Mine Owners and Mines of the Colorado Gold Rush

Laurel Michele Wickersheim
and Rawlene LeBaron

HERITAGE BOOKS
2009

HERITAGE BOOKS
AN IMPRINT OF HERITAGE BOOKS, INC.

Books, CDs, and more—Worldwide

For our listing of thousands of titles see our website at
www.HeritageBooks.com

Published 2009 by
HERITAGE BOOKS, INC.
Publishing Division
100 Railroad Ave. #104
Westminster, Maryland 21157

Copyright © 2007 Laurel Michele Wickersheim
and Rawlene LeBaron

Other books by the authors:

Colorado on the Eve of Statehood: An Edited Business Directory of the Pioneers Who Built the Centennial State

The Lost Cities of Colorado

All rights reserved. No part of this book may be reproduced or transmitted in any form or by any means, electronic or mechanical, including photocopying, recording or by any information storage and retrieval system without written permission from the author, except for the inclusion of brief quotations in a review.

International Standard Book Numbers
Paperbound: 978-0-7884-3135-7
Clothbound: 978-0-7884-8118-5

TO LEADVILLE

This book is dedicated to the Town of Leadville.

It was a cold December 1879 when Ralph Bayard, a reporter with the *St. Louis Post Dispatch*, packed his grip-sack and journeyed west to Leadville, Colorado. Drawn by the excitement and turbulence of the Colorado gold rush, he had accepted a job as city editor with Leadville's *Carbonate Weekly Chronicle*. His friends drafted cartoons depicting the dangers of life at the center of the gold rush, along with the following poem to which each contributed a verse, and mailed the package to him at the Leadville newspaper.

To Bayard
By his brother pencil pushers of the St. Louis press

T.J. Meek – *Times*:
    The rain was falling thick and fast
    As through the Four Courts [reporters' room] building passed
    A youth who held on high a flag
    Inscribed, "My carcass I will drag
        To Leadville."

Burke Waterloo – *Globe Democrat*:
    "Oh, stay!" the clerks and peelers cried,
    "The sender of the message lied
    A salary of thirty-five
    You could not earn to save your life
        In Leadville."

Frank Brooks – *Globe Democrat*:
    "Oh, stay with me!" Katrina yelled,
    As on to Ralph's coat tail she held.
    But Bayard pulled out through the door,
    And said: "I've heard that cry before,"
        Leadville

W.V. Byars – *Times*:
    He gave the maid one parting glance,
    Then onward he, as in a trance,
    The Union Depot reached at last –
    And Bayard's lot henceforth is cast
            In Leadville

D.L. Reid – *Post Dispatch*:
    "Mind not the pass," the old man said,
    Upon this road you're no dead-head."
    But still the youth went wildly on, - -
    Wild-eyed, rain-soaked, and quite forlorn
            To Leadville.

Clarence Howell – *Republican*:
    The warning words no weak'ning wake;
    Rash rushes Ralph the town to shake;
    Files forth, fierce to get in his work
    'Mid mad'ning, murd'ring mobs that lurk
            In Leadville.

Frank O'Neil – *Republican*:
    And when one day our Ralph shall spin
    One of his yarns so rich, but I'thin,'
    (That is to say – lacking in the elements
        which go to constitute reliable history)
    Some disapproving mining gent,
    On Western satisfaction bent,
    Will take his pop and bowie-knife
    And seek our little Bayard's life;
    In which event his heart will yearn
    For home, and quick his steps will turn
            From Leadville.

And Coroner Anlur begs to add:
    Unawed by bruisers, follow reason's plan,
    Assert thy quill, or quit Leadville, if you can.
    In spite of bowies, and in spite of whips,
    If to thyself thou canst thyself acquit,
    Stand steadfast, tho' alone in all conscious pride,
    Rather than err with Leadville on thy side.

Thos. J. Meek
Burke Waterloo
Frank Brooks
W.V. Byars
Bob Howell
   His
Davs X Reid
   Mark
Frank O'Neil
Hugo Auler

[Next to the signatures of Ralph Bayard's friends was a skull and crossbones, the 'seal of the gang.' Imagine the glee of Reid's companions as they prankishly substituted his signature with the "X" of one who could not read or write!]

Carbonate Weekly Chronicle
Saturday, 20 December 1879

Early Leadville, Colorado, with mines and mills to the left beyond the town. Mount Massive is in the background. This photo was taken about 1910 – 1915.

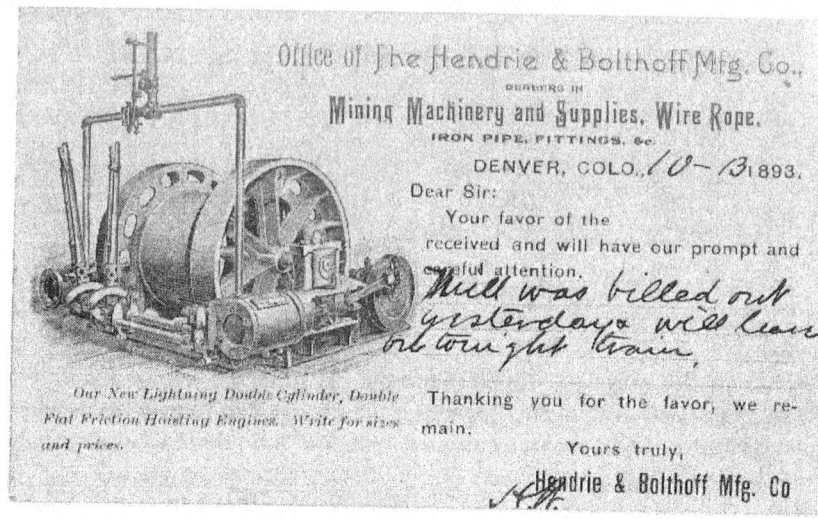

The Hendrie & Bolthoff Mfg. Company used this 1893 postal card to notify A.A. McGovney of Colorado Springs that the milling equipment they ordered would be shipped the next day via train.

Until 1893 postal cards in the U.S. were used primarily by companies to notify their customers of shipments and other events. This changed in 1893, when picture postcards were introduced at the 1893 Columbia Exposition in Chicago and soon became very popular.

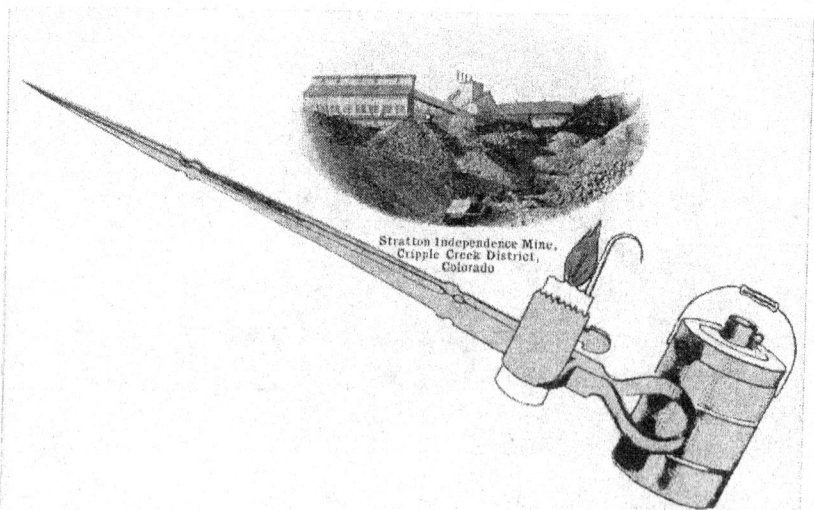

Miners had to warm dynamite before using it. This vintage postcard depicts a warming device for dynamite. Above the dynamite warmer is an inset of the Independence Mine that Winfield Scott Stratton discovered in 1891.

FOREWORD

In 1879, Thomas Corbett compiled a directory, organized by county, of Colorado's known gold, silver, and coal mines and ore mills. This was only three years after Colorado was granted statehood, and many of these jurisdictional lines have continued to shift to reflect an expanding population. Many new counties have since been carved out of the 1879 map of Colorado.

This book brings to life Thomas Corbett's classic out-of-print mining directory, with edits, annotations, and a full name index for today's reader.

This invaluable source of information about Colorado's turbulent gold rush names the owners, officers, and key employees for each mine. It also details the mine location, type of ore, mineral veins, yield, and other information of interest to historians, geologists, and modern prospectors.

Teams of horses hauled heavy mining machinery to the mines above Idaho Springs. This photo was taken about 1908.

Miners faced the constant danger of breaking through to underground water that would explode in an uncontrollable torrent through the mine walls, fill the tunnel, and wash them away. The Durant Mine [not named in this directory] near Aspen is shown with a comparatively gentle waterfall. This photo was taken between 1901 and 1907.

CONTENTS

Dedication iii

Foreword vii

Colorado Mines, by County
    Arapahoe County 1
    Boulder County 3
    Clear Creek County 77
    Custer County 153
    Gilpin County 165
    Grand County 231
    Hinsdale County 233
    Huerfano County 253
    Jefferson County 255
    La Plata County 259
    Lake County 263
    Ouray County 293
    Park County 337
    Pueblo County 361
    Rio Grande County 363
    San Juan County 369
    Summit County 439
    Weld County 463

Name Index (mine owners, officers, and key employees) 465

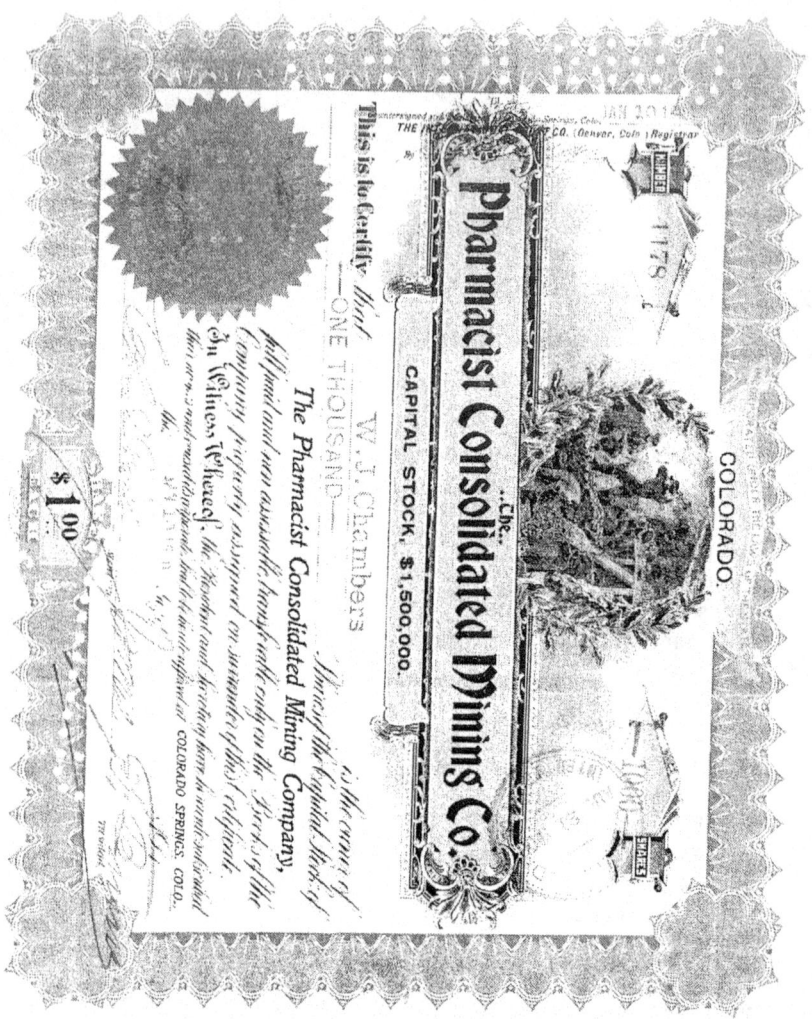

Stock certificate issued in 1900 for 1,000 shares of the Pharmacist Consolidated Mining Co. These old mining certificates have become highly collectible. Two pharmacists from Colorado Springs with more enthusiasm than prospecting skill discovered the Pharmacist Mine. Trusting their luck, they threw a hat in the air, marked the spot where it landed, and began digging for gold. The mine produced over a half million dollars in gold.

ARAPAHOE COUNTY

Thomas B. Corbett reported that Arapahoe County had a population of 32,000 in 1878. Corbett did not itemize specific Arapahoe County mines in the 1879 directory.

Boston and Colorado Smelting Co. Organized under the laws of the State of Massachusetts, May 11, 1867; capital paid up stock, $750,000, in 7,500 shares, $100 each; offices Boston, Massachusetts, Denver and Black Hawk, Colorado; J.W. Converse, president, J. Warren Merrill, treasurer, Prof. Nathaniel P. Hill, general manager, Henry R. Wolcott, assistant manager, Richard Pearce, metallurgist, A. von Schulz, assayer in charge. The Black Hawk Works were commenced in 1867, and consist of a group of buildings covering nearly 5 acres of ground. They are located on the Colorado Central Railroad and on North Clear Creek within the limits of Black Hawk, 1 mile from Central City. They have a capacity for treating 60 tons of ore daily, and employ 100 men.

Alma, Branch Works. Erected in 1874; Henry Williams, superintendent; located in the town of Alma in the county of Park, 63 miles from Black Hawk, and 100 miles from Denver. These works have a capacity for treating 25 tons of ore daily and employ 10 men.

Boulder Sampling and Crushing Works. Erected in 1876; C.G. Duncan, agent; located corner Pearl and Eighth streets in the city of Boulder, 43 miles from Denver, and 28 miles from Black Hawk; have a capacity for treating 25 tons of ore daily, and employ 6 men.

Argo. The Denver works, located 2 miles from Denver on the Colorado Central Railroad, was commenced in June 1878, and erected under the supervision of the general manager, Prof. Nathaniel P. Hill. This property comprises 80 acres of land, on which are build the main works of the company, consisting of a refinery 64 feet in width by 293 feet in length with a wing 40 by 77 feet, a smelting house 283 feet in length by 38 feet I width; an ore house 119 feet in width by 224 feet in length with a wing 33 by 91 feet; a calcining house 119 feet in width by 224 feet in length; an office, counting room, and assay laboratory 42 by 81 feet; and a coal shed 51 by 178 feet. These works are constructed of stone, and covered with a superior quality of corrugated iron roofing, and every department is equipped with the latest and most complete machinery selected with especial reference to the extraction of the precious metals in a manner profitable alike to the miner and the mill management. In addition to the foregoing buildings there are 5 two-story brick dwelling houses of 6 tenements each, a large hotel with accommodations for 100 persons, together with stables, sheds and other houses of minor importance. The works are surrounded by a substantial stone wall, have a capacity for treating 100 tons of ore daily, and give employment to 150 men.

Bunker Hill Placer Mining Co. Incorporated July 28, 1877, capital stock, $20,000, in 200 shares $100 each; office Littleton; trustees: John G. Lilley, Julius D. Hill, and Hiram S. Leach.

Colorado Dry Placer Mining Co. Incorporated April 16, 1877, capital stock, $350,000, in 3,500 shares, $100 each; office Denver; trustees: James C. Langhorne, John J. Van Demoer, Rufus W. Hedges, John Ermerins, Constant Duhem, Erastus S. Bennett, Lucy J. Bennett, Bertha Ermerins, and _____ Gortmans.

Denver Concentrating and Smelting Co. Incorporated December 8, 1874; capital stock, $100,000, in 100 shares, $1,000 each; office Denver; organized by C.F.A. Fischer, Francis M.F. Cazin, and Eugene P. Jacobson.

Denver Diamond Company. Incorporated November 18, 1872; capital stock, $5,000,000, in 50,000 shares, $100 each; office Denver; organized by William S. Walker, William N. Byers [the founder and publisher of *The Rocky Mountain News*], Benjamin F. Woodward, A. Cameron Hunt, Isaac N. Dawley, James H. Jones, Levin C. Charles, Samuel H. Elbert, and Henry Crow.

Denver Dry Ore Reduction Works. Scheyer & Co., proprietors; located in the City of Denver, convenient to the Denver and South Park Railroad depot; has a capacity for treating 6 tons of ore daily.

Denver Smelting and Refining Works. Incorporated December 20, 1872; capital stock, $100,000, in 1,000 shares, $100 each; located in Denver; organized by Joseph S. Miner, Hiram G. Bond, and Joseph E. Bates.

BOULDER COUNTY

Thomas Corbett reported that Boulder County had a population of 13,000 in 1878, and a bullion product of $713,642 for that year. In 1878, the mining districts were: Albion, Central, Gold Hill, Grand Island, James or Jim Creek, Magnolia, Sugar Loaf, and Ward.

Many of the Boulder area ores are tellurium, a rare, tin-colored metal that is easily scratched and very brittle. It is also easily melted. In 1880 Frank Hall, editor of the Mining Department of *Inter-Ocean*, a journal of Colorado politics, society, and mining, reported that the Boulder County tellurium belt starts in the mountains about six or seven miles west of Boulder. The belt is about five miles wide and about 20 miles from north to south. At first, area prospectors did not recognize this rare metal, just as prospectors did not recognize the black sands in Leadville as silver. The Red Cloud gold mine was discovered in Gold Hill in 1872. Ore from this promising mine assayed very high in gold, but miners had problems extracting the gold in the stamps and could not identify the tin-colored metal that seemed so abundant in the area. The miners sent ore samples to Professors Davis and Schirmer of the Denver Mint and to J. Alden Smith. They forwarded the ore samples to Professor F.A. Genth of the University of Pennsylvania, who identified the mystery metal as tellurium and informed them that Boulder was only the third known location where this rare metal was found. Tellurium was later discovered in other areas.

Some key mining camps and communities that boomed with the discovery of gold are now ghost towns: Balarat, Cardinal, Caribou, Crisman, Delphi [later renamed Wall Street], Magnolia, Sunshine, and Wagner's Camp.

Adda. William G. Pell, Capt. Ebenezer Rowland, Charles G. Van Fleet, and D.D. Hendrick, proprietors; office Boulder; claim 150 by 1,500 feet; discovered 1878; located near Springdale, Central mining district, 12 miles from Boulder, course of vein, northeast and southwest, average width 2 ½ feet, pitch 5 degrees north; nature of ore, tellurium and free gold; average value, $60 per ton; main shaft 25 feet deep. [Charles G. Van Fleet was an attorney from Pennsylvania who, in 1875, traveled to California for his health. On his trip West, he passed through Colorado and decided to settle in Boulder. He had interests in several area mines and briefly served as mayor of Boulder.]

Alamakee. Joseph Steppler, proprietor, office Denver; claim 50 feet by 1,200 feet, patented; discovered 1868; located on Gold Hill, 10 miles from Boulder; course of vein, northeast and southwest, average width 3 ½ feet, pitch 10 degrees north, pay vein 10 inches; nature of ore, free gold; average value, $165 per ton; main shaft 40 feet deep, 1 level 50 feet in length; yield to date, $4,000.

Alcrona. Chicago and Colorado Mining Co., proprietors; offices Boulder, Colorado, and Chicago, Illinois; claim 150 by 1,500 feet, patented; discovered 1875; located on Left Hand Creek, Ward mining district, 4 miles from town of Gold Hill, and 16 miles from Boulder; course of vein, east and west, average width 4 feet, pitch north, pay vein 6 to 15 inches; nature of ore, iron pyrites and free gold; average value, $122 per ton; main shaft 50 feet deep, 1 level 40 feet in length, and 50 feet from surface.

Alice. Owned by Henry and Ira F. Monell and others; office Salina; claim 150 by 1,500 feet; discovered 1878; located on Melvina Hill, Gold Hill mining district, 8 miles from Boulder; course of vein, northwest and southeast, average width 2 ½ feet, pitch 30 degrees north, pay vein 1 to 8 inches; nature of ore, tellurium; average value, $100 per ton; main shaft 60 feet deep.

Alps. John Williams and Matthew Holt, of Jamestown, proprietors; claim 150 by 1,500 feet; located in Central mining district, 1 mile from town of Springdale and 11 miles from Boulder; course of vein, northeast and southwest, average width 1 foot, pitch 5 degrees north; nature of ore, iron pyrites and free gold; main shaft 25 feet deep.

Alturas. John Williams, George Newcomb, and Daniel Drummond, proprietors; office Gold Hill; claim 50 by 1,400 feet; discovered 1871; located on Gold Hill, Gold Hill mining district, 12 miles from Boulder; course of vein, northeast and southwest, average width 6 feet, pitch 5 degrees northwest, pay vein 2 feet; nature of ore, tellurium, free gold and native silver; average value, $100 per ton; main shaft 25 feet, and 1 other shaft 15 feet deep, 1 drift 77 feet in length; total yield, $500.

American and Ajax. American Consolidated Gold and Silver Mining Co., of Connecticut, proprietors; incorporated 1877; capital stock, $1,000,000, in 100,000 shares, of $10 each; H.G. Angle, president, C. Goddard, secretary and treasurer, J. Alden Smith, superintendent; offices, Sunshine, Colorado, and 31 Broad Street, New York; claim 250 by 1,500 feet; discovered 1874; located in town of Sunshine, Gold Hill mining district, 6 miles from Boulder; course of vein, northeast and south-west, average width 2 ½ feet, pitch 6 degrees east, pay vein 2 inches to 2 feet; nature of ore, tellurium, of gold and silver, sylvanite, petzite, calaverite, coloradoite and free gold; average value of first class ore, $4,000, second class ore, $800, and third class ore, $200 per ton; main shaft 400 feet deep; five levels, 40, 100, 140, 200, and 300 feet from surface respectively, aggregating 1,000 feet in length; shaft house, 30 by 60 feet containing a hoisting apparatus, 20-horse power engine, machinery for crushing and sampling ore and 1 steam pump; assay laboratory in connection with office. [J. Alden Smith was born in Kennebec County, Maine. He studied geology, and came to Colorado in 1864 to report on mining properties. He stayed in Colorado, became editor of the *Miner's Register*, and opened an assay office. In 1872 he was appointed Territorial Geologist and two years later, in 1874 he was appointed superintendent of the American mine.]

American Eagle. James R. Davidson, William P. Clayton, J. Bedal, and G.B. Green, proprietors; office Gold Hill; claim 150 by 1,500 feet; discovered 1876; located in Gold Hill mining district, 10 miles from Boulder; course of vein, northeast and southwest, width 4 feet, pitch 5 degrees north, pay vein 12 inches; nature of ore free gold; average value, $75 per ton; main shaft 40 feet, and 1 other shaft 12 feet deep.

American Flag. John Holmes and S.B. Mills, proprietors; office Magnolia; claim 150 feet by 1,500 feet; discovered 1875; located on Magnolia Hill, Magnolia mining district, 8 miles from Boulder; course of vein, east and west, width 4 feet, pitch 30 degrees north, pay vein 4 to 8 inches; nature of ore, tellurium; average value, $75 per ton; main shaft 85 feet deep.

Americus. Owned by Hugh and Thomas F. Harkins and L.T. Haak; office Sugar Loaf; claim 150 by 1,500 feet; discovered 1875; located in Forest Gulch, Sugar Loaf mining district; 10 miles from Boulder; course of vein, northeast and southwest, width 5 feet, pitch northwest, pay vein 3 feet; nature of ore, tellurium; average value, $100 per ton; main shaft 35 feet deep.

Arlington. John W. James and P.J. Werley, proprietors; office Caribou; claim 50 by 1,500 feet, patented; discovered 1872; located at Cardinal, Grand Island mining district, 18 miles from Boulder; course of vein, east and west, width 5 feet, pay vein 5 inches to 2 feet, pitch 20 degrees north; nature of ore, iron pyrites, carbonates of copper and galena; average value, $200 per ton; main shaft 100 feet deep, 1 tunnel 150 in length.

Atchison. Owned by the Atchison Mining Co., Dundee Scotland; George G. Philip, superintendent; office Salina; claim 50 by 1,500 feet, patented; discovered 1874; located in Gold Hill mining district, near town of Salina, 8 miles from Boulder; course of vein, northeast and southwest, width 21 feet, pay vein 4 feet, pitch 25 degrees northeasterly; nature of ore, tellurium and gray copper; main shaft 100 feet deep, 1 tunnel 300 feet in length, intersecting vein at a depth of 150 feet, 1 level 25 feet in length; 10-stamp mill run by steam and water power.

B.F. Smith. Owned by William G. Pell, Capt. Ebenezer Rowland, and Charles Van Fleet; office Boulder; claim 150 by 1,500 feet; discovered 1876; located on Golden Age Hill, Elkhorn mining district, 2 ½ miles from Jamestown and 14 miles from Boulder; course of vein, northeast and southwest, average width 2 to 4 feet; pitch 25 degrees north, pay vein 4 to 16 inches; nature of ore, free gold; average value 10 ounces gold per cord; main shaft 100 feet deep.

Badger State. James H. Jones and D. Mahan, of Sunshine, proprietors; claim 150 by 1,500 feet; located in Gold Hill mining district, half mile from town of Sunshine, and 6 miles from Boulder; course of vein, northeast and southwest, width 6 feet; pitch 10 degrees north, pay vein 1 foot; nature of ore, tellurium and sylvanite; average value, $97 per ton; main shaft 90 feet deep, 1 level 25 feet in length.

Balarat. August Paddock and Frederick R. Luce, proprietors; office Boulder; claim 150 by 1,500 feet; discovered 1876; located on Balarat Hill, Central mining district, 19 miles from Boulder; vertical vein trending northeast and southwest, width 4 feet, pay vein 1 to 6 inches; nature of ore, tellurium; value $1,052 per ton; main shaft 100 feet deep, 1 tunnel 55 feet in length and 160 feet from surface, 1 level 25 feet in length; yield to date, $2,000.

Baron. Owned by C.C., F., and Clem C. Eddy, and A.R. West; office Salina; claim 150 by 1,500 feet, patented; discovered 1875; located on

Melvina Hill, Gold Hill mining district, half mile from town of Salina and 8 miles from Boulder; course of vein, northeast and southwest, width 3 feet, pay vein 10 inches; nature of ore, tellurium; average value, $200 per ton; main shaft 70 feet deep.

Bay State. Owned by Frederick W. Strout and James Stevens, of Sunshine; claim 150 by 1,500 feet; located in Wagner's Camp, Gold Hill mining district, 2 miles from town of Magnolia and 7 miles from Boulder; course of vein, northeast and southwest, width 4 feet, pitch 5 degrees southeast, pay vein 2 feet; nature of ore, tellurium of silver, average value, $255 per ton; main shaft 10 feet deep.

Bay State and Colorado Mining Co. Incorporated November 11, 1872; capital stock, $100,500, in 335 shares, $300 each, office Caribou; trustees: Theodore O. Saunders, E.W. Houghton, O.O. Austin, and W.H. Elliott.

Belcher. John A. Ellet, proprietor; office Boulder; claim 150 by 1,500 feet, patented; discovered 1873; located near town of Caribou, Grand Island mining district, 23 miles from Boulder; course of vein, northeast and southwest, width 4 feet, pay vein 6 to 18 inches; nature of ore, sulphurets and galena; average value 50 ounces in silver per ton; main shaft 110 feet deep; two levels aggregating 130 feet in length.

Belden. Frank C. Garbutt, and Thomas H. Baker, proprietors; office Denver; claim 150 by 1,500 feet; discovered 1874, located on Gold Run, Gold Hill mining district, 10 miles from Boulder; vertical vein running east and west, width 4 feet, pay vein 18 inches; nature of ore, gray copper and tellurium; assay value $500 per ton; main shaft 30 feet, and 1 other shaft 10 feet deep.

Belleview. Alex. Van Wendt and Joseph Steppler, proprietors, office Gold Hill; claim 150 by 1,500 feet; discovered 1872; located on Gold Hill, Gold Hill mining district, 12 miles from Boulder; vertical vein running northeast and southwest, width 6 feet, pay vein 2 feet; nature of ore, gray copper; assay value $6,500 per ton; main shaft 65 feet deep, 1 adit 200 feet in length; total yield, $2,500.

Ben C. Lowell. J.V. Pomeroy and John S. Reid, proprietors; office Magnolia; claim 50 by 1,500 feet, discovered 1875; located on Magnolia Hill, Magnolia mining district, 8 miles from Boulder; vertical vein trending east and west, average width 7 feet, pay vein 2 feet; nature of ore, free gold; average value, $55 per ton; one shaft 103 feet deep.

Bendigo. Owned by the Smuggler Mining Co.; claim 150 by 1,500 feet, discovered 1876; located in Long's Gulch, Central mining district, 17 miles from Boulder; course of vein, northeast and southwest, width 4 feet, pay vein 12 inches, pitch 5 degrees north; nature of ore, tellurium; operated through the main workings of the Smuggler mine.

Bettie. Granville M. Button and A. Gilmore, proprietors; office Boulder; claim 150 by 1,500 feet; discovered 1878; located on Jim Creek, Jim Creek mining district, 16 miles from Boulder; course of vein, northeast and southwest, width 3 feet, pitch north, pay vein 18 inches; nature of ore, tellurium; average value, $675 per ton; main shaft 50 feet deep; yield to date, $3,000.

Big Blossom. Claim 150 by 1,400 feet; located in Castle Gulch, Central mining district, 1 mile from Springdale and 12 miles from Boulder; course of vein, northeast and southwest; width 30 feet, pitch 25 degrees southeast; nature of ore, tellurium and free gold; value $50 to $20,000 per ton; main shaft 125 feet deep, 1 level 60 feet in length and 50 feet from surface; yield to date, $7,500.

Big John. William Haardt and John Hasselbacher, proprietors, office Springdale, claim 150 by 1,500 feet, discovered 1877; located in Castle Gulch, Central mining district, half mile from town of Springdale, and 12 miles from Boulder; vertical vein running north and south, width 2 feet, pay vein 10 inches; nature of ore, gray copper and sulphurets; average value, $491 per ton; main shaft 20 feet deep.

Big Thing. Anthony Arnett and Thomas J. Graham, proprietors; office Boulder; claim 50 by 2,775 feet, patented; discovered 1863; located in Central mining district, 14 miles from Boulder; course of vein, northwest and southeast, width 6 feet, pitch 5 degrees north, pay vein, 2 feet; nature of ore, galena and copper pyrites; average value, $40 per ton; main shaft 110 feet deep; yield to date, $3,000.

Birdie Emma. Turner Tunnel Mining and Milling Co., proprietors; office Boulder, Colorado, and Lexington, Kentucky; claim 150 by 1,500 feet; discovered 1877; located at the head of Weare Gulch, Grand Island mining district, 2 miles from Caribou and Nederland, and 20 miles from Boulder; course of vein, northeast and southwest, width 3 feet, pitch 17 degrees north, pay vein 14 inches; nature of ore, iron pyrites and free gold; reached by the Turner Tunnel at a distance of 50 feet from the mouth, and intersected at a depth of 30 feet.

Bismark. William Haardt and John Hasselbacher, proprietors; office, Springdale; claim 150 by 1,500 feet; located in Elkhorn Gulch, Central mining district, 2 miles from Springdale and 12 miles from Boulder; vertical vein running northeast and southwest, width 12 feet, pay vein 3 feet; nature of ore, tellurium; main shaft 15 feet deep.

Black Cloud. Black Cloud Gold and Silver Mining Company, of Michigan, proprietors; incorporated 1876; capital stock, $500,000; John B. Maas, president, John Richardson, vice president, John Mitchell, secretary and treasurer, Ebenezer Rowland, general superintendent; office Boulder; claim 50 by 1,500 feet; patented; discovered 1871; located on the west side of Bighorn Mountain, Gold Hill mining district, 2 miles from Salina, 1 mile from Gold Hill, and 11 miles from Boulder; course of vein, northeast and southwest, average width, 4 feet, pitch 10 degrees north, pay vein 6 to 30 inches; nature of ore, gray copper, zinc blende, rosin zinc, sulphurets of copper, iron and copper pyrites, lead and free gold; average value, $50 per ton; main shaft 200 feet deep, 3 levels, aggregate length 900 feet. [Ebenezer Rowland was born in Monmouthshire, England on March 16, 1831, and was very young when his family immigrated to America, settling in Portage County, Ohio. He moved to Boulder, Colorado, in 1875.]

Black Prince. Daniel Drummond, John Williams, and M.J. Holt, proprietors; office Left Hand; claim 150 by 1,500 feet; discovered 1876; located on Gold Hill, Gold Hill mining district, 10 miles from Boulder; vertical vein trending northeast and southwest, average width 2 feet, pay vein 6 inches; nature of ore, tellurium and free gold; average value, $50 per ton; main shaft 30 feet deep.

Bonanza. Silas T. Tumbleson, proprietor. Office Gold Hill; claim 150 by 1,500 feet; discovered 1877; located on Bonanza Hill, Central mining district, 5 miles from town of Jamestown, and 13 miles from Boulder; vertical vein running northwest and southeast, average width 4 feet, pay vein 22 inches; nature of ore free gold; average value, $80 per cord; main shaft 50 feet deep; aggregate depth of other shafts, 200 feet, 1 level 50 feet in length, and 35 feet from surface; yield to date, $10,000.

Bondholder. Cyrus Strock and Frederick C. Bucherdee, of Jamestown, proprietors; claim 150 by 1,500 feet, discovered 1878; located in Gillaspie Gulch, Central mining district, 2 ½ miles from Jamestown, 18 miles from Boulder; course of vein, northeast and southwest, average width, 4 feet, pay vein, 1 foot; nature of ore, tellurium and free gold; main shaft 50 feet deep.

Boulder. Porter T. Hinman, proprietor; office Modoc; claim 150 by 1,500 feet, patented; discovered 1874; located on Four Mile Creek, Sugar Loaf mining district, 10 miles from Boulder; vertical vein running northeast and southwest, average width 6 feet; nature of ore, iron pyrites and mispickle; average value, $30 per ton; main shaft 50 feet deep, and 1 level 20 feet in length.

Boulder Canon Mining Co. Incorporated March 28, 1877; capital stock, $200,000, in 2,000 shares, $100 each; office Boulder; organized by James P. Maxwell, A.W. Bash, E.H. Dimick, and M.A. Lathrop.

Boulder County. Owned by Boulder County Mining, Tunneling, Milling, Land and Town Co.; claim 150 by 1,400 feet, patented; discovered 1870, located on Boulder County Hill, Grand Island mining district, 2 miles from Nederland and 18 miles from Boulder; vein nearly vertical, average width 4 ½ feet; nature of ore, galena, free gold and silver; average value, $50 per ton; main shaft 150 feet deep, 1 tunnel 360 feet in length.

Boulder County Mining, Tunneling, Milling, Land and Town Co., Colorado. Incorporated June 12, 1877; capital stock, $2,000,000, in 200,000 shares, of $10 each; James A. Austin, president; Ira W. Pendleton, secretary and treasurer, Samuel P. Conger, superintendent; office Denver; property comprises a mill, run by water power and constructed of stone, containing 100 stamps and 40 Frue vaners, 6 lodes or mineral veins, 150 by 1,500 feet each, and 1,200 acres of mineral and agricultural land; located in Grand Island mining district, 2 miles from Nederland and 18 miles from Boulder.

Bramblett. John N., George W., and John W. Gillaspie, proprietors; office Rockville; claim 150 by 1,500 feet; discovered 1878; located in Gray Eagle Gulch, 2 miles from Gold Hill and 12 miles from Boulder; course of vein, east and west, average width 5 feet, pitch 10 degrees south, pay vein 2 feet; nature of ore, tellurium; average value, $30 per ton; 1 shaft 25 feet deep, 1 drift 263 feet in length.

Buckeye. Owned by Thomas Coleman and Charles W. Hamill, of Boulder; claim 150 by 1,500 feet; discovered 1876; located in Middle Boulder Gulch, Grand Island mining district, 2 ½ miles from Nederland and 15 miles from Boulder; vertical vein running northeast and southwest, average width 4 feet, pay vein 10 inches; nature of ore, galena and gray copper; average value, $50 per ton; 1 drift 25 feet in length.

Buckeye. Owned by Henry Deitz of Boulder, claim 150 by 1,500 feet; discovered 1876; located in Bummer Gulch, Sugar Loaf mining district, 9 miles from Boulder; course of vein, northeast and southwest, pitch 25 degrees north, average width 3 feet, pay vein 15 inches; nature of ore, tellurium; average value, $104 per ton; 1 shaft 25 feet deep. [Henry Deitz was born in Lewis County, New York, on March 3, 1841, and briefly studied law, then medicine, then teaching, before moving west to work with railroads – at one time he was the local treasurer, auditor and paymaster for the Trans-Continental Railroad in Texas. He moved to Colorado in 1876 because of his wife's failing health.]

Buckeye Mining and Tunneling Co. Incorporated February 4, 1876; capital stock, $500,000, in 50,000 shares, $10 each; office Denver; organized by Alexander M. Cassiday, D.R. Cassiday, J.V. Harbour, and W.S. Hinman.

Buck Horn. Anthony Arnett, proprietor; office Boulder; claim 50 by 3,000 feet, patented; discovered 1863; located in Jim Creek mining district, 12 miles from Boulder; vertical vein trending northeast and southwest, average width 2 feet, pay vein 1 foot; nature of ore, galena; average value, $100 per ton; main shaft 100 feet deep, yield to date, $10,000.

Buenos Ayres. Owned by Joseph McPherson, Noah Terry, and John Williams, of Jamestown; claim 150 by 1,500 feet discovered 1873; located at Jamestown, Central mining district, 15 miles from Boulder; course of vein, north and south, average width 3 feet, pitch 5 degrees west; nature of ore, iron and copper pyrites, and antimony; main shaft 80 feet deep, 1 tunnel 50 feet in length.

California. Daniel Sutfon and Frederick A. Squires, of Boulder, proprietors; claim 150 by 1,500 feet, patented; discovered 1863; located on Left Hand Creek, Ward mining district, 18 miles from Boulder; vertical vein running northeast and southwest, average width 4 feet, pay vein, 20 inches; nature of ore, free gold; average value, $400 per ton; main shaft

173 feet deep. [Frederick A. Squires was born in Granville, Massachusetts, on May 19, 1819. He was a tinner. He moved to Boulder in 1860.]

Careless Boy. Owned by the Smuggler Mining Co.; claim 150 by 1,500 feet, patented; discovered 1876; located in Long's Gulch, Central mining district, 17 miles from Boulder; course of vein, east and west, width 3 feet, pitch south; nature of ore, copper pyrites; 1 tunnel 180 feet, and 1 level 125 feet in length.

Caribou. Hon. Jerome B. Chaffee, proprietor; Eben Smith, superintendent; offices, Caribou and Denver, Colorado; claim 50 by 1,500 feet, patented; discovered 1870; located in town of Caribou, on the north side of Caribou Hill, Grand Island Mining District, 22 miles from Boulder; course of vein, east and west, width 5 feet, pitch 10 degrees north; nature of ore, native, ruby and brittle silver; average value 60 ounces silver per ton; developed by a main shaft 800 feet and other shafting aggregating 1,000 feet in depth, 1 tunnel 600 feet in length, intersecting vein at a depth of 316 feet; shaft house 70 by 175 feet, containing a 60-horse power engine and 2 boilers, 4 Wood machine drills, a Knowles pump, and a duplex air compressor with an 8-inch cylinder; yield to date, $1,225,000.

Caribou Debenture Syndicate. Incorporated Nov. 11, 1875; capital stock, $300,000, in 3,000 shares, of $100 each; office Denver; organized by Marc A. Shaffenburg, James F. Watson, and William H. Salisbury.

Carlisle. Thomas J. Graham, proprietor; office Boulder; claim 50 by 3,000 feet; located in Central mining district, near Jamestown, 14 miles from Boulder; vertical vein running northeast and southwest, average width 5 feet, pay vein 16 inches; nature of ore, iron and copper pyrites and galena; average value, $27 per ton; main shaft 50 feet long.

Carmack. Albion Mining and Milling Co., proprietors; claim 150 by 1,500 feet; discovered 1866; located on North Boulder Creek, Albion mining district, 6 miles from Nederland and 23 miles from Boulder; course of vein, northeast and southwest, average width 54 feet, pitch 20 degrees northwest; nature of ore, galena and carbonate of copper; average value, 20 ounces silver per ton; main shaft 40 feet deep.

Cash. Owned by the Cash Gold and Silver Mining Company of Colorado, incorporated June 25, 1875; capital stock, $250,000, in 2,500 shares, $100 each; office Denver; James H. Boyd, president, A.J. Bean, vice president and superintendent, Corydon W. Sanborn, secretary; claim

150 by 1,500 feet, patented; discovered 1872; located on Gold Hill, Gold Hill Mining District, 10 miles from Boulder; vertical vein trending northeast and southwest, width 5 feet, pay vein 6 to 18 inches; nature of ore, tellurium; average value, $150 per ton; main shaft 115 feet deep, 1 level 300 feet in length and 60 feet from surface.

Castle Rock. Frederick C. Dorl and Thomas Coleman, proprietors; office Boulder; claim 150 by 1,500 feet; discovered 1877; located on Castle Rock Mountain, Grand Island mining district, 3 miles from town of Nederland and 15 miles from Boulder; vertical vein trending east and west, width 4 feet, pay vein 10 inches; nature of ore, brittle silver; average value, $450 per ton; 1 adit 450 feet in length.

C.B.A. Cyrus B. Ayers and Charles McKinley, proprietors; office Boulder, claim 150 by 1,500 feet; discovered 1877; located in Long's Gulch, Central mining district, 4 miles from Jamestown and 20 miles from Boulder; vertical vein running northeast and southwest, average width 6 feet, pay vein 14 inches; nature of ore, tellurium; assay value $780 per ton; main shaft 34 feet deep.

Celestial. Anthony Arnett, proprietor; office Boulder; claim 150 feet by 1,500 feet, with a mill site of 5 acres, all of which is patented; discovered 1873; located in Spring Gulch, near town of Ward, Ward mining district, 20 miles from Boulder; vertical vein trending east and west, width 20 inches, pay vein 8 to 20 inches, nature of ore, iron pyrites; value $5,000 per ton; main shaft 60 feet deep, aggregate depth of other shafts 175 feet; good shaft house; 15-horse power engine, and other machinery; yield to date, $20,000.

Celestial Mining Co. Incorporated January 15, 1876; capital stock, $300,000, in 60,000 shares, $5 each; office Denver; trustees: Peter Winne, W.W. Pierce, Oliver P. Hamilton, and George W. Stone.

Champion. Henry Deitz of Boulder, proprietor; claim 150 by 1,500 feet; discovered 1876; located in Bummer Gulch, Sugar Loaf mining district 9 miles from Boulder; course of vein, northwest and southeast, pitch northeast, width 7 feet, pay vein 20 inches; nature of ore, free gold; average value, $62 per ton; 1 shaft 45 feet deep.

Chance. James Martin, William McKnight, and Thomas Skelton, proprietors; office, Boulder; claim 150 by 1,500 feet; discovered 1876; located between Sunbeam and Parker Gulches, Sugar Loaf mining district,

6 miles from Boulder; vertical vein running northeast and southwest; width 5 feet; pay vein 18 inches; nature of ore, tellurium and free gold; average value, $400 per ton; main shaft 80 feet deep; 1 level 20 feet in length; yield to date, $600.

Charter Oak. Owned by Ellen Rollins and W. Sellars; office Gold Hill, claim 150 by 1,500 feet; discovered 1876; located in Gold Hill mining district, 10 miles from Boulder; course of vein, northeast and southwest, average width 30 inches; pitch 5 degrees north, pay vein 20 inches; nature of ore, free gold; average value, $200 per cord; main shaft 30 feet deep; yield to date, $3,000.

Chicago and Colorado Mining Co. Incorporated April 25, 1876; capital stock, $250,000, in 25,000 shares, $10 each; offices, Boulder, Ward mining district, Colorado, and Chicago, Illinois; trustees: Frank J. Weist, Mancel Talcott, William G. Shedd, Nelson Morris, and Wesley Brainard.

Chimney. Owned by George Tanner, Peter Davis, and W. Worthing; office Salina; claim 150 by 1,500 feet; discovered 1874; located on Melvina Hill, Gold Hill mining district, 9 miles from Boulder; course of vein, east and west, width 3 feet, pitch 25 degrees north, pay vein 12 inches; nature of ore, free gold; assay value $24 per ton; main shaft 55 feet deep.

Cincinnati. Owned by Samuel Garret, A.M. Barckhamer, and John Maxwell; office Salina; claim 150 by 1,500 feet; discovered 1877; located on Melvina Hill, Gold Hill mining district, 1 mile from town of Salina and 8 miles from Boulder; course of vein, east and west, average width 7 feet, pitch 45 degrees north, pay vein 16 inches; nature of ore, galena, brittle silver and free gold; average value, $325 per ton; main shaft 60 feet, 1 other shaft 20 feet deep, 1 level 12 feet in length and 40 feet from surface; yield to date, $5,000.

Clarence. John B. McFall and David J. Rollins, proprietors; office Gold Hill; claim 150 by 750 feet; discovered 1876; located in Gold Hill mining district, 10 miles from Boulder; vertical vein running northeast and southwest, average width 3 feet, pay vein 4 to 8 inches; nature of ore, tellurium; average value, $266 per ton; main shaft 25 feet deep yield to date, $300.

Columbia. James Pomeroy of Boulder, proprietor; claim 150 by 300 feet, patented; discovered 1859; located on Niwot Hill, Ward mining district, 19 miles from Boulder; course of vein, east and west, average width 3 feet; nature of ore, iron and copper pyrites; average value, $40 per ton; main shaft 112 feet deep, aggregate depth of other shafts 95 feet, 1 level 27 feet in length and 73 feet from surface; yield to date, $10,000.

Columbia Mining Co. Incorporated January 7, 1875; capital stock, $1,000,000, in 10,000 shares, $100 each; office at works in Ward mining district; trustees: Andrew W. Gill, William A. Davidson, and David H. Moffat, Jr.

Cold Spring. Truman Whitcomb, proprietor; office Gold Hill; claim 50 by 1,400 feet; discovered 1871; located in Gold Hill mining district, 12 miles from Boulder; course of vein, north and south, width 7 feet, pay vein 15 inches; nature of ore, iron pyrites and free gold; average value, $425 per ton; main shaft 400 feet deep, 1 tunnel 480 feet in length, 4 levels, 80, 140, 200, and 290 feet from surface respectively, aggregating 1,400 feet in length; good shaft house; 15-horse power engine and other machinery.

Colorado Comstock. Frank C. Garbutt and Thomas A. Baker, proprietors; office Denver; claim 150 by 1,500 feet; discovered 1874; located on Hoosier Hill, Gold Hill mining district, 10 miles from Boulder; vertical vein running northwest and southeast; average width 15 feet, pay vein 2 feet; nature of ore, gray copper and sulphurets; average value, $200 per ton; main shaft 40 feet deep.

Columbus. Owned by the Columbus Mining Co., incorporated May 19, 1875; capital stock, $50,000, in 2,000 shares, $25 each; office Denver; claim 150 by 1,500 feet, patented; discovered 1865; located in Gold Hill mining district, 10 miles from Boulder; course of vein, northeast and southwest, average width 2 feet, pitch north, pay vein 10 inches; nature of ore, tellurium and iron pyrites; average value, $125 per ton; main shaft 100 feet deep, aggregate depth of other shafting 100 feet, 1 adit 100 feet in length, 2 levels 50 and 70 feet from surface respectively, aggregating 150 feet in length.

Comstock No. 2. Owned by H.J. Van Wetering and N.C. Peterson, of Sunshine; claim 150 by 1,500 feet; discovered 1875; located on Lee Hill, Gold Hill mining district, 3 miles from town of Sunshine, and 9 miles from Boulder; course of vein, east and west, width 5 feet, pitch 3 degrees north, pay vein 6 inches; nature of ore, tellurium; average value, $200 per ton; main shaft 50 feet deep.

Conger. Ulysses Pugh and Samuel Mishler, proprietors; office Caribou; claim 50 by 1,400 feet; discovered 1864; located on Caribou Hill, Grand Island mining district, 23 miles from Boulder; vertical vein trending northeast and southwest, width 2 feet; nature of ore, galena, iron and copper pyrites; average value, $84 per ton; main shaft 30 feet deep.

Consolidated Caribou Belt Mining Co. Incorporated February 9, 1876; capital; stock $500,000, in 50,000 shares, $10 each; office Denver; organized by W.H.J. Nichols, Leonard Cutshaw, Theron W. Johnson, Isaac H. Rosencrans, Jeremiah Waldron, Jean Rouilliard, George Duggan, Walter B. Jenniss, and George M. Lahaye.

Consolidated Fourth of July Mining Co. Incorporated October 14, 1875; capital stock, $1,000,000, in 10,000 shares, $100 each; office Denver, Colorado; organized by M.A. Lathrop, H.B. Blakeslee, J.M. Wilson, and J.B. Johnson.

Consolidated North Slope and Evans Tunnel Co. Incorporated November 27, 1876; capital stock, $500,000, in 50,000 shares, $10 each; offices Boulder, Colorado, and Cincinnati, Ohio; trustees: John S. Crawford, A.P.W. Skinner, and W.T. Forrest.

Consolidated Tellurium Belt Mining Co. Incorporated July 15, 1876; capital stock, $1,000,000, in 100,000 shares, $10 each; office Denver; trustees: Jeremiah Waldron, Jean B. Rouilliard, L.T. Stewart, N.V. Mahan, M.M. Hug, A.E. Rhinehart, and J.W. Lowerre.

Copper. Owned by David Evans and Frederick E. Sturns, of Cincinnati; claim 150 by 1,500 feet; discovered 1863; located in Gold Hill mining district 1 mile from town of Salina, and 8 miles from Boulder; course of vein, northeast and southwest, width 3 ½ feet, pitch 20 degrees northwest, pay vein 2 feet; nature of ore, galena and iron pyrites; average value, $60 per ton; main shaft 115 feet deep, 1 adit 52 feet in length.

Corning. Anthony Arnett and Horace Wolcott, proprietors, office Boulder; claim 50 by 1,500 feet, patented; discovered 1870; located on Niawot Hill, Ward mining district, 20 miles from Boulder; course of vein, east and west, width 4 feet, pitch 40 degrees north, pay vein 6 inches; nature of ore, iron and copper pyrites; 1 shaft 25 feet deep.

Corning Tunnel. Owned by the Corning Tunnel Mining and Reduction Co., incorporated December 10, 1875; capital stock, $200,000, in 2,000 shares, $100 each; office Boulder; organized by George C. Corning, Jarvis Gilbert, Frederick A. Squires, Daniel A. Robinson, Frederick Kehler, Edward A. Phillips, John P. Thompson, Albert G. Soule, John Morrison, Charles Y. McClure, John A. Ellet, and Corydon W. Sanborn; claim 1,500 by 3,000 feet, with 2 mill sites of 5 acres each; discovered 1872; located on Left Hand Creek, Gold Hill mining district, 2 miles from town of Gold Hill and 12 miles from Boulder; length of tunnel 1,100 feet. [Col. John A. Ellet was born near Alton, Illinois, on June 22, 1838, and went to California when he was 15 years old to join his father. He taught public school in Santa Clara until 1862, when he returned east to serve in the Civil War as Lt. Colonel of the Mississippi River Ram Fleet under the command of his uncle. He went into business in Vicksburg, Mississippi, before moving to Colorado in 1875. He was elected mayor of Boulder in 1880.]

Credit Mobilier. Webster City Gold and Silver Mining Co., proprietors; office Boulder; claim 50 by 1,500 feet; located in McKnight Gulch, Gold Hill mining district, 11 miles from Boulder; course of vein, northeast and southwest, width 3 ½ feet, pitch 5 degrees north, pay vein 14 inches; nature of ore, gray copper, galena, iron and copper pyrites; main shaft 50 feet deep, 1 tunnel 100 feet in length.

Crown Prince. Crown Prince Mining and Milling Co., proprietors; office Boulder; claim 1,500 by 3,000 feet; located on Four Mile Creek, Sugar Loaf mining district, 1 mile from town of Delphi, and 10 miles from Boulder; vertical vein running northwest and southeast; average width 25 feet; nature of ore, iron and copper pyrites; average value, $13 per ton; 1 shaft 25 feet deep.

Crown Prince Tunnel Mining Co., Colorado. Incorporated June 22, 1877; capital stock, $500,000, in 500,000 shares, $1 each; office Boulder; trustees: Platt Rogers, William Burke, and John Tierney.

Dale Owen. James K. Jones, proprietor; office Sunshine; claim 150 by 1,500 feet; discovered 1874; located in Sunshine Gulch, Gold Hill mining district, 1 mile from Sunshine, and 7 miles from Boulder; vertical vein running north and south; width 3 feet, pay vein 8 inches; nature of ore, gray copper, galena and iron; average value, $126 per ton; main shaft 25 feet deep, 1 tunnel 75 feet in length.

Dana or Hoosier. Squires, Neikirk & Co., proprietors; office Boulder; claim 50 by 1,500 feet, patented; discovered 1867; located on Hoosier Hill, near Gold Run, Gold Hill mining district, 12 miles from Boulder; vertical vein running north and south; average width 25 feet; nature of ore, gray copper and sulphurets of silver; average value, $250 per ton; main shaft 150 feet deep, 1 tunnel 100 feet in length; shaft house 40 by 60 feet; yield to date, $200,000.

Dead Medicine. Owned by Thomas Danford, O.P Patterson, and others; office Boulder; claim 150 by 1,500 feet, patented; discovered 1874; located in town of Sunshine, Gold Hill mining district, 6 miles from Boulder; vertical vein running north and south; average width 5 feet, pay vein 6 to 20 inches; nature of ore, tellurium; 1 shaft 50 feet deep.

Deadwood. Owned by Henry Dwyer, Frederick Pocock, and John Alderson; office Salina; claim 150 by 1,500 feet; discovered 1877; located on Melvina Hill, 8 miles from Boulder; course of vein, northeast and southwest; average width 3 feet, pitch 10 degrees northwest, pay vein 12 inches; nature of ore, tellurium; assay value, $477 per ton; main shaft 875 feet deep, 1 tunnel 70 feet in length; yield to date, $6,000,

December. Owned by Charles S. Davis, Allen J. and Noah Terry, of Jamestown; claim 150 by 1,500 feet; discovered 1875; located on Golden Age Hill; Central mining district, 2 miles from Jamestown, and 12 miles from Boulder; course of vein, northeast and southwest; average width 6 feet; pitch 15 degrees south, pay vein 20 inches; nature of ore, free gold; average value, $30 per ton; main shaft 100 feet deep, 1 level 35 feet in length and 50 feet from surface.

Delphi Consolidated Gold and Silver Mining Co. Incorporated December 24, 1877; capital stock, $1,000,000, in 100,000 shares, $10 each; offices Boulder, Colorado, and Cambridge, Ohio. The property of this company is patented and consists of 5 lode claims, 150 by 1,500 feet each, and a mill site of 5 acres; located in town of Delphi, Sugar Loaf mining district, 10 miles from Boulder; nature of ore, tellurium of gold and silver; average value, $50 silver, and $250 gold per ton; main shaft 100 feet deep, 1 tunnel 100 feet in length, 2 levels 20 feet each; have in store 50 tons of ore which will average $50 per ton.

Dexter. Owned by Alexander G. Montreuil, Ely Bruner, and Burdell & Platt; office Boulder; claim 150 by 1,500 feet; discovered 1877; located in Sunshine Gulch, Gold Hill mining district, 1 mile from Sunshine and 6 miles from Boulder; vertical vein running northeast and southwest, average width 4 feet, pay vein 2 to 4 feet; nature of ore, sylvanite and tellurium;

average value, $323 per ton; main shaft 50 feet deep, 1 tunnel 40 feet in length; yield to date, $300.

Dexter. Owned by Samuel Baeder, Jacob Hummel, and _____ Robbins; office Boulder; claim 150 by 1,500 feet; discovered 1875; located in Silver City Gulch, Gold Hill mining district, 9 miles from Boulder; vertical vein running northeast and southwest, average width 10 feet, pay vein 6 inches; nature of ore, tellurium and free gold; value $1,500 per ton; main shaft 40 feet deep, 1 level 12 feet in length.

Dexter. Owned by Scott Bros., Joseph Shaw, Samuel Wood, and E.A. Hupper; office Caribou; claim 50 by 1,400 feet; discovered 1872; located on Caribou Hill, Grand Island mining district, 23 miles from Boulder; course of vein, east and west, average width 2 feet, pitch 10 degrees north, pay vein 10 inches; nature of ore, galena and sulphurets; average value, $75 per ton; main shaft 30 feet deep; shaft house 14 by 20 feet.

Dime. Benjamin F. Smith and J. McCadon, proprietors; office Golden; claim 150 by 1,500 feet; discovered 1875; located on French Mountain, in town of Crisman, Sugar Loaf mining district, 8 miles from Boulder; course of vein, north and south, average width 5 feet, pitch 10 degrees northwest; nature of ore, tellurium; value $1,700 per ton, 1 tunnel 220 feet and 1 adit 25 feet in length.

Dolly Vardin. Lexington and Colorado Gold and Silver Mining Co., proprietors; capital stock, $300,000; John W. Weare, superintendent; claim 150 by 300 feet; discovered 1874; located in Pennsylvania Gulch, 5 miles from Nederland and 15 miles from Boulder; vertical vein, running northeast and southwest, average width 10 feet, pay vein 6 feet; nature of ore, iron and copper pyrites and free gold; average value, $40 per ton; main shaft 40 feet deep, 1 tunnel site 1,500 by 3,000 feet, with a tunnel 290 feet in length.

Dom Pedro. James K. Jones, proprietor; office Sunshine; claim 150 by 1,500 feet; discovered 1877; located in Sunshine Gulch, Gold Hill mining district, 1 mile from Sunshine and 7 miles from Boulder; course of vein, northeast and southwest, average width 4 feet, pay vein 3 inches; nature of ore, tellurium; average value, $30 per ton; 1 tunnel 28 feet in length.

Dorchester. Ferdinand C. Albright, of Caribou, proprietor; claim 150 by 1,500 feet; discovered 1872; located on Idaho Hill, Grand Island mining district, 23 miles from Boulder; course of vein, east and west, average width 4 to 5 feet, pitch 25 degrees north, pay vein 8 inches; nature of ore, galena, iron and copper pyrites; assay value $35 per ton, main shaft 60 feet, and 1 other 10 feet deep; shaft house 12 by 18 feet.

Doss. Orris and Frank O. Blake, George and Jacob Hepner, proprietors; office Boulder; claim 150 by 1,500 feet; discovered 1875; located on the south slope of Hoosier Hill, near town of Delphi, Sugar Loaf mining district, 9 miles from Boulder; course of vein, northeast and southwest, average width 4 feet, pitch 10 degrees northwest, pay vein 6 to 12 inches; nature of ore, tellurium and free gold; average value, $100 per ton; main shaft 150 feet deep, 2 levels 20 and 100 feet in length respectively, reached by tunnel at a distance of 270 feet from the mouth, and intersected at a depth of 250 feet; shaft house and 2 dwelling houses. [Major Orris Blake was born in Syracuse, New York, on March 25, 1830, studied law, and served in the Civil War, eventually serving on the staff of Major General McCook. He moved to Denver in 1872, and then to Boulder, where he divided his time between his law practice and his mining interests.]

Dunraven. John L. Schellenger, proprietor; office Magnolia; claim 150 by 1,500 feet; discovered 1876; located in Jackson's Camp, Magnolia mining district, 8 miles from Boulder; vertical vein trending east and west, width 2 feet; nature of ore, tellurium; value $2,000 per ton; 1 shaft 80 feet deep.

East Idaho. Owned by Charles B. Buckingham, J.L. Herzinger, S.B. Harker, Charles Haynor, and W.B. Williams; office Boulder; claim 150 by 750 feet, patented; located on Idaho Hill, Grand Island mining district, 20 miles from Boulder; course of vein, east and west, width 3 ½ feet, pitch 5 degrees north, pay vein 4 to 30 inches; nature of ore, silver glance and sulphurets; assay value $1,000 to $10,000 per ton; main shaft 90 feet deep; shaft house 12 by 20 feet.

Eastern Slope. Henry Deitz, proprietor, office Boulder; claim 150 by 1,500 feet; discovered 1876; located in Bummer Gulch, Sugar Loaf mining district, 9 miles from Boulder; course of vein, northeast and southwest, width 2 feet, pitch northwest; nature of ore, tellurium; average value, $45 per ton; 1 shaft 30 feet deep, 1 level 140 feet in length; yield to date, $3,500.

Eclipse. Owned by George Moore, James F. Clark, and others; office Gold Hill; claim 150 by 1,500 feet; discovered 1875; located on Four Mile Creek, Sugar Loaf mining district, 7 miles from Boulder; vertical vein running north and south, average width 2 feet, pay vein 4 inches; nature of ore, tellurium; average value, $500 per ton; main shaft 87 feet deep, aggregate length of adits 35 feet; total yield, $3,000.

El Dorado. Owned by Silas S. Kennedy, Benjamin H. Eaton, William H. Lessig, Jacob Hawke, and William A. Christian; office Balarat; claim 150 by 1,500 feet, patented; discovered 1876; located in Long's Gulch, near town of Balarat, Central mining district, 17 miles from Boulder; course of vein, north 15 degrees west, average width 3 feet, pitch 45 degrees east, pay vein 6 inches; nature of ore, tellurium; 1 tunnel 125 feet in length and 65 feet from surface, 1 level 160 feet in length and 65 feet from surface.

El Dorado No. 2. Owned by H.J. Van Wetering and N.D. Peterson, of Sunshine; claim 150 by 1,000 feet; located on Lee Hill, Gold Hill mining district, 3 miles from town of Sunshine and 9 miles from Boulder; course of vein, northwest and southeast, average width 4 feet, pitch 5 degrees east, pay vein 10 inches; nature of ore, tellurium; average value, $200 per ton; main shaft 80 feet deep, 1 drift 75 feet in length.

Elephant. Samuel Mishler and Iswold Lunsford, proprietors; office Caribou; claim 50 by 1,400 feet; discovered 1870; located on Idaho hill, near town of Caribou, Grand Island mining district, 23 miles from Boulder; course of vein, east and west, average width 9 feet, pay vein 5 feet, pitch 40 degrees north; nature of ore, silver glance and brittle silver; average value, $117 per ton; main shaft 125 feet deep, 2 levels 80 feet aggregate; yield to date, $3,000.

Ellen. Owned by Henry Laffer, Charles F. Pease, Frederick L. Higbee, and William McKay; office Springdale; claim 135 by 1,500 feet; discovered 1875; located near Jim Creek, Central mining district, 12 miles from Boulder; course of vein, northeast and southwest, average width 6 feet, pitch 10 degrees south, pay vein 6 to 12 inches; nature of ore, free gold; average value, $100 per ton; main shaft 100 feet deep; yield, $20,000.

Emmet. Anthony Arnett, of Boulder, proprietor; claim 50 by 1,700 feet, patented; discovered 1859; located on Gold Hill, Gold Hill mining district, 10 miles from Boulder; vertical vein running northeast and southwest; average width 4 feet; pay vein 2 feet; nature of ore, free gold; average value, $100 per ton; main shaft 100 feet deep; yield, $20,000.

Empress, Extension of Magnolia. Owned by James Stevens, Amos Bixby, Eugene Wilder, J. Alden Smith, and F.W. Strout; office Boulder; claim 150 by 1,500 feet; located in town of Magnolia, Gold Hill mining district, 7 miles from Boulder; course of vein, northeast and southwest; average width 6 feet, pitch 5 degrees southeast, pay vein 4 feet; nature of ore, tellurium; average value, $50 per ton; main shaft 20 feet deep.

Encounter. James D. Smith, William A. and Charles J. Christian, proprietors; office Denver; claim 150 by 1,500 feet; discovered 1877; located in Central Gulch, 2 miles from Balarat and 15 miles from Boulder; course of vein, east and west, average width 7 feet, pitch 35 degrees south; nature of ore, free gold; main shaft 15 feet deep.

Erie. Henry Taylor and Albert Bills, proprietors; office Boulder; claim 150 by 1,500 feet; discovered 1867; located near Gold Lake, Central mining district, 3 miles from Gold Hill and 15 miles from Boulder; vertical vein trending east and west; average width 5 feet, pay vein 10 inches; nature of ore, free gold; average value, $85 per ton; main shaft 30 feet deep.

Esta Bueno. Owned by Theodore O. Saunders; office Sunshine; claim 150 by 1,500 feet; discovered November 1874; located in Gold Hill mining district, 1 mile south of Sunshine and 5 miles from Boulder; course of vein, northeast and southwest, width 5 feet, pay vein 1 foot, pitch 20 degrees north; nature of ore, tellurium and free gold; average value, $50 per ton; developed by a shaft 50 feet deep and a surface opening 200 feet in length.

Eugene. Anthony Arnett, proprietor; office Boulder; claim 50 by 1,500 feet, patented; discovered 1859; located on Gold Hill, Gold Hill mining district, 8 miles from Boulder; vertical vein running northeast and southwest, average width 3 feet, pay vein 2 feet; nature of ore, free gold; average value, $100 per ton; main shaft 25 feet deep, aggregate depth of other shafts 150 feet; yield, $10,000.

Eureka. Owned by George H. Batchelder, James H. Guise, and James G. Rutter; office Left Hand; claim 150 by 1,500 feet, patented; discovered 1872; located on Gold Hill, Gold Hill mining district, 10 miles from Boulder; course of vein, east and west, width 5 feet, pay vein 12 inches, pitch north; nature of ore, galena and gray copper, value $80 to $300 per ton; main shaft 100 feet deep, 1 tunnel 20 feet in length.

Eureka. Owned by Dr. Shoup and John M. Maxwell; office Boulder; claim 150 by 6,000 feet; discovered 1876; located on North Boulder Creek, Albion mining district, 22 miles from Boulder; course of vein, northeast and southwest, average width 6 feet, pitch 10 degrees north; nature of ore, galena; average value 6 ounces silver per ton, 50 per cent lead and 20 per cent copper; 1 tunnel 100 feet from surface and 120 feet in length.

Evening Star. Capt. Ebenezer Rowland, John B. Maas, and George Thomas, proprietors; office Boulder; claim 150 by 1,500 feet; discovered 1877; located in McKnight Gulch, 2 miles from Gold Hill and 11 miles from Boulder; vertical vein running east and west, average width 3 feet; nature of ore, tellurium, and free gold; average value, $30 per ton; main shaft 25 feet deep, 1 drift 60 feet in length.

Evening Star. Benjamin F. Smith and Thomas Brady, proprietors; office Golden; claim 150 by 1,500 feet; discovered 1876; located near Sunbeam Gulch; Sugar Loaf mining district, 7 miles from Boulder; vertical vein running northeast and southwest, average width 5 feet, pay vein 12 inches; nature of ore, tellurium and free gold; average value, $700 per ton; main shaft 50 feet, and 1 other shaft 40 feet deep.

Extension. John A. Ellet & Co., proprietors; office Boulder; claim 50 by 1,400 feet, patented; discovered 1872; located on Caribou Hill, Grand Island mining district, near town of Caribou, 23 miles from Boulder; course of vein, east and west, average width 6 feet, pitch 10 degrees north, pay vein 1 to 20 inches; nature of ore, sulphurets, and native silver, average value 200 ounces silver per ton; main shaft 32 feet deep.

Fanny Mott. Armstrong, Ankeny & Co., proprietors; offices, Jamestown, Colorado, and Xenia, Ohio; claim 150 by 1,500 feet; discovered 1876; located on Golden Age Hill, Central mining district, 14 miles from Boulder; course of vein, northeast and southwest, pay vein 8 to 9 inches; nature of ore, free gold; average value, $40 per ton; main shaft 85 feet, and 1 other shaft 10 feet deep.

Fifty-Nine. Gold Hill Mining Co., proprietors; office Boulder; claim 50 by 1,200 feet, patented; discovered 1859; located near town of Gold Hill, 10 miles from Boulder; course of vein, northeast and southwest, average width 6 feet, pitch north, pay vein 2 feet; nature of ore, iron and copper pyrites; average value, $600 per cord; main shaft 230 feet deep, aggregate length of levels 1,000 feet; total yield, $100,000.

First National. Owned by The National Gold and Silver Mine and Mining Co.; claim 150 by 2,417 feet; discovered 1874; located in Weare Gulch, 2 miles from Caribou and 21 miles from Boulder; vertical vein running southeast and northwest, average width 10 feet, pay vein 10 inches to 3 feet; nature of ore, gray copper and free gold; average value, $60 per ton; main shaft 154 feet deep, aggregate length of other shafting 150 feet, 1 adit 104 feet from surface and 550 feet in length, 1 level 400 feet in length and 50 feet from surface; total yield, $5,000; on this property is erected a 25-stamp mill, containing all machinery necessary to the successful treatment of ore.

Forrest. Owned by John W. King, Henry C. Brown, George W. Brown, B.D. Spencer, and John Stuart; office Denver; claim 150 by 1,500 feet, patented; discovered 1871; located on Four Mile Creek, Sugar Loaf mining district, 10 miles from Boulder; course of vein, northeast and southwest; width 4 feet; pay vein 27 inches, nature of ore, tellurium and free gold; average value, $77 per ton; main shaft 145 feet deep; shaft house 15 by 18 feet; yield to date, $5,000.

Fourth of July. Owned by William J. and William E. Mann and Charles H. Taylor; claim 150 by 4,500 feet, discovered in 1874; located on Middle Boulder Creek, Grand Island mining district, 23 miles from Boulder; vertical vein running east and west; average width 50 to 300 feet; nature of ore, galena, carbonates, and gray copper; assay value, $500 per ton; main shaft 90 feet deep.

Fourth of July Consolidated Mining Co. Organized under the laws of the State of New York, March 4, 1878; capital stock, $3,000,000, in 300,000 shares, $10 each; offices Boulder, Colorado, and New York City; George Earle, President, Robert J.S. White, Secretary.

Frankfort. Heber F. and Horace A. Learnard, proprietors; office Rockville; claim 150 by 1,500 feet, discovered 1876; located in Rockville Camp, Gold Hill mining district, near town of Rockville, 10 miles from Boulder; course of vein, northeast and southwest, width 7 feet, pitch 10 degrees south, pay vein 6 inches; nature of ore, tellurium; average value, $30 per ton; 1 drift 36 feet in length.

Frieburg.. Harry C. Crary, proprietor; office Gold Hill; claim 50 by 1,450 feet; discovered 1873; located in Gold Hill mining district 10 miles from Boulder; vertical vein running northeast and southwest; average width 3 feet, pay vein 8 inches; nature of ore, tellurium and free gold; average value, $75 per ton; main shaft 55 feet deep, 1 adit 78 feet in length; yield, $1,500.

Frieburg Mining and Tunnel Co. Incorporated October 7, 1875; capital stock, $10,000, in 100 shares, $10 each; office Denver; organized by Loudon Mullin, George L. Aggers, and Gustave Korner.

Gage. Chicago and Colorado Mining Co., proprietors; offices Boulder, Colorado, and Chicago, Illinois; claim 150 by 1,500 feet, patented; discovered 1876; located on Left Hand Creek, Ward mining district, 4 miles from town of Gold Hill and 16 miles from Boulder; course of vein, east and west; average width 4 feet, pitch 45 degrees north, pay vein 10 inches; nature of ore, iron pyrites; average value, $84 per ton; main shaft 40 feet deep, 1 tunnel 200 feet in length.

General Grant. William and James Pell, proprietors; office Boulder; claim 150 by 1,500 feet, patented; located on Gold Hill, Gold Hill mining district, 10 miles from Boulder; vertical vein running northeast and southwest, average width 6 feet; nature of ore, iron pyrites; main shaft 60 feet deep.

George Washington. L.J. Butzel and William Howell, proprietors; office Boulder; claim 150 by 1,500 feet; discovered 1876; located in town of Salina, Sugar Loaf mining district, 10 miles from Boulder; vertical vein running north and south; average width 6 feet; pay vein 22 inches; nature of ore, galena and gray copper; average value, $175 per ton; 1 tunnel 85 feet in length.

Georgia Mining Co. Incorporated May 1, 1876; capital stock, $50,000, in 5,000 shares, $10 each; office Denver; organized by Herman Fischer, Paul C. Glave, and John Kiefer.

Gertrude. Henry Taylor and Albert Bills, proprietors; office Boulder; claim 150 by 1,500 feet; discovered 1877; located near Gold Lake, Central mining district, 3 miles from town of Gold Hill and 15 from Boulder; course of vein, northwest and southeast; average width 4 ½ feet, pitch 45 degrees north, pay vein 15 inches; nature of ore, free gold and carbonates; average value, $65 per ton; main shaft 18 feet deep.

Giant. Owned by John S. Reid, John S. Moore and L. Ford; office Magnolia; claim 150 by 1,500 feet; discovered 1876; located on Giant Hill, Magnolia mining district, nine miles from Boulder; course of vein, east and west, width 13 feet, pitch 5 degrees south, pay vein 2 to 22 inches; nature of ore, free gold; average value, $60 per ton; main shaft 87 feet deep, 2 adits 140 and 25 feet in length respectively, 2 levels, aggregate length 130 feet; shaft house 16 by 20 feet.

Gillaspie Gold and Silver Mining and Milling Co.. Incorporated October 1, 1875; capital stock, $125,000, in 12,500 shares, $10 each; office Denver; organized by James H. Ralston, D.F. Brown, and H.B. Gillaspie.

Gilpin County. Andrew G. Olson, proprietor; office Springdale; claim 150 by 1,500 feet; discovered 1875; located near town of Springdale, Central mining district, 12 miles from Boulder; course of vein, northeast and southwest; average width 4 feet, pitch 10 degrees south, pay vein 18 inches; nature of ore, pyrites and free gold; main shaft 40 feet deep.

Gladiator. Cyrus Taylor, proprietor; office Jamestown; claim 150 by 1,500 feet; discovered 1875; located on Jim Creek, in town of Springdale, Central mining district, 12 miles from Boulder; course of vein, northeast and southwest; average width 4 feet, pitch 20 degrees south, pay vein 5 inches to 3½ feet; nature of ore, tellurium, iron and copper pyrites; average value of first-class ore, $8,000, and second-class, $700 per ton; main shaft 125 feet deep, 1 adit 25 feet in length, 1 level 60 feet from surface and 25 feet in length.

Glide. William and S.W. Thompson, proprietors; office Rockville; claim 150 by 1,500 feet; located in town of Rockville, 2 miles from Sunshine and 10 miles from Boulder; course of vein, northeast and southwest; average width 6 feet, pitch 10 degrees south, pay vein 6 inches; nature of ore, tellurium and free gold; average value, $30 per ton; main shaft 50 feet deep, 1 tunnel 70 feet in length.

Gold Coin Mining Company of Colorado. Incorporated January 12, 1877; capital stock, $100,000, in 4,000 shares, $25 each; office Georgetown; trustees: Charles R. Fish, John G. Roberts, Albert R. Forbes, Charles H. Morris, Philip E. Morehouse, Adolph H. Borman, and E.R. Pierson.

Gold Hill Mining Co. Incorporated October 26, 1875; capital stock, $125,000, in 12,500 shares, $10 each; offices Denver and Sunshine, Colorado; organized by Almerson Wilson, George H. Batchelder, and Edward F. Wallace.

Gold Hill Tunnel Co. Incorporated June 15, 1875; capital stock, $250,000, in 2,500 shares, $100 each; office Boulder; organized by Edward A. Phillips, Frederick A. Squires, George H. Batchelder, John Morrison, and Joseph Roper, Jr.

Gold Nugget. Heber F. and Horace A. Learnard, of Rockville, proprietors; claim 150 by 1,500 feet; discovered 1877; located in Rockville Camp, Gold Hill mining district, 10 miles from Boulder; course of vein, northeast and southwest; average width 6 feet, pitch 10 degrees south, pay vein 6 inches; nature of ore, tellurium and free gold; average value, $30 per ton; 1 shaft 12 feet deep.

Gold Ring. Alexander Von Wendt and Gustave Rutar, proprietors; office Gold Hill; claim 50 by 1,500 feet; discovered 1875; located on Gold Hill, Gold Hill mining district, 12 miles from Boulder; vertical vein running northeast and southwest; average width 6 feet, pay vein 2 ½ feet; nature of ore, tellurium; average value, $70 per ton; main shaft 100 feet deep, 1 tunnel 166 feet in length, 1 level 60 feet in length and 65 feet from surface; yield, $8,000.

Golden. B.F. Smith, of Golden, proprietor; claim 150 by 1,500 feet; discovered 1876; located in Sunbeam Gulch, Sugar Loaf mining district, 7 miles from Boulder; course of vein, east and west; average width 4 feet, pitch 5 degrees north, pay vein 22 inches; nature of ore, tellurium; value $3,900 per ton; main shaft 42 feet deep.

Golden Age. Harlan P. Walker, proprietor; office Jamestown; claim 150 by 1,500 feet; discovered 1875; located on Golden Age Hill, Central mining district, 2 miles from Jamestown and 12 miles from Boulder; course of vein, northeast and southwest, average width 5 feet, pitch 45 degrees south, pay vein 30 inches; nature of ore, free gold; average value, 8 ounces gold per cord; main shaft 200 feet deep, other shafting aggregating 200 feet, 2 levels 100 feet from surface, 70 and 40 feet in length respectively; shaft house and blacksmith shop 30 by 47 feet; yield, $40,000. In connection with this property is a mill 40 by 70 feet, with a capacity for treating 15 tons of ore daily.

Golden Crown Mining Co. Incorporated January 29, 1876; capital stock, $300,000, in 30,000 shares, $10 each; office Denver; organized by D.F. Brown, J.H. Jones, and E.W. Pierce.

Golden Eagle. Henry Deitz, proprietor; office Boulder; claim 150 by 7,500 feet; discovered 1876; located in Bummer Gulch, Sugar Loaf mining district, 9 miles from Boulder; course of vein, northeast and southwest, average width 8 feet, pitch 25 degrees north; pay vein 2 to 3 feet; nature of ore, tellurium; average value, $150 per ton; main shaft 140 feet deep, 2 levels aggregating 140 feet in length; yield, $20,000.

Golden Era. Henry Taylor and Albert Bills, proprietors; office Boulder; claim 150 by 1,500 feet; discovered 1867; located near Gold Lake, Central mining district, 3 miles from Gold Hill and 15 miles from Boulder; course of vein, east and west; average width 3 feet, pitch south, pay vein 20 inches; nature of ore, carbonate of copper and free gold; average value, $50 per ton; main shaft 50 feet deep, 1 level 25 feet from surface and 20 feet in length.

Golden Gem. Owned by Joseph P. McIntosh, John A. Gilman, Allen J. and Noah Terry, and Robert A. McBride, of Jamestown; claim 150 by 1,500 feet; located in Central Gulch, Central mining district, 1 ½ miles from town of Jamestown and 12 miles from Boulder; course of vein, northeast and southwest; average width 5 feet, pitch 10 degrees south, pay vein 12 inches; nature of ore, free gold; main shaft 40 feet deep.

Government. Wright W. and Horace F. Hall, proprietors; office Left Hand; claim 150 by 1,500 feet; discovered 1877; located in Long's Gulch, Gold Hill mining district, 9 miles from Boulder; course of vein, northeast and southwest; average width 2 feet, pitch 18 degrees north, pay vein 18 inches; nature of ore, tellurium, gray copper and free gold; average value, $165 per ton; main shaft 25 feet deep.

Grand Central. Major Orris Blake, Thomas Reilly, William L. Crooks, proprietors; office Boulder; claim 150 by 1,500 feet; discovered 1874; located on Boulder Creek, Magnolia mining district, 6 miles from Nederland and 13 miles from Boulder; vertical vein running southeast and northwest; average width 20 feet; nature of ore, gray copper; average value, $50 per ton; main shaft 100 feet deep, 1 tunnel 20 feet in length.

Grand Republic. Victor Roux and Henry Chmal, proprietors; office Boulder; claim 150 by 1,500 feet; discovered 1873; located in Sunbeam

Gulch, Sugar Loaf mining district, 1 mile from Salina and 8 miles from Boulder; vertical vein trending east and west, width 6 feet, pay vein 18 inches; nature of ore, tellurium and free gold; average value, $100 per ton; main shaft 50 feet deep.

Grand Trunk. Webster Gold and Silver Mining Co., proprietors; Capt. Ebenezer Rowland, superintendent; office Boulder; claim 150 by 1,500 feet; discovered 1873; located in Gold Hill mining district, 8 miles from Boulder; course of vein, northeast and southwest, average width 6 feet, pitch 15 degrees north, pay vein 8 inches; nature of ore, tellurium; average value, $30 per ton; main shaft 50 feet deep.

Grand View. Owned by the Grand View Mining Co., of Cincinnati; organized under the laws of the state of Ohio; J.H. Clemmer, president, W.P. Powell, secretary, Cornelius C. Howell, manager; offices Sunshine, Colorado, and Cincinnati, Ohio; claim 150 by 1,500 feet, patented; discovered May, 1874; located in town of Sunshine, Gold Hill Mining district, 7 miles from Boulder; course of vein, northeast and southwest, average width 2 ½ feet; pitch 25 degrees west; nature of ore, tellurium, sylvanite, petzite, calaverite, coloradoite, and free gold, value $70 to $10,000 per ton; main shaft 350 feet deep, 4 levels, 50, 100, 166, and 225 feet from surface respectively, aggregating 500 feet in length; shaft house, 24 by 70 feet, with iron roof, containing a 20-horse power engine, and an apparatus for hoisting ore; assay laboratory 18 by 24 feet, and a large ore house; yield to date, $60,000. Since coming under the management of Mr. C.C. Howell, other property in the immediate vicinity is being purchased with a view of enlarging and reorganizing the old company.

Grand View Silver Mining Co., of Caribou, Colorado. Incorporated July 14, 1876, capital stock, $150,000, in 15,000 shares, $10 each; office Denver; organized by George M Lahaye, H.J. De Bruyn Prince, and Kemp G. Cooper.

Grant. Edward W. Henderson, of Central City, proprietor; claim 150 by 1,500 feet; discovered 1864; located at Ward, Ward mining district, 14 miles from Boulder; vertical vein running east and west, average width 2 feet; pay vein 4 to 18 inches; nature of ore, tellurium, aggregate value $250 pr cord; main shaft 25 feet deep; open surface 150 feet in length.

Gray Copper. Geo. R. Williamson, proprietor; office Crisman; claim 150 by 1,500 feet, patented; discovered 1876; located in Sunshine Gulch, Sugar Loaf mining district, 9 miles from Boulder; vertical vein running northeast and southwest, average width 2 feet, pay vein 18 inches; nature of ore, iron pyrites and carbonate of copper; average value, $450 per ton; 5 adits 50, 75, 90, 100, and 125 feet in length respectively; yield, $10,000.

Gray Eagle. S.F. Huddleston and John N. Gillaspie, proprietors; office Rockville; claim 50 by 2,800 feet, patented; discovered 1868; located on Gray Eagle Gulch, Left Hand Creek, 2 miles from Gold Hill and 12 miles from Boulder; course of vein, northeast and southwest, average width 5 feet, pitch 5 degrees south; pay vein 2 feet; nature of ore, tellurium, galena, zinc blende, iron pyrites and free gold; average value, $40 per ton; main shaft 80 feet deep, 2 tunnels 120 and 200 feet in length respectively, 1 level 50 feet in length.

Great Eastern. Owned by T.R. and Charles N. Stinson, Albert Brown, and L.R. Chittenden; office Denver; claim 150 by 1,500 feet; located on Hoosier Hill, Gold Hill mining district, near town of Salina, 10 miles from Boulder; course of vein, northwest and southeast, width 4 feet, pay vein 6 inches, pitch 15 degrees northwest; nature of ore, tellurium and free gold; average value, $200 per ton; main shaft 210 feet deep; shaft house 12 by 20 feet.

Great Eastern Mining Property. Cornelius C. Howell, manager; office Boulder; comprises 8 claims, 150 by 1,500 feet each; located on the same vein as the celebrated Melvina, near town of Salina, Gold Hill mining district, 10 miles from Boulder; course of vein, northeast and southwest, width 2 feet, pay vein 6 to 8 inches, pitch 25 degrees northwest; nature of ore, tellurium and free gold; average value, $250 per ton; main shaft 200 feet deep, 2 levels 90 feet each; shaft house 30 by 40 feet, containing an engine of 15-horse power; yield to date, $10,000.

Great Republic Gold and Silver Mining Co. Incorporated December 11, 1877; capital stock, $300,000, in 30,000 shares, $10 each; office Black Hawk; organized by Thomas J. Oyler, Frederick Eggert, and Edward O'Neil.

Greenback. Anthony Arnett and A.J. Kimber, proprietors; office Boulder; claim 150 by 1,500 feet; discovered 1877; located on Bonanza Hill, Central mining district, 15 miles from Boulder; course of vein, east and west, average width 3 feet, pitch north, pay vein 18 inches; nature of ore, free gold; average value, $80 per ton; main shaft 35 feet deep, aggregate depth of other shafts 60 feet, yield, $5,000.

Green Island. David McCormick, proprietor; office Gold Hill; claim 150 by 1,500 feet; discovered 1878; located in Gold Hill mining district, 11 miles from Boulder; vertical vein running northeast and southwest, average width 4 feet, pay vein 10 inches; nature of ore, free gold; 1 shaft 18 feet deep.

Grenada. Owned by William Sellers and David J. Rollins; office Gold Hill; claim 150 by 1,500 feet; discovered 1875; located in Gold Hill mining district 10 miles from Boulder; course of vein, northeast and southwest, average width 5 ½ feet, pitch 45 degrees south; nature of ore, tellurium; average value, $100 per ton; main shaft 30 feet deep.

Hanging Rock Group of Mines. Chicago and Colorado Mining Co., proprietors; located in Ward mining district, at head of the North Boulder, 4 miles from Gold Hill.

Happy Jack. Owned by Samuel Baeder, Jacob Hummel, and _____ Robbins; office Boulder; claim 150 by 1,500 feet; discovered 1875; located in Silver City Gulch, Gold Hill mining district, 9 miles from Boulder; vertical vein running northeast and southwest, average width 16 feet; pay vein 8 inches; nature of ore, tellurium and free gold; average value, $600 per ton; main shaft 45 feet deep, one tunnel 85 feet and 1 level 35 feet in length.

Harriet N. Sylvester Hill, of Boulder, proprietor; claim 150 by 1,500 feet; discovered 1877; located on Jim Creek, Jim Creek mining district, 15 miles from Boulder; vertical vein running east and west, average width 7 feet, pay vein 5 inches; nature of ore, iron pyrites; assay value, $100 per ton; main shaft 20 feet deep.

Hart. Eva Fee and Samuel L. Duston, proprietors; office Cincinnati, Ohio; claim 150 by 1,500 feet; discovered 1876; located on Hoosier Hill, Gold Hill mining district, 10 miles from Boulder; course of vein, northeast and southwest, average width 3 feet; pitch northwest, pay vein 20 inches; nature of ore, iron pyrites; average value, $50 per ton; main shaft 27 feet deep.

Hecla. Extension of Big Blossom, William G. Pell, Capt. E. Rowland, and D.D. Hendrick, proprietors; office Boulder; claim 150 by 1,500 feet; discovered 1878; located near Springdale, Central mining district, 12 miles from Boulder; course of vein, northeast and southwest, average width 3 ½ feet; pitch 5 degrees north; nature of ore, tellurium and free gold; average value, $60 per ton; main shaft 20 feet deep.

Hercules. Owned by Orris and Frank O. Blake and Jacob Cherney; office Boulder; claim 150 by 1,500 feet; discovered 1878; located on the south slope of Hoosier Mountain, Sugar Loaf mining district, near town of Delphi [later renamed Wall Street], 9 miles from Boulder, course of vein, northeast and southwest, average width 4 feet, pitch 10 degrees northwest; pay vein 6 to 12 inches; nature of ore, tellurium; average value, $220 per ton; main shaft 30 feet deep.

Hillsboro. Owned by Elbridge Forsyth, William Hobson, and Charles N. Hockaday; office Magnolia; discovered 1875; located at Magnolia, Magnolia mining district, 8 miles from Boulder; vertical vein trending east and west, width 4 feet; assay value $50 per ton; 1 shaft 25 feet deep.

Hoodo. William and James Pell, proprietors; office Boulder; claim 50 by 1,500 feet, patented; discovered 1859; located on Gold Hill, Gold Hill mining district, 9 miles from Boulder; vertical vein running northeast and southwest; average width 5 feet; nature of ore, iron and copper pyrites; average value, $60 per ton; main shaft 60 feet deep.

Hoosier, or Dana. Squires, Neikirk & Co., proprietors; office Boulder; claim 50 by 1,500 feet, patented; discovered 1867; located on Hoosier Hill, near Gold Run, Gold Hill mining district, 12 miles from Boulder; vertical vein running north and south; average width 25 feet; nature of ore, gray copper and sulphurets of silver; average value, $250 per ton; main shaft 150 feet deep, 1 tunnel 100 feet in length, shaft house 40 by 60 feet; yield to date, $20,000.

Hopewell. Owned by the National Gold and Silver Mill and Mining Co., Kentucky; capital stock, $750,000, Samuel Clay, Jr., president, Norman Whitney, vice-president, T.S. Hamilton, secretary, Nathan Weare, superintendent; claim 150 by 4,500 feet; discovered 1875; located on Caribou Hill, Grand Island mining district, 2 miles from Caribou and 21 miles from Boulder; vertical vein running east and west; average width 3 feet, pay vein 8 to 20 inches; nature of ore, gray copper, carbonates and sulphurets; average value, $200 to $400 per ton; main shaft 125 feet deep, 1 tunnel 300 feet in length and 70 feet from surface, 1 level 80 feet in length and 30 feet from surface.

Horsefall. Richard Blower, of Boulder, proprietor; claim 50 by 1,000 feet, patented; discovered 1859; located in Gold Hill mining district, near town of Gold Hill, 12 miles from Boulder; vertical vein running northeast and southwest; average width 9 feet, pay vein 2 to 16 inches; nature of ore, iron pyrites and free gold; average value, $70 per ton; main shaft 240 feet deep, aggregate depth of other shafts 1,000 feet, 1 level 300 feet in length and 90 feet from surface.

Huberty. Owned by Henry Huberty, Henry Mack, and M.J. Holt; office Gold Hill; claim 150 by 1,500 feet; discovered 1875; located in Gold Hill mining district; discovered 11 miles from Boulder; vertical vein running east and west, average width 5 feet, pay vein 6 inches; nature of ore, tellurium and free gold; main shaft 42 feet deep, 1 adit 60 feet in length.

Humboldt Property. Owned by Edward J. Binford and John J. Ellingham; office Denver; comprised of 5 patented claims 150 by 1,500 feet each; located in the town of Ward, Ward mining district, 18 miles from Boulder; course of vein, east and west, width 10 feet, pay vein 1 to 2 feet; nature of ore, iron pyrites and free gold; average value, $50 per ton; 2 shafts 50 and 150 feet deep respectively, shaft house 26 by 60 feet; 10 stamp mill, with a capacity for treating 10 tons of ore daily, containing a 20-horse power engine and a slack belt hoisting machine; yield to date, $3,000.

I.O.U. Henry N. Coffee, Sylvester Hill, and Anthony Arnett, proprietors; office Balarat; claim 150 by 1,500 feet; discovered 1874; located in Four Mile Gulch, Gold Hill mining district, 6 miles from Boulder, course of vein, northeast and southwest, average width 4 feet, pitch 10 degrees west; nature of ore, tellurium and gray copper; average value, $40 per ton; main shaft 70 feet, 1 tunnel 75 feet in length.

Illinois. Henry E. Wood, J. Alden Smith, and E.M. Hawkins, proprietors; office Sunshine; claim 150 by 1,500 feet; located in Sunshine Gulch, half mile from town of Sunshine, Gold Hill mining district, 7 miles from Boulder; course of vein, northeast and southwest, average width 2 feet, pitch 5 degrees east, pay vein 5 inches; nature of ore, tellurium, petzite, and sylvanite; average value, $108 per ton; main shaft 50 feet deep.

Illinois and Colorado Mining Co. Organized under the laws of the State of Illinois June 26, 1877; capital stock, $750,000, in 7,500 shares,

$100 each; offices Jacksonville, Illinois, and Boulder, Colorado; organized by Benjamin F. Bergen, Barnett T. Napier, and Isaac Cooper.

Ingram. Owned by Hiram Fuller, Thomas Shires, and C.G. Buckingham; office Boulder; claim 75 by 1,500 feet; located in Salina, 9 miles from Boulder; course of vein, northeast and southwest, average width 4 feet, pitch 10 degrees north; pay vein 6 inches; nature of ore, tellurium; average value of first class ore, $500 per ton; main shaft 100 feet deep.

Iron. Owned by Ulysses Pugh and Samuel Mishler, of Caribou, proprietors; claim 50 by 1,400 feet; discovered 1864; located on Caribou Hill, Grand Island mining district, 23 miles from Boulder; vertical vein running east and west; average width 8 feet, pay vein 6 feet; nature of ore, iron and copper pyrites; average value, $100 per ton; 1 shaft 25 feet deep.

Iron Mountain. Owned by George Fanner, Peter Davis, and W. Worthing; office Salina; claim 150 by 1,500 feet; discovered 1877; located on Melvina Hill, Gold Hill mining district, 9 miles from Boulder; course of vein, northeast and southwest, average width 10 inches, pitch 5 degrees northwest, pay vein 4 inches; nature of ore, tellurium; assay value $60 per ton; main shaft 20 feet deep.

Isabelle. Owned by Anthony Arnett, John Weese, and D. Whittier; office Boulder; claim 50 by 1,500 feet, patented; discovered 1868; located on Caribou Hill, Grand Island mining district, 22 miles from Boulder; vertical vein running northeast and southwest, average width 4 feet; nature of ore, sulphurets; average value, $850 per ton; main shaft 60 feet deep; yield, $10,000.

Jefferson. Owned by Capt. Ebenezer Rowland, George C. Corning, John P. Mitchell, and John B. Maas, office Boulder; claim 50 by 1,500 feet, patented; discovered 1871; located in McKnight Gulch, 2 miles from Gold Hill and 11 miles from Boulder; vertical vein running northeast and southwest, average width 4 feet, pay vein 16 inches; nature of ore, gray copper, free gold, zinc blende, and sulphurets of copper; average value, $50 per ton; main shaft 50 feet deep, 1 tunnel 150 in length.

John Jay. Archibald J. Van Deren, proprietor; office Boulder; claim 150 by 1,500 feet, patented; discovered 1875; located 3 miles from Jamestown, on James Creek, Central mining district, 15 miles from Boulder; course of vein, northeast and southwest, average width 3 ½ feet, pitch 22 degrees northwest; nature of ore, tellurium and sylvanite, average value, $100 per ton; main shaft 255 feet deep, and 2 other shafts 45 and 56 feet respectively, 6 levels, 40, 50, 80, 110, 167, and 250 feet respectively, aggregating 385 feet in length; 2 shaft houses, 1 ore house and blacksmith shop, hoisting apparatus and a 15-horse power engine.

John M. Kerns. Owned by William A. Meralls, Hugh D. McGraw, and Samuel Force, of Salina; claim 150 by 1,500 feet; discovered 1875; located on Hoosier Hill, Sugar Loaf mining district, half mile from town of Salina, and 8 miles from Boulder; course of vein, northeast and southwest, average width 4 ½ feet, pitch 10 degrees west, pay vein 14 inches; nature of ore, tellurium, sylvanite, and petzite, average value, $128 per ton; main shaft 110 feet deep, aggregate depth of other shafts 586 feet, 1 level 22 feet in length.

Juniata. Owned by Homer H. Smith, Cornelius Van Keuren, Van Nest Talmadge, and Charles Abbey; office Sunshine; claim 150 by 1,500 feet; discovered 1874; located in town of Sunshine, Gold Hill mining district, 7 miles from Boulder; vertical vein running northeast and southwest, average width 3 feet, pay vein 1 foot; nature of ore, galena, gray copper, and tellurium; average value, $200 per ton; main shaft 90 feet deep, 1 drift 30 feet in length.

Kekonga. Owned by Frederick W. Strout and John Pughe, of Sunshine; claim 150 by 1,475 feet, patented; discovered 1875; located on Magnolia Hill, Gold Hill mining district, 5 miles from Boulder; course of vein, northeast and southwest, average width 4 feet, pitch 5 degrees southeast, pay vein 2 feet; nature of ore, tellurium of silver; average value, $45 per ton; main shaft 70 feet deep, 1 level 30 feet in length and 60 feet from surface.

Keno. Thomas J. Graham and John W. Smith, proprietors; office Boulder; claim 50 by 3,000 feet, patented; discovered 1865; located on Hoosier Hill, near town of Gold Hill, 10 miles from Boulder; vertical vein running northeast and southwest, average width 20 feet; nature of ore, galena and gray copper; average value, $40 per ton; main shaft 50 feet deep.

Keystone. Owned by the Keystone Mining Co., Colorado. Incorporated Dec. 16, 1876; capital stock, $150,000, in 1,500 shares, $100 each; Jacob Carvell, president, Wm. A. Hardenbrook, vice president; office Boulder; claim 150 by 1,550 feet; discovered 1875; located on Magnolia

Hill, Magnolia mining district, 8 miles from Boulder; course of vein, northeast and southwest, average width 5 feet, pitch 45 degrees north, pay vein 1 to 20 inches; nature of ore, tellurium; average value, $350 per ton; main shaft 250 feet deep, aggregate length of other shafting 300 feet, 5 levels, 40, 70, 90, 155, and 200 feet from surface respectively, aggregating 600 feet in length; shaft house 30 by 40 feet, 1 engine 20-horse power, and 1 steam pump.

Keystone Mineral and Land Association, of Colorado. Thomas J. Graham, agent. The property of this company is located 9 miles west from Boulder, in the great mineral belt, which passes through the country in a northeasterly and southwesterly course, from the mining regions of Gilpin County on the southwest, and the Balarat district in the northeast part of Boulder County. It comprises over 2,000 acres of mineral land, surrounded by the richest mining districts in the county. Caribou adjoins it on the west, Magnolia on the east and south, and Sugar Loaf on the north. The north and middle Boulder creeks, and the stage road leading from Boulder to Nederland, Caribou and Central City passes through this property. Twenty lodes have been discovered on this land, some of which are rich in gold, silver, copper, and lead; they are very wide and only partially developed; shafts varying in depth from 10 to 70 feet have been sunk and many other prominent mineral bearing loads have not been worked; being discovered below the snow belt, mining operations can be conducted all winter; the water power is unlimited, and the surface is well supplied with timber suitable for all purposes. The title is patented from the government to the original purchasers; it was then conveyed by warrantee deeds to the present owners. Thomas J. Graham, of Boulder, Colorado, is the agent for this property, and to whom all communications should be addressed.

Key West. Owned by Harlan P. Walker, T.J. Griffin, John Miller, and Thomas F. Hendershot; office Jamestown; claim 150 by 1,500 feet; discovered 1878; located on Golden Age Hill, Central mining district, 2 miles from Jamestown, and 10 miles from Boulder; course of vein, northeast and southwest, average width 4 feet, pitch 5 degrees south, pay vein 8 inches; nature of ore, tellurium and free gold; main shaft 25 feet deep.

King William. Andrew G. Olson and J.V. Pomeroy, proprietors; office Jamestown; claim 150 by 1,500 feet; discovered 1875; located in King William Gulch, Central mining district, near town of Springdale, 12 miles from Boulder; course of vein, northeast and southwest, average width 5 feet, pitch 10 degrees south, pay vein 18 inches; nature of ore, tellurium and iron pyrites; average value of first-class ore, $10,000, second-class ore, $200, and third-class ore, $15 per ton; total yield, $3,000.

Kit Carson. Owned by Enos K. Baxter, A.M. Graves, and S.W. Leedon, of Central City; Claim 150 by 1,400 feet; discovered 1875; located on Magnolia Hill, Magnolia mining district, 8 miles from Boulder; vertical vein, trending east and west, average width 4 feet, pay vein 4 inches; nature of ore, tellurium; average value, $150 per ton, main shaft 50 feet deep.

Lafayette. Major Orris Blake, proprietor; office Boulder; claim 150 by 1,500 feet; located on the south slope of Hoosier Hill near town of Delphi [renamed Wall Street], Sugar Loaf mining district, 9 miles from Boulder; course of vein, northeast and southwest, average width 4 feet, pitch northwest, pay vein 1 foot; nature of ore, tellurium; average value of first-class ore $3,000 and second-class ore $450 per ton; main shaft 25 feet deep, reached by tunnel at a distance of 45 feet from the mouth, and intersected at a depth of 30 feet.

Lafayette. Owned by Cornelius Van Keuren and Charles North, of Sunshine; claim 150 by 1,350 feet; located on Lee Hill, Gold Hill mining district, 3 miles from the town of Sunshine, and 9 miles from Boulder; course of vein, northeast and southwest, average width 5 feet, pitch 10 degrees east, pay vein 15 inches; nature of ore, tellurium and free gold; average value, $700 per ton; main shaft 55 feet deep, aggregate depth of other shafts 30 feet, 1 drift 20 feet in length.

Last Chance. Owned by O.H. Harker, William Marr, George J. Foster, William McConnaughey, and B.H. Frisbee; office Providence; claim 150 by 1,500 feet; discovered 1875; located on north side of Jim Creek, Central mining district, 3 miles from Jamestown and 16 miles from Boulder; course of vein, northeast and southwest, average width 2 ½ feet, pitch 30 degrees northwest, pay vein 1 to 14 inches; nature of ore, tellurium; average value, $600 per ton; main shaft 100 feet, and 1 other shaft 50 feet deep, 2 levels 30 and 25 feet in length respectively; yield, $9,000.

Last Chance and Great Eastern. Owned by James M. Farmer, Joseph Collier, and others; office Salina; claim 150 by 1,514 feet; discovered 1875; located on Melvina Hill, Gold Hill mining district, 9 miles from Boulder; course of vein, northeast and southwest, average width 20 inches, pitch 20 degrees, pay vein 8 inches; nature of ore, tellurium; average value, $300 per ton; main shaft 200 feet deep; aggregate length of levels 180 feet; shaft house 14 by 20 feet; yield, $10,000.

Licon. Samuel Baeder, of Boulder, proprietor; claim 150 by 1,500 feet; discovered 1874; located in Sunbeam Gulch, Sugar Loaf mining district, 2 miles from town of Salina, and 8 miles from Boulder; vertical vein running northeast and southwest, average width 8 feet, pay vein 1 foot; nature of ore, tellurium and brittle silver; average value, $30 per ton; main shaft 60 feet deep.

Lily of the West. Owned by Mat. Heiser, Louis Coffman, Emory C. Nutt, and others; office Sunshine; claim 150 by 1,500 feet; located in Castle Gulch, Central mining district, near town of Springdale, 11 miles from Boulder; course of vein, northeast and southwest, average width 2 ½ feet, pitch 10 degrees south, pay vein 10 inches; nature of ore, tellurium of silver; average value, $58 per ton; main shaft 60 feet deep, 1 drift 25 feet in length.

Lindley. John C. Blake, proprietor; office Boulder; claim 75 by 1,100 feet, patented; discovered 1871; located on Four Mile Creek, Sugar Loaf mining district, 1 mile from town of Delphi [later renamed Wall Street], and 10 miles from Boulder; vertical vein running northwest and southeast, average width 30 feet; nature of ore, copper pyrites and gray copper; average value, $30 per ton; main shaft 115 feet deep shaft house 20 by 20 feet.

Little Alice. J.B. Joslin, Austin and William Boden, and Wenzel Tucker, proprietors; office Rockville; claim 150 by 1,500 feet; discovered 1873; located on Left Hand Gulch, Gold Hill mining district 1 ½ miles from Sunshine, and 9 miles from Boulder; course of vein, northeast and southwest, average width 5 feet, pitch 5 degrees south; nature of ore, tellurium, sylvanite, ruby silver, and free gold; average value, $250 per ton; main shaft 150 feet deep, aggregate length of levels 50 feet.

Little Annie. Owned by H.C. and E.P. Thompson and Charles Davis, of Boulder; claim 150 by 1,500 feet; discovered 1876; located in Bummer Gulch, Sugar Loaf mining district, 6 miles from Boulder; course of vein, east and west; width 3 feet, pitch 5 degrees north; pay vein 3 inches; nature of ore, tellurium; average value, $600 per ton; main shaft 50 feet deep; yield, $1,000.

Little Belle. Owned by Henry Deitz and Benjamin Hayman, of Boulder; claim 150 by 1,500 feet; discovered 1866; located in Bummer Gulch, Sugar Loaf mining district, 9 miles from Boulder, course of vein, east and west; average width 6 feet, pitch 40 degrees north; pay vein 30 inches; nature of ore, tellurium; average value, $65 per ton; 2 shafts, aggregate length 130 feet; 1 tunnel 100 feet in length, 1 shaft house 20 by 24 feet; yield, $5,000.

Little Brittain [sic]. Marcus Bronson and Henry C. Thompson, proprietors; office Boulder; claim 150 by 1,500 feet; discovered 1875; located near Hoosier Hill, 1 mile from Salina and 10 miles from Boulder; course of vein southeast and northwest; average width 3 feet; pay vein 6 inches, pitch 5 degrees north; nature of ore, gray copper; average value 100 ounces silver per ton; main shaft 90 feet deep.

Little Daisey. Orris and Frank O. Blake and Jacob Cherney, proprietors; office Boulder; claim 150 by 1,500 feet, patented; discovered 1875; located on the south slope of Hoosier Mountain, Sugar Loaf mining district, near town of Delphi [later renamed Wall Street], 9 miles from Boulder; course of vein, northeast and southwest, average width 3 feet, pitch 10 degrees northwest, pay vein 8 inches; nature of ore, tellurium and iron pyrites; average value, $225 per ton; main shaft 75 feet deep.

Little Dorrit. Owned by Charles B. Buckingham, J. Jay Joslin, and the Selkirk Estate; claim 150 by 1,500 feet, patented; discovered 1875; located on Keystone Hill, Magnolia mining district, 8 miles from Boulder; course of vein, northeast and southwest, width 3 feet, pay vein 6 inches, pitch 40 degrees north; nature of ore, tellurium; average value, $200 per ton; 3 shafts, aggregate depth 235 feet, 1 level 75 feet in length; yield, $10,000.

Little Eddie. Silas S. Kennedy, Benjamin H. Eaton, William H. Lessig, Jacob Hawke, and William A. Christian, proprietors; office, Balarat; claim 150 by 1,500 feet; discovered 1876; located in Long's Gulch, Central mining district, in town of Balarat, 17 miles from Boulder; course of vein, northeast and southwest, average width 2 feet, pitch east, pay vein 3 inches; nature of ore, tellurium; average value, $300 per ton; main shaft 25 feet deep, 1 drift 35 feet in length and 20 feet from surface.

Little Giant. Owned by Dr. Harrison Goodwin, William W. Birch, and Giles Blevin, of Longmont; claim 75 by 900 feet; located in Sunshine, Gold Hill mining district, 7 miles from Boulder; course of vein, northeast and southwest, average width 5 feet, pitch 5 degrees southwest, pay vein 10 inches; nature of ore, tellurium, sylvanite, petzite, and free gold; average value, $60 per ton; main shaft 100 feet deep.

Little Giant. Owned by Elbridge Forsyth, David Waldron, and William Blake; office Magnolia; claim 150 by 1,500 feet; discovered 1877; located on Magnolia Hill, Magnolia mining district, 8 miles from Boulder; course of vein, northeast and southwest, width 3 feet, pay vein 14 inches, pitch 45 degrees north; nature of ore, tellurium, assay value $200 per ton; main shaft 60 feet deep.

Little Giant. Owned by H.J. Van Wetering and P.H. Van Diest; office Sunshine; claim 50 by 600 feet; located in town of Sunshine, Gold Hill mining district, 7 miles from Boulder; vertical vein running northeast and southwest, average width 5 feet, pay vein 14 inches; nature of ore, tellurium; average value, $50 per ton; main shaft 26 feet deep.

Little Rebel. Cyrus B. Ayres, proprietor, office Boulder; claim 150 by 1,500 feet; discovered 1877; located near Gifford Gulch, Central mining district, 2 miles from Jamestown and 18 miles from Boulder; course of vein, northeast and southwest, average width 5 feet, pitch 45 degrees northeast, pay vein 15 inches; nature of ore, tellurium; average value, $200 per ton; one shaft 11 feet deep.

Little Sue. John Borgeson, proprietor; office Gold Hill; claim 150 by 1,500 feet; discovered 1875; located on Gold Hill, Gold Hill mining district, 12 miles from Boulder; vertical vein running northeast and southwest; average width 3 feet; pay vein 18 inches; nature of ore, galena and tellurium; average value, $100 per ton; main shaft 40 feet deep, aggregate depth of other shafts 20 feet.

Logan, Quincy and Croesus, Consolidated. Aaron W. Kellogg & Co., proprietors; office Sunshine; claim 150 by 4,000 feet; discovered 1875; located in Sunbeam Gulch, Sugar Loaf mining district, 7 miles from Boulder; course of vein, north and south; average width 5 to 12 feet; pitch 15 degrees, pay vein 4 to 12 inches; nature of ore, tellurium and free gold; average value, $625 per ton; main shaft 106 feet deep, aggregate depth of other shafts 100 feet, aggregate length of levels 240 feet; yield, $21,000.

Longfellow. James D. Smith, William A. and Charles J. Christian, proprietors; office Denver; claim 150 by 1,500 feet; discovered 1877; located in Long's Gulch, in town of Balarat, Central mining district 17 miles from Boulder; course of vein, east and west; average width 5 feet, pitch 15 degrees south, pay vein 12 inches; nature of ore, sulphurets; average value, $105 per ton; main shaft 100 feet deep, 1 tunnel 60 feet in length and 35 feet from surface, 1 level 65 feet in length and 60 feet from surface; total yield, $1,200.

Louis. Henry C. Thompson, proprietor; office Boulder; claim 150 by 1,500 feet; discovered 1876; located on Louis Mountain, Central mining district half mile from Springdale and 13 miles from Boulder; course of vein, northeast and southwest, average width 3 ½ feet, pitch 5 degrees south, pay vein 14 to 18 inches; nature of ore, tellurium and free gold; average value, $300 per ton; main shaft 100 feet deep, aggregate length of other shafting 150 feet; this property is now leased to the Turner Tunnel Mining and Milling Co., of Louisville, Kentucky; total yield, $5,000.

M.M. Pomeroy. Owned by William Mann, Ferdinand C. Albright, and E.J. Katzenmayer; office Caribou; claim 50 by 1,500 feet; discovered 1872; located on Caribou Hill, Grand Island mining district, 23 miles from Boulder; vertical vein running east and west, width 3 to 5 feet, pay vein 5 to 18 inches; nature of ore, galena and sulphurets; average value, $80 per ton; main shaft 40 feet deep.

Mack. Anthony Arnett, proprietor; office Boulder; claim 50 by 1,500 feet, patented; discovered 1859; located on Gold Hill, Gold Hill mining district, 8 miles from Boulder; course of vein, northeast and southwest, average width 3 feet, pitch south, pay vein 20 inches; nature of ore, tellurium; average value, $200 per ton; main shaft 30 feet deep, aggregate depth of other shafts 110 feet, 1 level 90 feet in length and 30 feet from surface; yield, $5,000.

Magnetic. Daniel Drummond and Joel T. St. Clair, proprietors; office Left Hand; claim 150 by 1,500 feet; discovered 1877; located on Gold Hill, Gold Hill mining district, 10 miles from Boulder; vertical vein trending northeast and southwest, width 3 feet, pay vein 6 inches; nature of ore, iron pyrites; average value, $125 per ton; main shaft 25 feet deep.

Magnolia. Owned by Charles G. Buckingham, C.A. Stuart, and others; office Magnolia; claim 150 by 3,000 feet, patented; discovered 1875; located on Magnolia Hill, Magnolia mining district, 8 miles from Boulder; course of vein, east and west, width 3 feet, pitch 5 degrees north, pay vein 10 inches, nature of ore, tellurium, silver glance and ruby silver; value $150 per ton; main shaft 100 feet deep, 1 level 45 feet in length.

Magnolia Consolidated Gold Mining and Concentration Co. Incorporated March 23, 1876; capital stock, $300,000, in 3,000 shares, $100 each; office Boulder; organized by John Landon, Leland W. Green, Daniel A. McPherson, J.H. Salisbury, E.L. Hubbard, B.F. Pine, and Alfred Tucker.

Malava. Owned by George Tanner, Peter Davis, and W. Worthing; office Salina; claim 150 by 1,500 feet; discovered 1876; located on Melvina Hill, Gold Hill mining district, 9 miles from Boulder; vertical vein running north and south; average width 10 inches, pay vein 3 inches; nature of ore, tellurium; assay value $25 per ton; main shaft 60 feet deep.

Marblehead. Owned by Frederick Pocock, A. Wilder, Lawrence Singer, and John Alderson; office Salina; claim 150 by 1,500 feet; discovered 1874; located on Bighorn Mountain, Gold Hill mining district, 8 miles from Boulder; course of vein, northeast and southwest, average width 4 feet, pitch 10 degrees northwest, pay vein 1 foot to 18 inches; nature of ore, tellurium and sylvanite; average value, $375 per ton; main shaft 38 feet deep; aggregate depth of other shafts 69 feet, 1 level 15 feet in length, and 25 feet from surface; yield, $30,000.

Mary. Owned by Henry Monell and others; office Salina; claim 150 by 1,500 feet; claim 150 by 1,500 feet; discovered 1878; located on Melvina Hill, Gold Hill mining district, 8 miles from Boulder, course of vein, northeast and southwest, average width 1 foot, pitch 15 degrees north, pay vein 1 inch; nature of ore, tellurium; assay value, $800 per ton; main shaft 13 feet deep.

Maryland. Owned by Joseph Pasco, H.C. Cameron, and D.H. Sayler; office Salina; claim 150 by 1,500 feet; discovered 1876; located on Gold Run, Gold Hill mining district, 8 miles from Boulder; course of vein, northeast and southwest, average width 3 feet, pitch 5 degrees southeast, pay vein 18 inches; nature of ore, galena and zinc blende; assay value, $60 per ton; 1 adit 75 feet in length.

Mastodon Mining Co. Incorporated October 30, 1878; capital stock, $400,000, in 40,000 shares, $10 each; office Denver; organized by Gabriel Netter, E. Winslow Cobb, and Clarence P. Elder.

May Flower. John L. Butzel and Andrew J. Mackey, proprietors; office Boulder; claim 150 by 1,500 feet, patented; discovered 1876; located near Lee Hill, Gold Hill mining district, 8 miles from Boulder; vertical vein trending northeast and southwest, width 15 feet, pay vein 4 to 12 inches; nature of ore, tellurium; average value, $125 per ton; main shaft 140 feet deep.

McCarty Mining Co. Incorporated September 4, 1874; capital stock, $50,000, in 500 shares, $100 each; office Boulder; organized by Alexander Boyd, Leander McCarty, and Henry H. Curtis.

Meland. Owned by Victor Roux, Henry Chmal, and Henry Lions, of Boulder; claim 150 by 1,500 feet; discovered 1874; located in Boulder Canyon, Grand Island mining district, 8 miles from Boulder; vertical vein running east and west, average width 12 feet, pay vein 3 feet; nature of ore, tellurium; average value, $100 per ton; main shaft 60 feet deep, 1 tunnel 150 feet in length.

Melvina. Owned by Henry Neikirk, Henry Meyring, Melvin Bailey, and Marion Kissler; office Boulder; claim 150 by 1,500 feet, patented; discovered 1875; located on Melvina Hill, near town of Salina, Gold Hill mining district, 9 miles from Boulder; course of vein, northeast and southwest, average width 3 feet, pitch 10 degrees northeast, pay vein 8 inches; nature of ore, tellurium; average value, $500 per ton; main shaft 300 feet deep; aggregate length of levels 1,000 feet; engine and ore house 40 by 40 feet; one engine, 12-horse power; yield, $200,000.

Memphis Gold and Silver Mining Co. Incorporated May 1877; capital stock, $300,000, in 60,000 shares, $5 each; offices Sunshine, Colorado, and Broadway, New York; A.R. Chisolm, president, J. Harvey Jones, vice president, William S. Walker, secretary, William McClure, treasurer, Owen E. LeFevre, managing director; claim 150 by 3,000 feet; discovered 1871; located in Left Hand Gulch, near town of Sunshine, Gold Hill mining district, 7 miles from Boulder; course of vein, northeast and southwest, width 4 feet; pay vein 1 foot, pitch 45 degrees northwest; nature of ore, tellurium and free gold; average value, $65 per ton; developed by 4 shafts, aggregating 215 feet in depth, and 3 levels, aggregating 325 feet in length.

Merchants' and Miners' Tunnel Co. Incorporated July 21, 1876; capital stock, $500,000, in 50,000 shares, $10 each; office Denver; organized by J. Jay Joslin; Augustus B. Ingols, and A.J. Bean.

Mexico. Owned by the Santa La Saria Mining Co., Colorado; claim 150 by 1,500 feet; discovered 1856; located on White Cloud Mountain,

Grand Island mining district, 2 ½ miles from Nederland and 15 miles from Boulder; course of vein, northeast and southwest, average width 5 feet, pitch 30 degrees southwest, pay vein 1 foot; nature of ore, galena, gray copper, native silver, and sulphurets of silver; average value, 100 ounces silver per ton; will be reached by the Santa La Saria tunnel at a distance of 225 feet from the mouth and intersected at a depth of 400 feet, 4 levels aggregate length 500 feet; yield, $30,000.

Milwaukee. Owned by the Sunnyside Mining Co., of Milwaukee, Wisconsin; claim 150 by 2,500 feet, patented; discovered 1871; located on Bald Mountain, Ward mining district, 8 miles from town of Gold Hill and 20 miles from Boulder; course of vein, east and west, width 3 to 10 feet, pitch 35 degrees north, pay vein 1 to 3 feet; nature of ore, iron pyrites, and carbonate of copper; average value, $120 per cord; main shaft 235 feet deep, aggregate depth of other shafts 197 feet, aggregate length of levels 200 feet; yield, $11,000.

Mineral Point. Joseph Steppler and Joseph Steppler, Jr., proprietors; office Denver; claim 150 by 1,500 feet, patented; located on Big Horn Mountain, Gold Hill mining district, 10 miles from Boulder; course of vein, northeast and southwest; width 5 feet, pitch 25 degrees north, pay vein 30 inches; nature of ore, tellurium and gray copper; value, $450 per ton; main shaft 85 feet deep, 1 tunnel 40 feet, and 1 level 45 feet in length.

Miner's Hope. Owned by Hiram Fuller and Thomas Shires; office Salina; claim 150 by 1,500 feet, patented; discovered 1874; located in Silver City Gulch, Gold Hill mining district, 8 miles from Boulder; course of vein, east and west, width 2 feet, pitch 20 degrees north, pay vein 4 to 6 inches; nature of ore, tellurium; average value, $550 per ton; main shaft 40 feet deep, aggregate length of adits 33 feet; yield, $1,600.

Mining Company of New Jersey. Organized under the laws of the State of New Jersey, February 10, 1876; capital stock, $500,000, in 5,000 shares, $100 each; office Camden, New Jersey; trustees: A.G. Curtin, Alexander G. Cattell, Aden Alexander, D.S. Stetson, Thomas H. Dudley, Dell Noblit, Jr., Constance Curtin, John J. Curtin, and Edward Dudley.

Minneapolis. Owned by Joseph Steppler and Joseph Steppler, Jr., of Denver; claim 150 by 1,500 feet, patented; discovered 1874; located on Big Horn Mountain, Gold Hill mining district, 8 miles from Boulder; course of vein, northeast and southwest; width 3 feet, pay vein 6 inches, pitch 45 degrees north; nature of ore, free gold and tellurium; value, $500 per ton; 1 shaft 50 feet deep, 1 tunnel 30 feet in length.

Missouri Valley. Owned by Samuel Markert, George Spencer, George Fairburn, and William Martin; office Caribou; claim 50 by 1,400 feet, discovered 1873; located on Caribou Hill, Grand Island mining district, 23 miles from Boulder; course of vein, northeast and southwest, average width 2 feet, pitch 15 degrees north, pay vein 12 inches; nature of ore, galena, iron and copper pyrites; average value, $350 per ton; main shaft 600 feet deep, 1 level 60 feet in length and 50 feet from surface; shaft house 16 by 20 feet; yield, $12,000

Modoc. Henry Deitz, of Boulder, proprietor; claim 150 by 1,500 feet; discovered 1876; located in Bummer Gulch, Sugar Loaf mining district, 9 miles from Boulder; vertical vein running southeast and northwest, average width 3 feet; nature of ore, copper pyrites; 1 shaft 20 feet deep, one tunnel 100 feet in length.

Mollie Mullen. Silas S. Kennedy, Benjamin H. Eaton, William H. Lessig, Jacob Hawke, and William A. Christian, proprietors; office Balarat; claim 150 by 1,500 feet, patented; discovered 1876; located in Long's Gulch, Central mining district, 17 miles from Boulder; vertical vein running northwest and southeast, average width 4 feet, pay vein 6 inches; nature of ore, tellurium; average value, $40 per ton; main shaft 50 feet deep.

Moltke. Chicago and Colorado Mining Co., proprietors; offices Boulder, Colorado, and Chicago, Illinois; claim 900 by 7,500 feet, patented; discovered 1874; located on Moltke Hill, Ward mining district, 16 miles from Boulder; vertical vein running east and west, average width 3 feet, pay vein 5 inches; nature of ore, galena; average value, $350 per ton; 1 tunnel 200 feet in length, which intersects the vein at a depth of 18 feet, aggregate length of levels 800 feet.

Monitor. Monitor Mining Co., proprietors; office Rochester, New York, H.P. Everest, president, E.A. Hupper, superintendent, claim 150 by 1,500 feet, patented; discovered 1870; located on Idaho Hill, Grand Island mining district, 23 miles from Boulder; course of vein, east and west, width 3 ½ feet, pitch 20 degrees north; nature of ore, galena and sulphurets, average value, $70 per ton; main shaft 265 feet deep; aggregate length of levels 230 feet; shaft, engine and ore-house combines, 16-horse power engine and hoisting apparatus.

Monongahela. Owned by Richard Veale, C.C. Bradbury, and William G. Shedd, of Sunshine; claim 150 by 1,500 feet; discovered 1874; located in town of Sunshine, Gold Hill mining district, 7 miles from Boulder; vertical vein, trending northeast and southwest, average width 4 feet, pay vein 8 inches; nature of ore, tellurium; main shaft 40 feet deep.

Monongahela. Located on Caribou Hill, near town of Caribou, 22 miles from Boulder; main shaft 100 feet deep.

Montana. Owned by Samuel and William Garret, of Salina; claim 150 by 1,500 feet; discovered 1877; located on Big Horn Mountain, Gold Hill mining district, near town of Salina, 8 miles from Boulder; course of vein, north and south, average width 6 feet, pitch 10 degrees west, pay vein 14 inches; nature of ore, sylvanite and free gold; value $1,800 per ton; main shaft 85 feet deep, aggregate depth of other shafts 60 feet, 1 adit 30 feet in length; yield, $5,000.

Morning Star. Paul P. Miller, proprietor; claim 150 by 1,500 feet; discovered 1873; located in Spring Gulch, Ward mining district, near town of Columbia, 19 miles from Boulder; course of vein, northeast and southwest, average width 10 feet, pitch north, pay vein 8 inches; nature of ore, free gold; average value, $300 per ton; main shaft 30 feet deep, aggregate depth of other shafts 70 feet, 1 tunnel 100 feet in length and 30 feet from surface.

Mount Pleasant. Owned by Thomas J. Graham and John W. Smith; office Boulder; claim 50 by 1,400 feet, patented; discovered 1863; located in Central mining district, half mile from Jamestown and 14 miles from Boulder; course of vein, northwest and southeast, average width 4 feet, pitch 5 degrees north, pay vein 4 to 15 inches; nature of ore, galena and copper pyrites; average value, $50 per ton; main shaft 75 feet deep, aggregate depth of other shafts 140 feet, 1 level 60 feet in length.

Mount Sterling. Owned by John W. McFall and William Sellers; office Gold Hill; claim 150 by 1,500 feet; discovered 1875; located in Gold Hill mining district, 10 miles from Boulder; vertical vein running northeast and southwest, average width 4 feet, pay vein 13 inches; nature of ore, tellurium; average value, $275 per ton; main shaft 25 feet deep.

Mount Vernon. Located on Caribou Hill, near town of Caribou, Grand Island mining district, 22 miles from Boulder; developed by 5 shafts, 20, 30, 40, 60, and 100 feet in depth respectively; value of ore, 40 to 80 ounces silver per ton.

Mountain Chief. Charles E. Wiswall and W.A. Hammond, proprietors; office Gold Hill; claim 150 by 1,500 feet; discovered 1875; located in Graham Gulch, 2 miles from town of Gold Hill and 10 miles from Boulder; course of vein, east and west, average width 4 feet, pitch 25 degrees south, pay vein 1 foot; nature of ore, decomposed granite; average value, $1,000 per ton; main shaft 125 feet deep, 2 drifts 100 and 150 feet in length respectively.

Mountain Chief. Mountain Chief Mining Co., proprietors; incorporated 1877; capital stock, $20,000; B. Frank Wood, president, John H. Wood, secretary and treasurer, Thomas L. Wood, superintendent; offices, Sunshine, Colorado, and Broad Street, New York; claim 50 by 1,400 feet, patented; discovered 1872; located in Silver City, Gold Hill mining district, near town of Sunshine, 7 miles from Boulder; course of vein, northeast and southwest, average width 3 feet, pitch 30 degrees northwest; nature of ore, galena, gray copper, and iron pyrites; average value, $35 per ton; main shaft 136 feet deep.

Mountain Lion. Owned by Frederick R. Luce and Augustus Paddock, of Boulder; claim 150 by 1,450 feet, patented. Discovered 1875; located on Magnolia Hill, Magnolia mining district, 9 miles from Boulder; course of vein, east and west, width 4 feet, pay vein 1 to 12 inches, pitch 45 degrees north; nature of ore, tellurium, value $520 per ton; main shaft 200 feet deep, 3 levels aggregate length 600 feet; shaft house 22 by 60 feet; yield, $35,000.

Mountain Lion Mining Co. Incorporated August 11, 1876; capital stock, $250,000, in 10,000 shares, $25 each; office Boulder; organized by Andrew W. Gill, Phillip W. Ver Plank, and Charles P. Berdell.

Mountain Ram. Granville M. Bottoms, proprietor; office Boulder; claim 150 by 1,500 feet; discovered 1876; located on James Creek, James Creek mining district, 16 miles from Boulder; course of vein, northeast and southwest, width 6 feet, pay vein 12 inches, pitch 20 degrees north; nature of ore, tellurium; average value, $47 per ton; main shaft 115 feet deep; yield, $2,000.

Mountain Tiger. John W. McFall and J.M. Miller, proprietors; office Gold Hill; claim 150 by 1,500 feet; discovered 1876; located in Gold Hill mining district, 10 miles from Boulder; vertical vein running northeast and southwest, average width 3 feet, pay vein 5 inches; nature of ore, tellurium; assay value $500 per ton; main shaft 60 feet deep.

Mountain View. Mountain Chief Mining Co., William E. Cook and James A. Thomason, proprietors; offices Sunshine, Colorado, and New York City; claim 150 by 1,500 feet; discovered 1874; located in town of Sunshine, Gold Hill mining district, 7 miles from Boulder; vertical vein running northeast and southwest, average width 2 ½ feet, pay vein 8 inches; nature of ore, galena, gray copper, and iron pyrites; average value of first-class ore $900, second-class ore $200, and third-class ore $60 per ton; main shaft 125 feet deep, 2 levels, aggregate length 160 feet; shaft house 30 by 46 feet; yield, $2,000.

Mountain View, East Extension. Owned by Benjamin F., Thomas L., and John R. Wood; office Sunshine; claim 150 by 375 feet; main shaft 25 feet deep, average value of ore $150 per ton.

Mouth. Owned by Alex. L. Gravelle, Burrell McPherson, and Isaac N. Loper; office Gold Hill; claim 150 by 1,500 feet; discovered 1877; located on North Boulder Creek, Sugar Loaf mining district, 3 miles from Nederland and 20 miles from Boulder; vertical vein running north and south, average width 100 feet, pay vein 2 feet; nature of ore, gray copper; average assay value $200 per ton; 1 tunnel 75 feet from surface and 45 feet in length.

Mystic. Owned by Anne Bixby, Eugene Wilder, James Stevens, and Gorham Foster, of Boulder; claim 150 by 1,500 feet; discovered 1875; located in Wagner's Camp, Gold Hill mining district 1 miles from town of Magnolia and 7 miles from Boulder; course of vein, northeast and southwest, average width 4 feet, pitch 5 degrees southeast; pay vein 2 feet; nature of ore, tellurium; average value, $255 per ton; main shaft 50 feet deep.

National. Angelo Ruvera, proprietor; office Salina; claim 150 by 1,500 feet; discovered 1876; located in Ringold Gulch, Sugar Loaf mining district, 2 miles from town of Gold Hill and 11 miles from Boulder; course of vein, northeast and southwest, average width 10 feet, pitch south, pay vein 1 foot; nature of ore, galena and gray copper; 1 shaft 12 feet deep, 1 tunnel 54 feet in length, 1 level 35 feet in length and 31 feet from surface.

National Consolidated Gold Mining and Milling Co., Lexington, Kentucky. Organized under the laws of the State of Kentucky, May 8, 1878; capital stock, $750,000, in 7,000 shares, $100 each, stock assessable; office Lexington, Kentucky; trustees: Samuel Clay, Jr., Nathan Weare, R.W. Hocker, and F.H. Brown.

Native Silver. Owned by the _____ Mining Co., of New Jersey; office Caribou; claim 150 by 2,000 feet, patented; discovered 1870; located on Caribou Hill, Grand Island mining district, 23 miles from Boulder; course of vein, east and west, average width 5 ½ feet, pitch 5 degrees north, pay vein 2 inches to 3 feet; nature of ore, galena; average value, $120 per ton; main shaft 200 feet deep; aggregate length of levels 600 feet; one engine and steam pump; yield, $175,000.

Nazarene. James K. Jones, proprietor; office Sunshine; claim 150 by 1,500 feet; discovered 1877; located in Sunshine Gulch, Gold Hill mining district, 1 mile from Boulder; course of vein, northeast and southwest, average width 3 ½ feet, pitch 10 degrees north, pay vein 1 foot; nature of ore, iron pyrites; average value 2 ounces gold per ton; 1 tunnel 75 feet, and 1 drift 25 feet in length.

Neggaunee. Webster City Gold and Silver Mining Co., proprietors; Ebenezer Rowland, superintendent; office Boulder; claim 150 by 1,500 feet; discovered 1874; located near Springdale, 12 miles from Boulder; course of vein, northeast and southwest, average width 3 feet, pitch 5 degrees north, pay vein 6 inches; nature of ore, gray copper and tellurium; average value, $40 per ton; main shaft 125 feet deep, 1 drift 50 feet in length.

Nelly Kleine. H.J. Van Wetering and P.H. Van Diest, proprietors; office Sunshine; claim 150 by 1,500 feet; discovered 1875; located in Sunshine, Gold Hill mining district, 7 miles from Boulder; vertical vein running northeast and southwest, average width 3 ½ feet, pay vein 5 inches; nature of ore, tellurium; average value, $50 per ton; main shaft 60 feet deep.

New York. William A. Davidson, proprietor; office Sunshine; claim 150 by 1,500 feet; discovered 1874; located in Gold Hill mining district, in town of Sunshine, 7 miles from Boulder; course of vein, northeast and southwest, average width 13 feet, pitch 15 degrees south, pay vein 3 feet; nature of ore, tellurium, calaverite, sylvanite, and free gold; main shaft 80 feet deep, aggregate depth of other shafts 150 feet.

Niawot. William A. Davidson, proprietor; office Valmont; claim 75 by 1,500 feet, patented; discovered 1859; located on Niawot Hill, Ward mining district, 18 miles from Boulder; course of vein, east and west, width 15 feet, pitch 40 degrees north; nature of ore, iron pyrites, gray copper and native silver, average value, $125 per cord; main shaft 420 feet deep, aggregate depth of other shafting 300 feet, 1 level 400 feet in length and 225 from surface; 50-stamp mill in good order; yield to date, $600,000.

Nicholls. George W. Gregg and M. Granville Pease, proprietors; office Gold Hill; claim 50 by 1,500 feet, patented; discovered 1868; located on Hoosier Hill, Gold Hill mining district, 10 miles from Boulder; vertical vein running north and south, average width 4 feet, pay vein 17 inches; nature of ore, gray copper, assay value $50 per ton; main shaft 63 feet deep; shaft house 12 by 14 feet.

Nil Desperandum. Oscar B. Force, James E. and M. Munford, proprietors; office Sunshine; claim 150 by 1,500 feet; discovered 1874; located in Sunshine Gulch, Gold Hill mining district, 7 miles from Boulder; course of vein, northeast and southwest, average width 6 feet, pitch 5 degrees northwest, pay vein 1 inch to 4 feet; nature of ore, tellurium, sylvanite, and free gold; average value, $300 per ton; main shaft 140 feet in length; shaft house 16 by 18 feet; yield, $500.

No Name. William Fullerton and R. G. Dun, proprietors; office New York City; claim 50 by 1,400 feet patented; discovered 1869; located on Caribou Hill, Grand Island mining district 23 miles from Boulder; course of vein, northeast and southwest, average width 3 feet, pay vein 3 inches to 3 feet; nature of ore, gray copper, iron and copper pyrites; average value, $410 per ton; main shaft 530 feet deep, aggregate length of levels 970 feet; shaft 30 by 46 feet, one engine 25-horse power; yield, $300,000.

North. Owned by C.A. Stewart, H. Fullan, and Thomas Shires; office Magnolia; claim 150 by 1,500 feet; discovered 1875; located on Magnolia Hill, Magnolia mining district, 8 miles from Boulder; vertical vein trending east and west, width 3 feet, pay vein 12 inches; nature of ore, tellurium; average value, $50 per ton; 1 shaft 50 feet deep, 1 tunnel 25 feet in length.

Northwestern. Wm. J. Mann, proprietor; office Caribou; claim 150 by 1,400 feet; discovered 1871; located in town of Caribou, Grand Island mining district, 23 miles from Boulder; vertical vein trending northeast and southwest, width 3 to 7 feet, pay vein 4 inches to 2 feet; nature of ore, galena, brittle silver and sulphurets; average value, $250 per ton; 1 shaft 100 feet deep.

Osceola. Owned by William G. Shedd, N.S. Culver, C.C. Bradbury, F.L. Martin, and L.H. Bartlett; office Sunshine; claim 150 by 1,500 feet, patented; discovered 1874; located in town of Sunshine, Gold Hill mining district, 7 miles from Boulder; course of vein, northeast and southwest, width 5 feet pay vein 8 to 20 inches; nature of ore, tellurium, sylvanite, and free gold; value $200 per ton; main shaft 105 feet deep, aggregate length of levels 125 feet; shaft house 16 by 18 feet.

Ogallalah. Owned by John C. Henry, Major Orris Blake, R.H. Whitely, and Edward S. Walker; office Boulder; claim 75 by 1,100 feet; discovered 1861; located on Four Mile Creek, Sugar Loaf mining district, one mile from town of Delphi [later renamed Wall Street] and 10 miles from Boulder; course of vein, northeast and southwest; average width 5 feet, pitch 10 degrees northwest, pay vein 3 feet; nature of ore, galena, gray copper and zinc blende; average value of first-class ore 400 ounces and second-class 36 ounces of silver per ton; 1 tunnel 425 feet in length, in which there is a shaft 50 feet deep, 1 level 45 feet in length; two-story ore house 20 by 30 feet.

Ogallalah Mining Co. Incorporated March 22, 1876; capital stock, $250,000, in 5,000 shares, $50 each; office Boulder; organized by Henry and John C. Blake, Edward S. Walker, and Frances E. Ames.

Ohio-Colorado Reduction and Mining Co. Incorporated January 6, 1875; capital stock, $150,000, in 150 shares, $1,000 each; office Salina, Colorado; organized by J.F. Saunders, Aaron Ferneon, S.M. and Elijah G. Penn.

Oro Cash. Owned by George H. Batchelder and F.A. Squires, of Boulder; claim 150 by 1,500 feet, patented; discovered 1868; located on Gold Hill Gold Hill mining district, 10 miles from Boulder; vertical vein running northeast and southwest; average width 4 feet; nature of ore, tellurium and free gold; average value, $125 per cord; main shaft 150 feet deep, 1 level 150 feet in length and 150 feet from surface; yield, $20,000.

Ozark Tunnel Mining Property. Owned by D. Mortimore, M.D., of Denver; comprises 6 lodes, or mineral veins, 150 by 1,500 feet each, a tunnel site 1,500 by 2,310 feet, and a mill site containing 5 acres of ground, on which is erected a good dwelling house. This property was discovered in 1874, and is located in Gold Hill mining district, in the center of the tellurium ore belt, between the mines of Gold Hill and Salina, being 1 ½ miles from each place, and 6 miles from the city of Boulder, Colorado. It is surrounded by the most prominent mines in the state, and may well be considered one of the largest mining estates, owned in 1 body. The owner, who is largely interested in other mines, considers this property the most valuable in the state.

Palatine. Horace Wolcott, proprietor; office Left Hand; claim 150 by 1,500 feet; discovered 1876; located on Gold Hill, Gold Hill mining district, 10 miles from Boulder; vertical vein running northeast and southwest, width 3 feet, pay vein 15 inches; nature of ore, galena and iron pyrites; average value, $35 per ton; 1 shaft 22 feet deep.

Parallel. C.B. Farwell & Co., proprietors; office Gold Hill; claim 50 by 1,450 feet, patented; discovered 1873; located in Gold Hill mining district, 10 miles from Boulder; course of vein, northeast and southwest; width 4 ½ feet, pitch 25 degrees north, pay vein 15 inches; nature of ore, galena and iron pyrites; average value, $75 per ton; main shaft 75 feet deep.

Parstele. Victor Roux, proprietor; office Boulder; claim 150 by 1,500 feet; discovered 1872; located in Sunbeam Gulch, Sugar Loaf mining district, 1 mile from town of Salina and 8 miles from Boulder; vertical vein running east and west; width 5 feet, pay vein 12 inches; nature of ore, tellurium; average value, $50 per ton; main shaft 20 feet deep, 1 tunnel 57 feet in length.

Pay-master. Owned by Benjamin F. Smith and Frederick Wilkins; office Golden; claim 150 by 1,500 feet; discovered 1875; located in Sugar Loaf mining district, near town of Crisman, 7 miles from Boulder; vertical vein running east and west; average width 5 feet; pay vein 30 inches, pitch 30 degrees west; nature of ore, tellurium, sylvanite, petzite, and free gold; average value, $278 per ton; main shaft 80 feet deep; yield, $1,000.

Pay-master. Henry A. Barrows and William A. Meralls, proprietors; office Sunshine; claim 150 by 1,500 feet; discovered 1874; located near town of Sunshine, Gold Hill mining district, 6 miles from Boulder; course of vein, northeast and southwest; width 5 feet, pay vein 30 inches, pitch 30 degrees west; nature of ore, tellurium, sylvanite, petzite, and free gold; average value, $278 per ton; main shaft 80 feet deep; yield, $1,000.

Perrigo. Samuel Mishler & Co., proprietors; office Central City; claim 150 by 1,500 feet; discovered 1864; located on Caribou Hill, Grand Island mining district, 23 miles from Boulder; vertical vein trending southeast and northwest; width 4 feet; pay vein 18 inches; nature of ore, iron and copper pyrites; average value, $75 per ton; 1 shaft 60 feet deep, shaft house 12 by 20 feet.

Phil Sheridan. Owned by the Phil Sheridan and Gillaspie Consolidated Mining Co., incorporated October 9, 1875; capital stock, $250,000, in 2,500 shares, $10 each; office Denver; claim 150 by 1,400 feet; located in town of Sunshine, Gold Hill mining district, 7 miles from Boulder; course of vein, east and west, width 4 feet, pay vein 4 to 6 inches; nature of ore, tellurium and free gold; average value, $250 per ton; main shaft 100 feet deep; shaft house 17 by 24 feet.

Point of Rocks Mining, Milling and Tunneling Co. Incorporated October 21, 1875; capital stock, $250,000, in 2,500 shares, $100 each; office Boulder, Colorado; organized by James L. Wilson, William B. Fowler, William H. Fisher, James H. Decker, and William A. Hardenbrook.

Poison. Owned by William Sellers, Robert Brown, and David J. Rollins, office Gold Hill; claim 150 by 1,500 feet; discovered 1875; located in Gold Hill mining district, 10 miles from Boulder; vertical vein running north and south; width 4 feet, pay vein 20 inches; nature of ore, tellurium; assay value $500 per ton; main shaft 30 feet deep.

Poor Man. Owned by Neil D. McKenzie, G.I. and W.R. Stebbins, Edward J. Binford, and Mrs. A. Arnold; office Caribou; claim 150 by 1,400 feet, patented; discovered 1869; located in town of Caribou, Grand Island mining district, 20 miles from Boulder, course of vein, east and west, width 3 feet, pay vein 18 inches; nature of ore, native silver and sulphurets; value, $160 per ton; main shaft 220 feet deep, 3 levels, aggregate length 550 feet; shaft house 30 by 60 feet; yield, $35,000.

Poor Man. Owned by Moses Hallett and John S. Reid; office Boulder; claim 150 by 1,500 feet; discovered 1876; located in town of Magnolia, Magnolia mining district, 8 miles from Boulder, course of vein, northeast and southwest, average width 6 feet, pay vein 3 inches, pitch 5 degrees south; nature of ore, tellurium; value, $2,750 per ton; main shaft 125 feet deep, 1 level 40 feet in length and 65 feet from surface.

Potosi. Owned by Wyke & Ellison and William Davidson; office Sunshine; claim 150 by 1,500 feet; discovered 1874; located in Gold Hill mining district, half mile from town of Sunshine and 6 miles form Boulder; vertical vein running northeast and southwest, width 4 feet; nature of ore, tellurium and petzite; main shaft 60 feet deep.

Potosi Silver Mining Co. Incorporated June 18, 1877; capital stock, $600,000, in 60,000 shares, $10 each; offices Denver, Colorado, and Alexandria, Virginia; organized by Charles C. Tompkins, T.B. Wilson, W.H. Marbury, B.F. Nalle, F.T. Hawkes, Orville L. Grant, Peyton Randolph, L. Wilbur Reid, and B. Thompkins.

Prescott. Granville M. Button and John Miner, proprietors; office Boulder; claim 150 by 1,500 feet; discovered 1877; located on James Creek, 2 miles from Jamestown and 16 miles from Boulder; vertical vein trending northeast and southwest, width 3 feet, pay vein 18 inches; nature of ore, tellurium; value $85 per ton; main shaft 25 feet deep; yield, $1,000.

Pride of the Mountain. Thomas Danford & Co., proprietors; office Boulder; claim 150 by 1,500 feet, patented; discovered 1875; located on Four Mile Creek, 3 miles from Boulder; course of vein, east and west, width 7 feet, pay vein 5 inches, pitch 15 degrees north; nature of ore, tellurium; value $60 per ton; 1 shaft 30 feet deep, 1 adit 50 feet in length.

Rambler. Owned by John and James J. Miller, Thomas F. Hendershot, and George Kountze; office Jamestown; claim 150 by 1,500 feet; discovered 1878; located on Golden Age Hill, Central mining district, 2 miles from town of Springdale and 10 miles from Boulder; course of vein, northeast and southwest, width 18 feet, pitch 18 degrees south, pay vein 14 inches; nature of ore, tellurium; main shaft 40 feet deep.

Rapid Rhone. James D. Smith, William A. and Charles J. Christian, proprietors; office Denver; claim 150 by 1,500 feet; discovered 1877; located in Central Gulch, 2 miles from Balarat and 15 miles from Boulder; course of vein, east and west, average width 4 feet, pitch 25 degrees south, pay vein 12 inches; nature of ore, free gold; main shaft 30 feet deep.

Ready Cash. Owned by C. Eschler, John Flykiger, and John Kesler; office Caribou; claim 150 by 1,400 feet; discovered 1870; located on Idaho Hill, Grand Island mining district, 20 miles from Boulder; course of vein, east and west, width 4 feet, pay vein 5 inches; nature of ore, iron pyrites and carbonate of copper; assay value $244 per ton; one shaft 70 feet deep, 1 level 20 feet in length; shaft house 14 by 16 feet.

Ready Cash. Owned by Emory C. Nutt, A.H. McClure, J.N. Davis, and W.W. Burch, of Sunshine; claim 150 by 635 feet; discovered 1876; located head of Sunshine Gulch, Gold Hill mining district, near town of Sunshine, 7 miles from Boulder; course of vein, northeast and southwest, average width 10 feet, pitch 10 degrees north, pay vein 14 inches; nature of ore, tellurium, sylvanite, and free gold; average value, $650 per ton; main shaft 60 feet deep, 1 tunnel 40 feet in length.

Ready Cash. Owned by James H. and James W. Robinson; office Gold Hill; claim 150 by 1,500 feet; discovered 1875; located in Gold Hill mining district, 10 miles from Boulder; vertical vein running northeast and southwest, average width 4 feet, pay vein 18 inches; nature of ore, tellurium, assay value $800 per ton; main shaft 30 feet deep, 1 adit 30 feet in length.

Ready Cash. Owned by Silas T. Tumbleson; office Gold Hill; claim 150 by 1,500 feet; discovered 1878; located on Bonanza Hill, Central mining district, 3 ½ miles from town of Gold Hill and 13 miles from Boulder; course of vein, northwest and southeast, average width 6 feet, pitch 20 degrees east, pay vein 2 feet; nature of ore, free gold; average value, $120 per cord; main shaft 10 feet deep.

Rebecca. Henry C. Thompson, William T. and Joseph M. Grout, proprietors; office Boulder; claim 150 by 1,500 feet, discovered 1875; located in town of Magnolia, Magnolia mining district, 7 miles from Boulder; course of vein southeast and northwest, average width 3 ½ feet, pitch 10 degrees north, pay vein 18 to 20 inches; nature of ore, tellurium; average value of first-class ore $300 per ton, second-class ore 5 ounces gold per cord; main shaft 150 feet and 1 other 80 feet deep, 2 levels, aggregate length 83 feet; shaft house 10 by 36 feet, and 1 engine 20-horse power; yield, $2,000.

Red Cloud. Cyrus Strong and John J. Riethmann, proprietors; office Denver; claim 50 by 1,400 feet, patented; discovered 1872; located on the west slope of Gold Hill, 10 miles from Boulder; course of vein, northeast and southwest; average width 3 feet, pitch 10 degrees northwest, pay vein 6 inches; nature of ore, tellurium; average value, $750 per ton; main shaft 400 feet, and 1 other shaft 100 feet deep, 1 tunnel 225 feet in length, and 100 feet from surface, aggregate length of levels 1,000 feet; yield, $60,000.

Red Cloud, No. 2. John W. Corser, proprietor; office Gold Hill; claim 150 by 1,500 feet, discovered 1875; located on Gold Hill, 12 miles from Boulder; course of vein, northeast and southwest, average width 6 feet, pitch 25 degrees north, pay vein 2 ½ feet; nature of ore, free gold; assay value, $250 per ton; main shaft 24 feet deep.

Ringgold. Owned by George Reemer, D.H. Sayler, Harry Carter, John Pierce, and G.F. Mayer; office Salina; claim 150 by 1,500 feet, discovered 1875; located on Hoosier Hill, Gold Hill mining district, 8 miles from Boulder; course of vein, north and south, average width 30 inches, pitch 25 degrees west, pay vein 1 to 20 inches; nature of ore, tellurium; average value, $50 per ton; main shaft 115 feet deep, aggregate length of levels 70 feet, being 90 feet from surface, shaft house 16 by 20 feet.

Rip Van Dam. Fred L. Higbee, Charles F. Pease, and Britton A. Hill, proprietors; office Springdale; claim 150 by 1,500 feet; discovered 1875; located in Castle Gulch, near town of Springdale, 12 miles from Boulder; course of vein, northeast and southwest, average width 20 feet, pitch 20 degrees southeast, pay vein 6 inches to 3 feet; nature of ore, tellurium and free gold; 1 shaft 50 feet deep, 1 level 40 feet in length.

Robert E. Lee. Owned by John Corser, John Simpson, and John Mitchell; office Gold Hill; claim 150 by 1,500 feet, discovered 1875; located on 7:30 Hill, Gold Hill mining district, 12 miles from Boulder; course of vein, northeast and southwest, average width 3 feet, pitch 5 degrees north, pay vein 4 to 8 inches; nature of ore, free gold; average value, $125 per ton; main shaft 40 feet deep.

Rocky Mountain Gold and Silver Mining Co. Incorporated January 20, 1876; capital stock, $100,000, in 1,000 shares, $100 each; office Boulder, organized by M.A. Lathrop, H.M. Spalding, and James P. Maxwell.

Rocky Mountain Mammoth. Owned by Caleb S. Stowell, A.J. Emrick, H. Summers, and G.N. Eaton; office Magnolia; claim 150 by 1,500 feet, discovered 1876; located on Magnolia Hill, 8 miles from Boulder; course of vein, east and west, width 5 feet; pay vein 20 inches, pitch 20 degrees south; nature of ore, tellurium and silver glance; value, $105 per ton; 1 shaft 50 feet deep.

Rocky Mountain Mining and Reduction Co. Incorporated May 29, 1877; capital stock, $1,000,000, in 100,000 shares, $10 each; office Denver; trustees: Matthew D. Brett, Robert Connely, John M. Lord, William F. Shanks, and George W. Parker.

Romance. Cyrus Taylor, proprietor; office Jamestown; claim 150 by 1,500 feet; discovered 1877; located in Central mining district, 2 miles from town of Springdale and 14 miles from Boulder; course of vein, northeast and southwest, average width 3 feet, pitch 25 degrees south, pay vein 18 inches; nature of ore, galena, iron and copper pyrites; 1 tunnel 18 feet in length.

Rough Ridge. Clifford E. Sherwood and Charles P. Slade, proprietors; office Gold Hill; claim 150 by 1,500 feet; discovered 1865; located on Big Horn mountain, Gold Hill mining district, 10 miles from Boulder; course of vein, northeast and southwest; average width 2 ½ feet, pitch south, pay vein 2 inches; nature of ore, tellurium; assay value, $150 per ton; 1 shaft 20 feet deep.

Royal Mining Co. Incorporated December 18, 1872; capital stock, $500,000, in 5,000 shares, $100 each; office Boulder; organized by Edwin N. Beach, Stephen Green, and Nathaniel Hayden.

Sac and Fox. Owned by John May and Mrs. William Millsbaugh; office Cincinnati, Ohio; claim 150 by 1,500 feet; discovered 1875; located at Jackson's Camp, near town of Magnolia, 8 miles from Boulder; course of vein, east and west, width 2 feet, pay vein 3 to 4 inches, pitch 5 degrees north, nature of ore, tellurium; value $300 per ton; 1 shaft 50 feet deep, 1 tunnel 45 feet in length.

Salina Consolidated Mining Co. Incorporated December 7, 1875; capital stock, $200,000, in 20,000 shares, $10 each; office Denver; organized by E.W. Pierce; J.H. Ralston, and Peter Winne.

Salina Mining and Milling Co. Incorporated December 27, 1875; capital stock, $60,000, in 6,000 shares, $10 each; office Salina, Boulder County, Colorado; organized by Jacob Eador, John R. Stearns, and Lamartine Alderman.

San Antonio. Owned by Rufus Linderman, J.E. Rice, James and Alexander Gilmore; office Sunshine; claim 150 by 1,500 feet; discovered 1875; located near town of Sunshine, 8 miles from Boulder; course of vein, northeast and southwest, width 5 feet, pay vein 12 inches, pitch 5 degrees south; nature of ore, tellurium and galena; value, $100 per ton; 1 tunnel 160 feet in length, from which a shaft is sunk 75 feet deep.

Santa Fe. Owned by Ebenezer Rowland, George Davy, Alexander McCall, and George Thomas; office Boulder; claim 150 by 1,500 feet; discovered 1873; located on Gold Hill, 2 miles from town of Gold Hill and 11 miles from Boulder; course of vein, northeast and southwest, average width 4 feet, pitch 5 degrees north, pay vein 24 inches; nature of ore, tellurium, free gold, and sulphurets; average value, $40 per ton; one shaft 30 feet deep, 1 drift 80 feet in length.

Santa La Saria Mining Co., Colorado. Incorporated January 25, 1877; capital stock, $3,000,000, in 300,000 shares, $10 each; Albro L. Parsons, president, Curtis R. Parsons, vice president, Charles W. Ellis, Secretary, D. Mortimore, M.D., assistant secretary and superintendent; offices Denver, Colorado, and 40 Broad Street, New York. This property comprises 6 lodes, or mineral veins, 150 by 1,500 feet each; tunnel site 1,500 by 3,000 feet, 4 mill sites, containing 5 acres of ground each; dwelling and boarding houses, blacksmith shop, ore house and stable; located on White Cloud Mountain, Grand Island mining district, 2 ½ miles from Nederland and 15 miles from Boulder. The North Boulder Creek runs through this property, affording ample waterpower for milling and other purposes.

Savannah. Andrew J. Mackey and James A. Carr, proprietors; office Boulder; claim 150 by 1,500 feet, patented; discovered 1859; located had of Dock's Gulch, near town of Gold Hill, 12 miles from Boulder; course of vein, northeast and southwest, average width 4 feet, pitch northwest, average width 4 feet, pitch northwest, pay vein 12 inches; nature of ore, tellurium; average value, $100 per ton; main shaft 50 feet deep; yield, $10,000.

Scott. George W. Day and William Iliff, proprietors; office Denver; claim 75 by 100 feet; discovered 1876; located on Four Mile Creek, Sugar Loaf mining district, 10 miles from Boulder; course of vein, northeast and southwest, width 3 feet, pay vein 18 inches; nature of ore, gray copper and iron pyrites; value $65 per ton; 1 shaft 40 feet deep, 1 tunnel 130 feet in length.

Scott. Owned by Jarred Wood, J.T. Clark, and S. McBarnes; office Gold Hill, claim 150 by 1,200 feet; discovered 1877; located in Gold Hill mining district, 10 miles from Boulder; course of vein, northeast and southwest; average width 4 feet, pitch 10 degrees south, pay vein 8 inches; nature of ore, free gold; assay value $35 per ton; 1 shaft 10 feet deep.

Segengottes. Owned by H.J. Van Wetering and P.H. Van Deist, of Sunshine; claim 150 by 650 feet; discovered 1876; located in Gold Hill mining district, within the limits of Sunshine, 7 miles from Boulder; course of vein, northeast and southwest, average width 4 feet, pitch 10 degrees south; pay vein 10 inches; nature of ore, tellurium; average value, $60 per ton; main shaft 12 feet deep.

Seven-Thirty. Owned by Samuel Moore, Perry Brewer, E.A. Hupper, and Berger Bros.; office Caribou; claim 150 by 3,000 feet, patented; discovered 1870; located on Caribou Hill, Grand Island mining district, 23 miles from Boulder; course of vein, east and west, average width 3 feet, pitch 10 degrees north, pay vein 1 to 22 inches; nature of ore, silver glance and sulphurets, average value, $143 per ton; main shaft 185 feet, and 1 other shaft 40 feet deep; aggregate length of levels 295 feet; shaft house 28 by 50 feet, containing a 20-horse power engine; yield, $25,000.

Sherman. Located near town of Caribou, Grand Island mining district, 22 miles from Boulder; the main shaft is sunk 45 feet below the level of the Caribou tunnel, which intersects the vein at the depth of 210 feet; value of ore, 60 ounces silver per ton.

Shields. B.F. Smith, proprietor; office Golden; claim 150 by 1,500 feet; discovered 1878; located in Sunbeam Gulch, Sugar Loaf mining district, 8 miles from Boulder; vertical vein running northeast and southwest, average width 3 feet, pay vein 3 to 8 inches; nature of ore, copper pyrites; average value, $100 per ton; 1 adit 60 feet in length.

Silver Cloud Mining Co. Incorporated December 6, 1875; capital stock, $30,000, in 300 shares, $100 each; office North Boulder, Grand Island mining district; organized by Charles T. Naylor, John Collier, and Charles H. Conklin.

Silverdale. O.H. Harker, Geo. J. Foster, and Robert D. Kenney, proprietors; office Boulder; claim 150 by 1,500 feet; discovered 1874; located north side of Four Mile Canyon, near town of Sunshine, 6 miles from Boulder; course of vein, northeast and southwest, average width 2 feet, pitch 10 degrees northwest, pay vein 1 to 6 inches; nature of ore, tellurium; average value, $300 per ton; main shaft 60 feet deep, one level 20 feet in length; yield, $1,500.

Silver Point. Thomas J. Graham and William E. Beck, proprietors; office Boulder; claim 50 by 1,500 feet, patented; discovered 1872; located on Idaho Hill, Grand Island mining district, near town of Caribou, 20 miles from Boulder; course of vein, east and west, average width 5 feet, pitch 5 degrees north, pay vein 6 to 20 inches; nature of ore, gray copper, galena, and brittle silver; average value, $125 per ton; main shaft 100 feet deep; 3 levels, aggregate length 100 feet; yield, $6,000.

Silver Point. Owned by H.J. Van Wetering and P.H. Van Deist, of Sunshine claim 150 by 1,500 feet; discovered 1874; located in Gold Hill mining district, near town of Sunshine, 7 miles from Boulder; vertical vein running northeast and southwest, average width 4 feet, pay vein 6 inches; nature of ore, tellurium; average value, $50 per ton; main shaft 55 feet deep.

Sir John Franklin. John L. Schellenger, proprietor; office Magnolia; claim 150 by 1,500 feet; discovered 1877; located at Jackson's Camp, Magnolia mining district, 8 miles from Boulder; course of vein, east and west, width 2 feet, pay vein 5 inches, pitch 5 degrees north; nature of ore, tellurium; 1 shaft 12 feet deep.

Sivyer Placer Claim. Owned by George J. Sivyer & Co., of Boulder; claim comprises 160 acres of ground; located on Four Mile Creek, Ward mining district, 14 miles from Boulder.

Slide. Owned by the American Consolidated Gold and Silver Mining Co., of Connecticut; J. Alden Smith, superintendent; claim 150 by 1,500 feet; discovered 1876; located on Gold Hill, 2 ½ miles from Sunshine and 8 miles from Boulder; vertical vein running northeast and southwest, average width 4 feet, pay vein 18 inches; nature of ore, tellurium, petzite, calaverite, altaite, and free gold; average value of first-class ore $2,500, second-class ore $500, and third-class ore $150 per ton; main shaft 250 feet deep; 4 levels, aggregating 650 feet in length; shaft house 30 by 40 feet.

Smuggler. Owned by Silas S. Kennedy, Benjamin H. Eaton, William H. Lessig, Jacob Hawke, and William A. Christian; office Balarat; claim 150 by 1,500 feet, patented; discovered 1876; located in town of Balarat, Central mining district, 17 miles from Boulder; course of vein, north 2 degrees west; average width 6 feet, pitch 15 degrees, pay vein 8 inches; nature of ore, tellurium and sylvanite; average value, $300 per ton; main shaft 180 feet and 2 other shafts 135 feet each, 3 tunnels, 65, 110, and 75 feet from surface respectively, aggregating 375 feet in length, 3 levels, 54, 130, and 170 feet from surface respectively, aggregating 1,037 feet in length; 20-stamp concentration mill with a capacity for treating 14 tons of ore in 24 hours; 20-horse power engine, 2 ore houses and 2 store and boarding houses; yield to date, $100,000.

South Caribou Tunneling and Mining Co. Incorporated August 31, 1874; capital stock, $5,000,000, in 50,000 shares, $100 each; office Denver; organized by S.P. Conger, Daniel C. Stover; Jerry N. Hill, and C.C. Alvord.

Sovereign People. Owned by Ulysses Pugh and William Woods; office Caribou; claim 150 by 1,400 feet, patented; discovered 1869; located on Caribou Hill, Grand Island mining district, 23 miles from Boulder; course of vein, east and west; width 5 feet, pay vein 18 inches, pitch 10 degrees north; nature of ore, iron and copper pyrites; value $300 per ton; 2 shafts, aggregate depth 172 feet; yield to date, $10,000.

Spencer. Owned by William Donald, Joseph Shaw, and L.K. Smith; office Caribou; claim 50 by 1,400 feet, patented; discovered 1870; located on Caribou Hill, Grand Island mining district, 23 miles from Boulder; course of vein, northeast and southwest, width 3 feet, pay vein 1 to 18 inches, pitch 10 degrees northwest; nature of ore, sulphurets and carbonate of copper; 1 shaft 116 feet; 1 level 100 feet in length; yield to date, $10,000.

Spurgeon. West extension of John Jay. Owned by Enos K. Baxter and Seth B. Bowker, M.D. [Dr. Bowker, a practicing physician in Sunshine, was listed in *The Colorado Business Directory* for 1878]; office Sunshine; claim 150 by 1,500 feet; located in Central mining district, 4 miles from Jamestown and 14 miles from Boulder; course of vein, northeast and southwest, average width 2½ feet, pitch 15 degrees northeast, pay vein 8 inches; nature of ore, tellurium; average value, $150 per ton; one shaft 10 feet deep, 1 tunnel 100 feet in length.

Standard. John G. Evans and E.M. Tarvin, proprietors; office Springdale; claim 150 by 1,500 feet; discovered 1877; located in Elkhorn Gulch, Central mining district, 1 mile from town of Springdale and 12 miles from Boulder; course of vein, northeast and southwest; average width 6 feet, pitch 35 degrees west, pay vein 36 inches; nature of ore, iron and copper pyrites and free gold; average value, $360 per cord; main shaft 75 feet; and 1 other shaft 20 feet deep.

Standard. Owned by Mrs. J.B. Shaw and Harry C. Crary; office Gold Hill; claim 50 by 1,500 feet; discovered 1873; located in Gold Hill mining district, 10 miles from Boulder; vertical vein running northeast and southwest, average width 4 feet, pay vein 10 inches; nature of ore, tellurium and free gold; average value, $250 per ton; main shaft 45 feet deep.

Sterling. Sterling Mining Co., proprietors; offices Gold Hill, Colorado, and Philadelphia, Pennsylvania; claim 150 by 1,500 feet; discovered 1874; located on Gold Hill, Gold Hill mining district, 10 miles from Boulder; course of vein, northeast and southwest, average width 3 ½ feet, pitch 10 degrees north, pay vein 8 inches; nature of ore, tellurium; average value, $345 per ton; main shaft 80 feet deep, 1 tunnel 75 feet, and 1 level 25 in length.

St. Joe. Owned by James B. Gould, J.M. Carnahan, and George W. Chambers; office Gold Hill; claim 150 by 1,500 feet, discovered 1875; located on Gold Hill, Gold Hill mining district, 10 miles from Boulder; course of vein, northeast and southwest, average width 3 feet, pitch 10 degrees northwest, pay vein 10 inches; nature of ore, tellurium; average value, $400 per ton; main shaft 100 feet deep, 2 levels, 60 and 70 feet from surface respectively, aggregating 155 feet in length; shaft house 20 by 32 feet; total yield, $10,000.

Stone. Owned by George W. Grigg and M. Granville Pease; office Gold Hill; claim 150 by 1,500 feet; discovered 1876; located on Hoosier Hill, Gold Hill mining district, 10 miles from Boulder; course of vein, northeast and southwest, average width 12 feet, pitch 10 degrees, pay vein 2 ½ feet; nature of ore, gray copper; assay value $90 per ton; main shaft 30 feet deep.

Stonewall Jackson. Ferdinand C. Albright, proprietor; office Caribou; claim 150 by 1,500 feet; discovered 1870; located on Pomeroy Mountain, Grand Island mining district, 1 mile from Caribou and 23 miles from Boulder; course of vein, east and west, average width 6 feet, pay vein 3 inches, pitch 5 degrees north; nature of ore, galena, and carbonates of copper; average value, $375 per ton; 1 shaft 80 feet deep.

St. Paul. Owned by Isaac Bedal, Cornelius Van Keuren, and William Collier; office Sunshine; claim 150 by 1,500 feet; discovered 1875; located on Grand View Hill, Gold Hill mining district, 8 miles from Boulder; vertical vein running northeast and southwest, average width 3 feet, pay vein 6 inches; nature of ore, tellurium; assay value $1,700 per ton; main shaft 50 feet deep.

St. Vrain Mining Co. Incorporated January 8, 1877; capital stock, $500,000, in 20,000 shares, $25 each; office Boulder; organized by Samuel P. Ely, W.E. Westbrook, Edward B. Gay, and Charles G. Van Fleet.

Sugar Loaf Mining Co. Incorporated March 28, 1878; capital stock, $300,000, in 30,000 shares, $10 each; offices Boulder, Colorado, and New York City; organized by Thomas B. Read, Benjamin G. Bloss, Nathan W. Riker, George C. Ames, and Charles N. Morgan.

Sugar Loaf and Anchor. Owned by W.W. Sleight, of Denver; claim 150 by 1,500 feet; discovered 1874; located in Sugar Loaf mining district, 10 miles from Boulder; course of vein, northeast and southwest, average width 3 feet, pitch 25 degrees northwest, pay vein 6 inches; nature of ore, tellurium; average value, $40 per ton; main shaft 85 feet deep, 1 level 15 feet in length; shaft house 30 by 46 feet.

Sultana. Owned by Jacob Hummel and Samuel Baeder; office Boulder; claim 150 by 1,500 feet, discovered 1875; located in Silver City Gulch, Gold Hill mining district, near town of Salina, 9 miles from Boulder; course of vein, northeast and southwest, average width 4 feet, pitch north, pay vein 2 inches; nature of ore, tellurium; average value, $2,000 per ton; main shaft 40 feet deep, 1 tunnel 85 feet in length and 200 feet from surface.

Sunshine. Thomas Danford, proprietor; office Boulder; claim 150 by 2,200 feet, patented; discovered 1873; located in town of Sunshine, Gold Hill mining district, 6 miles from Boulder; course of vein, northeast and southwest, width 5 feet, pay vein 1 to 4 inches, pitch 15 degrees northwest; nature of ore, tellurium; average value, $160 per ton; one shaft 130 feet deep, 1 adit 50 feet in length.

Swansea Coal and Iron Mining Co. Incorporated February 15, 1872; capital stock, $589,875 in 7,865 shares, $75 each; office Denver; organized by Charles B. Kountze, John Harper, A. Davidson, Charles C. Welch, Jonathan S. Smith, Low P. Cook, and William B. Berger.

Sylvan. Owned by Mary G. Arnett, Sylvester Hill, and Daniel Flynn; office Boulder; claim 150 by 1,500 feet; discovered 1872; located on Four Mile Creek, Gold Hill mining district, 1 mile from town of Salina and 10 miles from Boulder; vertical vein running northeast and southwest, average width 4 feet, pay vein 10 inches; nature of ore, tellurium; average value, $35 per ton; main shaft 33 feet deep and 1 other shaft 22 feet deep, 1 tunnel 40 feet in length.

Tellurium Crown Mining Co. Incorporated August 14, 1876; capital stock, $600,000, in 60,000 shares, $10 each; office Denver; organized by Jean B. Rouilliard, W.H.J. Nichols, Leonard Cutshaw, Edgar Tawdey, and Stephen Lashuer.

Tellurium Tunnel Mining Co. Hal. Sayer, superintendent; claim 1,500 by 3,000 feet; discovered 1876; located at the base of Sunshine Hill, Gold Hill mining district, 8 miles from Boulder; length of tunnel 238 feet.

Tenbroek. James D. Smith, William A. and Charles J. Christian, proprietors; office Denver; claim 150 by 1,500 feet; discovered 1877; located in Central Gulch, 2 miles from Balarat and 15 miles from Boulder; vertical vein running east and west, average width 4 feet; nature of ore, sulphurets; average value, 80 ounces silver per ton; main shaft 17 feet deep.

Ten-Forty. Located on Caribou Hill, Grand Island mining district, 22 miles from Boulder; developed by a shaft 100 feet deep.

Three Brothers. Owned by Henry Carter, John Pierce, and others; office Salina; claim 150 by 1,500 feet; discovered 1876; located on Melvina Hill, Gold Hill mining district, 8 miles from Boulder; course of vein, north and south, average width 1 foot, pitch 20 degrees west, pay vein 1 inch to 1 foot; nature of ore, tellurium; assay value, $150 to $5,000 per ton; 1 tunnel 80 feet in length.

Tilden. Owned by John Williams, George Newcomb, and John Campbell; office Gold Hill; claim 150 by 1,400 feet; discovered 1877; located near town of Gold Hill, Gold Hill mining district, 12 miles from Boulder; course of vein, northeast and southwest, width 1 foot; nature of ore, iron pyrites; main shaft 40 feet deep.

Tillie Butzel. Andrew J. Mackey and John L. Butzel, proprietors; office, Boulder, Colorado, and Saugerties, New York; claim 150 by 1,500 feet, patented; discovered 1874; located in Gold Hill mining district, 1 mile from Sunshine and 7 miles from Boulder; vertical vein, trending northeast and southwest, average width 5 feet, pay vein 8 inches; nature of ore, tellurium, petzite, and free gold; average value, $250 per ton; discovery shaft 19 feet, two other shafts 45 and 85 feet deep respectively, 1 tunnel 200 feet in length, 1 level 45 feet in length and 128 feet from surface.

Trade Dollar. Owned by the Trade Dollar Silver Mining Co., of Colorado; incorporated under the laws of the State of New York January 25, 1875; capital stock, $500,000, in 5,000 shares, $100 each; offices New York City and Caribou, Colorado; organized by R. Ogden Doremus, George K. Listance, Christopher Miller, Leonard S. Root, and B.S. Budd; claim 500 by 5,000 feet, patented; located on Caribou Hill, Grand Island mining district, 23 miles from Boulder.

Triune. Owned by Joseph Airy, George L. Aggers, B.E. Hale, and Henry Meyring; office Denver; claim 150 by 1,500 feet, patented; discovered 1875; located on Gold Run, Gold Hill mining district, 8 miles from Boulder; course of vein, northeast and southwest, average width 12 feet, pitch 5 degrees south, pay vein 1 to 2 feet; nature of ore, free gold and tellurium; average value, $500 per ton; main shaft 80 feet deep, 1 level 55 feet in length; total yield, $1,000.

Trojan. Located in Grand Island mining district, near town of Caribou, 22 miles from Boulder; developed by 2 shafts, 60 and 130 feet deep respectively.

Tryon Mining and Concentrating Co. Incorporated October 4, 1875; capital stock, $20,000, in 800 shares, $25 each; office Hunt's Mill, Boulder County, Colorado; organized by Thomas P. Gallup, William A. Hardenbrook, and Radcliffe B. Lockwood.

Turner Tunnel Mining and Milling Co. Incorporated under the laws of the State of Kentucky, November 14, 1877; capital stock, $275,000, in 2,750 shares, $100 each; B.F. Farrar, president, S.L. Shivel, secretary, William H. Cassel, treasurer, Professor Harry Turner, superintendent; offices, Boulder, Colorado, and Lexington, Kentucky; claim 1,500 by 3,000 feet; discovered 1877; located at the head of Weare Gulch, Grand Island mining district, 2 miles from Caribou and Nederland and 20 miles from Boulder; the course of the tunnel is south 21 degrees 10 minutes west, is now 175 feet in length, and was designed to intersect 7 lodes or mineral veins.

Twinn. Owned by H.C. and E.P. Thompson, H.H. Tuttle, and C.J. Spalding; office Boulder; claim 150 by 1,500 feet; discovered 1877; located at Camp Rebecca, Magnolia mining district, 8 miles from Boulder; course of vein, east and west, pitch 5 degrees north, width 15 feet, pay vein 8 to 10 inches; nature of ore, tellurium; average value, $50 per ton; main shaft 30 feet deep.

Union. Owned by Angelo Ruvero, of Salina; claim 150 by 1,500 feet, discovered 1877; located in Ringgold Gulch, Sugar Loaf mining district, 2 miles from Salina and 11 miles from Boulder; course of vein, northeast and southwest, average width 3 feet, pitch south, pay vein 18 inches; nature of ore, galena; average value, $125 per ton; one tunnel 59 feet in length.

Union. Owned by Dr. N.S. Culver, of Colorado Springs; claim 150 by 1,500 feet, discovered 1875; located in Gold Hill mining district, near town of Sunshine, 7 miles from Boulder; course of vein, east and west, average width 3 feet; pitch 5 degrees north, pay vein 5 inches; nature of ore, tellurium; main shaft 35 feet deep.

Union. Owned by Samuel Baeder and Robert McKee of Boulder; claim 150 by 1,500 feet, discovered 1877; located on Boulder Creek, Grand Island mining district, 2 ½ miles from town of Salina and 12 miles from Boulder; vertical vein running northeast and southwest, average width 10 feet; pay vein 18 inches; nature of ore, tellurium; average value, $84 per ton; main shaft 25 feet deep.

United States Mint. Owned by Whipple & Henderson, and George D. Cook; office Left Hand; claim 150 by 1,500 feet, discovered 1876; located on Gold Run, Gold Hill mining district, 10 miles from Boulder; course of vein, northeast and southwest, width 3 feet, pitch 5 degrees south, pay vein 12 inches; nature of ore, free gold; average value, $72 per ton; 1 tunnel 50 feet in length.

Victor. Owned by Charles E. Buckingham and others; office Boulder; claim 150 by 1,500 feet, patented; discovered 1874; located in Somerville, Gold Hill mining district, 10 miles from Boulder; vertical vein, width 3 feet, pay vein 1 to 10 inches; nature of ore, gray copper and tellurium; average value, $250 per ton main shaft 235 deep; total yield, $1,500.

Victoria Mining Co. Incorporated September 28, 1875; capital stock, $250,000, in 25,000 shares, $10 each; offices Boulder and Denver, Colorado; organized by Edward F. Wallace, Henry Moore, Judson B. Shaw, James W. Robinson, Isaac H. Locke, and George H. Batchelder.

Virginia. Owned by Joseph Steppler and Joseph Steppler, Jr.; office Denver; claim 150 by 1,500 feet, patented; located on Big Horn Mountain, Gold Hill mining district, near town of Salina, 8 miles from Boulder; course of vein, northeast and southwest, width 5 feet, pay vein 15 inches, pitch 35 degrees north; nature of ore, tellurium and free gold; average value, $450 per ton; main shaft 50 feet deep.

Wamego. Owned by the Smuggler Mining Co., claim 150 by 1,500 feet, patented; discovered 1876; located in Long's Gulch, Central mining district, 17 miles from Boulder; vertical vein trending northeast and southwest, width 4 feet, pay vein 2 inches; nature of ore, tellurium; average value, $300 per ton; 1 tunnel 125 feet in length.

Wandering Jew. Owned by John Verdew, Joseph Begsbee, and Frederick C. Bucherdee; claim 150 by 1,500 feet; discovered 1877; located on Golden Age Hill, Central mining district, 1 mile from Jamestown, and 18 miles from Boulder; course of vein, northeast and southwest, average width 2 feet, pitch 20 degrees southeast, pay vein 2 inches; nature of ore, tellurium and free gold; average value, $153 per ton; main shaft 50 feet and 1 other shaft 10 feet deep.

Ward Mining Company, Massachussets. Incorporated April 20, 1877; capital stock, $30,000; organized by Oliver Ames, William A. Hayes, John J. Dixwell, John R. Brewer, and others.

Warsaw. Owned by George Walton, Elmer White, and Charles W. Smith; office Sunshine; claim 150 by 1,500 feet; discovered 1874; located on Warsaw Hill, Gold Hill mining district, 1 mile from town of Sunshine, and 6 miles from Boulder; vertical vein running northeast and southwest; average width 4 feet, pay vein 10 inches; nature of ore, tellurium; average value, $300 per ton; main shaft 70 feet deep aggregate length of drifts 380 feet.

Washington. Heber A. and Horace F. Learnard, proprietors; office Rockville; claim 150 by 1,500 feet; located in Rockville, 2 miles from Sunshine, and 10 miles from Boulder; course of vein, east and west;

average width 5 feet, pitch 5 degrees north, pay vein 6 inches; nature of ore, tellurium and iron; average value, $80 per ton; main shaft 25 feet deep.

Washington. Owned by Frank Touser, Anthony Wilde, and Lawrence Singer; office Salina; claim 150 by 1,500 feet; discovered 1874; located on Sunshine Hill, Gold Hill mining district, 1 mile from town of Salina, and 8 miles from Boulder; vertical vein running northeast and southwest; average width 5 feet, pay vein 18 inches; nature of ore, tellurium; assay value $14,000 per ton; main shaft 70 feet deep, aggregate depth of other shafts 64 feet; yield, $1,000.

Washington Avenue. Owned by the Washington Avenue Silver Mining Co., of Cincinnati; claim 150 by 4,500 feet, patented; discovered 1870; located in Grand Island mining district, 5 miles from Nederland, and 16 miles from Boulder; course of vein, north and south; average width 5 feet, pitch 10 degrees west; nature of ore, galena, zinc blende and iron pyrites; average value, $100 per ton; main shaft 320 feet, and 1 other shaft 300 feet deep; shaft house 30 by 80 feet, ore house 26 by 60 feet; concentration mill 44 by 130 feet, with all necessary machinery for the successful treatment of ore; total yield, $50,000.

Washington Irving. Owned by Jacob Ellison, Frederick Eberhardt, John Jordie, and Martin Hosang; office Sunshine; claim 150 by 1,500 feet; discovered 1875; located in Sunshine Gulch, Gold Hill mining district, half mile from town of Sunshine, and 7 miles from Boulder; vertical vein running northeast and southwest average width 4 feet; pay vein 12 inches, nature of ore, tellurium; average value, $400 per ton; main shaft 35 feet deep; 1 drift 100 feet in length.

Waverly. Jacob Ellison, proprietor; office Boulder; claim 150 by 1,500 feet; discovered 1875; located on Lee Hill, Gold Hill mining district, 1 mile from Sunshine, and 8 miles from Boulder; course of vein, northeast and southwest, average width 3 feet, pitch 20 degrees east; pay vein 1 to 6 inches; nature of ore, tellurium and free gold; average value, $75 per ton; main shaft 50 feet deep, 1 level 30 feet in length.

Webster City. Webster City Gold and Silver Mining Co., Michigan; incorporated 1876; capital stock, $500,000, Capt. Ebenezer Rowland, of Boulder, superintendent; claim 150 by 1,500 feet; discovered 1873; located in Pennsylvania Gulch, Sugar Loaf mining district, 5 miles from Gold Hill, and 13 miles from Boulder; course of vein, northeast and southwest, average width 12 feet, pitch 10 degrees north, pay vein 14 inches; nature of ore, iron and copper pyrites, decomposed galena, and sulphurets; average value, $50 per ton; main shaft 100 feet deep.

Webster City Placer Claim. Owned by the Webster City Gold and Silver Mining Co., of Michigan; Capt. Ebenezer Rowland, superintendent; comprises 35 acres of patented ground; located in Pennsylvania Gulch on Four Mile Creek, Sugar Loaf mining district, 5 miles from Gold Hill and 12 miles from Boulder.

Weist. Owned by the Chicago and Colorado Mining Co., offices Boulder, Colorado, and Chicago, Illinois; claim 150 by 1,500 feet, patented; discovered 1875; located on Left Hand Creek, Ward mining district, 4 miles from town of Gold Hill and 16 miles from Boulder; course of vein, east and west, width 3 feet, pitch 5 degrees north; pay vein 7 inches; nature of ore, iron pyrites; average value, $15 per ton main shaft 40 feet deep; aggregate depth of other shafts 110 feet.

Welcome. Victor Roux, proprietor; office Boulder; claim 150 by 1,500 feet; discovered 1871; located in Sunbeam Gulch, Sugar Loaf mining district, 1 mile from town of Salina and 8 miles from Boulder; vertical vein running east and west; average width 5 feet, pay vein 12 inches; nature of ore, tellurium; average value 54 ounces silver per ton; main shaft 22 feet deep.

Welfare. Wright W. and Horace F. Hall, proprietors; office Left hand; claim 150 by 1,500 feet; discovered 1876; located in Long's Gulch, Gold Hill mining district, 10 miles from Boulder; course of vein, northeast and southwest; width 6 feet, pay vein 4 feet; nature of ore, galena and gray copper; value $70 per ton; 1 shaft 50 feet deep; 1 level 17 feet in length.

Western Slope. Santa La Saria Mining Company, Colorado, proprietors; offices Denver, Colorado, and 41 Broad Street, New York; claim 150 by 1,500 feet; discovered 1874; located on White Cloud mountain, Grand Island mining district 2 ½ miles from Nederland and 13 miles from Boulder; course of vein, northeast and southwest; average width 5 feet; pitch 30 degrees south; pay vein 10 inches; nature of ore, gray copper, native silver, and sulphurets of silver; average value 50 ounces silver per ton; will be reached by the Santa La Saria tunnel at a distance of 250 feet from the mouth and intersected at a depth of 400 feet; 1 drift 100 feet in length.

White Crow. John Morrison and others, proprietors; office Boulder; claim 150 by 1,500 feet, patented; discovered 1874; located in Gold Hill mining district, 1 mile from Sunshine and 7 miles from Boulder; nature of ore, tellurium; average value, $300 per ton; main shaft 70 feet deep.

White Pilgrim. Black Cloud Gold and Silver Mining Co., proprietors; office Boulder; claim 150 by 1,500 feet; located in McKnight Gulch, 2 miles from Gold Hill and 11 miles from Boulder; course of vein, northeast and southwest; average width 2 ½ feet; pitch 3 degrees south; pay vein 6 inches; nature of ore, iron and copper pyrites, free gold and zinc blende; 1 drift 20 feet in length; operated through the Black Cloud tunnel.

Wild Boy. Owned by Henry Wood; office Sugar Loaf; claim 150 by 1,500 feet; discovered 1876; located in Long's Gulch, Sugar Loaf mining district, 11 miles from Boulder; course of vein, northeast and southwest; average width 18 inches; pitch 10 degrees north; pay vein 1 foot; nature of ore, iron pyrites; assay value $200 per ton; main shaft 15 feet deep.

Winona. Owned by Hatch, Davidson & Co. [Israel B. Hatch, David Davidson, and A.D. Dunbar]; office Denver; claim 50 feet by 1,342 feet, patented; discovered 1858; located on Gold Hill, Gold Hill mining district, 8 miles from Boulder; course of vein, northeast and southwest; width 5 feet, pay vein 8 to 26 inches; pitch 5 degrees west; nature of ore, tellurium and free gold; average value, $150 per ton; main shaft 100 feet deep; other shafting aggregating 400 feet; 1 drift of 50 feet and a surface opening 100 feet in length; shaft house 14 by 24 feet, containing a 20-horse power engine and hoisting apparatus; this property is located on the same vein as the well known "Slide Mine," and is reached by the Slide tunnel at a distance of 4,500 feet from the mouth, and intersected at a depth of 300 feet; yield to date, $25,000.

Winona, Gold Hill Mining Co. Incorporated December 13, 1875; capital stock, $250,000, in 25,000 shares of $10 each; offices Denver, Gold Hill, and Boulder, Colorado; organized by James H. Ralston, Francis H. La Grave, Joseph Steppler, and Judson B. Shaw.

Worchester. Samuel Baeder, proprietor; office Boulder; claim 150 by 1,500 feet; discovered 1876; located in Sunbeam Gulch, Sugar Loaf mining district, 1 mile from town of Salina and 8 miles from Boulder; vertical vein running northeast and southwest; average width 14 feet, pay vein 2 inches; nature of ore, tellurium; average value, $84 per ton; main shaft 20 feet deep.

Wren. Andrew J. Beasley & Co., proprietors; office Salina; claim, 150 by 1,500 feet; discovered 1875; located on Melvina Hill, Gold Hill mining district, half mile from town of Salina and 8 miles from Boulder; course of vein, northeast and southwest; average width 14 inches; pitch north; pay vein 6 inches; nature of ore, tellurium; average value, $200 per ton; main shaft 185 feet deep; aggregate length of tunnels 150 feet; shaft house 16 by 20 feet.

W.W. Consolidated Mining Co. Incorporated December 27, 1875; capital stock, $200,000, in 20,000 shares, $10 each; office Denver; organized by James H. Ralston, Peter Winne, and E.W. Pierce.

Xenia. Armstrong, Ankeny & Co., proprietors; offices Jamestown, Colorado, and Xenia, Ohio; claim 150 by 1,500 feet; discovered 1877; located on Golden Age Hill, Central mining district, 2 ½ miles from Jamestown, and 14 miles from Boulder, vertical vein running northeast and southwest; pay vein 8 inches; nature of ore, free gold; value $30 per ton; main shaft 45 feet deep.

Xerxes. Homer H. Smith and Emory C. Nutt, proprietors; office Sunshine; claim 150 by 600 feet; discovered 1876; located in town of Sunshine, Gold Hill mining district, 7 miles from Boulder; course of vein, northeast and southwest; average width 5 feet; nature of ore, tellurium; main shaft 25 feet deep

Yellow Boy. Owned by Henry Taylor and Albert Bills, of Boulder; claim 150 by 1,500 feet, discovered 1877; located near Gold Lake, Central mining district, 3 miles from town of Gold Hill, and 15 miles from Boulder; vertical vein running east and west; average width 5 feet; pay vein 16 inches; nature of ore, free gold; 1 shaft 12 feet deep.

Young America, Extension of American. Owned by William Davidson, C.C. and David Strock; office Sunshine; claim, 150 by 1,500 feet, patented; discovered 1874; located in town of Sunshine, Gold Hill mining district, 7 miles from Boulder; course of vein, northeast and southwest, width 5 feet; pay vein 2 feet; pitch 5 degrees north; nature of ore, tellurium, calaverite, sylvanite, and free gold; average value of first-class ore $4,000, second-class ore $250, and third-class ore $100 per ton; main shaft 170 feet deep; 1 level 82 feet in length.

Boulder County Ore Mills

Atchison Mining Co.'s Mill. Located at the mouth of Gold Run, 8 miles from Boulder; has a capacity for treating 10 tens of ore daily; the main building is 50 feet wide by 60 feet in length.

Bates Mill. Owned by Hon. Jerome B. Chaffee, located on North Boulder Creek, Grand Island mining district, 3 miles from Caribou, and 23 miles from Boulder; has a capacity for treating 10 tons of ore daily.

Black Cloud Mill. Located in town of Summerville, 1 mile from Salina, and 11 miles from Boulder; the main building is 36 feet wide by 102 feet long, and contains a 25-horse power engine, Blake crusher, Cornish rolls, and all other necessary machinery; has a capacity for treating 25 tons of ore daily.

Boulder County Mill. Owned by the Boulder County Mining, Tunneling, Milling, Land and Town Co., located on North Boulder Creek, Grand Island mining district, 2 miles from Nederland, and 18 miles from Boulder; contains 100 stamps and 40 Frue vaners.

Boulder Sampling and Crushing Works. Owned by the Boston and Colorado Smelting Co.; C.G. Duncan, agent; located corner Pearl and Eighth streets, in the city of Boulder; has a capacity for treating 25 tons of ore daily.

Boyd's Smelting Works. James H. Boyd, proprietor; located near the limits of the town of Boulder, at the mouth of Boulder Canyon; main building is 60 feet wide by 90 feet long, and contains a Blake crusher, Cornish rollers, 2 turbine water wheels, 17 and 35 inches in diameter, respectively, blast and separating furnaces and apparatus; has a smelting capacity for treating 15 tons of ore in 24 hours, and a crushing capacity of 25 tons daily.

Brett's Mill. Located on Left Hand Creek, near mouth of James Creek, 8 miles from Boulder.

Cincinnati Concentration Mill. Owned by John R. Davey and Robert Simpson; located on James Creek, near Jamestown, 14 miles from Boulder.

Corning Mill. Owned by the Corning Tunnel, Mining and Reduction Co.; located on Four Mile Creek, 2 miles from Sunshine and 9 miles from Boulder; size of building 40 by 100 feet.

Evert's Mill. Located on Four Mile Creek, 2 miles from Boulder; the main building is 40 feet in width and 50 feet in length, with an engine house 25 by 35 feet; contains 10 stamps, 4 Frue vaners, a 20-horse power engine, crushers, etc.

Golden Age Mill. Harland P. Walker, proprietor; located on James Creek, in Jamestown, 12 miles from Boulder; has a capacity for treating 15 tons of ore in 24 hours; the building is a wooded structure, 40 feet in width by 70 feet in length, and contains 25 stamps, weighting 550 pounds each; a 25-horse power engine, 4 Bartola pans, Austrian percussion tables, water wheel, etc.; run by steam and water power.

Humboldt Mill. Owned by Edward J. Binford and John J. Ellingham; located at the head of Indiana Gulch, near town of Ward, in Ward mining district, 18 miles from Boulder; has a capacity for treating 10 tons of ore daily, and contains 10 stamps, a 20-horse power, engine, etc.; main building is 36 feet wide by 40 feet in length.

Long's Peak Mining Company's Mill. Located in town of Ward, 18 miles from Boulder.

National Mill. Owned by the National Gold and Silver Mining Co., of Lexington, Kentucky; Nathan Weare, superintendent; located in Weare Gulch, 2 miles from Caribou and 18 miles from Boulder; contains 25 stamps and all other necessary machinery.

Nederland Mill. Owned by Hon. Jerome B. Chaffee, Eben Smith, superintendent; located in town of Nederland, Grand Island Mining district, 19 miles from Boulder; the building is a wooden structure, 100 feet in width by 165 feet in length, and contains 15 stamps, a Blake crusher, 4 cylinders, 120-horse power engine, amalgamating pans, agitators, etc.; has a capacity for treating 10 tons of ore daily.

New Jersey Mill. Owned by the Mining Company of New Jersey; located at Caribou, Grand Island mining district, 23 miles from Boulder; has a capacity for treating 10 tons of ore daily.

Ni Wot Mill. Owned by William A. Davidson, of Valmont; located in Ward mining district, 18 miles from Boulder; has 50 stamps.

Orodelfan Smelting Works. Hunt, Barber & Co., proprietors; located in Boulder canyon, at the mouth of Four Mile Creek, 3 miles from Boulder; has a capacity for treating 10 tons of ore daily.

Pomeroy's Mill. J.V. Pomeroy, proprietor; located in town of Ward, 18 miles from Boulder; has a capacity for treating 10 tons of ore daily.

Rust's Concentration Mill. Owned by George W. Rust; located in the city of Boulder.

Smuggler Mill. Owned by Charles Van Fleet, of Boulder; located near town of Balarat, 17 miles from Boulder; main building is 28 feet in width by 30 feet in length, with a wing 30 by 45 feet, contains 20 stamps and a 20-horse power engine, etc.; has a capacity for treating 14 tons of ore daily.

Stewart Mill. Operated by James Martin & Co., located on Pearl Street, Boulder; contains 10 stamps, concentration tables, 1 furnace, large size Bartola pan, 1 Burr mill, and 2 engines.

Washington Avenue Mill. Owned by the Washington Avenue Silver Mining Co., of Cincinnati, Ohio; located in Grand Island mining district, near Nederland, 16 miles from Boulder. The main building is 44 feet in width by 130 feet in length, and contains all necessary machinery.

Williard's Mill. Owned by Z.A. Williard; located on James Creek, near Jamestown, 13 miles from Boulder; main building is 45 feet in width by 60 feet in length, with an engine house 15 by 40 feet.

Boulder County Coal Mines

Davidson Coal and Iron Mining Co. Incorporated April 28, 1873; capital stock, $160,000, in 32,000 shares, $50 each; office Denver; organized by William A. Davidson, Jonathan S. Smith, George W. Smiley, Charles B. Kountze, and William B. Berger.

Marshall Coal Mining Co. Incorporated November 17, 1877; capital stock, $30,000, in 300 shares, $100 each; office Denver; trustees: Augustine G. and Nathaniel P. Langford and Samuel S. Davidson.

Rob Roy. Henry T. West, proprietor; property comprises 160 acres of coal land; located in the town of Canfield, on Boulder Valley Railroad, 25 miles from Denver; developed by a main shaft 125 feet, and an air shaft of the same depth, 6 entries aggregating 2,400 feet, and 7 air courses aggregating 2,400 feet in length; has a capacity for raising 200 tons of coal daily.

Star. Owned by Joseph C. Dresser, Benjamin M. Williams, and William O. Wise; property comprises 160 acres of coal land, located in town of Canfield, on Boulder Valley Railroad, 1 mile from town of Erie, and 25 miles from Denver; developed by a main shaft 103 feet deep, and an air shaft of same depth, 6 entries aggregating 1800 feet, and 5 air courses aggregating 1,200 feet in length; has a capacity for raising 200 tons of coal daily, and has produced to date 20,000 tons.[William O. Wise, a leading citizen of Canfield, was born in Dodgeville, Wisconsin, on 28 October 1848, and moved to Colorado in the spring of 1870 for his health. He may have had tuberculosis. Penniless when he arrived, Wise worked as a farmhand in the Canfield area and later homesteaded a farm in Canfield. When he discovered coal on an adjacent property, he formed a company to purchase and develop the coal mine, which they named the Star. He was superintendent of the Star mine, postmaster of Canfield, and was elected Colorado State Representative.]

Superior. Owned by David Hill and James Thompson; office Denver; this property comprises 80 acres of coal land, located 1 mile from town of Erie, and 22 miles from Denver, on the Boulder Valley Railroad; developed by 2 shafts 80 and 100 feet, respectively; has a capacity for producing 50 tons of coal daily.

Welch Coal Co. Incorporated August 26, 1878; capital stock, $50,000, in 1,000 shares, $50 each; office Golden; directors, Charles C. Luther J., and Frank L. Welch.

Remnants of Boulder County's Wall Stgreet Mine and Mill rise from the hillside, resembling an English castle out of place and out of time.

Unidentified abandoned mine in Boulder County with the wreckage of an old sluice leaning against the enclosure.

CLEAR CREEK COUNTY

Thomas Corbett reported that Clear Creek County had a population of 12,000 in 1878, and the bullion produced that year was valued at $2,250,000. In 1878, the mining districts were Argentine, Cascade, Downieville, East Argentine, Geneva, Grass Valley, Griffith, Idaho, Independent, Montana, Morris, Queen, Spanish Bar, Trail Creek, Virginia, and Upper Union.

Adelia. George L. Cannon, proprietor; office, Idaho Springs; claim 50 by 1,500 feet; discovered 1872; located on Schaffter Mountain, Idaho mining district, half mile from town of Idaho Springs; vertical vein, trending northeast and southwest, width 5 feet, pay vein 12 inches; nature of ore, free gold; average value, $104 per ton; main shaft 50 feet deep.

Adrian. George L. Cannon, proprietor; office, Idaho Springs; claim 150 by 1,500 feet; discovered 1872; located on Schaffter Mountain, near Idaho Springs, 33 miles from Denver; vertical vein, trending northeast and southwest, width 4 feet, pay vein, 10 inches; nature of ore, free gold; average value, 20 ounces gold per cord; 1 shaft 60 feet deep.

Advance. Owned by the Burleigh Mining Co., Ivers Phillips, superintendent; claim 150 by 3,000 feet; discovered 1871; located on Sherman Mountain, Griffith mining district, near town of Silver Plume, 3 miles from Georgetown; course of vein; east and west, width 4 feet, pay vein 4 inches, pitch 20 degrees north; nature of ore, galena; value, 200 ounces silver per ton; reached by the Burleigh Tunnel at a distance of 1,000 feet from the mouth and intersected at a depth of 550 feet; 1 level 40 feet in length.

Aetna. Owned by George E. Marsh and William J. Gilchrist; office, Georgetown; claim 50 by 3,000 feet, patented; located on Griffith Mountain, Griffith mining district, 1 mile from Georgetown; course of vein, northeast and southwest, pitch 5 degrees north, width 20 feet, pay vein 11 inches; nature of ore, galena and sulphurets; value, 300 ounces silver per ton; 1 shaft 20 feet deep; 1 level 140 feet in length.

Akron Mining Co. Incorporated April 24, 1878; capital stock, $100,000, in 1,000 shares, $100 each; office Akron, Ohio; trustees: William W. Moulton, Joseph A. Beebe, Richard S. Elkins, Edward Mize, Joy H. Pendleton, and David L. King.

Akron Tunnel. Owned by the Akron Tunnel Co., claim 1,500 by 3,000 feet; discovered 1876; located on Columbia Mountain, Montana mining district, near town of Lawson, 6 miles from Georgetown; length of tunnel 400 feet.

Alabama. Owned by C.N. Shipman; claim 50 by 1,600 feet, patented; discovered 1865; located on Leavenworth Mountain, Griffith mining district, 3 miles from Georgetown; width of vein 5 feet; pay vein 1 to 4 inches; value of first-class ore 500 to 2,000 ounces, and second-class ore 350 to 450 ounces silver per ton; developed by 2 tunnels on the vein, 1 of which is 200 feet in length, and 2 shafts sunk from the tunnel level, 50 and 100 feet, respectively.

Albro. Located on Silver Creek near Georgetown; width of vein 4 feet; value of ore $40 per ton; developed by a shaft 200 feet deep and a tunnel on the vein, which intersects the shaft at a depth of 100 feet.

Allen Mining Property. Samuel P. Allen, proprietor; office Empire; claim 50 by 1,500 feet; located on Silver Mountain, Upper Union mining district, 2 miles from Georgetown; course of vein northeast and southwest, width 4 feet, pitch 15 degrees northwest, nature of ore, iron and copper pyrites; value $150 per cord; main shaft 150 feet deep, 1 tunnel 900 feet in length, which intersects the vein at a depth of 450 feet; yield, $13,000.

Alpine. Horatio E. and Morris Hazard and David Rousseau, proprietors; office Central City; claim 50 by 1,500 feet; discovered 1872; located on McClelland Mountain, East Argentine mining district, 7 miles from Georgetown; vertical vein, trending northeast and southwest, width 6 feet, pay vein 10 inches; nature of ore, galena; average value of first-class ore, 300 ounces, and second-class ore, 150 ounces silver per ton; 2 levels, aggregate length 125 feet.

Alps. Lebanon Mining Co., proprietors, Julius G. Pohle, superintendent; office Georgetown; claim 50 by 1,500 feet, patented; discovered 1870; located in Cascade mining district, 4 miles from Idaho Springs; vertical vein, trending northeast and southwest, width, 20 feet, pay vein 14 inches; nature of ore, galena and gray copper; average value, 150 ounces silver per ton; 1 shaft, 40 feet deep.

Amazon. Owned by John McIntosh, James Ray, and David Hill; office Georgetown; claim 150 by 1,500 feet; discovered 1876; located on Sherman Mountain, near town of Silver Plume, Griffith mining district, 2 miles from Georgetown; course of vein, north and south, width 6 feet, pitch 10 degrees east; 1 tunnel 180 feet in length.

American Chief. Owned by George Westman, C.S. Abbot, and others; office Lawson; claim 150 by 1,500 feet; discovered 1877; located on Columbia Mountain, Montana mining district, near town of Lawson; vertical vein, running east and west, width 4 feet, pay vein 6 inches; nature of ore, galena; average value, $125 per ton; main shaft 50 feet deep.

American Eagle. Francis L. Andre, proprietor; office Empire; claim 150 by 1,500 feet; discovered 1875; located on Silver Mountain, Upper Union mining district, 2 miles from Georgetown; vertical vein, running north and south, average width 4 feet, pay vein 18 inches; nature of ore, iron and copper pyrites and free gold; average value, $150 per cord; main shaft 90 feet and 1 other shaft 40 feet deep.

American Mining Co. Incorporated June 24, 1870; capital stock, $2,000,000, in 20,000 shares, $100 each; office Washington, D.C.; organized by Sayles J. Bowen, Charles H. Parsons, Thomas Antisell, Edward G. Ross, Alpheus S. Williams, Thomas T. Crittenden, Stephen K. Kane, Alfred B. Mullet, and William Thorpe.

Anderson Prospecting Co. Organized under the laws of the Commonwealth of Kentucky, July 19, 1872; capital stock, $55,000, in 550 shares, $100 each; office Louisville, Kentucky; organized by W.H. Anderson, H.H. Sale, James Ruddle, John F. Babbitt, Benjamin S. Weller, William H. Sale, R.J. Crawford, J.T. Murphy, and James McGrain.

Anglo Saxon. Located on Griffith Mountain, Griffith mining district, near Georgetown; nature of ore, silver glance; average value 1,000 ounces silver per ton; developed by levels and tunnels.

Annoka County. Owned by Jas. Kinkead and Thomas H. Thatch; office Fall River; claim 50 by 800 feet, patented; discovered 1860; located on Donaldson Hill, Spanish Bar mining district, near town of Fall River; course of vein, northeast and southwest, average width 4 feet, pay vein 15 inches, pitch 25 degrees south; nature of ore, iron and copper pyrites; average value, $110 per ton; main shaft 50 feet deep; aggregate depth of other shafting 75 feet.

Aorta Tunnel. Owned by the Bay State Mining Co., and Charles P. Baldwin; offices Georgetown, Colorado, and Boston, Massachusetts; claim 1,500 by 3,000 feet; located on Silver Mountain, Upper Union mining district, near town of Empire, 2 miles from Georgetown; length of tunnel 960 feet, designed to cut the following named lodes: Arta, Coyote, Erie County, Fountain, Harden County, Livingston County, Must Cash, Sandusky, Siegel, Silver Mountain, Tempest, Tenth Legion. These lodes are intersected by the tunnel at a depth from 450 to 800 feet.

Argentine. Located on Leavenworth Mountain, Griffith mining district, 3 miles from Georgetown; discovered 1875; width of vein 7 feet, pay vein 2 feet; value of ore, 400 ounces silver per ton; developed by a shaft 100 feet deep, two levels 150 and 285 feet in length, respectively, and 1 tunnel, which intersects the vein at a depth of 60 feet.

Argentum Company. Incorporated June 21, 1876; capital stock, $2,000, in 20 shares, $100 each; offices Denver; organized by Aaron B. Robbins, Henry A. Bagley, and William F. Perkins.

Argonaut. Nathan White & Co., proprietors; office Leadville; claim 150 by 1,500 feet; discovered 1875; located on Leavenworth Mountain, Griffith mining district, 2 miles from Georgetown; course of vein, northeast and southwest, width 4 feet, pay vein 10 inches, pitch 20 degrees east; nature of ore, sulphurets, galena, and iron pyrites; average value, 150 ounces silver per ton; 1 shaft 45 feet deep.

Argus. West Argentine Mining Co., proprietors; John A. Stacey, president, William H. Baldwin, secretary, Charles P. Baldwin, superintendent; office Georgetown; claim 50 by 3,000 feet, patented; discovered 1865; located on Kelso Mountain, above timber line, Argentine mining district, 4 miles from Bakerville, and 12 miles from Georgetown; course of vein, east and west, pitch northwest, width 5 feet, pay vein 1 to 5 inches; average value of ore 225 to 400 ounces silver per ton; 2 shafts 16 and 20 feet in depth, respectively; 1 level 40 feet in length.

Aspinwall Mining and Milling Co. Incorporated February 3, 1877; capital stock $600,000, in 60,000 shares, $10 each; office Denver; trustees: Calvin C. Davidson, Eli M. Ashley, Aaron Gove, James Lawson, and J.

Walling Root. [Aaron Gove was born in East Hampton Falls, New Hampshire, on 26 September 1839, and moved to Colorado in 1874, after serving in the Civil War. He was the superintendent of public schools in East Denver, Colorado, and helped establish the Denver Public Library.]

Astor, Extension. J.H. Rogers, J. Shields, John Shillito, and W.C. Hicock, proprietors; office Georgetown; claim 50 by 1,400 feet, patented; discovered 1868; located on Democrat Mountain, Griffith mining district, 2 miles from Georgetown; course of vein, northeast and southwest, average width 18 feet, pitch 10 degrees, pay vein 5 inches; nature of ore, galena and sulphurets, average value 300 ounces silver per ton; main shaft 35 feet deep, 1 level 20 feet in length. (See W.B. Astor Mine.)

Atlantic. Owned by James Peck and Sons; office Empire; claim 150 by 1,500 feet, patented; discovered 1864; located on Silver Mountain, Upper Union mining district, 1 mile from Georgetown; vertical vein, running northeast and southwest, average width 8 feet, pay vein 7 feet; nature of ore, iron and copper pyrites; average value, $40 per ton; main shaft 230 feet deep; 1 level 50 feet in length; yield, $5,000.

Atlantic. Owned by T.F. Simmons, H.O. and Schuyler Button; office Georgetown; claim 100 by 3,000 feet; discovered 1867; located on Sherman Mountain, Griffith mining district, 3 miles from Georgetown; course of vein, northeast and southwest, width, 11 feet, pitch 45 degrees north; nature of ore, galena; value 60 ounces silver per ton; 1 shaft 50 feet deep, 1 tunnel 200 feet in length.

Ayshire Tunnel. Owned by John W. Edwards & Co.; office Idaho Springs; claim 1,500 by 3,000 feet; discovered 1874; located on Montgomery Mountain, Grass Valley mining district, 33 miles from Denver; length of tunnel 105 feet.

Bald Eagle. Owned by David Green; located in Virginia Canyon, Idaho mining district, 3 miles from Central City; pay vein 15 inches; nature of ore, galena and iron pyrites; average value, $50 per ton; main shaft 75 feet deep.

Ball's Placer. Owned by David J. Ball, comprises 17 acres of patented property; located at the mouth of Lion Gulch, 3 miles north from Georgetown.

Baker. Owned by the Baker Silver Mining Co.; office Georgetown; claim 1,500 by 3,000 feet, patented; discovered 1860; located on Kelso mountain, East Argentine mining district, 13 miles from Georgetown; course of vein, east and west, average width 15 feet, pitch 45 degrees north; nature of ore, sulphurets and galena; average value, $200 per ton; main shaft, 480 feet deep.

Baltic. Owned by Edward Bowden & Co., office Mill City; claim 150 by 1,500 feet, patented; discovered 1875; located on Columbia Mountain, Montana mining district, 2 miles from Lawson and 6 miles from Georgetown; course of vein, east and west, average width, 10 feet, pay vein 2 to 3 feet, pitch 45 degrees north; nature of ore, gray copper, galena, and native silver; average value, $150 per ton; main shaft 200 feet deep, 3 adits, aggregate length 420 feet.

Baltic. Owned by the Revenue Mining Co., located in Geneva mining district, 4 miles from Montezuma and 13 miles southwesterly from Georgetown.

Baltimore Tunnel Mining Co. John Tomay, agent; office Georgetown; claim 50 by 1,500 feet, patented; located on Brown Mountain, Griffith mining district, 3 miles from Georgetown; width of crevice, 3 feet; average value of ore, 300 ounces silver per ton; main shaft, 120 feet deep, 1 tunnel 750 feet in length; 3 levels, aggregate length 460 feet.

Bank of England. George L. Cannon, proprietor; office Idaho Springs; claim 150 by 1,500 feet; discovered 1878; located on Russell Mountain, 2 miles from Idaho Springs and 33 miles from Denver; course of vein, east and west, width 4 feet, pay vein 14 inches; nature of ore, free gold; average value 7 ounces gold per cord; main shaft 40 feet deep.

Baxter. Owned by Church Bros. and Enos K. Baxter; office Georgetown claim 50 by 100 feet, patented; located on Sherman Mountain, near town of Silver Plume, Griffith mining district, 3 miles from Georgetown; course of vein, northeast and southwest; width 4 feet, pay vein 6 inches, pitch 10 degrees north; nature of ore, zinc blende and galena; average value, 180 ounces silver per ton; 3 shafts, aggregate depth 380 feet; 1 tunnel 75 feet in length; intersected by the Diamond tunnel.

Belleview Tunnel. E.T. Carr, Charles McKee, and J. Jay Joslin, proprietors; office Central City; claim 1,500 by 3,000 feet; discovered 1877; located on Belleview Mountain, Virginia mining district, 3 miles from Central City; length of tunnel 210 feet.

Ben Adams. Owned by the Lebanon Mining Co., Julius G. Pohle, superintendent; office, Georgetown; claim 50 by 1,600 feet, patented; discovered 1865; located on Republican Mountain, Griffith mining district, 2 miles from Georgetown; course of vein, northeast and southwest, width 4 to 6 feet, pay vein 4 inches, pitch 10 degrees north; nature of ore, gray copper; average value, 160 ounces silver per ton; main shaft 65 feet deep; 1 level 40 feet in length.

Benjamin Harding. Owned by Lebanon Mining Co., Julius G. Pohle superintendent; office Georgetown; claim 50 by 1,600 feet, patented; discovered 1865; located on Republican Mountain, Griffith mining district, 1 mile from Georgetown; vertical vein, trending northeast and southwest, width 20 feet, pay vein 4 inches, pitch 5 degrees south; nature of ore, galena; average value, 500 ounces silver per ton; 3 shafts, aggregate depth 240 feet, 3 levels, aggregate length 800 feet; cross cut tunnel, 300 feet in length.

Benton. Owned by the Benton Gold and Silver Mining Co., of Chicago; claim 50 by 1,600 feet, patented; discovered 1868; located on Brown and Sherman mountains, Queen mining district, 3 miles from Georgetown; course of vein, northeast and southwest; width 4 to 18 feet; pay vein, 4 to 10 inches; nature of ore, galena, gray copper, and zinc blende; average value, 100 ounces silver per ton; main shaft 80 feet deep; 1 tunnel 70 feet in length, 2 levels, 30 feet each.

Big Gun. David M. De Witt, proprietor; office Georgetown; claim 150 by 1,500 feet; discovered 1876; located on Sherman Mountain, Griffith mining district, 3 miles from Georgetown; width of vein 4 feet, pay vein 2 to 4 inches; 1 shaft 30 feet deep, reached by tunnel, at a distance of 300 feet from the mouth and intersected at a depth of 165 feet.

Blue Jacket. Owned by Cashier Silver Mining Co., and Cushman Estate; office Georgetown; claim 50 by 900 feet; discovered 1867; located on Red Elephant Mountain, 1 mile from Lawson; course of vein, east and west; average width 15 feet, pitch 40 degrees north, pay vein 6 to 18 inches; nature of ore, galena and native silver; assay value, 40 ounces silver per ton; 1 shaft 60 feet deep; 1 adit 60 feet from surface and 100 feet in length.

Blue Jacket, West Extension. Owned by J.H. Adams, A.D. Blodgett, Justin E. Dubois, F.H. Allison, and the Cushman Estate; claim 50 by 2,100 feet; discovered 1867; located near town of Lawson, on Red Elephant Mountain, 46 miles from Denver; curse of vein, northeast and southwest, width 15 feet, pay vein 6 to 18 inches, pitch 45 degrees north; 1 shaft 32 feet deep.

Blue Wing. Owned by George H. Barrett & Co., of Georgetown; claim 150 by 1,500 feet; discovered 1877; located on Red Elephant Mountain, 1 mile from town of Lawson, Downieville mining district, 6 miles from Georgetown; course of vein, northeast and southwest, width 12 feet, pitch 20 degrees north, pay vein 10 inches; nature of ore, galena; average value, $50 per ton; 1 shaft 42 feet deep.

Bobtail. Owned by George M. Cummings and David B. Nash; office Silver Plume; claim 150 by 1,500 feet; discovered 1873; located on Sherman Mountain, Griffith mining district, 3 miles from Georgetown; vertical vein, trending northeast and southwest, width 4 feet, pay vein 2 to 6 inches; value of ore, 240 ounces silver per ton; main shaft 25 feet deep; 1 tunnel 20 feet in length.

Bonanza Tunnel. Owned by Patrick McNulty, John Caine, and Anthony Joslin; claim 1,500 by 3,000 feet; located on Democrat Mountain, Griffith mining district.

Boulder Nest. Owned by Joseph Reynolds, James I. Gilbert, Charles R. Fish, A.H. Borman, and G.H. Barrett; office Georgetown; claim 150 by 1,100 feet, patented; discovered 1877; located on Red Elephant Mountain, Downieville mining district, 1 mile from Lawson; course of vein, northeast and southwest, average width 4 feet, pitch 10 degrees north, pay vein 1 to 3 feet; nature of ore, galena, gray copper, and sulphurets; average value, $140 per ton; main shaft 275 feet deep and 3 other shafts aggregating 575 feet; 3 levels, 140, 220, and 290 feet in length, respectively; 1 steam pump, 2 engines, 10- and 20-horsepower, respectively, yield, $200,000.

Britannia Tunnel. Owned by the Revenue Mineral Co.; office Grant, Park County, Colorado; claim 1,500 by 3,000 feet; located on Revenue Mountain, above timberline, Geneva mining district, 15 miles from Georgetown; course of tunnel north and south; length 1,000 feet.

Broadway Tunneling and Silver Mining Co. Incorporated May 18, 1875; capital stock, $500,000, in 5,000 shares, $100 each; office New York City and Georgetown, Colorado; organized by Andrew J. Cropsey, A.S. Palmer, George W. Hoagland, Arthur J. Fitch, Frank J. Marshall, and W.H. Moore.

Brother Jonathan. Owned by Carver J. Gross, Francis Gallup, and Edward E. Hartwell; office Georgetown; claim 150 by 1,500 feet; discovered 1867; located on Democrat Mountain, Griffith mining district, 1 mile from Georgetown; course of vein, east and west, pitch north, width 6 feet, pay vein, 4 inches; nature of ore, galena; average value, $50 per ton; 1 tunnel, 40 feet in length.

Buchanan. Owned by M.W. Latson and others; office Georgetown; claim 150 by 1,500 feet; discovered 1875; located on Leavenworth Mountain, Griffith mining district, 2 miles from Georgetown; course of vein, north and south, average width 20 feet, pitch 20 degrees, pay vein 4 inches; nature of ore, galena and gray copper; average value, $175 per ton; main shaft 50 feet and 1 other shaft 35 feet deep, 1 tunnel, 100 feet in length.

Buchanan Silver Mining Co. Incorporated December 2, 1876; capital stock $100,000, in 1,000 shares, $100 each; offices Georgetown, Colorado, and Indianapolis, Indiana; organized by Norman M. and Morris M. Ross, Austin W. Morris, James Buchanan, and M.W. Latson.

Buckskin. Owned by Alexander Lawson, Joseph I. Gilbert, and Nils Larsen; office Georgetown; claim 150 by 1,000 feet; discovered 1877; located on Red Elephant Mountain, Downieville mining district, 45 miles from Denver; vertical vein, trending east and west, width 6 feet, pay vein 7 inches; nature of ore, galena; average value, 84 ounces silver per ton; 1 shaft 30 feet deep, 2 tunnels aggregate length 240 feet, shaft house, 12 by 20 feet.

Bulldozer Tunnel. Alexander Lawson, S.B. Bishop, and B. Willis, proprietors; office Lawson; claim 1,500 by 3,000 feet; discovered 1877; located on Red Elephant Mountain, 1 mile from town of Lawson, and 45 miles from Denver; length of tunnel, 100 feet; designed to cut the White, Free America, Carbon, Apex, and other lodes. [William Alexander Lawson's inn, the Six Mile House, formed the nucleus for the town of Lawson when silver was discovered in 1875 and 1876.]

Burleigh Tunnel. Owned by the Burleigh Mining Co.; organized under the laws of the State of Massachusetts; common stock $400,000; offices Georgetown, Colorado, and Fitchburg, Massachusetts; Charles H. Burleigh, president, Col. Ivers Philips, superintendent; this property comprises a tunnel claim, 1,500 by 3,000 feet, with 30 acres of ground for mill and building sites, all of which are patented; length of tunnel 2,300 feet; located at the base of Sherman Mountain, Griffith mining district, near town of Silver Plume, 3 miles from Georgetown.

Bush. Perry Kalbaugh, proprietor; office Golden; claim 50 by 1,400 feet, patented; discovered 1869; located on Sherman Mountain, near town of Silver Plume, 3 miles from Georgetown; vertical vein, trending northeast and southwest, width 5 feet, pay vein 10 inches; nature of ore, galena, zinc blende, and sulphurets; value 500 ounces silver per ton; 2 shafts, aggregate depth, 150 feet, 1 level 40 feet in length, reached by the Burleigh Tunnel [see separate listing] 1,000 feet from the mouth and intersected at a depth of 350 feet.

Caledonia. Owned by Caledonia Silver Mining Co.; office Georgetown; claim 50 by 750 feet, patented; discovered 1865; located on Republican Mountain, Griffith mining district, 3 miles from Georgetown; course of vein, northeast and southwest, width 3 feet, pitch 10 degrees north, pay vein 14 inches; nature of ore, gray copper and galena; average value, 375 ounces silver per ton; aggregated depth of shafts 375 feet, aggregated length of levels 400 feet.

Cannon. George L. Cannon, proprietor; office Idaho Springs; claim 150 by 1,500 feet; located on Schaffter Mountain, Idaho mining district, 1 mile from town of Idaho Springs and 33 miles from Denver; course of vein, northeast and southwest, average width 5 feet, pitch 10 degrees north, pay vein 10 inches; nature of ore, free gold; main shaft 50 feet deep.

Capital. A.C. Carpenter, proprietor; office, Lawson; claim 50 by 3,000 feet; discovered 1868; located on Capital Mountain, Montana mining district, 1 mile from Lawson and 43 miles from Denver; course of vein east and west, width 6 feet, pay vein 15 inches, pitch 35 degrees north; nature of ore, galena, gray copper, and native silver; average value, 100 ounces silver per ton; main shaft 180 feet deep, 1 level 200 feet in length; yield, $4,000.

Captain Wells. Owned by Charles P. Baldwin and William T. Reynolds; office Georgetown; claim 50 by 1,000 feet, patented; located on

Sherman Mountain, Griffith mining district, 3 miles from Georgetown; course of vein, northeast and southwest, width 8 feet, pay vein 1 to 15 inches; nature of ore, galena, porphyry, and zinc blende; average value of first-class ore 4,920 ounces, second-class ore 244 ounces, and third-class ore 61 ounces silver per ton; main shaft 105 feet deep, aggregate depth of other shafts 120 feet, 1 level 50 feet in length.

Carbonate. Owned by William N. Dickerson, William H. Doe, and J.F. Bearce; office Lawson; claim 150 by 1,500 feet; discovered 1878; located on Columbia Mountain, Montana mining district, 2 miles from Lawson; vertical vein, running east and west, width 4 feet, pay vein 10 inches; nature of ore, galena and copper pyrites; average value, $75 per ton; 1 shaft 25 feet deep, 1 tunnel 30 feet in length.

Cascade. Cascade Mining Co., proprietors; office Idaho Springs; claim 50 by 2,100 feet; discovered 1864; located on Ute Creek, Cascade mining district, 6 miles from Georgetown; course of vein, northeast and southwest, width 5 feet, pitch 30 degrees north, pay vein 4 feet; nature of ore, galena; average value, 300 ounces silver per ton; main shaft 65 feet and 1 other shaft 40 feet deep; 1 adit 690 feet in length.

Cascade. Owned by the Young America Silver Mining Co.; office Georgetown; claim 50 by 700 feet; discovered 1866; located on Sherman Mountain, Griffith mining district, 3 miles from Georgetown; course of vein, northeast and southwest, average width 5 feet, pitch 10 degrees north, pay vein 10 inches; nature of ore, gray copper; average value, 300 ounces silver per ton; aggregate depth of shafts, 130 feet, 1 tunnel 75 feet in length.

Cash La Poudre Tunnel. William Baker, proprietor; office, Georgetown; claim 1,500 by 3,000 feet; discovered 1872; located on Leavenworth Mountain, Griffith mining district, near Silver Dale, 3 miles from Georgetown; length of tunnel 200 feet.

Cashier. Zadock Kalbaugh, W.W. Wrigley, and Erskine McClelland, proprietors; office, Georgetown; claim 50 by 700 feet, patented; discovered 1866; located on Sherman Mountain, Griffith mining district, 3 miles from Georgetown; vertical vein, running northeast and southwest, average width 4 feet, pay vein 2 to 12 inches; nature of ore, galena and sulphurets; average value, 200 ounces silver per ton; aggregate length of levels, 300 feet.

Cashier, West. Cashier Silver Mining Co., Massachusetts, proprietors; offices Georgetown, Colorado, and Boston, Massachusetts; Horatio G. Parker, president, Arthur W. Hoyt, secretary and treasurer; claim 700 linear feet, patented; discovered 1866; located near town of Brownville, on Sherman Mountain, 3 miles from Georgetown; course of vein, east and west, average width 10 feet, pitch 10 degrees north, pay vein 8 inches; nature of ore, galena, gray copper, and zinc blende; average value, $99 per ton; 2 shafts, aggregate depth 280 feet; 4 levels, aggregate length 368 feet.

Centennial. Owned by R. Majors, A. Horn, and John Thompson; located in Geneva mining district, 4 miles from Montezuma and 13 miles from Georgetown, width of vein 18 inches; nature of ore, galena, gray copper, and copper pyrites; value, 140 ounces silver per ton; operated through a tunnel from an adjoining claim.

Centennial. Owned by Horatio E. and Morris Hazard, John Collier, and J.S. Boreham; office Central City; claim 150 by 1,500 feet; discovered 1875; located at the head of Virginia Canyon, Virginia mining district, 3 miles from Central City; course of vein, northeast and southwest, width 4 feet; nature of ore, iron pyrites; average value, 4 ounces gold per cord; 6 shafts, aggregate depth 200 feet.

Centennial. Samuel Tobin, proprietor; office Idaho Springs; claim 150 by 1,500 feet; discovered 1876; located on Hukill Gulch, Spanish Bar mining district, 2 miles from Idaho Springs; vertical vein, trending northeast and southwest, width 4 feet, pay vein, 12 inches; nature of ore, galena, iron and copper pyrites; average value, $100 per ton; 1 shaft 60 feet deep.

Centennial Star. Owned by C.F. Kercher and O.P. Fluke; located on Kelso Mountain, West Argentine mining district, 8 miles from Georgetown; developed by a surface opening, 60 feet in length; value of ore, 600 ounces silver per ton.

Center Tunnel. Owned by C.J. Johnson, August Nilson, Nils Frohm, and August Peterson; office Georgetown; claim 1,500 by 3,000 feet; discovered 1874; located on Democrat Mountain, Griffith mining district, 3 miles from Georgetown; nature of ore, galena, gray copper, wire silver, iron and copper pyrites; average value, $450 per ton; this tunnel is now 360 feet in length, and has cut the Montana, Woodstock, and Juniper lodes.

Central Iowa Silver Mining Co., Iowa. Incorporated March 17, 1877; capital stock, $100,000, in shares of $10 each; offices Georgetown, Colorado, and Marshalltown, Iowa; organized by Frank L. Downend, A.H. Neidig, E.C. McMillan, William H. Thomas, Nathan Worley, Jr., J.B. Stattler, and E.C. Rice.

Champion. Edward W. Williams, William and Edward Jones, and Owen Hughes, proprietors; office Central City; claim 150 by 3,000 feet; discovered 1878; located on Bellevue Mountain, Idaho mining district, 3 miles from Central City; course of vein, northeast and southwest, width 4 feet, pay vein 1 foot, pitch 25 degrees north; value of ore, 20 ounces gold per cord; main shaft 60 feet deep; 1 drift 40 feet in length, and 55 feet from surface.

Champion. J. Alden Smith and Arthur Gorham, proprietors; office Boulder; claim 50 by 200 feet, discovered 1859; located on Champion Mountain, Trail Run mining district, 3 miles from Idaho Springs; course of vein, northeast and southwest, width 2 feet, pitch 5 degrees northwest, pay vein 10 inches; nature of ore, iron and copper pyrites, and free gold; average value, $132 per cord; main shaft, 70 feet deep; 1 drift 30 feet, and 1 level 50 feet in length.

Champion. Owned by the Lebanon Mining Co., of Colorado and New York; Julius G. Pohle, superintendent; offices Georgetown, Colorado, and New York City; claim 50 by 1,500 feet, patented; discovered 1870; located on Republican Mountain, Griffith mining district, 2 miles from Georgetown; course of vein, northeast and southwest, width 150 feet, pitch southeast, pay vein 3 to 4 inches; nature of ore, galena; average value, 500 ounces silver per ton main shaft 40 feet deep, 2 tunnels, 60 and 40 feet, respectively; aggregate length of levels 700 feet.

Charter Oak. Lebanon Mining Co., proprietors; Julius G. Pohle, superintendent; office Georgetown; claim 50 by 1,500 feet, patented; discovered 1870; located in Cascade mining district, 3 miles from Idaho Springs; vertical vein, trending northeast and southwest, width, 20 feet; nature of ore, galena, gray copper, and sulphurets; average value, 150 ounces silver per ton; 1 shaft 50 feet deep.

Cincinnati. Owned by Charles C. Welch, David C. Crawford, James F. Devere, and R.W. Chinn; office Denver; claim 150 by 1,500 feet; discovered 1877; located on Columbia Mountain, Montana mining district, 1 mile from town of Lawson and 46 miles from Denver.

Clara. Owned by H.E. Lane; office Baltimore; claim 150 by 1,300 feet, patented; discovered 1873; located on Griffith Mountain, Griffith mining district, 2 miles from Georgetown; course of vein, east and west, pitch 10 degrees north, width 9 feet, pay vein 6 inches; nature of ore, sulphurets and silver glance; average value, 300 ounces per ton; 2 tunnels, aggregate length 250 feet.

Clarissa. William H. Bush, proprietor; office Central City; claim 150 by 1,500 feet, patented; located on Virginia Hill, Virginia mining district, 2 miles from Central City.

Clear Creek Co. Incorporated December 28, 1875; capital stock, $100,000, in 2,000 shares, $50 each; offices Georgetown, Colorado, and New York City; organized by H. Augustus Taylor, Frank M. Taylor, and N.T. Porter.

Clear Creek Mining and Improvement Co. Incorporated April 5, 1873; capital stock, $100,000, in 1,000 shares, $100 each; offices, Georgetown, Colorado, and New York City; organized by Reuben S. Middleton, Locke W. Winchester, Jay C. Wemple, and J. Warren Brown.

Cold Stream. Owned by William Clinn; office Baltimore; claim 150 by 1,500 feet; discovered 1871; located on Republican Mountain, near Silver Plume, Griffith mining district, 3 miles from Georgetown; course of vein, east and west, width 6 feet, pay vein 18 inches; nature of ore, galena, zinc blende, and sulphurets; average value, 135 ounces silver per ton; main shaft 200 feet deep, 1 tunnel 350 feet in length, 3 levels, aggregate length, 250 feet.

Cold Stream Consolidated Mining Co. Incorporated June 6, 1878; capital stock, $1,000,000, in 100,000 shares, $10 each; office Georgetown; trustees: John Glenn, Henry Crow, William T. Reynolds, Jacob Fillius, and Charles R. Fish.

Collom Idaho Ore Dressing Co. Incorporated April 17, 1874; capital stock $50,000, in 500 shares, $100 each; office Idaho Springs, Colorado, and Trenton, New Jersey; organized by John and Charles Collom, Daniel Peters, Frederick R. Williamson, and William Hancock.

Colorado. Located on the eastern slope of the main range, west of Geneva Valley, 4 miles from Montezuma and 13 miles from Georgetown;

developed by a tunnel 100 feet in length, which intersects the vein at a depth of 40 feet.

Colorado Central. Owned by Charles H. Marshall and G.W. Hall; office, Georgetown; claim 50 by 1,500 feet, patented; discovered 1872; located on Leavenworth Mountain, Griffith mining district, 2 miles from Georgetown; vertical vein, trending northeast and southwest, width, 75 feet; pay vein, 18 inches; nature of ore, galena and gray copper; average value, 500 ounces silver per ton main shaft, 150 feet deep, 1 tunnel 250 feet in length.

Colorado Corporation. Incorporated May 17, 1873; capital stock $20,000,000, in 200,000 shares, $100 each; office New York City; organized by Arthur A. Brown, Robert Edwards, L.M. Lawson, John H. King, John W. Brown, Seth B. Cole, John A. Biggs, Drake DeKay, Samuel W. Canfield, Edward H. Nichols, S.M. Blatchford, H.L. Johnson, William Thorpe, Charles Medary, V. Whitcomb, John E. Risley, Henry C. Brown, and Mathias Goetzel.

Colorado Diamond Tunnel Silver Mining Co. John A. Fish, superintendent; office Georgetown; the property of this company is patented, and comprises a tunnel claim 1,500 by 3,000 feet, and 3 acres of ground, suitable for building and other purposes; length of tunnel, 1,400 feet; located on Republican Mountain, Griffith mining district, 1 mile from Georgetown.

Colorado State. Owned by G.W. Cowles & Co.; office Empire; claim 150 by 1,500 feet; discovered 1876; located on Silver Mountain, Upper Union mining district, 2 miles from Georgetown; vertical vein, running northeast and southwest, width 5 feet, pay vein 30 inches; nature of ore, free gold, iron and copper pyrites; average value, $150 per cord; main shaft 30 feet deep, 1 adit, 30 feet in length.

Colorado United Mining Co., Limited. Organized under the laws of England, June 1877; capital stock 325,000 English pounds, in 65,000 shares, 5 English pounds each; offices Brownville, Colorado, and London, England; Cecil Beden, chairman, Francis Andrews, secretary, Hon. William A. Hamill, manager. The property of this company is located on Brown and Sherman mountains at town of Brownville, 3 miles from Georgetown. It comprises the Terrible, Silver Ore, Brown, Oneida, Last Chance, and other lodes of minor importance, together with a new concentration mill, the capacity of which is 35 tons of ore daily. These mines are patented and well developed and are among the best producers in the state.

Combs. Owned by Perry Kalbaugh, of Georgetown; claim 50 by 1,400 feet, patented; discovered 1866; located on Sherman Mountain, Griffith mining district, 3 miles from Georgetown; course of vein, northeast and southwest, average width 4 feet, pitch north, pay vein 2 to 7 inches; nature of ore, galena; average value, 6 ounces silver per ton; aggregate depth of shafts, 200 feet, aggregate length of levels, 125 feet.

Comet. Owned by the Ohio Comet Silver Mining Co.; incorporated 1876; J.A. Hawkes, president, O. Ballard, Jr., secretary and treasurer, Phillip E. Morehouse, superintendent; capital stock, $100,000; offices Georgetown, Colorado, and Circleville, Ohio; claim 50 by 1,500 feet, patented; discovered 1867; located on Griffith Mountain, Griffith mining district, 2 miles from Georgetown; width of crevice, 100 feet; nature of ore, galena, zinc blende, iron and copper pyrites; average value, 180 ounces silver per ton; 4 shafts, aggregate dept 455 feet, 1 tunnel, 250 feet, and 1 level 80 feet in length.

Comstock. Owned by James I. Gilbert, J.H. Western, and Joseph Reynolds; office Georgetown; claim 50 by 3,000 feet; discovered 1872; located on Red Elephant Mountain, near town of Lawson, Downieville mining district, 6 miles from Georgetown; course of vein, east and west, width 30 feet, pay vein 3 feet; main shaft 120 feet deep, 1 tunnel 340 feet, and 1 adit 400 feet in length.

Confidence. David M. DeWitt, proprietor; office Georgetown; claim 150 by 1,500 feet; discovered 1877; located on Sherman Mountain, near Brownville, Griffith mining district, 3 miles from Georgetown; width of vein 3 ½ feet; open cut, 25 feet in length and 15 feet in depth; 1 tunnel 50 feet in length.

Congress. Owned by William Mendenhall, and others; located in Geneva mining district, 4 miles from Montezuma and 13 miles from Georgetown; developed by a shaft 75 feet deep and a drift 80 feet long; value of ore, 200 ounces silver per ton.

Conmorn. Owned by Frederick Bunholzer, of Georgetown; claim 150 by 1,500 feet; discovered 1877; located on Red Elephant Mountain, Downieville mining district, 1 mile from Georgetown; course of vein, northeast and southwest, width 40 feet, pitch north, pay vein 18 inches;

nature of ore, galena and carbonate of copper; average value 130 ounces silver per ton; main shaft 16 feet deep.

Conqueror. Owned by the Conqueror Gold Mining Co.; Park Disbrow, superintendent and vice president; office Empire; claim 50 by 1,500 feet; discovered 1864; located on Silver Mountain, Upper Union mining district, near town of Empire, 3 miles from Georgetown; course of vein, northeast and southwest, width 4 to 26 feet, pitch 10 degrees north; nature of ore, iron pyrites; average value, $300 per cord; main shaft 400 feet deep, aggregate depth of other shafting 175 feet; 1 tunnel 800 feet in length and 265 feet from surface; 1 level 240 feet in length and 265 feet from surface; yield, $75,000.

Consolidated Aspinwall and Durango Mining Co. Incorporated August 1, 1876; capital stock, $600,000, in 60,000 shares, $10 each; office Denver; organized by James Lawson, Richard C. Reily, J.W. Boot, Talmadge Norwood, C.C. Davidson, R.L. Martin, and J.S. Smith.

Consolidated Hercules and Roe Silver Mining Co. Incorporated February 2, 1876; capital stock, $1,000,000, in 100,000 shares, $10 each; offices New York City, Georgetown and Denver, Colorado; G.W.E. Griffith, president, K.G. Cooper, vice president, Charles M. Williams, secretary, H.M. Griffin, treasurer and superintendent. The property of this company is located on Brown and Sherman Mountains, Griffith and Queen mining district, 2 miles from Silver Plume and 4 miles from Georgetown. It comprises the East Roe, Hercules, John E. McClung, and J.M. Wilson lodes, and a mill site containing 5 acres of ground. This property is patented, surrounded by timber, is well developed and has yielded to date, $500,000.

Cook. James Kinkead and Thomas H. Thatch, proprietors; office Fall River; claim 50 by 500 feet, patented; discovered 1860; located on Donaldson Hill, Spanish Bar mining district, near town of Fall river, 33 miles from Denver; course of vein, northeast and southwest, average width 4 feet, pitch 30 degrees south, pay vein 30 inches; nature of ore, copper and iron pyrites; average value, $160 per cord; main shaft 110 feet deep, 2 adits 140 and 170 feet in length, respectively; 12-stamp mill, with all necessary machinery for the successful treatment of ore.

Cope. Owned by William H. Latshaw; office Idaho Springs; claim 50 by 2,000 feet, patented; discovered 1868; located in Virginia mining district, on the Continental Divide, between Virginia Canyon and Fall

River, 3 miles from Central City; course of vein, northeast and southwest, pitch 45 degrees north, width 4 ½ feet, pay vein 2 feet; nature of ore, iron and copper pyrites; average value, $118 per ton; main shaft 50 feet deep, other shafting aggregating 50 feet, 1 drift 40 feet in length.

Corry City. Charles R. Fish & Co., proprietors; office Georgetown; claim 50 by 1,000 feet, patented; discovered 1861, located on Sherman Mountain, Griffith mining district, 2 miles from Georgetown; vertical vein, trending northeast and southwest, width 12 feet; nature of ore, galena; average value, $23 per ton; main shaft 80 feet, and 1 other shaft 30 feet deep 1 tunnel 170 feet in length; total yield, $5,000.

Corry City. Colorado Diamond Tunnel Silver Mining Co., proprietors; office Baltimore, Maryland; claim 1,000 [linear?] feet, patented; located on Republican Mountain, Griffith mining district, half mile from Georgetown; course of vein, east and wet, width 40 feet, pitch 10 degrees north; nature of ore, galena and gray copper; main shaft 120 feet deep, 1 tunnel 1,500 feet in length.

Crown Point and Virginius. George and Robert Martin, Job V. Kimber, W.H. Cheatley, Harley B. Morse, John Scudder, and William Armor, proprietors; office Central City; claim 150 by 1,500 feet, patented; discovered 1859; located at the head of Virginia Canyon, 2 miles from Central City; course of vein, northeast and southwest, width 4 feet, pitch 20 degrees north; nature of ore, iron and copper pyrites; average value of mill ore, 10 ounces gold per cord, smelting ore, 100 ounces per ton; main shaft 125 feet deep, aggregate depth of other shafts, 300 feet; total yield, $60,000.

Crown Silver Mining Co. Incorporated May 31, 1877; capital stock, $500,000, in 50,000 shares, $10 each; office Denver; trustees: Robert Connely, John M. Lord, William F. Shanks, William Braden, and Horace W. Hibbard.

Coupon Placer. Owned by William Moore, H.M. Kline, and E. R. Abodie; claim consists of 86 acres of ground, located within the limits of the town of Empire, Upper Union mining district, 1 mile from Georgetown.

Daniel Peters. Owned by the Snow Drift Silver Mining and Reduction Co., England; George Teal, agent; office Georgetown; claim 30 by 1,400 feet, patented; discovered 1863; located on Republican Mountain,

Griffith mining district, near Silver Plume, 2 miles from Georgetown course of vein, northeast and southwest; width 16 feet, pitch north, pay vein 8 inches; nature of ore, sulphurets; average value, 250 ounces silver per ton; main shaft 80 feet deep, 2 levels, aggregate length 100 feet.

Denver. Owned by H.O., J.S., and Schuyler Button and the McMurdy Estate; office Georgetown; claim 150 by 1,500 feet, patented; discovered 1864; located on Sherman Mountain, Griffith mining district, 3 miles from Georgetown; course of vein, northeast and southwest, width 18 feet, pay vein 8 to 12 inches, pitch 15 degrees north; nature of ore, galena and sulphurets; average value, $450 per ton, 3 shafts, aggregate depth 180 feet, aggregate length of levels 1,100 feet.

Detroit. Harry Almer & Co., proprietors; office Georgetown; claim 150 by 1,500 feet; discovered 1872; located on Kelso Mountain, West Argentine mining district, 9 miles from Georgetown; vertical vein, running northeast and southwest; width 4 feet, pay vein 5 inches; nature of ore, galena and sulphurets; average value, $500 per ton; 1 adit 100 feet in length.

Dexter. John Coburn, Charles P. Baldwin, and R.H. Coe, proprietors; claim 150 by 1,500 feet, patented; discovered 1877; located on Red Elephant Mountain, Downieville mining district, 6 miles from Georgetown; course of vein, east and west, pitch north, width 16 feet, pay vein 2 to 18 inches; nature of ore, galena, gray copper, iron and copper pyrites; aggregate length of levels, 225 feet.

Diamond. Owned by W.N. Webster and Daniel Diamond; claim 150 by 1,500 feet; discovered 1877; located on Columbia Mountain, Montana mining district, 2 miles from Lawson and 6 miles from Georgetown; 1 shaft 40 feet deep.

Diamond Joe. Owned by James I. Gilbert and Joseph [nicknamed "Diamond Joe"] Reynolds; office Georgetown; claim 50 by 3,000 feet, patented; located on Kelso Mountain, West Argentine mining district, 10 miles from Georgetown; course of vein, northeast and southwest, width 12 feet, pay vein 4 to 18 inches, pitch 5 degrees north; nature of ore, gray copper and sulphurets of silver; average value, 400 ounces silver per ton; 2 shafts, aggregate depth 500 feet; 3 levels, aggregate length 913 feet, intersected by tunnel at a depth of 800 feet.

Diamond Joe, East Extension. Owned by J.H. Boun, C. Kelley, and R. Carruth; claim 150 by 1,500 feet; discovered 1877; located on Kelso Mountain, West Argentine mining district, 9 miles from Georgetown; course of vein, east and west, width 5 feet, pay vein 8 inches; nature of ore, galena and sulphurets; average value, 180 ounces silver per ton; 1 adit 105 feet in length; yield to date, $2,000.

Diamond Tunnel. Owned by the Colorado Diamond Tunnel Silver Mining Co., Baltimore, John A. Fish, superintendent; office Georgetown; tunnel claim 1,500 by 3,000 feet, and 3 acres of ground at the mouth of tunnel, all of which is patented; length of tunnel, 1,400 feet; located on Republican Mountain, Griffith mining district, 1 mile from Georgetown.

Dives. Owned by the McMurdy Estate, Purdue Silver Mining Co., and others, located on Sherman Mountain, near town of Silver Plume, 3 miles from Georgetown; this property was discovered in the year 1868, is well developed, and has produced to date $1,500,000.

Douglas. Lebanon Mining Co., proprietors, Julius G. Pohle, superintendent, office Georgetown; claim 50 by 1,500 feet; discovered 1870; located on Lebanon Mountain, Cascade mining district, 4 miles from Georgetown; vertical vein, running northeast and southwest, width 6 feet, pay vein 1 inch; nature of ore, galena; average value, 150 ounces silver per ton; 1 shaft, 32 feet deep.

Douglas Mining Co. Incorporated October 19, 1870; capital stock, $1,000,000, in 10,000 shares, $100 each; office New York; organized by Mathias Goetzel, Edward L. Lynch, Frederick Leposin, Alexander McDonald, Patrick McDonald, Jacob Ritter, and William A. Neschke.

Dragon. Owned by Charles E. Robinson & Co.; office Spanish Bar; claim 150 by 1,500 feet; discovered 1865; located on Hukill Gulch, 2 miles from Idaho Springs, Spanish Bar mining district, 43 miles from Denver; vertical vein, running northeast and southwest, width 25 feet; nature of ore, galena, iron and copper pyrites; 1 tunnel 125 feet in length.

Drummond. Owned by W.N. Webster and Daniel Drummond; claim 150 by 1,500 feet; discovered 1877; located on Columbia Mountain, Montana mining district, near town of Lawson; 1 shaft 40 feet deep.

Dubuque. A.S. Benne, W.G. Wakefield, and John C. Easley, proprietors; office Spanish Bar; claim 150 by 1,500 feet; discovered 1878;

located on Dubuque Mountain, Spanish Bar mining district, near town of Fall River, 41 miles from Denver; course of vein, northwest and southeast, width 12 feet, pitch 30 degrees northeast, pay vein 4 feet; nature of ore, copper and iron pyrites; average value, $100 per ton; main shaft 25 feet deep.

Dunderburg. Owned by James Peck & Sons; office Empire; claim 150 by 1,500 feet; discovered 1863; located on Silver Mountain, Upper Union mining district, 1 mile from Georgetown; vertical vein, running northeast and southwest; width 3 feet; nature of ore, iron and copper pyrites; average value, $60 per ton; main shaft 20 feet deep, 1 tunnel 80 feet in length; total yield, $3,000.

Dunderburg. Owned by Thos. C. Old; office Georgetown; claim 150 by 3,000 feet, patented; discovered 1868; located on Sherman Mountain, near town of Brownville, Griffith mining district, 3 miles from Georgetown; course of vein, northeast and southwest, width 3 feet, pitch 5 degrees north, 3 pay veins averaging 6, 12, and 20 inches, respectively; nature of ore, galena, gray copper, and ruby silver; average value, $248.60 per ton; main shaft 160 feet and 1 other shaft 120 feet deep, 1 tunnel 40 feet in length at the end of which is a shaft 50 feet deep, 2 adits 180 and 280 feet in length, respectively; 2 winzes connecting adits, 60 feet each; shaft house 25 by 35 feet; total yield from December 8, 1877, to June 1, 1878, $147,000, which was taken out, at a cost of 20 percent of the yield.

Dunkirk. Owned by the Herman Mining Co.; capital stock, $200,000, none of which has been sold for less than par value; John D. Dix, president, Benjamin F. Morris, vice president, George W. Dix, secretary, Charles H. Morris, superintendent; claim 150 by 1,400 feet, patented; discovered 1874; located on Republican Mountain, Griffith mining district, near town of Silver Plume, 3 miles from Georgetown; course of vein, nearly east and west, width 100 feet, pay vein 1 to 24 inches, pitch 35 degrees north; nature of ore, iron and copper pyrites, galena, gray copper, zinc blende, wire silver, and silver glance; main shaft, 400 feet deep, aggregate depth of other shafting, 200 feet; 3 levels, aggregating 600 feet in length, and 300 feet of cross-cutting; shaft and engine house, 20 by 70 feet, containing a 20-horse power reversible engine, and a steam pump having a capacity for raising 40 gallons of water per minute for a depth of 600 feet.

E. Pluribus Unum. Owned by W.T. Reynolds and James A. Wilson; office Georgetown; claim 50 by 3,000 feet; discovered 1868; located on

Sherman Mountain, Griffith mining district, 2 miles from Georgetown; vertical vein running east and west, width 3 feet, pay vein 12 inches; nature of ore, galena; average value, $300 per ton; 1 tunnel, 165 feet in length, 3 adits aggregating 180 feet in length; total yield, $12,000.

Eagan Tunnel. Located on Republican Mountain, near town of Silver Plume, Griffith mining district, 3 miles from Georgetown.

Eagle Bird. Owned by Lewis C. Rockwell; office Central City; claim 50 by 1,400 feet, patented; discovered 1869; located on Sherman Mountain, Griffith mining district, 3 miles from Georgetown; course of vein, northeast and southwest, width 9 feet, pay vein 7 inches; nature of ore, galena and zinc blende; average value, 150 ounces silver per ton; 2 shafts, aggregate depth 256 feet.

East Roe Silver Mining Co. Incorporated June 23, 1875; capital stock, $150,000, in 1,500 shares, $10 each; offices Georgetown and Denver, Colorado; organized by Samuel Watson, William H. Cushman, William M. Clark, Benjamin F. Napheys, and Aaron B. Robbins.

Eclipse. Owned by the Goetzel Silver Mining Co.; office Georgetown; claim 50 by 1,500 feet, patented; discovered 1865; located in Griffith mining district near Georgetown; course of vein, northeast and southwest, width 2 feet, pay vein 10 inches; nature of ore, galena; average value, 240 ounces silver per ton; main shaft 140 feet deep, 1 tunnel 300 feet in length.

Edgar. Owned by the Edgar Mining Co.; office Louisville, Kentucky; claim 150 by 1,500 feet; located in Hukill Gulch, Spanish Bar mining district, 2 miles from Idaho Springs; vertical vein running north and south, width 4 feet, pay vein 10 inches; nature of ore, iron pyrites and galena; main shaft 50 feet deep, 2 adits, 105 and 200 feet in length, respectively; yield, $10,000.

Edgar. Owned by David N. Smith, of Georgetown; claim 150 by 1,500 feet, patented; discovered 1871; located on Democrat Mountain, Griffith mining district, 1 mile from Georgetown; course of vein, northeast and southwest, width 5 feet, pitch north, pay vein 1 to 8 inches; nature of ore, galena and sulphurets; main shaft 125 feet deep, 1 level 100 feet in length.

Edwards' Placer. Owned by John W. Edwards, of Idaho Springs; claim consists of 24 acres of patented ground; located on Chicago Creek, near Idaho Springs, 33 miles from Denver. It is worked by hydraulic pressure, and has yielded to date $100,000; also a claim of 500 by 2,500 feet, which is worked by an engine of 16-horse power, located in Spring Gulch, 1 mile from Idaho Springs.

Edweni. Owned by Carver J. Goss, of Denver; claim 50 by 1,500 feet; discovered 1867; located near the town of Brownville, on Brown Mountain, Griffith mining district, 3 miles from Georgetown; vertical vein, running northeast and southwest; width 16 feet; pay vein 3 inches; nature of ore, galena and sulphurets; average value 725 ounces silver per ton; 1 shaft 50 feet deep, 1 tunnel 75 feet in length and 50 feet from surface, 20 feet in length.

Eli Courtney. Owned by Julius G. Pohle, H.S. Morrison, Samuel P. Bishop, C.C. Burnett, and the McMurdy Estate; claim 150 by 800 feet, patented; discovered 1864; located on Leavenworth Mountain, Griffith mining district, 1 mile from Georgetown; vertical vein, trending northeast and southwest, width 6 feet; nature of ore, gray copper and galena; average value, 300 ounces silver per ton.

Elijah Hise. Owned by the Lebanon Mining Co., Julius G. Pohle, superintendent; office Georgetown; claim 50 by 1,600 feet, patented discovered 1865; located on Republican Mountain, Griffith mining district, 1 mile from Georgetown; course of vein, northeast and southwest, width 9 feet, pay vein 6 inches, pitch 10 degrees north; nature of ore, galena; average value, 300 ounces silver per ton; main shaft 110 feet deep, 3 levels aggregating 500 feet in length.

Elliot. Owned by B.R. Elliot and E. Staples; office Georgetown; claim 150 by 1,500 feet; discovered 1872; located on Republican Mountain, Griffith mining district, 2 miles from Georgetown; course of vein, northeast and southwest, width 4 feet, pay vein 10 inches; nature of ore, galena; average value, $215 per ton; 1 shaft 35 feet deep.

Emma. Office, Georgetown; claim 50 by 1,500 feet, patented; discovered 1867; located on Democrat Mountain, Griffith mining district, 3 miles from Georgetown; course of vein, east and west, width 35 feet, pay vein 6 to 14 inches; nature of ore, galena; average value, 125 ounces silver per ton; main shaft 48 feet deep.

Emma No. 2. Owned by Samuel Linscott, E. Harrison, and H. Marcus Bronson; office Georgetown; claim 150 by 1,500 feet, patented; located on Democrat Mountain, Griffith mining district, 2 miles from Georgetown; course of vein, southwest and northwest, width 10 feet, pay vein 2 inches, pitch 25 degrees north; nature of ore, galena and gray copper; average value, $1,000 per ton; 1 shaft 46 feet deep; total yield, $5,000.

Empire City. Owned by William Moore and others; office Empire; claim 150 by 1,500 feet; discovered 1877; located on Cavode Mountain, Upper Union mining district, 1 mile from Georgetown; vertical vein, running northeast and southwest, width 4 feet, pay vein 10 inches; nature of ore, free gold; value $400 per ton; 4 adits 14, 43,82, and 163 feet in length, respectively.

Empire Ditch and Placer Co. Reorganized March 18, 1878; capital stock, $200,000, in 4,000 shares of $50 each; office Georgetown; organized by Frank M. Taylor, Timothy G. Negus, and Robert S. Morrison.

Empire Mining Co., of Colorado. Incorporated June 3, 1863; capital stock, $10,000,000, in 100,000 shares, $100 each; office New York City; organized by Elijah Weston, Samuel W. Canfield, W.W. Vanderbilt, A.M. Wilcox, Seth B. Cole, B. Franklin Clark, C.H. Wessell, William Thorpe, and Robert Edwards.

Enterprise Tunnel and Mining Co. Incorporated July 9, 1877; capital stock, $100,000, in 1,000 shares, $100 each; office Georgetown; trustees: Charles J. Morrill, Albert Townsend, Wiley E. Morrill, Henry Anderson, Henry Seifried, Henry E. Newcomb, Henry Thompson, and John S. Holms; claim 1,500 by 3,000 feet; discovered 1877; located on Red Elephant Mountain, Downieville mining district, 1 mile from town of Lawson and 45 miles from Denver; length of tunnel 300 feet.

Equator. Owned by James Peck and Sons; office Empire; claim 150 by 1,500 feet, patented; discovered 1864; located on Silver Mountain, Upper union mining district, near town of Empire, 1 mile from Georgetown; vertical vein, trending northeast and southwest, width 4 feet, pay vein 1 foot; nature of ore, iron and copper pyrites and free gold; average value, $75 per ton; main shaft 75 feet deep, 1 tunnel 100 feet in length; 12-stamp mill with a capacity for treating 10 tons of ore daily; yield to date, $6,000.

Equator. Owned by the Equator Mining and Smelting Co.; organized under the laws of the State of Illinois, October 29, 1878; capital stock, $300,000, in 3,000 shares, $100 each; offices Georgetown, Colorado, and Chicago, Illinois; William O. Carpenter, president, John Turck, vice president, James B. Goodman, secretary and treasurer, H.S. Kearney, agent and superintendent; the property of this company is located on Leavenworth Mountain, Griffith mining district, 2 miles from Georgetown. It comprises the Equator, Munsell, Halcyon, Hunter, and other lodes of minor importance, together with a good water power and 15 acres of ground, suitable for building and other purposes. The Equator claim is 50 by 1,400 feet, patented; discovered July 28, 1866; the vein is nearly vertical, trends northeast and southwest, is from 100 to 200 feet wide and has pay veins from 1 to 18 inches; the ore is gray copper and galena, and averages $400 per ton. It is developed by a main shaft 450 feet deep, drifts aggregating 4,000 feet in length, and a cross cut tunnel 100 feet in length, which will intersect the vein at a distance of 1,000 from the mouth and at a depth of 550 feet fro the surface; the surface improvements consist of a shaft house 30 by 80 feet, a 25-horse power engine and hoisting apparatus, and an ore house 15 by 40 feet; total yield, $1,000,000.

Essex. Owned by William Howe, William Shehan, William and Edward Horrigan; claim 150 by 1,500 feet; discovered 1877; located on Brown Mountain, above timberline, 3 miles from Georgetown; width of vein 8 feet, pay vein 6 inches; main shaft 40 feet deep.

Excelsior. Owned by Marshall Hollister and William Thompson; office Georgetown; claim 150 by 1,500 feet; located on Red Elephant Mountain, Downieville mining district, 6 miles from Georgetown; course of vein, east and west, width 25 feet, pay vein 1 to 18 inches; nature of ore, galena and copper pyrites; value 75 ounces silver per ton; 1 shaft 35 feet deep; 1 tunnel 140 feet in length.

Excelsior. William Baker, proprietor; office Georgetown; claim 150 by 1,400 feet, discovered 1870; located on Leavenworth Mountain, Griffith mining district, near Silver Dale, three miles from Georgetown; course of vein, east and west, width 20 feet, pay vein 2 feet, pitch north; nature of ore, sulphurets; average value, 300 ounces silver per ton; main shaft 50 feet deep; 1 tunnel 200 feet in length.

Fairmount. Owned by the estate of John W. Thackeray; claim 50 by 3,000 feet, patented; discovered 1865; located on Schaffter Mountain,

Spanish Bar mining district, 2 miles from Idaho Springs; vertical vein, running northeast and southwest, width 5 feet, pay vein 18 inches; nature of ore, iron and copper pyrites; average value, $200 per ton; 1 shaft 115 feet deep, 2 levels aggregate length 380 feet; total yield, $18,000.

Fall River Placer Mining Co. John M. Osborn, A.S. Bennett & Co.; claim 160 acres of ground, located in Fall River Gulch, near town of Fall River, 41 miles from Denver.

Federal. Charles R. Fish & Co., proprietors; office Georgetown; claim 50 by 1,400 feet, patented; discovered 1870; located on Griffith Mountain, Griffith mining district, 1 mile from Georgetown; vertical vein, running northeast and southwest, width 4 feet, pay vein 10 inches; nature of ore, sulphurets and silver glance; average value, $500 per ton; aggregate depth of shafts 295 feet, 4 adits, 50, 70, 100, and 130 feet in length, respectively; yield, $10,000.

Fenian Star. Owned by John E. Morris, M.J. McKinley, Robert Morris, and W. Hart; office Idaho Springs; claim 150 by 1,400 feet; discovered 1866; located on Hukill Gulch, Spanish Bar mining district, 2 miles from Idaho Springs; course of vein, northeast and southwest, width 4 feet, pay vein 16 inches; 1 shaft 40 feet deep, 2 tunnels aggregating 115 feet in length.

Flat Iron. Owned by George Lange, Thomas Renwick, and George Arnold; office Lawson; claim 150 by 900 feet, located on Columbia Mountain, Downieville mining district, 45 miles from Denver; course of vein, east and west; width 30 feet, pay vein 24 inches; nature of ore, galena and silver glance; average value, $300 per ton; main shaft 40 feet deep, 1 adit 250 feet in length, ore house 10 by 15 feet.

Forest Queen. R.W. Chin & Co., proprietors; office Lawson; claim 150 by 1,500 feet; discovered 1878; located on Columbia Mountain, Montana mining district, 1 mile from Lawson and 46 miles from Denver; 1 shaft 40 feet deep.

Fortunatus. West Argentine Mining Co., proprietors, John A Stacey, president, William H. Baldwin, secretary, Charles P. Baldwin, superintendent; office Georgetown; claim 50 by 1,500 feet, patented; discovered 1865; located on Kelso Mountain, Argentine mining district, 4 miles from Bakerville and 12 miles from Georgetown; width of vein 6 feet, pay vein 1 to 6 inches; nature of ore, galena and gray copper; average value, 285 ounces silver per ton; main shaft 25 feet deep, 1 level 30 feet long.

Fountain. Owned by William Baker, of Georgetown; claim 150 by 1,400 feet; discovered 1872; located on Leavenworth Mountain, Griffith mining district, 3 miles from Georgetown; course of vein, east and west, pay vein 1 to 5 feet; nature of ore, galena; average value, 200 ounces silver per ton; main shaft 65 feet deep.

Free America. James I. and Joseph Gilbert, proprietors; office Georgetown; claim 150 by 1,500 feet, patented; discovered 1877; located on Red Elephant Mountain, Downieville mining district, 1 mile from Lawson; course of vein, northeast and southwest, average width 35 feet, pitch 10 degrees north, pay vein 4 to 20 inches; nature of ore, galena, gray copper and native silver; average value, 150 ounces silver per ton; main shaft 300 feet deep, 4 levels, 100, 160, 220, and 280 feet from surface, respectively, aggregating 700 feet in length, total yield, $135,000.

Freeland. John M. Dumont and Edwin S. Platt, proprietors; office Spanish Bar; claim 1,500 by 2,600 feet, patented; discovered 1860; located 2 ½ miles from the mouth of Trail Creek, 2 ½ miles from Fall River, and 40 miles from Denver; course of vein, northeast and southwest, average width 7 feet, pitch 45 degrees, pay vein 18 inches; nature of ore, iron and copper pyrites; average value of first-class ore $100, second-class $60, and third-class $40 per ton; main shaft 100 feet deep; 4 adits 15, 400, 650, and 750 feet in length, respectively; shaft house 30 by 45 feet; in the ore house of this mine is stored 1,000 tons of assorted ore, ready for market.

French Mining Co. Incorporated April 1, 1872; capital stock, $25,000, in 500 shares, $50 each; office Georgetown; organized by Antoine Roy, Pierre Beaupre, Saline Virginia, Napoleon Leduc, Onesine Chamberland, John Dupuis, and others.

Frostburgh. Owned by Zadock Kalbaugh and W.W. Wrigley; office Georgetown; claim 50 by 1,400 feet, patented; discovered 1866; located on Sherman Mountain, near Silver Plume, Griffith mining district, 3 miles from Georgetown; course of vein, northeast and southwest, width 4 feet, pitch 10 degrees north, pay vein 1 to 10 inches; nature of ore, galena and sulphurets; average value, 375 ounces silver per ton; 1 shaft 40 feet deep, 1 level 50 feet in length.

G.M. Church. Charles E. Robinson and Gilbert Wakefield, proprietors; office Spanish Bar; claim 150 by 1,500 feet; discovered 1878; located on Red Elephant Mountain, near town of Lawson, 46 miles from Denver; vertical vein, running northeast and southwest, average width 4 feet, pay vein 1 to 7 inches; nature of ore, galena; average value, $100 per ton; main shaft 70 feet deep. (See also H.H. Church Mine.)

Galic. William Teller, proprietor; office Denver; claim 50 by 1,500, patented; located on Griffith Mountain, Griffith mining district, 2 miles from Georgetown; width of vein 4 feet; average value of ore, $75 per ton; main shaft 80 feet deep, 1 tunnel 150 feet in length.

Galena. Owned by Gilbert Wakefield and Simon Tyre; office Fall River; claim 150 by 1,500 feet; discovered 1878; located in Spring Gulch, Morris mining district, 2 miles from Idaho Springs; course of vein, northeast and southwest, width 4 feet, pay vein 5 inches, pitch 30 degrees north; nature of ore, native copper and galena; average value, $50 per ton; 1 shaft 40 feet deep, 1 tunnel 45 feet in length.

Garno. Located on Kelso Mountain, 8 miles from Georgetown; developed by a shaft 50 feet in depth, which shows a vein 3 feet in width; the nature of the ore is carbonate of copper, gray copper, and galena; average value, 200 ounces silver per ton.

General J.H. Harney. Owned by the Lebanon Mining Co., Julius G. Pohle, superintendent; office Georgetown; claim 50 by 1,600 feet, patented; located on Republican Mountain, Griffith mining district, 1 mile from Georgetown; vertical vein, trending northeast and southwest, width 6 feet, pay vein 6 inches; nature of ore, galena; average value, 80 ounces silver per ton; main shaft 50 feet deep, 1 tunnel 40 feet in length.

General Marion. Owned by W.H. and H.B. Johnson, Dr. Vandervort, W.F. and G.O. Huff, office, Georgetown; claim 150 by 700 feet; discovered 1867; located on Leavenworth Mountain, Griffith mining district, 2 miles from Georgetown; course of vein, northeast and southwest, width 8 inches, pay vein 3 to 8 inches, pitch 5 degrees north; nature of ore, galena; average value, 150 ounces silver per ton; main shaft 40 feet deep, 1 level 30 feet in length.

General Marshall. Owned by the Lebanon Mining Co., Julius G. Pohle, superintendent; office Georgetown; claim 50 by 1,600 feet; discovered 1865; located on Republican Mountain, 1 mile from

Georgetown; course of vein, northeast and southwest, width 6 feet, pay vein 1 to 12 inches, pitch 10 degrees north; nature of ore, galena; average value, 300 ounces silver per ton; 1 shaft 40 feet deep.

General Scott. Owned by J.H. Rogers, J. Shields, and W.C. Hicock; office Georgetown; claim 50 by 1,500 feet, patented; discovered 1866; located on Democrat Mountain, 2 miles from Georgetown; course of vein, northeast and southwest, width 22 feet, pitch north, pay vein 1 to 8 inches; nature of ore, sulphurets; average value, 145 ounces silver per ton; main shaft 20 feet deep, 1 tunnel 25 feet in length.

George D. Prentiss. Owned by the Lebanon Mining Co.; Julius G. Pohle, superintendent, office Georgetown; claim 50 by 1,600 feet, patented; discovered 1865; located on Republican Mountain, Griffith mining district, 1 mile from Georgetown; vertical vein, trending northeast and southwest, width 20 feet, pay vein 10 inches to 3 feet; nature of ore, galena; average value, 200 ounces silver per ton; 2 shafts, aggregate depth 100 feet; aggregate length of levels 270 feet; reached by the Lebanon Tunnel at a distance of 400 feet from the mouth, and intersected at a depth of 185 feet.

German. John Tubb, proprietor; office Central City; claim 100 by 1,200 feet; discovered 1865; located on Belleview Mountain, Virginia mining district, 3 miles from Central City; course of vein, northeast and southwest, width 2 ½ feet; nature of ore, galena and iron pyrites; average value, 5 ounces gold per cord; main shaft 120 feet deep, 1 level 60 feet in length and 90 feet from surface.

Gilman. Owned by the Planet Silver Mining Co., office Georgetown; claim 900 linear feet; located head of Geneva Gulch, Argentine mining district, 15 miles from Georgetown; vertical vein, running northeast and southwest, width of pay vein, 1 to 18 inches; nature of ore, galena, gray copper and iron and copper pyrites; average value, 100 ounces silver per ton; main shaft 40 feet deep, 2 levels, 40 and 100 feet deep, respectively, aggregating 500 feet in length; total yield, $10,000.

Glasgow. Located on Brown Mountain, Griffith and Queen mining districts, 3 miles from Georgetown.

Globe. Owned by J.B. Johnson, O.E. Clark, David and George Meyers; office Georgetown; claim 150 by 450 feet; discovered 1877; located on Leavenworth Mountain, East Argentine mining district, 9 miles from Georgetown; vertical vein, running north and south; width 5 feet, pay vein 18 inches; nature of ore, galena; average value, $85 per ton; 1 shaft 20 feet deep.

Goetzel Silver Mining Co. Incorporated October 19, 1870; capital stock $1,000,000, in 10,000 shares, $100 each; office New York City; organized by Mathias Goetzel, Edward L. Lynch, C.C. Schiefendecker, Frederick Leposin, Louis Gaiser, G.M. Van Buren, and Joseph Boh.

Gold Dirt. Owned by James Peck & Sons; office Empire; claim 150 by 1,000 feet; discovered 1864; located on Silver Mountain, Upper Union mining district, 1 mile from Georgetown; vertical vein, running north and south, width 6 feet, pay vein 20 inches; nature of ore, iron and copper pyrites and free gold; average value, $80 per cord; main shaft 40 feet deep, 1 level 30 feet, and 1 tunnel 100 feet in length.

Gold Eagle. Owned by William Moore and others; office Empire; claim 150 by 1,500 feet; discovered 1860; located on Silver Mountain, Upper Union mining district, 2 miles from Georgetown; vertical vein, running northeast and southwest, width 15 feet, pay vein 18 inches; nature of ore, free gold; average value, $75 per ton, 3 adits, 48, 65, and 80 feet in length, respectively; total yield, $65,000.

Gold King. Owned by G.W. Cowles & Co.; office Empire; claim 150 by 1,500 feet; discovered 1876; located on Silver Mountain, Upper Union mining district, 3 miles from Georgetown; vertical vein, running north and south, width 4 feet, pay vein 8 to 12 inches; nature of ore, free gold, iron and copper pyrites; average value, $244 per cord; aggregate depth of shafts 110 feet, aggregate length of adits 140 feet; total yield, $12,000.

Golden Era. Owned by G.W. Cowles & Co.; office Empire; claim 150 by 1,500 feet; discovered 1876; located on Silver Mountain, Upper Union mining district 3 miles from Georgetown; vertical vein, running northwest and southeast, width 3 feet, pay vein 16 to 20 inches; nature of ore, iron and copper pyrites and free gold; average value, $200 per cord; main shaft 40 feet deep; aggregate length of levels 90 feet, 1 tunnel 60 feet in length.

Golden Wing. Edward W. Williams, William and Edward Jones, and Owen Hughes, proprietors; office Central City; claim 150 by 1,500 feet; discovered 1877; located on Belleview Mountain, Idaho mining district, 3 miles from Central City; course of vein, northeast and southwest, width 4 feet, pitch 25 degrees north, pay vein 1 foot; nature of ore, iron and copper pyrites, average value of mill ore, 20 ounces gold per cord; main shaft 40 feet deep.

Good Enough. Owned by N.J. Moss and W.A. Doyle; office Lawson; claim 150 by 1,500 feet; discovered 1878; located on Columbia Mountain, Montana mining district, 6 miles from Georgetown; 1 shaft 25 feet deep.

Goshen. Owned by Frank Hood, William S. Smith, and B. Devotie; office Georgetown; claim 150 by 3,000 feet; discovered 1870; located on Columbia Mountain, 1 mile from Lawson; course of vein, northeast and southwest, width 4 feet, pitch 40 degrees north, pay vein 12 inches; nature of ore, galena and gray copper; average value, 150 ounces length 100 feet, 1 tunnel 60 feet in length.

Goss. Owned by Carver J. Goss of Denver; claim 50 by 1,500 feet; discovered 1867; located on Brown Mountain, Griffith mining district, 3 miles from Georgetown; course of vein, northeast and southwest, pitch 10 degrees north, width 3 ½ feet, pay vein 2 inches; nature of ore, sulphurets; average value, 1,200 ounces silver per ton; 1 shaft 45 feet deep, 2 tunnels aggregate length 100 feet; 1 level 50 feet in length.

Grand Belcher Silver Mining Co. Incorporated October 8, 1877; capital stock, $3,750,000, in 375,000 shares, $10 each; office New York City; trustees: Frederick Leporieu, Mathias Goetzel, Thomas Baer, Levy Bernstein, and Joseph Koch.

Grand View. Owned by H.C. Cowles, D. and C. Cross; office Empire; claim 150 by 150 feet; discovered 1876; located on Silver Mountain, Upper Union mining district, 2 miles from Georgetown, vertical vein, running northeast and southwest, width 5 feet, pay vein 1 foot; nature of ore, free gold; average value, $62 per cord; main shaft 25 feet deep, 2 adits 30 feet in length, respectively.

Grant. Lebanon Mining Co., proprietors; Julius G. Pohle, superintendent; claim 150 by 1,500 feet, patented; discovered 1870; located in Cascade mining district, 4 miles from Idaho Springs; vertical vein, running northeast and southwest, width 20 feet, pay vein 14 inches; nature of ore, galena and gray copper; average value, 150 ounces silver per ton; 1 shaft 50 feet deep.

Grant. Joseph Reynolds and James I. Gilbert, proprietors; office Georgetown; claim 150 by 1,500 feet; discovered 1872; located on Red Elephant Mountain, Downieville mining district, 1 mile from Lawson; course of vein, northeast and southwest, width 8 feet, pitch 15 degrees north, pay vein 1 to 20 inches; nature of ore, galena, gray copper, and zinc blende; average value, $100 per ton; main shaft 100 feet, and 1 other shaft 60 feet deep, 2 tunnels 240 and 400 feet in length, respectively; total yield, $1,400.

Graves. Owned by Richard Baxter, of Baltimore; claim 150 by 750 feet; discovered 1872; located on Griffith Mountain, Griffith mining district, 2 miles from Georgetown; course of vein, east and west, pitch 10 degrees north; nature of ore, sulphurets; average value, 300 ounces silver per ton; 1 shaft 50 feet deep.

Gray. Eli Gray, proprietor; office Silver Dale; claim 150 by 1,500 feet; discovered on Grays Mountain, Griffith mining district, 3 miles from Georgetown; width of vein 4 feet, pay streak 2 feet; average value of ore, $200 per ton; main shaft 18 feet deep.

Great Eastern. Owned by J. Mills, of Milwaukee; claim 50 by 1,500 feet, patented; discovered 1860; located on Trail Creek, 3 miles from Idaho Springs, in Trail Creek mining district; course of vein, northeast and southwest; 1 shaft 100 feet deep.

Great Eastern. Owned by B. S. Stafford and J. Pratt; office Georgetown; claim 50 by 1,500 feet, patented; discovered 1872; located on Republican Mountain, 3 miles from Georgetown; course of vein, northeast and southwest, width 12 feet, pay vein 1 to 20 inches, pitch 10 degrees south; nature of ore, galena and sulphurets; value, 200 to 700 ounces silver per ton; 1 shaft 125 feet deep, 1 tunnel 300 feet in length.

Great Eastern. Owned by B.S. Stafford and J. Pratt, of Georgetown; claim 150 by 3,000 feet; discovered 1868; located on Sherman Mountain, 3 miles from Georgetown, in Griffith mining district; course of vein, northeast and southwest, pitch 45 degrees north, width 7 feet, pay vein 10 inches; nature of ore, galena and gray copper; average value, 200 ounces silver per ton; main shaft 60 feet deep, 1 tunnel 90 feet in length, 1 level 100 feet long.

Great Mogul. Owned by George L. Cannon, of Idaho Springs; claim 150 by 1,500 feet; discovered 1872; located on Belleview Mountain, Fall

River mining district, 2 miles from Idaho Springs and 33 miles from Denver; course of vein, east and west, width 5 feet, pay vein 3 feet, pitch 30 degrees north; nature of ore, sulphurets of silver; average value, 100 ounces silver per ton; 1 shaft 130 feet deep, 1 adit 60 feet in length.

Great Republic. Owned by G.W. Cowles & Co.; office Empire; claim 150 by 1,500 feet; discovered 1876; located on Silver Mountain, Upper Union mining district, 2 miles from Georgetown; vertical vein, running north and south, width 7 feet; nature of ore, iron and copper pyrites and free gold; average value, $60 per cord; aggregate depth of shafts 150 feet, 1 adit 80 feet in length; yield to date, $10,000.

Greeley Mining Co. Incorporated August 7, 1875; capital stock, $25,000, in 500 shares, $50 each; office Greeley; organized by A.Z. Salomon, James F. Benedict, Daniel Hawkes, J.M. Freeman, Silas S. Kennedy, and J.S. Barrett.

Greenback. Owned by Park Disbrow, of Empire; claim 150 by 1,500; discovered 1877; located on Silver Mountain, Upper Union mining district, near town of Empire, 45 miles from Denver; course of vein, northeast and southwest, width 4 feet, pay vein 12 inches; nature of ore, free gold; average value, $175 per cord; 1 shaft 69 feet deep, 1 tunnel 65 feet in length, 1 level 80 feet in length; total yield, $1,500.

Grinnell. G.W. Cowles & Co. proprietors; office Empire; claim 150 by 1,500 feet; discovered 1875; located on Silver Mountain, Upper Union mining district, 2 miles from Georgetown; vertical vein, running east and west, width 4 feet, pay vein 10 to 14 inches; nature of ore, free gold; average value, $120 per ton; main shaft 30 feet deep, 1 tunnel 70 feet in length.

Gum Tree. Owned by Israel Stotts; claim 150 by 1,500 feet, patented; discovered 1860; located at Spanish Bar, Trail Creek mining district, 15 miles from Georgetown; course of vein, northeast and southwest, width 4 feet, pay vein 12 inches; nature of ore, iron and copper pyrites; average value, $100 per ton; 1 adit 200 feet in length.

Gunboat. Located on Brown Mountain, Griffith and Queen mining districts, 3 miles from Georgetown.

H. H. Church. Charles E. Robinson & Co., proprietors; office Spanish Bar; claim 150 by 3,000 feet; discovered 1878; located on Fall

River, Morris mining district, 2 ½ miles from Idaho Springs and 43 miles from Denver; main shaft 80 feet deep, 1 adit 114 feet in length. (See also G.M Church Mine.)

Halcyon. Owned by the Equator Silver Mining and Smelting Co.; H.S. Kearney, agent and superintendent; claim 50 by 1,500 feet; discovered 1867; located on Leavenworth Mountain, Griffith mining district, 2 miles from Georgetown; developed by a shaft 50 feet deep, and a tunnel 80 feet in length.

Hansbrough. Owned by P.M. Hansbrough, William Jones, and William Clark, of Georgetown; claim 150 by 1,400 feet, patented; discovered 1876; located on Democrat Mountain, Griffith mining district, 2 miles from Georgetown; vertical vein, trending northwest and southeast, width 6 feet, pay vein 2 feet; nature of ore, galena and gray copper; average value, 218 ounces silver per ton; 1 shaft 45 feet deep; 2 tunnels aggregate length 175 feet.

Hawkeye. Owned by William Moore and David Dulaney; office Georgetown; claim 150 by 1,500 feet; discovered 1873; located on Red Elephant Mountain, near town of Lawson, Downieville mining district, 45 miles from Denver; course of vein, northeast and southwest. [David Elkanah Dulaney was born July 1828 at Wyeth Court House, West Virginia, and moved west to Illinois with his parents as a youngster. He married his second wife, Swedish-born Sophie Nilson in 1862, and they moved to Colorado later that year, settling at a succession of mining camps: Payne's Bar, Trail Creek (where they opened a grocery store), Empire, and others. Their first child, Clara Dulaney died July 5, 1865, in Missouri Flats, a mining camp overlooking Central City. Today, little Clara's small grave is all that remains of the long vanished community. The growing Dulaney family moved to Georgetown, where David Dulaney worked as a carpenter, continued prospecting, and eventually made a number of strikes. Little Clara was not forgotten; her brother George later settled in Fort Collins and named his daughter Clara. David Dulaney committed suicide in July 1917, less than three years after the death of his wife Sophie.]

Helmick Silver Mining Co., of Colorado. Incorporated November 2, 1869; capital stock, $600,000, in 6,000 shares, $100 each; office Washington, D.C.; trustees: William Helmick, Samuel A. Pugh, Walter W. Burdett, Marks Lissburger, Robert S. Lacey, F. Howard, and Norman M Ross.

Helmick Tunnel Co. Incorporated January 20, 1876; capital stock, $400,000, in 4,000 shares, $100 each; offices Washington, D.C. and Georgetown, Colorado; organized by Flodvardo Howard, James W. Barker, William Helmick, Robert S. Lacey, B. Peyton Brown, Samuel A Pugh, and Charles A. Metcalf.

Henry. Owned by the Marshall Silver Mining Co., of Colorado; office Georgetown; claim 150 by 1,500 feet; discovered 1878; located on Leavenworth Mountain, Griffith mining district, 2 miles from Georgetown; course of vein, northeast and southwest, width 10 feet; nature of ore, galena and gray copper; 1 shaft 25 feet deep.

Henry Ward Beecher. Owned by James Rogers, John Shields, and W.C. Hicock; office Georgetown; claim 150 by 2,000 feet, patented; discovered 1860; located on Republican Mountain, Griffith mining district, 3 miles from Georgetown; course of vein, northeast and southwest; average width 9 feet, pitch north, pay vein 1 to 20 inches; nature of ore, galena; average value 40 ounces silver per ton; main shaft 25 feet deep; 1 level 20 feet in length.

Hercules. Owned by the Consolidated Hercules and Roe Silver Mining Co.; G.W.E. Griffith, president, H.M. Griffin vice president; office Denver; claim 50 by 3,000 feet, patented; discovered 1868; located on Brown Mountain, Griffith mining district, 3 miles from Georgetown; course of vein, east and west, width 6 feet, pay vein 6 inches, pitch 5 degrees north; nature of ore, galena and gray copper; average value, 300 ounces silver per ton.

Hercules and Seven-Thirty Consolidated Silver Mining Co. Incorporated October 2, 1875; capital stock, $150,000, in 1,500 shares, $10 each; office Georgetown; organized by Horace Watkins, John R. Hambel, Charles T. Bellamy, and William M. Clark.

Herkimer. Owned by Benjamin F., Egbert, and Albert Johnson; office Denver; claim 50 by 1,200 feet, patented; located on the south branch of South Clear Creek, Griffith mining district, 1 ½ miles from Georgetown; nature of ore, argentiferous galena; developed by a shaft 70 feet deep and a surface opening 60 feet in length; 500 pounds of ore from this mine yielded 180 ounces of pure silver, worth $241.20 in coin and at the rate of $964.80 per ton. This is a good tunnel site, and water power can be obtained on this property.

Herman Mining Co. Organized under the laws of the state of New York, capital stock, $200,000, John D. Dix, president, Benjamin F. Morris, vice president, George W. Dix, secretary, Charles H. Morris superintendent, offices Georgetown, Colorado, and New York City.

Hickman. Owned by the Silver Plume Mining Co., London, England; office Georgetown; claim 50 by 3,000 feet, patented; discovered 1865; located on Republican Mountain, Griffith mining district, 3 miles from Georgetown; course of vein, northeast and southwest, width 3 feet, pay vein, 5 inches, pitch 15 degrees north; nature of ore, galena and gray copper; 1 shaft 120 feet deep, 1 level 100 feet in length.

High Grade. George L. Cannon, proprietor; office Idaho Springs; claim 150 by 1,500 feet, discovered 1873, located on Schaffter Mountain, Idaho mining district, near town of Idaho Springs, 33 miles from Denver; vertical vein, running northeast and southwest, width 4 feet, pay vein 16 inches; nature of ore, galena; average value, $184 per ton; 1 adit 340 feet in length.

Highland Chief. Charles C. Welch, David C. Crawford, James F. Devere, and R.W. Chinn, proprietors; office Denver; claim 150 by 1,500 feet; discovered 1877; located on Columbia Mountain, Montana mining district, 1 mile from town of Lawson and 46 miles from Denver.

Hobart. Charles P. Baldwin and Samuel Swett, proprietors; office Georgetown; claim 150 by 1,500 feet; discovered 1877; located on Columbia Mountain, Griffith mining district, 2 miles from Georgetown; course of vein, north and south, pitch west; nature of ore, carbonate of copper and gray copper; average value, 225 ounces silver per ton; 2 shafts aggregate depth 50 feet, 1 level 40 feet in length.

Home Stake. Owned by James McIntosh, H. Hartzel, and J.L. Wilson; office Lawson; claim 150 by 1,500 feet; discovered 1878; located on Columbia Mountain, Montana mining district, 46 miles from Denver; vertical vein, running east and west, width 4 feet, pay vein 14 inches; nature of ore, galena and sulphurets; average value, 200 ounces silver per ton; 1 shaft 54 feet deep.

Home Mining Association of Colorado Springs. Incorporated February 29, 1876; capital stock, $25,000, in 2,500 shares, $10 each; office

Idaho Springs; organized by Matthew O. Coddington, James C. Wright, Abram M. Noxon, John A. Dory, and Romeo D. Strong.

Hoodon. Charles H. Myers and John W. Gilbert, proprietors; office Georgetown; claim 150 by 1,500 feet; discovered 1877; located on Kelso Mountain, West Argentine mining district, 9 miles from Georgetown; vertical vein, running north and south, width 6 feet, pay vein 8 inches; nature of ore, galena and copper pyrites; average value, $115 per ton; 1 adit 125 feet in length.

Howdah. Owned by William Riley and R. Cunningham; office Georgetown; claim 150 by 1,500 feet; discovered 1876; located on Democrat Mountain, Griffith mining district, 3 miles from Georgetown; width of vein 20 feet, pay streak 6 inches; average value of ore, 550 ounces silver per ton; main shaft 50 feet deep, 1 tunnel 330 feet, and 2 levels aggregating 180 feet in length.

Hugo. Charles C. Welch, David C. Crawford, James F. Devere, and R.W. Chinn, proprietors; office Denver; claim 150 by 1,500 feet, discovered 1877; located on Columbia Mountain, Montana mining district, 1 mile from town of Lawson and 46 miles from Denver; course of vein, northeast and southwest, width 4 feet, pitch 30 degrees north; nature of ore, galena and carbonate of copper; average value, $114 per ton; 1 shaft 25 feet deep.

Hukill. Owned by the Hukill Gold and Silver Mining Co. of Colorado; incorporated June 28, 1876; capital stock, $1,000,000, in 100,000 shares, $10 each; offices Idaho Springs, Colorado, and 17 Broad Street, New York; J.L. Brownell, president, John De Lano, secretary; claim 50 by 1,288 feet, patented; discovered 1865; located on Hukill Hill, Spanish Bar mining district, near town of Idaho Springs; course of vein, east and west, width 5 feet, pay vein 6 inches, pitch north; nature of ore, galena; average value, 115 ounces silver per ton; 10 shafts aggregate depth 800 feet, 1 tunnel 300 feet in length, 3 levels aggregate length 800 feet.

Huldah. West Argentine Mining Co.; proprietors John A. Stacey, president, William H. Baldwin, secretary, Charles P. Baldwin, superintendent; office Georgetown; claim 50 by 1,600 feet, patented; discovered 1865; located on Kelso Mountain, above timberline, Argentine mining district, 4 miles from Bakerville and 12 miles from Georgetown; width of vein 30 feet, pay vein 1 to 12 inches; nature of ore, galena, zinc blende, iron and copper pyrites; average value, 125 ounces silver per ton;

main shaft 150 feet deep, 1 tunnel 30 feet in length, 3 levels aggregate length 166 feet.

Hunter. Owned by the Equator Mining and Smelting Co.; H.S. Kearney, agent and superintendent; offices Georgetown, Colorado, and Chicago, Illinois; claim 150 by 1,400 feet; discovered 1878; located on Leavenworth Mountain, Griffith mining district, 2 miles from Georgetown; developed by a tunnel 100 feet in length and a shaft 60 feet deep.

Independent Silver Mining Co. Incorporated November 1, 1876; capital stock $150,000, in 1,500 shares, $100 each; office Georgetown; organized by James Oscar Stewart, Anthony J. August, and Charles W. Burdsal.

International. Owned by the Hukill Silver Mining Co.; office Georgetown; claim 50 by 1,500 feet, patented; discovered 1865; located on McClelland Mountain, Argentine mining district, 7 miles from Georgetown; width of vein 5 feet, pay streak 8 inches; 1 shaft 100 feet deep, 1 tunnel 900 feet in length.

Iowa. Owned by David N. Smith, W.M. Fletcher, and Peter Shively; office Georgetown; claim 150 by 1,500 feet, patented; discovered 1877; located on Red Elephant Mountain, Downieville mining district, 1 mile from Lawson and 6 miles from Georgetown; course of vein, northeast and southwest, width 4 feet, pay vein 2 feet, pitch 45 degrees northwest; nature of ore, galena; average value, 100 ounces silver per ton; main shaft 150 feet deep.

Ivers Phillips. Owned by the Burleigh Mining Co.; office Georgetown; claim 150 by 3,000 feet, patented; discovered 1877; located on Sherman Mountain, Griffith mining district, 3 miles from Georgetown; course of vein, east and west, width 120 feet, pitch 10 degrees north; nature of ore, zinc blende, galena, and gray copper; reached by the Burleigh Tunnel [see separate listing] at a distance of 1,900 feet from the mouth and intersected at a depth of 1,800 feet; 1 level 50 feet in length.

James. Owned by R.C. Moore, P. M. McCrellis, and A.F. James; office Georgetown; claim 150 by 1,500 feet; discovered 1877; located on Red Elephant Mountain, 1 mile from town of Lawson, Downieville mining district, 6 miles from Georgetown; course of vein, east and west, pitch 20 degrees north, width 20 feet, pay vein 8 inches; nature of ore, galena;

average value, $50 per ton; 1 shaft 130 feet deep, 1 level, 110 feet in length; shaft house 37 by 24 feet.

James A. Gage. Owned by the Herman Mining Co., New York; John D. Dix, president, Benjamin F. Morris, vice president, George W. Dix, secretary, Charles H. Morris, superintendent; capital stock, $200,000, none of which has been sold for less than par value; offices Georgetown, Colorado, and New York City; claim 50 by 1,500 feet, patented; discovered 1869; located on Republican Mountain, near town of Silver Plume, 3 miles from Georgetown; course of vein, east and west, width 80 feet, pay vein 1 foot, pitch 40 degrees north; nature of ore, zinc blende and galena; average value, 400 ounces silver per ton; main shaft 160 feet deep, 1 tunnel 20 feet, and 1 drift 40 feet in length.

James Guthrie. Owned by the Lebanon Mining Co., Julius G. Pohle, superintendent; office Georgetown; claim 50 by 1,200 feet, patented; discovered on Republican Mountain, Griffith mining district, 1 mile from Georgetown; course of vein, northeast and southwest, width 12 feet, pay vein 4 to 15 inches, pitch 10 degrees south; nature of ore, galena; average value, 500 ounces silver per ton; 2 shafts, aggregate depth 200 feet, 1 level 140 feet in length; reached by the Lebanon Tunnel at a distance of 450 feet from the mouth and intersected at a depth of 200 feet.

Joe Coaley. Edward C. Guibor, proprietor; office Empire; claim, 150 by 1,500 feet; discovered 1876; located on Silver Mountain, Upper Union mining district, 2 miles from Georgetown; vertical vein, running northeast and southwest, width 3 feet, pay vein 1 foot; nature of ore, iron pyrites; average value, $100 per cord; aggregate depth of shafts 124 feet; 1 tunnel 40 feet in length; yield, $1,800.

John Bull. Owned by Charles H. Marshall & Co., and McMurdy Estate; office Georgetown; claim 50 by 1,500 feet; discovered 1868; located on Leavenworth Mountain, Griffith mining district, 3 miles from Georgetown; course of vein, northeast and southwest, pitch 10 degrees north, pay vein 15 inches; nature of ore, sulphurets; average value 500 ounces silver per ton; 1 shaft 100 feet deep, 3 levels, aggregate length 200 feet.

John E. McClung. Consolidated Hercules and Roe Silver Mining Co., proprietors; H.M. Griffin, treasurer and superintendent; office Denver; claim 50 by 3,000 feet, patented; discovered 1868; located on Brown Mountain near town of Brownville, Griffith and Queen mining districts, 3 miles from Georgetown; width of vein 6 feet, pay vein 11 inches; nature of ore, galena; average value, 175 ounces silver per ton.

John J. Roe, West. Owned by Albert Johnson, of Georgetown; claim 50 by 1,500 feet, patented; discovered 1873; located on Brown Mountain, Queen mining district, 3 miles from Georgetown; vertical vein, trending northwest and southeast, width 4 feet, pay vein 8 inches; nature of ore, galena; average value, 300 ounces silver per ton; aggregate depth of shafts, 340 feet, 1 tunnel 440 feet, and 1 level 245 feet in length.

John M. Wilson. Owned by the Consolidated Hercules and Roe Silver Mining Co.; G.W.E. Griffith, president, H.M. Griffin, treasurer; office Denver; claim 50 by 3,000 feet, patented; discovered 1868; located on Brown Mountain, Griffith mining district, 3 miles from Georgetown; width of vein 3 feet, pay vein 4 inches; nature of ore, galena and sulphurets; average value, 300 ounces silver per ton.

John Paul Jones. Owned by S.H. White, of Idaho Springs; claim 150 by 1,400 feet; discovered 1866; located in Virginia Canyon, near Idaho Springs; vertical vein, running northeast and southwest; width 5 feet, pay vein 20 inches; nature of ore, iron and copper pyrites and galena; average value, $60 per ton; 1 shaft 60 feet deep, 1 adit 160 feet in length.

Jordan. Owned by Patrick McNulty, of Georgetown; claim 150 by 1,500 feet, patented; discovered 1875; located on Democrat Mountain, Griffith mining district, 3 miles from Georgetown; course of vein, northeast and southwest, width 10 feet, pay vein 6 inches; nature of ore, galena; average value, 250 ounces silver per ton; 1 tunnel 150 feet in length.

Joseph E. Johnston. Lebanon Mining Co., proprietors; Julius G. Pohle, superintendent; office Georgetown; claim 150 by 1,500 feet; discovered 1872; located on the north side of Leavenworth Mountain, Griffith mining district, 1 mile from Georgetown; vertical vein, running northeast and southwest, width 8 feet, pay vein 6 inches; nature of ore, galena; average value, 120 ounces silver per ton; 1 shaft 40 feet deep.

Joseph E. Johnson. Lebanon Mining Co., proprietors; Julius G. Pohle, superintendent; office Georgetown; claim 150 by 1,500 feet; discovered 1872; located on the north side of Leavenworth Mountain, Griffith mining district, 1 mile from Georgetown; vertical vein, running northeast and southwest, width 8 feet, pay vein 6 inches; nature of ore, galena; average value, 120 ounces silver per ton; 1 shaft 40 feet deep.

Joseph Reynolds Mining Property. Owned by Joseph Reynolds, James I. Gilbert, and James Dailey; office Georgetown; 4 claims 150 by 1,500 feet each, all of which are patented; vertical veins, trending east and west, width 15 feet each; nature of ore, galena, copper pyrites, sulphurets, and native silver; average value, 500 ounces silver per ton; main shaft 100 feet deep, 3 adits, aggregate length 300 feet; total yield, $8,000.

Judd and Crosby Silver Reduction and Mining Co. Incorporated April 7, 1873; capital stock, $160,000, in 1,600 shares, $100 each; offices Georgetown, Colorado, and Chicago, Illinois; organized by Norman B. Judd, William H. Cushman, F.W. Crosby, Leonard G. Calkins, and Robert O. Old.

Junction. Owned by William P. Lynn, of Georgetown; claim 150 by 1,500 feet; discovered 1877; located on Leavenworth Mountain, Griffith mining district, 2 miles from Georgetown; vertical vein, trending northeast and southwest, width 5 feet, pay vein 14 inches; nature of ore, sulphurets; average value 105 ounces silver per ton; 1 shaft 40 feet deep.

Kanawha. Owned by G.G. White & Co., of Georgetown; claim 150 by 1,500 feet; discovered 1878; located on Miller Mountain, 2 miles from Georgetown; course of vein, east and wet, width 5 feet, pitch 10 degrees north; nature of ore, galena; average value, $125 per ton; main shaft 66 feet deep.

Kansas and Colorado Mining and Milling Co. Incorporated November 4, 1876; capital stock, $50,000, in 2,000 shares, $50 each; office, Mill City; organized by Justin E. Dubois, C.B. Keith, Edwin Bowden, August Blackman, Thomas Pearce, and Simon Dingle.

Kansberg. Owned by E. Harrison and M.J. Morrison; office Georgetown; claim 150 by 1,500 feet, patented; discovered 1873; located on Democrat Mountain, Griffith mining district, 2 miles from Georgetown; course of vein, east and west, width 7 feet, pay vein 2 to 4 inches, pitch 5 degrees south; nature of ore, galena and gray copper; average value, $600 per ton; 1 shaft 100 feet deep, 1 adit 140 feet in length.

Kirtley Tunnel Mining Co. Jeremiah Kirtley, J.N. Roberts, A. Medley, Charles A. Martine, and E.S. Weaver; claim 150 by 1,500 feet;

patented; discovered 1877; located on Leavenworth Mountain, Griffith mining district, 3 miles from Georgetown; length of tunnel 700 feet; width of vein 6 feet; average value of ore, 275 ounces silver per ton; main shaft 30 feet deep, 1 level 100 feet in length.

L.W. Powell. Owned by the Lebanon Mining Co.; Julius G. Pohle, superintendent; office Georgetown; claim 50 by 1,400 feet; discovered 1865; located on Republican Mountain, Griffith mining district, 1 mile from Georgetown; course of vein, northeast and southwest, width 3 feet, pay vein 6 to 14 inches, pitch south; nature of ore, galena; average value, 150 ounces silver per ton; 1 shaft 25 feet deep.

La Crosse. Owned by W.N. Webster, W.H. Doyle, Daniel Drummond, R.W. Tallman, and N.J. Morse; office Lawson; claim 150 by 3,000 feet; discovered 1878; located on Columbia Mountain, Montana mining district, 2 miles from Lawson; course of vein, northeast and southwest, width 6 feet, pitch 10 degrees north, pay vein 5 inches; nature of ore, galena; average value, 105 ounces silver per ton; 3 shafts, aggregate depth 74 feet.

Lady Anna. Owned by the Lebanon Mining Co., Julius G. Pohle, superintendent; offices Georgetown, Colorado, and New York City; claim 50 by 1,500 feet; discovered 1870; located on Republican Mountain, Griffith mining district; 2 miles from Georgetown; course of vein, northeast and southwest; width 4 feet; nature of ore, galena; average value, 200 ounces silver per ton; 1 tunnel 100 feet in length.

Lady Augusta. Owned by the Lebanon Mining Co.; Julius G. Pohle, superintendent; offices Georgetown, Colorado, and New York City; claim 150 by 1,500 feet; discovered 1870; located on Republican Mountain, Griffith mining district, 2 miles from Georgetown; course of vein, northeast and southwest, width 4 feet, pitch northwest; nature of ore, galena; assay value, 600 to 800 ounces silver per ton; main shaft 10 feet deep, 1 tunnel 50 feet in length.

Lady Emma. Lebanon Mining Co., proprietors; Julius G. Pohle, superintendent; office Georgetown; claim 50 by 1,500 feet; discovered 1870; located in Cascade mining district, 4 miles from Idaho Springs; course of vein, northeast and southwest, width 8 feet, pitch northwest, pay vein 8 inches; nature of ore, galena and sulphurets; average value, 300 ounces silver per ton; 1 shaft 60 feet deep.

Lafayette. Owned by William Baker, of Georgetown; claim 150 by 1,500 feet; discovered 1873; located on Leavenworth Mountain, Griffith mining district, 3 miles from Georgetown; course of vein, east and west, width 10 feet, pitch 15 degrees north, pay vein 10 inches; nature of ore, sulphurets and gray copper; value, 1,170 ounces silver per ton; main shaft 45 feet deep, 1 level 25 in length.

Lafayette. Owned by George Berry and Co.; office New York City; claim 150 by 1,500 feet; discovered 1878; located on Clear Creek, Spanish Bar mining district, 1 mile from town of Idaho Springs; course of vein, northeast and southwest.

Lake Superior. Northwestern Silver Mining Co.; capital stock, $300,000; William L. Hadley, manager; office Georgetown; claim 50 by 800 feet, patented; discovered 1868; located on the west face of Leavenworth Mountain, Griffith mining district, ½ mile from Georgetown; width of vein 6 feet, pay vein 18 inches; average value of ore, 60 ounces silver and 15 ounces gold per ton; 1 shaft 150 feet deep.

Lancaster. Owned by John Tubb; office Central City; claim 150 by 1,100 feet; discovered 1863; located on Belleview Mountain, Virginia mining district, 3 miles from Central City; width of vein 18 inches; average value of ore, 6 ounces gold per cord; main shaft 70 feet deep, 1 level 50 feet in length and 40 feet from surface.

Lancaster. Owned by Du Bois Tooker & Co., of Georgetown; claim 150 by 1,500 feet; discovered 1877; located on Red Elephant Mountain, near town of Lawson, Downieville mining district, 6 miles from Georgetown; course of vein, southeast and northwest, width 5 feet, pitch 30 degrees north, pay vein 10 inches; nature of ore, galena and copper pyrites; average value, $100 per ton; 1 shaft 70 feet deep, 1 tunnel 75 feet in length.

Last Chance. Owned by William Clark, of Silver Plume; claims 150 by 1,500 feet, patented; discovered 1872; located on Republican Mountain, Griffith mining district, 2 miles from Georgetown; course of vein, northeast and southwest, pitch north, width 5 feet, pay vein 4 inches; nature of ore, sulphurets; average value, 125 ounces silver per ton; 1 shaft 12 feet deep, 1 tunnel 175 feet in length.

Last Chance. Owned by the Colorado United Mining Co.; located on Brown Mountain, near Brownville, 3 miles from Georgetown.

Leavenworth Mountain Mining and Tunneling Co. Incorporated June 12, 1873; capital stock, $1,000,000, in 10,000 shares, $100 each; office Georgetown; organized by Howard C. Chapin, Charles H. Utter, David T. Griffith, and F.J. Marshall.

Lebanon Tunnel. Owned by the Lebanon Mining Co., organized under the laws of the State of New York, October 24, 1870; capital stock, $1,000,000, in 10,000 shares, $100 each; Felix Stoiber, president, Thomas Oliver, treasurer, Julius G. Pohle, superintendent and resident director; offices Georgetown, Colorado, and 229 Grand Street, New York City; claim 150 by 3,000 feet; discovered 1871; located on Republican Mountain, Griffith mining district, 1 mile from Georgetown; course of tunnel, north 28 degrees west, length 830 feet; has cut 8 lodes at a depth of 80, 150, 150, 220, 250, 300, 400, and 450 feet respectively.

Leviathan Tunnel. Owned by the Leviathan Tunnel Co., M.N. Robertson, manager; office, Grant, Park County, Colorado; claim 1,500 by 3,000 feet; located on Revenue Mountain above timberline, Geneva mining district, 15 miles from Georgetown; course of tunnel, north and south; has cut 5 lodes and is 700 feet in length.

Lewis. B.N. Sanford, C. Strong, and Edward C. Guibor, proprietors; office Empire; claim 150 by 1,500 feet; discovered 1877; located on Cavode Mountain, Upper Union mining district, 1 mile from Georgetown; vertical vein, running northeast and southwest, width 4 feet, pay vein 14 inches; nature of ore, hematite iron; average value, $240 per ton; main shaft 20 feet deep, 1 adit 45 feet in length; total yield, $5,000.

Lincoln. Owned by John M. Dumont, of Spanish Bar; claim 150 by 2,450 feet, patented; discovered 1860; located on Clear Creek, 39 miles from Denver, in Spanish Bar mining district; course of vein, northeast and southwest, width 4 feet, pitch 40 degrees north; nature of ore, iron and copper pyrites; average value, $200 per ton; 3 shafts, aggregate depth 250 feet, 1 adit 225 feet in length.

Lynn Tunnel Co., of Georgetown, Colorado. Incorporated November 8, 1869; capital stock, $300,000, in 3,000 shares, $100 each; offices, Georgetown and Denver, Colorado; trustees: Bela M. Hughes, J.W. Horner, William P. Lynn, J. Harvey Jones, and F.A. Pope.

Little Emma. Owned by Patrick McNulty, Charles W. Burdsal, V.E. Steward, and John Cain; office Georgetown; claim 150 by 1,500 feet, patented; discovered 1877; located on Democrat Mountain, Griffith mining district, 3 miles from Georgetown; course of vein, east and west, width 4 feet, pay vein 6 inches, pitch 35 degrees north; nature of ore, galena, sulphurets of silver, and pyrites of copper; value, $800 per ton; 1 shaft 40 feet deep, 1 tunnel 430 feet, and 1 level 180 feet in length.

Live Yankee. Owned by H. Henderson and J.H. Kirkland; office Empire; claim 150 by 1,500 feet; discovered 1867; located on Columbia Mountain, Montana mining district, near town of Lawson; vertical vein, running east and west, width 3 feet; nature of ore, galena; average value, $140 per ton; main shaft, 90 feet deep, aggregate depth of other shafts 80 feet.

Live Yankee, Extension. Owned by the Akron Tunnel Co.; claim 150 by 1,500 feet, patented; discovered 1876; located on Columbia Mountain, Montana mining district, near town of Lawson, 6 miles from Georgetown; vertical vein, running east and west, width 5 feet, pay vein 10 inches; nature of ore, galena and gray copper; average value, $50 per ton; 1 shaft 25 feet deep, 1 tunnel 400 feet in length.

Livingston County. Owned by the Bay State Mining Co.; office Boston, Massachusetts; W.L. Candler, superintendent; claim 50 by 800 feet; discovered 1862; located on Silver Mountain, near town of Empire, Upper Union mining district, 2 miles from Georgetown and 50 miles from Denver; course of vein, northeast and southwest, width 4 feet, pitch 15 degrees northwest; nature of ore, iron and copper pyrites, average value, $150 per cord; main shaft 100 feet, and 1 other shaft 40 feet deep; 1 tunnel 950 feet in length; 12-stamp mill and machinery; yield to date, $200,000.

Loretta. Owned by the Central Iowa Silver Mining Co., Iowa; E.G. McMillan, president, E.C. Rice, vice president, J.B. Stattler, secretary, A.T. Burchard, treasurer, Frank L. Downend, superintendent; claim 150 by 750 feet; discovered 1872; located near Silver Plume, on Republican Mountain, 3 miles from Georgetown; course of vein, northeast and southwest, width 18 feet, pitch 32 degrees north, pay vein 2 ½ feet; nature of ore, galena, gray copper, and sulphurets of silver; main shaft 160 feet deep, 2 tunnels, aggregate length 175 feet, 1 level, 180 feet from surface and 106 feet in length.

Loretta, West Extension. Owed by Frank L. Downend and Emanuel Hirsch; office Georgetown; claim 150 by 750 feet; discovered 1872; located on Republican Mountain, Griffith mining district, 3 miles from

Georgetown; course of vein, northeast and southwest, width 18 feet, pay vein 2 feet, pitch north; nature of ore, galena, sulphurets of silver, and gray copper; 1 shaft 30 feet deep.

Lucky Hesperus. Claim 50 by 3,000 feet, discovered 1867; located on Democrat Mountain, Griffith mining district, 2 miles from Georgetown; vertical vein, running northeast and southwest, width 6 feet, pay vein 7 inches; nature of ore, galena and gray copper; average value, 200 ounces silver per ton; main shaft 489 feet deep, aggregate length of levels 450 feet.

Lynn Boyd. Owned by the Lebanon Mining Co., Julius G. Pohle, superintendent; office Georgetown; claim 50 by 800 feet; discovered 1867; located on Republican Mountain, Griffith mining district, 1 mile from Georgetown; course of vein, north and south, width 4 feet, pay vein 1 foot; nature of ore, galena; average value, 355 ounces silver per ton; main shaft 20 feet deep.

Magnet. Located on Griffith Mountain; nature of ore, galena and sulphurets; value, 200 ounces silver per ton; pay vein 1 inch to 2 feet.

Mahany Gold Mining Co. Incorporated July 26, 1876; capital stock, $300,000, in 6,000 shares, $50 each; office Georgetown; organized by James O. Steward, Jeremiah Mahany, and Louis C. Damarin.

Major Anderson. Owned by Benjamin F. Egbert and Albert Johnson; claim 50 by 2,700 feet, patented; located in Griffith mining district in the lower part of Georgetown, within 100 rods of the Stewart Reduction Works; course of vein, northeast and southwest, width 3 feet, pay vein 4 inches, pitch south; nature of ore, galena and sulphurets; has ranged in value from $45 to over $900 to the ton; developed by 2 shafts aggregating 140 in depth.

Mammoth and Glasgow Silver Mining Co. Incorporated May 29, 1874; capital stock, $500,00 in 5,000 shares, $100 each; office, Georgetown; organized by Frank W. Johns, Thomas C. Johns, and James R. Polk Prindle.

Mammoth. Located on Brown Mountain near Brownville, 4 miles from Georgetown.

Manhattan. David M. DeWitt proprietor; office Georgetown; claim 150 by 1,250 feet; discovered 1875; located on Sherman Mountain near

Brownville, Griffith mining district, 3 miles from Georgetown; width of vein 3 feet, pay vein 6 inches; 3 shafts aggregate depth 100 feet, 1 tunnel 50 feet in length.

Marseilles. Owned by William Baker, of Georgetown; claim 150 by 1,400 feet, discovered 1873, located on Leavenworth Mountain, Griffith mining district, 3 miles from Georgetown; course of vein, northeast and southwest, width 4 feet, pitch north, pay vein 5 inches; nature of ore, sulphurets; average value, 367 ounces silver per ton; 1 tunnel 12 feet in length.

Marshall Silver Mining Co., of Georgetown. Incorporated September 25, 1868; capital stock, $600,000, in 6,000 shares, $100 each; office Georgetown; F.J. Marshall, president, D. Ernest Foster, secretary and treasurer. The property of this company is located on Leavenworth Mountain, 2 miles from Georgetown. It comprises 13,000 linear feet of ground, nearly all of which is patented.

Mary Ann. Lebanon Mining Co., proprietors; Julius G. Pohle superintendent; office Georgetown; claim 50 by 1,500 feet, discovered 1871; located in Cascade mining district, 3 miles from Idaho Springs; vertical vein, running northeast and southwest, width 6 feet, pay vein 6 inches; nature of ore, galena; average value, 50 ounces silver per ton; 1 shaft 40 feet deep.

Mary Jane. Owned by James Peck & Sons; office Empire; claim 150 by 1,500 feet; discovered 1877; located on Eureka Hill, Upper Union mining district, 1 mile from Georgetown; vertical vein, running northeast and southwest, width 5 feet, pay vein 8 inches; nature of ore, iron and copper pyrites; assay value, $200 per cord; main shaft 12 feet deep.

Matilda Fletcher. Owned by Ruben Morseman, Joseph Powers, and William Clark; office Georgetown; claim 50 by 1,500 feet, patented; discovered 1870; located on Democrat Mountain, 3 miles from Georgetown; course of vein, east and west, width 12 feet, pay vein 24 inches, pitch southwest; nature of ore, galena and sulphurets of silver; average value, 180 ounces silver per ton; 1 shaft 100 feet deep, 2 tunnels aggregating 900 feet in length.

May Flower. Owned by George Berry & Co.; office New York; claim 150 by 1,500 feet; discovered 1878; located on Clear Creek, Spanish Bar mining district, 1 mile from Idaho Springs and 42 miles from Denver.

McDonald. George Westman, proprietor; office Lawson; claim 150 by 1,500 feet; discovered 1877; located on Columbia Mountain, Montana mining district, near town of Lawson; course of vein, east and west, width 8 feet, pitch 45 degrees south, pay vein 1 foot; nature of ore, galena and sulphurets; average value, $400 per ton; main shaft 50 feet deep, 1 tunnel 100 feet in length; total yield, $1,600.

McLelland. Owned by William Gorsline and others; office Georgetown; claim 50 by 3,000 feet; discovered 1868; located on Leavenworth Mountain, Griffith mining district, 3 miles from Georgetown; vertical vein, running northeast and southwest, width 5 feet, pay vein 3 feet; nature of ore, galena; average value, 150 ounces silver per ton; main shaft 350 feet deep, 1 level 100 feet in length.

Mendota. Claim 50 by 3,000 feet, patented; discovered 1865; located in Sherman Mountain, Griffith mining district, 2 miles from Georgetown; course of vein, northeast and southwest, width 5 feet, pay vein 8 inches, pitch 10 degrees north; nature of ore, gray copper and galena; average value, $275 per ton; main shaft 110 feet deep, 2 levels aggregate length 300 feet.

Mills Tunneling, Mining, and Ore Reduction Co. Incorporated June 14, 1878; capital stock, $300,000, in 30,000 shares, $10 each; office Georgetown; stock assessable; trustees: Stephen Decatur, William M. Clark, Charles R. Fish, E.H.N. Patterson, Joseph K. Mills, and George E. Kettle.

Minnie. Owned by Frank Downend, Emanuel Hirsch, and M. Fogarty; office Georgetown; claim 150 by 1,500 feet; discovered 1874; located on Republican Mountain, Griffith mining district, 3 miles from Georgetown; course of vein, northeast and southwest, width 18 feet, pay vein 2 feet; nature of ore, galena, gray copper and iron pyrites; average value, 40 ounces silver per ton; 2 shafts aggregate depth 200 feet, 1 tunnel 110 feet in length.

Monticello Silver Mining Co., of Colorado. Incorporated May 27, 1870; capital stock, $1,000,000, in 10,000 shares, $100 each; office Georgetown; organized by D.J. Morrell, H.L. Cake, J.W. Forney, W.W. Wrigley, Simon J. Stine, Edward M. McCook, and A.W. Barnard. [Edward M. McCook was twice the territorial governor of Colorado, serving from

1869 to1873, and again 1874 to 1875. He was born 15 June 1833 and died 9 September 1909.]

Monticello Tunnel. Located near Silver Plume, on Republican Mountain, Griffith mining district, 2 miles from Georgetown; length of tunnel 1,200 feet.

Morning Star. Alexander Lawson, proprietor; office Lawson; claim 150 by 1,500 feet; discovered 1878; located on Red Elephant Mountain, Downieville mining district, 6 miles from Georgetown; course of vein, northeast and southwest, width 7 feet, pitch 45 degrees north; nature of ore, galena and native silver; average value, 500 ounces silver per ton; 1 shaft 40 feet deep.

Morris Mining Co. Incorporated December 26, 1869; capital stock, $600,000, in 6,000 shares, $100 each; office Washington, D.C.

Multum in Parvo. Owned by F. Hood, W.S. Smith, and B. Devotie; office Georgetown; claim 150 by 1,500 feet; discovered 1878; located on Columbia Mountain, Montana mining district, 46 miles from Denver; course of vein, northeast and southwest, width 4 feet; 1 shaft 25 feet deep.

Munn and Loomis Placer. Claim consists of 14 acres of ground, located on Fall River, near town of Fall River, 1 mile from railroad and 40 miles from Denver; owned by Joseph A. Munn and M. Loomis & Co.

Munn and Miller Placer. Joseph A. Munn and Frank Miller, proprietors; office Spanish Bar, claim comprises 36 acres of ground; located on Fall River, near town of Fall River, 1 mile from railroad and 40 miles from Denver.

Munsell. Owned by the Equator Silver Mining and Smelting Co.; H.S. Kearney, superintendent; office Georgetown, Colorado, and Chicago, Illinois; claim 50 by 2,000 feet, patented; discovered 1867; located on Leavenworth Mountain, Griffith mining district, 2 miles from Georgetown; width of vein 8 feet, pay vein 1 to 18 inches; nature of ore, gray copper and galena.

Munsell. Owned by Hal. Sayer; office Georgetown; claim 50 by 3,000 feet, patented; discovered 1867; located on Leavenworth Mountain, Griffith mining district, 1 mile from Georgetown; vertical vein, running northeast and southwest, width 3 feet, pay vein 1 to 12 inches; nature of ore, galena and sulphurets; average value, 225 ounces silver per ton; main shaft 85 feet deep, aggregate length of tunnels 350 feet.

Murphy. G.W. Cowles & Co., proprietors; office Empire; claim 150 by 3,000 feet; discovered 1877; located on Silver Mountain, Upper Union mining district, 2 miles from Georgetown; course of vein, northeast and southwest, width 4 feet, pitch 45 degrees northeast, pay vein 12 to 18 inches; nature of ore, iron and copper pyrites, carbonates and free gold; average value, $400 dollars per ton; 1 shaft 24 feet, aggregate length of tunnels 125 feet; total yield, $1,000.

Murry. Owned by Dubois Tooker & Co.; office Lawson; claim 300 by 1,500 feet; discovered 1878; located in the town of Lawson, on Colorado Central Railroad, 40 miles from Denver; vertical vein, trending northeast and southwest, width 5 feet, pay vein 10 inches; nature of ore, galena and copper pyrites; average value, 90 ounces silver per ton; 2 shafts 90 feet each, 1 level 50 feet in length and 80 feet from surface.

Muscatine. Owned by R.H. Wood & Co., of Georgetown; claim 50 by 1,500 feet; discovered 1868; located on Kelso Mountain, Argentine mining district, 9 miles from Georgetown; course of vein, east and west, width 4 feet, pay vein 18 inches; nature of ore, sulphurets; average value, $500 per ton; 1 adit 12 feet in length.

Nat. Wickliff. Owned by the Lebanon Mining Co.; Julius G. Pohle, superintendent; office Georgetown; claim 50 by 1,600 feet, patented; discovered 1865; located on Red Elephant Mountain, Griffith mining district, 1 mile from Georgetown; course of vein, northeast and southwest, width 5 feet, pay vein 3 inches to 1 foot; nature of ore, galena and iron pyrites; average value, $200 per ton; main shaft 50 feet deep, 1 level 20 feet in length.

Nash Tunnel. Julius G. Pohle, proprietor; office Georgetown; claim 300 by 3,000 feet; discovered 1868; located on Leavenworth Mountain, Griffith mining district, 3 miles from Georgetown. This tunnel is 260 feet in length and has cut 2 lodes, 1 at a depth of 160 feet, and 1 at a depth of 200 feet.

Native American. Owned by R.W. Chinn & Co., of Lawson; claim 150 by 1,500 feet; discovered 1877; located on Columbia Mountain, near town of Lawson, Spanish Bar mining district, 46 miles from Denver; vertical vein, tending northeast and southwest, width 4 feet, pay vein 15 inches; nature of ore, galena and gray copper; average value, $175 per ton; 1 adit 40 feet in length.

Navajoe. Owned by George L. Cannon, of Idaho Springs; claim 150 by 1,500 feet, discovered 1874; located on Russell Mountain, Idaho mining district, 33 miles from Denver; course of vein, east and west, pitch 20 degrees north, width 4 feet, pay vein 4 inches; nature of ore, sulphurets; average value, $130 per ton; main shaft 90 feet deep, 1 tunnel 100 feet and 1 level 15 feet in length.

Neith. G. and S.H. Vivian, proprietors; office Empire; claim 150 by 1,500 feet; discovered 1875; located on Cavode Mountain, Upper Union mining district, half mile from Georgetown; vertical vein, running northeast and southwest, width 5 feet, pay vein 3 feet; nature of ore, copper and iron pyrites; average value, $70 per cord; 1 adit 60 feet in length.

Neurenberg. Owned by T.E. Schwarz, Edwin Bowden, and W.V. Stephens; office Lawson; claim 150 by 1,500 feet; discovered 1878; located on Columbia Mountain, Montana mining district, 2 miles from Georgetown; vertical vein, trending east and west, width 6 feet, pay vein 2 feet; nature of ore, sulphurets; average value, 200 ounces silver per ton; 1 adit 50 feet in length.

New Castle. James Kinkead and Thomas H. Thatch, proprietors; office Fall River; claim 150 by 800 feet; discovered 1860; located on Donaldson Hill, Spanish Bar mining district, near town of Idaho Springs; course of vein, northeast and southwest.

New Era. Owned by the Burleigh Mining Co., of New York; office Georgetown; claim 150 by 3,000 feet; discovered 1871; located on Sherman Mountain, Griffith mining district, 3 miles from Georgetown; course of vein, east and west, pitch 30 degrees north, width 12 feet, pay vein 3 feet; reached by the Burleigh Tunnel [see separate listing] at a distance of 920 feet from the mouth, and intersected at a depth of 535 feet; 3 levels aggregating 700 feet in length.

New Era. Owned by J. Mills, of Milwaukee; claim 150 by 1,500 feet, patented; discovered 1860; located on Trail Creek, 3 miles from Idaho Springs.

New Loan. Owned by the Yandes Silver Mining Company; office Georgetown; claim 50 by 3,000 feet; discovered 1865; located on Sherman

Mountain, Griffith mining district, 2 miles from Georgetown; vertical vein, running northeast and southwest, width 6 feet; nature of ore, galena; average value, $210 per ton; 1 adit 300 feet in length.

Nott Tunnel. Claim 1,500 by 3,000 feet; discovered 1873; located on Leavenworth Mountain, Griffith mining district, 2 miles from Georgetown; length of tunnel 135 feet.

Nuckolls. Owned by Hukill Silver Mining Co.; office Georgetown; claim 150 by 1,500 feet, patented; discovered 1863; located on Columbia Mountain, Griffith mining district, 1 mile from Georgetown; width of vein 3 feet, pay streak 8 inches; average value of ore, 162 ounces silver per ton; main shaft 240 feet deep.

Number Five Mining Co., of Georgetown, Colorado. Incorporated July 13, 1875; capital stock, $50,000, in 5,000 shares, $10 each; offices Georgetown and Denver, Colorado; organized by George Teal, John R. Hambel, Henry C. Harrington, Charles B. Patterson, and Eli M. Ashley.

O.K. Charles R. Fish & Co., proprietors; office Georgetown; claim 50 by 1,500 feet, patented; discovered 1865; located on Leavenworth Mountain, Griffith mining district, 2 miles from Georgetown; course of vein, northeast and southwest, width 3 feet, pitch 45 degrees northwest, pay vein 8 inches; nature of ore, sulphurets and galena, average value, $450 per ton; main shaft 200 feet deep, aggregated length of levels 300 feet, 3 tunnels 60, 300, and 400 feet in length, respectively; total yield, $50,000.

O.R.S. Theodore King, proprietor; office Georgetown; claim 50 by 3,000 feet; discovered 1868; located on Leavenworth Mountain, 3 miles from Georgetown; vertical vein, running east and west, pay vein 1 to 18 inches; nature of ore, carbonate of copper; average value, 150 ounces silver per ton; 1 shaft 40 feet deep.

Old Missouri. Located on Brown Mountain, near Brownville, 4 miles from Georgetown.

Oneida. Owned by the Colorado United Mining Co.; located on Brown Mountain, near town of Brownville, 3 miles from Georgetown.

Pacific. Owned by James Peck & Sons; office Empire; claim 150 by 1,500 feet; discovered 1877; located on Cavode Mountain, Upper Union mining district, 1 mile from railroad; vertical vein, running northeast and southwest, width 5 feet, pay vein 13 inches; nature of ore, iron and copper pyrites; average value, $125 per ton; 1 adit 30 feet in length.

Pacific. Owned by S.H. White, of Idaho Springs; claim 150 by 1,500 feet; discovered 1860; located in dry gulch near Idaho Springs, Spanish Bar mining district, 35 miles from Denver; course of vein, northeast and southwest, width 6 feet, pitch 15 degrees north, pay vein 25 inches; nature of ore, iron and copper pyrites; average value, $35 per ton; 1 shaft 25 feet deep.

Pay Rock. Owned by the Pay Rock Consolidated Mining Co., of Georgetown, Colorado; incorporated January 12, 1875; capital stock, $350,000, in 3,500 shares, $100 each; organized by George E. Noyes, Charles H. Morris, Thomas W. Ellis, Peter Ellis, and William C. Snow; George E. Noyes, president, Thomas W. Ellis, vice president and superintendent, Charles H. Morris, secretary and treasurer; claim 50 by 1,500 feet, patented; discovered 1871; located on Republican Mountain, Griffith mining district, near town of Silver Plume, 2 miles from Georgetown; course of vein, northeast and southwest, width from 4 to 150 feet, pay vein from 1 to 24 inches, pitch 15 degrees south; value of ore, 105 to 715 ounces silver per ton; 3 shafts aggregate depth 700 feet, aggregate length of levels 2,500 feet.

Pelican. Located on Sherman Mountain, near town of Silver Plume, 3 miles from Georgetown. This property is well developed and has produced to date $1,500,000.

Peralto. Owned by the Hermann Mining Co., New York; John D. Dix, president, Benjamin F. Morris, vice president, George W. Dix, secretary, Charles H. Morris, superintendent; capital stock, $200,000, none of which has been sold for less than par value; offices Georgetown, Colorado, and New York City; claim 150 by 1,500 feet, patented; discovered 1872; located on Republican Mountain, near Silver Plume, 3 miles from Georgetown; course of vein, east and west, pitch 30 degrees north, width 15 feet, pay vein 1 foot; 2 shafts aggregate depth 50 feet, 1 tunnel, 110 feet in length.

Philadelphia Tunnel Co., of Colorado. Incorporated October 13, 1869; capital stock, $500,000, in 5,000 shares, $100 each; offices, Georgetown, Colorado, and Philadelphia, Pennsylvania; trustees: John R. Hall, Ambrose W. Bernard, and James Henshall.

Phillips. Owned by E.T. Carr, Charles McKee, and Richard Phillips; office Central City; claim 150 by 1,500 feet; discovered in 1874; located on Belleview Mountain, Virginia mining district, 3 miles from Central City; width of vein 20 inches; average value of mill ore 5 ounces gold per cord, smelting ore, $91 per ton; 1 shaft 60 feet deep; 2 levels, aggregate length 60 feet.

Phoenix. Owned by James Kinkead and Thomas H. Thatch; office Fall River; claim 50 by 1,100 feet; discovered 1862; located on Donaldson Mountain, Spanish Bar mining district, near town of Fall River, 40 miles from Denver.

Phoenix. Owned by Zadock Kalbaugh, Henry Crow, and Erskine McClelland; office Georgetown; claim 50 by 3,000 feet, located on Sherman Mountain, 3 miles from Georgetown; width of vein 6 feet, pay streak 4 inches; average value of ore, 1,000 ounces silver per ton; 8 shafts aggregating 400 feet in depth, 6 levels aggregating 750 feet in length.

Pilot Knob. Lebanon Mining Co., proprietors; Julius G. Pohle, superintendent; office Georgetown; claim 50 by 1,500 feet; discovered 1871; located on Pilot Knob Mountain, Trail Creek mining district, 3 miles from Idaho Springs; course of vein; northeast and southwest, width 6 feet, pitch north, pay vein 5 inches; nature of ore, galena; average value, 300 ounces silver per ton; 1 shaft 40 feet deep.

Pioneer. Francis L. Andre, proprietor; office Empire; claim 150 by 1,500 feet, patented; discovered 1862; located on Silver Mountain, Upper Union mining district, 2 miles from railroad; course of vein, northeast and southwest, width 4 feet, pitch 10 degrees east, pay vein 6 inches to 3 feet; nature of ore, iron pyrites and free gold; average value, $65 per cord; main shaft 100 feet deep, 1 level 250 feet in length and 70 feet from surface, 1 tunnel 150 feet in length.

Pittsburgh. Samuel P. Allen and David J. Ball, proprietors; office Empire; claim 50 by 1,300 feet, patented; discovered 1865; located on Silver Mountain, Upper Union mining district, 2 miles from railroad; course of vein, northeast and southwest, width, 10 feet, pitch 15 degrees; nature of ore, iron and copper pyrites; average value, $125 per cord; main shaft 60 feet deep, aggregate length of levels 150 feet, 1 tunnel 500 feet I length; total yield, $20,000.

Planet. Owned by the Planet Silver Mining Co.; office Georgetown; claim 1,050 linear feet; located at head of Geneva Gulch, Argentine mining district, 15 miles from Georgetown; course of vein, north and south, width 3 feet, pitch west, pay vein, 1 to 14 inches; nature of ore, bismuth, silver, galena, iron and copper pyrites; average value, $100 per ton; 1 adit 140 feet in length.

Pocahontas. Owned by R. Woods & Co., of Georgetown; claim 150 by 1,500 feet; discovered 1868; located on Kelso Mountain, West Argentine mining district, 9 miles from Georgetown; course of vein, east and west, width 5 feet, pay vein 12 inches; nature of ore, galena; average value, $500 per ton; 2 adits aggregate length 285 feet.

Polar Star. Owned by the Polar Star Gold and Silver Mining Co., New York; incorporated April 1877, capital stock, $500,000, in 50,000 shares, $10 each; offices Denver, Colorado, and New York City; O.H. King, president, W.C. Boone, secretary, H.C. Donnell, superintendent; claim 50 by 1,500 feet, patented; discovered 1866; located on Leavenworth Mountain, Griffith mining district, 2 miles from Georgetown; course of vein, northeast and southwest, width 10 to 20 feet, pay vein 1 to 18 inches, pitch north; nature of ore, galena and zinc blende; average value, 200 ounces silver per ton; 3 shafts aggregate depth 300 feet, 2 tunnels aggregate length 1,400 feet, 3 levels aggregate length 275 feet.

Pride of the West. Owned by the Hukill Silver Mining Co.; office Georgetown; claim 50 by 1,500 feet, patented; discovered 1865; located on Sherman Mountain, Griffith mining district, 2 miles from Georgetown; width of vein 5 feet, pay vein 10 inches; value of ore, 260 ounces silver per ton; main shaft 75 feet deep.

Princess. Owned by William P. Lynn, of Georgetown; claim 50 by 1,400 feet; discovered 1871; located on Leavenworth Mountain, Griffith mining district, 2 miles from Georgetown; course of vein, northeast and southwest, width 4 feet, pay vein 8 inches, pitch north; nature of ore, galena; 1 shaft 50 feet deep.

Providence. Owned by David N. Smith, of Georgetown; claim 100 by 1,500 feet, patented; discovered 1868; located on Democrat Mountain, Griffith mining district, 1 mile from Georgetown; course of vein, northeast and southwest, width 4 feet, pay vein 2 inches to 2 feet; nature of ore, galena and sulphurets; main shaft 60 feet deep, 1 tunnel 300 feet in length, which intersects the vein at a depth of 125 feet.

Providence, Extension. Owned by H. Marcus Bronson, Andrew Wold, and E. Harrison; office Georgetown; claim 150 by 1,500 feet; discovered 1877; located on Democrat Mountain, Griffith mining district, 3 miles from Georgetown; course of vein, east and west, width 10 feet, pay vein 4 inches, pitch 5 degrees north; nature of ore, sulphurets; average value, $560 per ton, 1 shaft, 25 feet deep, 1 tunnel 50 feet in length.

Pulaski. Owned by the Pulaski Silver Mining Co.; office Washington Court House, Ohio; claim 150 by 2,100 feet, patented; discovered 1874; located in Griffith mining district, 2 miles from Georgetown; vertical vein, running east and west, width 20 feet; nature of ore, galena and free gold; average value, $25 per ton; 2 tunnels 100 and 200 feet in length respectively.

Purdue Silver and Gold Mining and Ore Reduction Co. Organized under the laws of the State of Indiana; capital stock, $85,000, in 17,000 shares, $50 each.

Puzzler. Owned by Alexander Lawson and A.S. Carpenter, of Lawson; claim 150 by 1,500 feet, patented; discovered 1877; located on Red Elephant Mountain, near town of Lawson, Downieville mining district, 45 miles from Denver; course of vein, northwest and southeast, width 10 feet, pitch 45 degrees north, pay vein 6 inches; nature of ore, galena and copper pyrites; average value, $200 per ton; main shaft 120 feet deep.

Queen of the West. Owned by the Burlington and Colorado Silver Mining Co.; located on Democrat Mountain; yield to date, $150,000.

R. A. Miner. Owned by J.F. Snodgrass, T. Warwick & Co.; office Georgetown; claim 150 by 2,000 feet, patented; discovered 1865; located on Sherman Mountain, Griffith mining district, 2 miles from Georgetown; vertical vein, running northeast and southwest, width 7 feet, pay vein 4 inches; nature of ore, galena; average value, $150 per ton; main shaft 30 feet deep, 1 tunnel 150 feet in length.

Rackett. John Tubb, proprietor; office Central City; claim 150 by 850 feet; discovered 1864; located on Belleview Mountain, Virginia mining district, 3 miles from Central City; width of vein 15 inches; average value of ore, 8 ounces gold per cord; 2 shafts aggregating 150 feet in depth, 1 level 50 feet in length and 70 feet from surface.

Revenue. Owned by the Revenue Mineral Co.; office Grant, Park County, Colorado; claim 150 by 1,500 feet, patented; discovered 1872; located on Revenue Mountain, Geneva mining district, 15 miles from Georgetown; vertical vein, trending northeast and southwest, width 5 feet, pay vein 15 inches; nature of ore, galena and gray copper; 2 shafts 300 feet each; 2 levels 250 feet in length each; stone shaft house 18 by 20 feet.

Richmond. Owned by John Turck; located on McClelland Mountain, near Georgetown; developed by a tunnel 250 feet in length, and 2 shafts 20 and 50 feet in depth, respectively; the pay vein is 7 inches wide, and the average value of the ore is 150 ounces silver per ton.

Rider. Owned by the Burleigh Mining Co., of New York; claim 150 by 3,000 feet; discovered 1873; located on Sherman Mountain, Griffith mining district, 3 miles from Georgetown; course of vein, east and west, pitch north, width 4 feet, pay vein 1 to 12 inches; nature of ore, galena and gray copper.

Rip Van Winkle, West. Owned by the West Rip Van Winkle Silver Mining Co.; incorporated October 29, 1878; capital stock, $150,000, in 15,000 shares, $10 each; office Denver; trustees: Carver J. Goss, William B. Slosson, and Rodney Curtis; claim 150 by 1,500 feet, discovered 1869; located on Brown Mountain, Griffith mining district, 3 miles from Georgetown; course of vein, northeast and southwest, width 5 feet; nature of ore, sulphurets; average value, 500 ounces silver pr ton; 2 shafts, 50 and 35 feet deep, respectively, 1 level 50 feet in length.

Roberson. John W. Edwards, proprietor; office Idaho Springs; claim 50 by 1,500 feet; discovered 1865; located on Roberson Hill, Idaho mining district, 2 miles from town of Idaho Springs; vertical vein, running east and west, width 4 feet, pay vein 12 inches; nature of ore, free gold and iron pyrites; average value, $67 per cord; main shaft 170 feet deep, aggregate depth of other shafts 70 feet, aggregate length of levels 239 feet, 1 adit 200 feet in length; total yield, $15,000.

Robert E. Lee. Lebanon Mining Co., proprietors, Julius G. Pohle, superintendent; office Georgetown; claim 150 by 1,500 feet; discovered 1872; located north side of Leavenworth Mountain, Griffith mining district, 1 mile from Georgetown; vertical vein, running northeast and southwest, width 4 feet, pay vein 4 to 5 inches; nature of ore. Sulphurets and gray copper; average value, 800 ounces silver per ton; 1 shaft 50 feet deep, 1 level 50 feet from surface and 40 feet in length.

Robinson Placer. Charles E. Robinson, proprietor; claim comprises 5 acres of patented ground; located on Clear Creek, in town of Fall River, Spanish Bar mining district, 45 miles from Denver.

Rockford. Clase [Chas.?] J. Johnson, August Nelson, Nils Frohm, and August Peterson, proprietors; office Georgetown; claim 150 by 1,500 feet; discovered 1876; located on Democrat Mountain, Griffith mining district, 3 miles from Georgetown; course of vein, east and west, width 12 feet, pitch south; nature of ore, galena and gray copper; 1 shaft, 25 feet deep, reached by a tunnel at a distance of 360 feet from the mouth, and intersected at a depth of 160 feet.

Roe, East. Owned by the Consolidated Hercules and Roe Silver Mining Co.; G.W.E. Griffith, president, H.M. Griffin, treasurer; office Denver; claim 50 by 1,500 feet, patented; discovered 1867; located on Brown Mountain, Griffith mining district, 3 miles from Georgetown; course of vein, northeast and southwest, width, 12 feet, pitch 5 degrees north, pay vein 1 to 10 inches; nature of ore, galena and gray copper; average value, 450 ounces silver per ton; aggregate depth of shafts 700 feet, aggregate length of levels 2,497 feet, 1 tunnel 240 feet in length.

Rogers, Fred. Owned by the Lincoln Mining Co.; W.A. Burr and Duncan McArthur, agents; office Georgetown; claim 50 by 1,500 feet; discovered 1869; located on Democrat Mountain, Upper Union mining district, 2 miles from Georgetown; nature of ore, gray copper and silver glance; value of first-class ore, 500 to 700 ounces, second-class ore, 200 to 350 ounces, and third-class, 150 to 200 ounces silver per ton; main shaft 200 feet deep; 12-horse power engine and a No. 3 Knowles pump; yield to date 175,000 ounces.

Rosencrans. Z.H. and Ann Willard, proprietors; office Boston, Massachusetts; claim 75 by 600 feet, patented; discovered 1862; located on Silver Mountain, Upper Union mining district, 2 miles from railroad; course of vein, east and west, pitch 10 degrees south, width 7 feet; nature of ore, free gold; average value, $140 per cord; main shaft 90 feet deep; total yield, $14,000.

Ruby Silver Mining Co. Incorporated October 17, 1874; capital stock, $400,000, in 4,000 shares, $100 each; offices Georgetown,

Colorado, and Chicago, Illinois; organized by John A. Logan, George Townsend, Andrew Meyers, Solomon Robeson, and David H. Mitchell.

Saco. Charles R. Fish, proprietor; office Georgetown; claim 50 by 900 feet, patented; discovered 1865; located on Leavenworth Mountain, Griffith mining district, 2 miles from Georgetown; vertical vein, trending northeast and southwest, width 3 feet, pay vein 10 inches; nature of ore, galena and gray copper; average value, $300 per ton; aggregate depth of shafts 380 feet, aggregate length of tunnels 530 feet; total yield, $250,000.

Salisbury. Owned by John M. Dumont, of Spanish Bar; claim 150 by 800 feet; discovered 1860; located on Clear Creek, Spanish Bar mining district, 2 miles from Idaho Springs; course of vein, east and west, pitch 45 degrees north, width 4 feet, pay vein 12 inches; nature of ore, iron and copper pyrites; average value, $60 per ton.

Sallie Ward. Owned by the Lebanon Mining Co.; Julius G. Pohle, superintendent; office Georgetown; claim 50 by 1,600 feet, patented; discovered 1865; located on Republican Mountain, Griffith mining district, 1 mile from Georgetown; width of vein 4 feet, pay streak 2 feet; average value of ore, 800 ounces silver per ton; main shaft 160 feet deep, 1 level 100 feet in length and 160 feet from surface; reached by the Lebanon Tunnel at a distance of 320 feet from the mouth and intersected at a depth of 160 feet.

Samuel J. Tilden. Owned by the Marshall Silver Mining Co.; F.J. Marshall, president and superintendent, W.C. Williams, secretary, G.W. Hall, treasurer; office Georgetown; claim 50 by 1,500 feet, patented; discovered 1876; located on Leavenworth Mountain, Griffith mining district, 2 miles from Georgetown; width of vein 15 feet, pay streak 10 inches; average value of ore, 600 ounces silver per ton; main shaft 150 deep, 3 levels, aggregate length 400 feet; reached by the O,K. Tunnel, 200 feet from the mouth, and intersected at a depth of 110 feet.

Schaffter. Owned by Peter, Theobald, and P.P. Schaffter; office Idaho Springs; claim 150 by 1,500 feet; discovered 1860; located on Schaffter Mountain, Spanish Bar mining district, 3 miles from town of Idaho Springs; vertical vein, trending northeast and southwest, width 4 feet, pay vein 18 inches; nature of ore, free gold; average value, $40 per ton; main shaft 100 feet and 1 other shaft 33 feet deep; total yield, $22,000.

Seaton. Owned by Willard Teller and Clinton Reed; office Denver; claim 50 by 100 feet, patented; located on Seaton Mountain, Idaho mining district, 1 mile from town of Idaho Springs; course of vein, northeast and southwest, pay vein 2 feet; nature of ore, galena; main shaft 170 feet deep, 1 tunnel 100 feet in length; shaft house 18 by 20 feet.

Selma. Lebanon Mining Co., proprietors; Julius G. Pohle, superintendent; office Georgetown; claim 150 by 1,500 feet; discovered 1872; located on north side of Leavenworth Mountain, Griffith mining district, 1 mile from Georgetown; vertical vein, trending northeast and southwest; average width 4 feet, pay vein 3 to 4 inches; nature of ore, galena; average value, 400 ounces silver per ton; 1 shaft 40 feet deep, 1 level 60 feet in length.

Seven-Thirty. Owned by the Consolidated Hercules and Roe Silver Mining Co.; capital stock, $1,000,000, 100,000 shares, $10 each; offices Georgetown and Denver, Colorado; G.W.E. Griffith, president, H.M. Griffin, treasurer and superintendent, C.M. Williams, secretary, K.G. Cooper, vice president; claim 50 by 3,000 feet, patented; discovered 1866; located on Brown Mountain, Griffith and Queen mining districts, 3 miles from Georgetown; course of vein, northeast and southwest, width 4 to 20 feet, pay vein 1 to 20 inches, pitch north; nature of ore, galena, gray copper, and zinc blende; average value, 300 ounces silver per ton; 2 shafts, aggregate depth 400 feet, 1 tunnel 400 feet in length, 4 levels [sic] 100, 200, and 250 feet from the surface, respectively, aggregating 600 feet in length.

Seven-Twenty. B.F. Wells, of Black Hawk, proprietor; claim 50 by 850 feet; discovered 1872; located on Seaton Mountain, Idaho mining district, 1 mile from Idaho Springs; course of vein, northeast and southwest, width 4 feet, pitch 10 degrees north, pay vein 6 inches to 2 feet; nature of ore, galena and gray copper; average value, $150 per ton; main shaft 250 feet, and 1 other shaft 100 feet deep; will be operated through the Seven-Twenty Tunnel, which is now 100 feet in length; hoisting apparatus and a 40-horse power engine.

Seventy-Seven. Owned by William H. Moore & Co.; office, Georgetown; claim 50 by 924 feet; discovered 1872; located on Leavenworth Mountain, Griffith mining district, 2 miles from Georgetown; course of vein, north and south, width 12 to 15 feet, pitch 33 degrees west; nature of ore, sulphurets; value, 1,300 ounces silver per ton; 2 shafts, aggregate depth 145 feet, 2 tunnels aggregate length 280 feet, 1 level 120 feet in length.

Shively. Owned by David N. Smith, of Georgetown; claim 150 by 1,500 feet, patented; discovered 1872; located on Brown Mountain, Queen mining district, near Brownville 2 miles from Georgetown course of vein, northeast and southwest, width 4 feet, pitch north, pay vein 10 inches; nature of ore, galena; average value, 500 ounces silver per ton; main shaft 75 feet deep, 1 tunnel 279 feet in length, 2 levels aggregating 455 feet in length.

Shoo Fly. Lebanon Mining Co., proprietors; Julius G. Pohle, superintendent; office Georgetown; claim 50 by 1,500 feet; discovered 1871; located on Pilot Knob Mountain, Trail Creek mining district, 3 miles from Idaho Springs; vertical vein, running northeast and southwest, width 6 feet, pay vein 1 to 3 inches; nature of ore, galena; average value, 80 ounces silver per ton; 1 shaft 50 feet deep.

Silver Belt. Charles C. Welch, David C. Crawford, James F. Devere, and R.W. Chinn, proprietors; office Denver; claim 150 by 1,500 feet; discovered 1877; located on Columbia Mountain, Montana mining district, 1 mile from town of Lawson and 46 miles from Denver.

Silver Bluff. Lebanon Mining Co., proprietors; Julius G. Pohle, superintendent; office Georgetown; claim 50 by 1,500 feet, patented; discovered 1870; located in Cascade mining district, 3 miles from Idaho Springs; course of vein northeast and southwest, width 10 feet, pitch north; nature of ore, galena; average value, 200 ounces silver per ton; 1 shaft, 40 feet deep, 2 tunnels aggregate length 100 feet.

Silver Cloud. Owned by J. Rogers, John Shields, and W.C. Hicock; office Georgetown; claim 150 by 1,500 feet, patented; discovered 1866; located on Democrat Mountain, 3 miles from Georgetown; course of vein, northeast and southwest, width 7 feet, pitch 5 degrees north, pay vein 5 inches; nature of ore, sulphurets and zinc blende; average value, 104 ounces silver per ton; main shaft 20 feet deep, 1 tunnel 25 feet in length.

Silver Cloud. Silver Cloud Mining Co., proprietors; office Georgetown; claim 50 by 3,000 feet, patented; discovered 1868; located on Hanna Mountain, Queen mining district, 3 miles from Georgetown; vertical vein, running northeast and southwest, width 4 feet, pay vein 4 inches; nature of ore, galena and gray copper; main shaft 70 feet deep, 2 tunnels 75 feet in length, respectively.

Silver Cloud. Owned by the Victor Silver Mining Co.; office Georgetown; claim 150 by 1,500 feet, patented; discovered 1874; located on Democrat Mountain, Upper Union mining district, 2 miles from Georgetown; course of vein, northeast and southwest, width 15 feet, pitch 25 degrees south, pay vein 10 inches; nature of ore, sulphurets; average value, $1,000 per ton; main shaft 90 feet deep, 3 tunnels 75, 100, and 290 feet in length, respectively; total yield, $40,000.

Silver Dale. Owned by Frank Bunholzer, Herman Housen, R.A. Ames, and Mrs. Phoebe Gilmore [one of the few Colorado women to own a mining interest]; office Georgetown; claim 1,500 by 3,000 feet, patented; discovered 1873; located on Leavenworth Mountain, Griffith mining district, in town of Silver Dale, 3 miles from Georgetown; length of tunnel 400 feet; has cut 4 lodes, all of which show good ore, averaging 1,000 ounces silver per ton.

Silver Glance. James Oscar Stewart, proprietor; office Georgetown; claim 150 by 500 feet, patented; discovered 1874; located on Democrat Mountain, Griffith mining district, 3 miles from Georgetown; width of vein 20 feet, pay vein 18 inches, average value of ore, 300 ounces silver per ton; 1 shaft 40 feet deep, aggregate length of tunnels 420 feet, aggregate length of levels 400 feet.

Silver Ore. Located on Brown Mountain, near Brownville, 4 miles from Georgetown.

Silver Glance Mining Co. Incorporated April 7, 1876; capital stock, $150,000, in 1,500 shares, $10 each; office Georgetown; organized by James Oscar Stewart, Charles W. Burdsal, A.J. August, John R. Hambel, and Nathan S. Hurd.

Silver King. Owned by Charles F. Hendrie and Henry Bolthoff; office Central City; claim 150 by 1,500 feet; discovered 1876; located in Spring Gulch, Independent mining district, 10 miles from Georgetown; course of vein, northeast and southwest, width 4 feet, pitch 25 degrees southeast; nature of ore, galena and gray copper; average value, $40 per ton; main shaft 40 feet deep, 1 tunnel 60 feet, and 1 drift 100 feet in length.

Silver Mountain. Francis F. Andre, proprietor; office Empire; Claim 50 by 600 feet; discovered 1862; located on Silver Mountain, Upper Union mining district, 2 miles from railroad; course of vein, northeast and southwest, width 4 feet, pitch west, pay vein 2 feet; nature of ore, iron and copper pyrites; average value, $40 per cord; main shaft 135 feet deep, 1 tunnel 600 feet in length, which intersects the vein at a depth of 135 feet; total yield, $150,000.

Silver Mountain. Owned by James Peck & Sons; office Empire; claim 150 by 40 feet; patented; discovered 1861; located on Silver Mountain, Upper Union mining district, 2 miles from railroad; vertical vein, running northeast and southwest, width 6 feet, pay vein 2 feet; nature of ore, iron and copper pyrites; average value, $90 per ton; main shaft 65 feet deep.

Silver Plume. Silver Plume Mining Co., proprietors; George Teal, agent; office Georgetown; claim 50 by 1,400 feet, patented; discovered 1863; located near town of Silver Plume, on Republican Mountain, 2 miles from Georgetown; course of vein, northeast and southwest, width 12 feet, pitch 20 degrees north, pay vein 1 to 18 inches; nature of ore, galena and sulphurets; average value, 600 ounces silver per ton; main shaft 100 feet deep, 5 tunnels aggregate length 1,800 feet.

Silver Queen Mining Co. Incorporated October 31, 1872; capital stock, $5,000, in 50 shares $100 each; office Georgetown; organized by Irving J. Pollock, R.W. Steele, Augustus H. Whitehead, Albert Townsend, J.B. O'Connor, Theodore King, Henry Boyer, John B. Snodgrass, Glaize, Bender and Co., and H. Clinton Cowles.

Snow Drift. Owned by the Snow Drift Silver Mining and Reduction Co., England; George Teal, agent; office Georgetown; claim 50 by 1,400 feet, patented; discovered 1863; located on Republican Mountain, Griffith mining district, near Silver Plume, 2 miles from Georgetown; course of vein, northeast and southwest, width 4 feet, pitch 70 degrees north, pay vein 9 inches; nature of ore, galena and sulphurets; average value, 1,000 ounces per ton; main shaft 550 feet deep, 1 tunnel 50 feet in length, which intersects the vein at a depth of 280 feet; 6 levels aggregated length 600 feet; total yield, $140,000.

Specie Payment. Owned by the Sunshine Mining Co., New York; John Newman, president and general manager, Thomas Farrell, vice president, George H. Sagendorf, treasurer, W.H. Poor, secretary; offices

Idaho Springs, Colorado, and Troy, New York; claim 150 by 1,500 feet, patented; discovered 1872; located on Belleview Mountain, Virginia mining district, 3 miles from Idaho Springs and 3 miles from Central City; course of vein, northwest and southeast, width 4 feet, pitch 30 degrees north, pay vein 12 inches; nature of ore, iron and copper pyrites; average value of mill ore, 8 ounces gold per cord, smelting ore, $175 per ton; main shaft 100 feet deep and 2 other shafts aggregating 180 feet in depth, 1 adit 360 feet and 1 level 65 feet in length; engine house and office, 20 by 30 feet; has yielded under the present management $40,000.

Specie Payment. Owned by N.J. Moss, W.A. Doyle, and J.J. Smith; office Lawson; claim 150 by 1,500 feet; discovered 1878; located on Columbia Mountain, Montana mining district, 2 miles from Lawson and 6 miles from Georgetown; width of vein, 6 feet; nature of ore, galena, iron and copper pyrites; 1 adit 25 feet in length.

Specie Tunnel. Located on the south side of Brown Mountain, near Brownville, four miles from Georgetown.

Steamboat. Owned by the Broadway Tunneling and Silver Mining Co.; office Georgetown; claim 150 by 1,500 feet; discovered 1873; located on Leavenworth Mountain, 2 miles from Georgetown; course of vein, northeast and southwest, width 4 feet, pay vein 6 inches; nature of ore, galena and gray copper; average value, 300 ounces silver per ton; main shaft 100 feet deep, 2 tunnels aggregating 500 feet in length, and 4 levels aggregating 410 feet in length.

Steamboat, West Extension. Owned by W.H. Moore, Nathan S. Hurd, A.L. Palmer, and A.D. Bullis; claim 150 by 1,426 feet, patented; discovered 1873; located on Leavenworth Mountain, 2 miles from Georgetown; vertical vein, trending northeast and southwest, width 4 feet, pay vein 5 inches; nature of ore, galena and gray copper; average value, 210 ounces silver per ton; main shaft 64 feet deep, 1 tunnel 165 feet in length, 2 levels aggregate length 165 feet.

Stephens' Placer. Owned by Andrew Stephens and others; office Idaho Springs; claim comprises 36 acres of patented ground located on Spanish Bar, 2 miles from town of Idaho Springs, and 36 miles from Denver.

Stevens. Owned by the Stevens Mining Co.; located on McClelland Mountain, near Georgetown; this property is well developed and has produced largely; the pay vein is 6 to 10 inches in width, and the average value of the ore is 200 ounces silver per ton.

Sunny South. H.C. Cowles, proprietor; office Empire; claim 150 by 1,500 feet; discovered 1876; located on Silver Mountain, Upper Union mining district, 2 miles from railroad; vertical vein, running east and west, width 3 feet, pay vein 6 inches; nature of ore, free gold; average value, $100 per cord; main shaft 25 feet deep, 1 adit 35 feet in length.

Sunshine Mining Co. Organized under the laws of the State of New York, July 15, 1875; capital stock, $50,000, in 500 shares, $100 each; offices Idaho Springs, Colorado, and Troy, New York; John Newman, president and agent, George H. Sagendorf, treasurer, W. H. Poor, secretary.

Taylor. Owned by W.W. Taylor, T.J. Huff, and W. Collins; office Idaho Springs; claim 150 by 1,500 feet; discovered 1878; located in Hukill Gulch, Spanish Bar mining district, 2 miles from Idaho Springs; vertical vein, running northeast and southwest, width 4 feet, pay vein, 5 inches; nature of ore, galena and iron pyrites; average value, $42 per ton; main shaft 25 feet deep.

Taylor Tunnel. Owned by E.C. McMillan and Frank L. Downend; office Georgetown; claim 1,500 by 3,000 feet; discovered 1877; located on Republican Mountain, near town of Silver Plume, 2 ½ miles from Georgetown; length of tunnel 100 feet; designed to intersect the Loretta Lode.

Tenth Legion. Owned by the Knickerbocker Mining Co.; office New York City; claim 50 by 400 feet, patented; located on Silver Mountain, Upper Union mining district, 2 miles from railroad; course of vein, northeast and southwest, width 3 to 30 feet, pay vein 14 inches, pitch 10 degrees north; nature of ore, iron and copper pyrites and carbonate of copper; average value, 100 ounces silver per ton; main shaft 265 feet deep, aggregate length of levels, 215 feet, being 90, 180, and 230 feet from surface, respectively; shaft house 30 by 60 feet, containing a 12-horse power engine; yield to date, $120,000.

Terrible. Located on Brown Mountain, near town of Brownville, 3 miles from Georgetown; course of vein, north 77 degrees east, pitch 16 degrees north, pay vein 9 inches; nature of ore, galena, zinc blende, and copper pyrites; has produced largely.

Tom Corwin. Owned by James Rogers, John Shields, and W.C. Hicock; office Georgetown; claim 50 by 1,500 feet, patented; discovered 1866; located on Democrat Mountain, Griffith mining district, 2 miles from Georgetown; course of vein, northeast and southwest, width 12 feet, pitch north, pay vein 14 inches; nature of ore, carbonate of lead; average value, $120 per ton; aggregate depth of shafts, 45 feet.

Trailor. Owned by Alexander Lawson; office Lawson; claim 150 by 1,500 feet; discovered 1878; located on Red Elephant Mountain, Downieville mining district, near town of Lawson, 45 miles from Denver.

Trailor. Owned by W.W. Trailor, T.J. Huff, and W. Collins; claim 150 by 1,500 feet; discovered 1878; located on Hukill Gulch, Spanish Bar mining district, 2 miles from Idaho Springs; course of vein, northeast and southwest, width 4 feet, pay vein 5 inches; nature of ore, galena, iron and copper pyrites, average value, $50 per ton; 1 shaft 40 feet deep.

Treasure. Owned by Nelson and John Tallman and K.P. Grove, of Georgetown; claim 150 by 1,500 feet; discovered 1877; located on Democrat Mountain, Griffith mining district, 2 miles from Georgetown; course of vein, southwest and northeast, width 20 feet, pay vein 20 inches, pitch 5 degrees south; nature of ore, sulphurets, average value, $60 per ton; 1 tunnel 75 feet in length.

Treasure. Owned by William Baker and Dr. Vandervort, of Georgetown; claim 150 by 1,600 feet; discovered 1872; located on Leavenworth Mountain, Griffith mining district, 3 miles from Georgetown; course of vein, northeast and southwest, width 55 feet, pitch 10 degrees north, pay vein 8 inches to 3 feet; nature of ore, gray copper and galena; average value, 377 ounces silver per ton; aggregate length of tunnels 300 feet.

Trenton Mining Co. Incorporated September 26, 1872; capital stock, $25,000, in 5,000 shares, $5 each; offices Idaho Springs, Colorado, and Trenton, New Jersey; organized by John Collom, Daniel Peters, and Charles Collom.

Trevellian Placer. Owned by James Trevellian & Co.; office Spanish Bar; claim comprises 3 acres of ground; located on Seser's Bar, near town of Idaho Springs, 41 miles from Denver.

Trio. William and Henry Helmer, I.N. Henry, and R.L. Rohm proprietors; office Central City; claim 150 by 950 feet; discovered 1875; located on the eastern slope of Belleview Mountain, Idaho mining district, 3 miles from Central City; course of vein, northeast and southwest, pitch 45 degrees north, width 3 feet, pay vein 8 inches to 1 foot; nature of ore, iron and copper pyrites, average value of mill ore, 18 ounces gold per cord, smelting ore, $175 per ton; main shaft 100 feet deep, 1 drift 100 feet in length.

Tropic. Owned by Turnbull & Co.; claim 150 by 1,500 feet, patented; discovered 1869; located on Columbia Mountain, Montana mining district, 2 miles from Lawson and 6 miles from Georgetown; vertical vein, trending east and west, width 7 feet, pay vein 12 inches; nature of ore, galena; average value, $110 per ton; 1 adit 180 feet in length.

Tunnel. Owned by Fletcher Kelso; located on Kelso Mountain, 8 miles from Georgetown; course of vein, east and west, pay vein 5 inches; nature of ore, galena and porphyry; value, 100 ounces silver per ton; developed by a shaft 100 feet deep and a tunnel, 40 feet in length.

Tycoon. Located on Brown Mountain, near Brownville, 4 miles from Georgetown.

Union Tunnel. Located on Brown Mountain, near Brownville, 4 miles from Georgetown.

United Pelican and Dives Mining Company of Colorado. Incorporated December 20, 1877; capital stock, $1,500,000, in 150,000 shares, $10 each; office Georgetown; trustees: Bela M. Hughes, Eli S. Streeter, Thomas McCunniff, Benjamin F. Napheys, Edward Y. Naylor, William A. Hamill, Frank M. Taylor, George G. Symes, and John L. Routt. [John L. Routt was born April 25, 1826 in Kentucky and died August 13, 1907. He owned the Morning Star silver mine in Leadville -- see listing in Lake County -- and worked as a miner while serving as governor. One of the territorial leaders to successfully pursue statehood for Colorado, Routt served three terms, once as the last territorial governor, once as the first governor of the State of Colorado and again as the seventh governor of the state.]

U.S. Coin. Located on Brown Mountain, near Brownville, 4 miles from Georgetown.

Vetra Madre Mining Co. Incorporated June 14, 1877; capital stock, $500,000, in 20,000 shares, $25 each; office Denver; trustees: James F. Welborn, John Stuart, Daniel Sayer, Jacob F. Cypher, and Milton Mills.

Washington. Located on Brown Mountain, near Brownville, 4 miles from Georgetown.

Wooster. Owned by William Mendenhall and others; located on Kelso Mountain, 8 miles from Georgetown; developed by a shaft 50 feet deep, a tunnel 175 feet, and 1 level 55 feet in length; value of ore, 200 ounces silver per ton.

Victor. Owned by D.S. Bailey, S.W. Nott, and Cushman Estate; office Georgetown; claim 50 by 1,500 feet; discovered 1866; located on Griffith Mountain, Griffith mining district, 1 mile from Georgetown; vertical vein, running northwest and southeast, width 2 feet; nature of ore, galena, zinc blende, and silver glance; average value, $350 per ton; main shaft 20 feet deep, 1 tunnel 50 feet in length.

Victor. Owned by E.T. Carr and William Warwick; office Central City; claim 100 by 1,000 feet; discovered 1873; located on Belleview Mountain, Virginia mining district, 3 miles from Central City; width of vein, 15 inches, average value of ore, 6 ounces gold per cord; main shaft 70 feet deep.

Victor. Victor Silver Mining Co., proprietors; Thomas J. Oyler, president, Ed. A. Seiwell, secretary, John Needham, superintendent; office Black Hawk; claim 150 by 1,400 feet, patented; discovered 1863; located on Seaton Hill, Idaho mining district, near town of Idaho Springs, 2 miles from railroad; course of vein, northeast and southwest.

Victor Silver Mining Co. Incorporated April 7, 1876; capital stock, $150,000, in 1,500 shares, $100 each; offices Georgetown, Colorado, and Indianapolis, Indiana; organized by Benjamin F. Claypool, Richard J. Bright, Edward F. Claypool, Morton W. Latson, and William R. McKeen.

Victor Silver Mining Co. Incorporated August 2, 1871; capital stock, $50,000, in 500 shares, $100 each; office Black Hawk; organized by Edward Bowden, George E. Congden, M.F. Bebee, Ed. A. Seiwell, Samuel P. Lathrop, John Needham, Frederick Leighton, George A. Patten, and S.W. Lincoln.

Virgin. Owned by Zadock Kalbaugh and W.W. Wrigley; office Georgetown; claim 50 by 1,400 feet, patented; discovered 1866; located on Sherman Mountain, Griffith mining district, 3 miles from Georgetown; course of vein, northeast and southwest; width 4 feet, pitch 5 degrees north, pay vein 3 to 5 inches; nature of ore, galena and sulphurets; average value, 375 ounces silver per ton; 1 shaft, 45 feet deep, aggregate length of levels 110 feet.

Vulcan. Owned by George and David Meyers and C.E. Pollard; office Georgetown; claim 150 by 1,500 feet, patented; located on Republican Mountain, Griffith mining district, 3 miles from Georgetown; vertical vein, trending north and south, width 3 feet, pay vein 6 inches; nature of ore, galena, gray copper, and sulphurets; average value, $400 per ton; 1 shaft 60 feet deep, 1 tunnel 95 feet, and 1 tunnel 150 feet in length.

W. B. Astor. Owned by the Astor Mining Co., Cincinnati; offices Georgetown, Colorado, and Cincinnati, Ohio; patented claim, discovered 1866; located on Democrat Mountain [one of Colorado's "fourteeners," at 14,148 feet], Griffith mining district, 2 miles from Georgetown; vertical vein, trending northeast and southwest, width 50 feet, pay vein 2 to 18 inches; nature of ore, silver glance, gray copper and galena; value, 500 ounces silver per ton; 5 shafts, aggregate depth 400 feet, 1 adit 750 in length. (See Astor, Extension)

Wall Street. Owned by Charles P. Baldwin and heirs of Jacob Snider; office Georgetown; claim 50 by 3,000 feet, patented; discovered 1867; located south of Silver Creek, Montana mining district, 2 miles from Lawson and 8 miles from Georgetown; width of vein 15 feet, pay vein 1 to 10 inches; nature of ore, silver glance; average value, 300 ounces silver per ton; 3 shafts aggregate depth 230 feet, 2 levels, 105 and 60 feet in length, respectively.

Walton. Owned by David N. Smith, of Georgetown; claim 150 by 1,500 feet; discovered 1872; located on Brown Mountain, Queen mining district, 1 mile from Brownville and 2 miles from Georgetown; course of vein, northeast and southwest, width 6 feet, pitch north, pay vein 4 inches; nature of ore, galena and gray copper; average value, 346 ounces silver per ton; main shaft 32 feet deep, 1 tunnel 324 feet, and 1 level 100 feet in length.

Well Bassett. Owned by the Lebanon Mining Co.; Julius G. Pohle, superintendent; office Georgetown; claim 150 by 1,500 feet; discovered 1868; located on Republican Mountain, Griffith mining district, 2 miles from Georgetown; course of vein, northeast and southwest; width, 20 feet, pitch northwest; nature of ore, galena; main shaft 20 feet deep; 1 adit, 60 feet, and 1 level 65 feet in length.

West Rip Van Winkle Silver Mining Co. [See Rip Van Winkle, West] Incorporated October 29, 1878; capital stock, $150,000, in 15,000 shares, $10 each; offices Denver, Colorado, and St. Joseph, Missouri; organized by William B. Slosson, Rodney Curtis, and Carver J. Goss.

Whale. Hukill Gold and Silver Mining Co., proprietors; office Idaho Springs; claim 150 by 1,500 feet, patented; discovered 1860; located on Clear Creek, Spanish Bar mining district, 1 mile from Idaho Springs and 41 miles from Denver; main shaft 225 feet deep, 1 adit 277 feet in length; ore mill 75 by 139 feet, capacity 20 tons per day.

White. Owned by Philip E. Morehouse, Carver J. Goss, Charles H. Marshall, and Henry Thompson; office Lawson; claim 150 by 2,750 feet, patented; discovered 1877; located on Red Elephant Mountain, near town of Lawson, 45 miles from Denver; course of vein, northwest and southeast, width 8 to 25 feet, pay vein 2 to 24 inches, pitch 45 degrees north; nature of ore, native silver, gray copper, and galena; average value, $225 per ton; main shaft 100 feet deep, aggregate depth of other shafting 250 feet, 4 adits 60, 80, 90, and 250 feet in length, respectively, 1 level 70 feet from surface and 80 feet in length; yield to date, $40,000.

White Silver Mining Co. Incorporated July 31, 1878; capital stock, $100,000, in 1,000 shares, $100 each; offices, Georgetown, Colorado, and New York City; directors, Charles E. Biglow, Joshua R. Biglow, Alexander McDonald, John H. Weston, and William H. Moore.

Windsor. Owned by George L. Cannon; office Idaho Springs; claim 150 by 1,500 feet; discovered 1873; located on Schaffter Mountain, Idaho mining district, near town of Idaho Springs; vertical vein, running northeast and southwest, width 5 feet, pay vein 10 inches; nature of ore, free gold; average value, 6 ounces gold per cord; main shaft 40 feet, and 1 other shaft 30 feet deep.

Winnebago. Horace H. Atkins, proprietor; office Georgetown; claim 150 by 1,500 feet, patented; located on Leavenworth Mountain, Griffith mining district, 1 mile from Georgetown; course of vein, northeast and southwest, width 3 ½ feet, pitch 15 degrees north, pay vein 9 inches; nature of ore, gray copper; average value, $275 per ton; main shaft 125 feet deep.

Wisconsin Central. Owned by Edward Riley & Co.; office Georgetown; claim 50 by 1,500 feet; discovered 1865; located on Kelso Mountain, West Argentine mining district, 9 miles from Georgetown; vertical vein, trending north and south, width 5 feet, pay vein 15 inches; nature of ore, galena; average value, $75 per ton; 3 adits, aggregate length 480 feet; yield to date, $3,000.

Wyandott. Owned by B.C. Catren and Thomas Fenney; office Georgetown; claim 150 by 1,500 feet; discovered 1878; located on Griffith Mountain, Griffith mining district, 1 mile from Georgetown; width of vein, 4 feet, pay streak 12 inches; average value of ore, 200 ounces silver per ton; main shaft 40 feet deep, 1 level, 30 feet in length.

Yandes Silver Mining Co. Incorporated April 7, 1876; capital stock, $100,000, in 1,000 shares, $100 each; offices Georgetown, Colorado, and Indianapolis, Indiana; organized by Benjamin F. and Edward F. Claypool, Morton W. Latson, William R. McKeen, and Richard J. Bright.

Yingling. Owned by Charles E. Robinson; office Spanish Bar; claim 150 by 1,500 feet; discovered 1877; located on Clear Creek, Spanish Bar mining district, 42 miles from Denver; vertical vein, trending north and south, width 2 feet; 1 adit 115 feet in length.

Young America. Enos K. Baxter Co., proprietors; office Central City; claim 150 by 1,500 feet; discovered 1867; located on Red Elephant Mountain, Downieville mining district, near town of Lawson, 45 miles from Denver; vertical vein, running east and west, width 3 feet, pay vein 6 to 12 inches; nature of ore, galena, gray copper, and native silver; average value, $200 per ton; main shaft 100 feet and 3 other shafts 30, 40, and 60 feet deep, respectively, 1 level 100 feet from surface and 40 feet in length.

Young America. Owned by the Young America Silver Mining Co.; office Georgetown; claim 150 by 1,500 feet, patented; discovered 1866; located on Downieville Hill, Downieville mining district, 6 miles from Georgetown; course of vein, northeast and southwest, width 6 feet, pitch north, pay vein 6 inches; nature of ore, galena and gray copper average value, 300 ounces silver per ton; aggregate depth of shafts 345 feet, 1 adit 520 feet in length.

Clear Creek County Ore Mills

Bay State Mill. Owned by the Bay State Mining Co., of Boston, Massachusetts; located near town of Empire, 2 miles from Georgetown; contains 12 stamps, with other machinery.

Church Bros. Sampling and Crushing Works. Located in Georgetown; have a crushing capacity of 40 tons daily; the main building is 50 by 80 feet, with a storehouse 25 by 30 feet.

Clear Creek Co. H. Augustus Taylor, president, Frank M. Taylor, secretary and treasurer; the works of this company are located in Georgetown, have a capacity for treating 50 tons of ore daily, consists of 3 buildings, 28 by 84, 18 by 76, and 40 by 120 feet, respectively, and contain all necessary machinery for the successful treatment of ore.

Collom Idaho Ore Dressing Co. Incorporated April 17, 1874; capital stock, $50,000, in 500 shares, $100 each; office Idaho Springs, Colorado, and Trenton, New Jersey; organized by Charles Collom, Daniel Peters, Frederick R. Williamson, and William Hancock.

Conqueror Mill. Owned by the Conqueror Gold Mining Co.; Park Disbrow, vice president and superintendent; located in town of Empire, 3 miles from Georgetown; the main building is 35 by 50 feet, and contains all necessary machinery.

Equator Mill. Owned by the Equator Mining and Smelting Co.; William O. Carpenter, president, John Turck, vice president, James B. Goodman, secretary and treasurer, H.S. Kearney, agent and superintendent; located 2 miles from Georgetown, in Griffith mining district.

Fall River Mill. Owned by Joseph A. Munn and John M. Osborn; located in town of Fall River; has a capacity for treating 50 tons of ore in 24 hours, and contains a 70-horse power water wheel and 20 stamps; main building 45 by 50 feet.

Farwell Mill. Owned by John V. Farwell, of Chicago; Samuel Learned, manager; has a capacity for treating 20 tons of ore daily, and is located in the upper part of Georgetown, on the west branch of Clear Creek; the main building is 90 by 140 feet, contains 15 stamps, Dodge crusher, 24-inch Cornish rollers, 3 Bruckner cylinders, 2 turbine water wheels, 50-horse power engine, &c.

Hall's Crushing and Sampling Works. Owned by G.W. Hall & Co.; located in Georgetown; has a capacity for treating 75 tons of ore in 24 hours; main building is 45 by 60 feet, and contains a 40-horse power turbine wheel, crushers, rollers, etc.

Lebanon Concentration and Sampling Mill. Owned by the Lebanon Mining Co., of New York; Julius G. Pohle, superintendent; located near Silver Plume, 2 miles from Georgetown.

McCann's Crushing and Sampling Works. Owned by T.J. McCann; located in town of Lawson, on Colorado Central Railroad, 45 miles from Denver; main building is 25 by 40 feet, and contains a 10-horse power engine, crushers, rollers, etc.

Miles Concentration and Sampling Works. Owned by Miles & Co.; located in town of Idaho Springs; capacity 25 tons of ore per day; main building is 45 by 60 feet, and contains a 40-horse power engine, crushers, rollers, etc.

Pelican Mill. Operated by Franklin Ballou and Benjamin F. Napheys; located at junction of Main and Alpine streets, Georgetown; run by steam and waterpower; has a capacity for treating 80 tons of ore in 24 hours; main building is 80 by 100 feet, and contains a 60-horse power engine, an amalgamating mill of 10 stamps, 8 1-ton barrels, 5 Bruckner cylinders, and a water wheel 12 feet in diameter.

Robinson Mill. Owned by Charles E. Robinson; located on Clear Creek, 2 miles from Idaho Springs and 42 miles from Denver; main building is 40 by 70 feet and contains a 25-horse power engine, a Robinson furnace, pair of steam stamps, a Frieberg pan, and one large crusher; has a capacity for treating 10 tons of ore daily.

Rocky Mountain Crushing and Sampling Works. Owned by Mathews, Morris & Co.; James F. Mathews, manager; located corner of 3rd and Argentine streets, Georgetown; has a capacity for treating 30 tons of ore in 10 hours; main building is 40 by 60 feet, with an office and assay laboratory attached; contains a 25-horse power engine, crusher, Cornish rollers, and all other necessary machinery.

Silver Plume Concentration and Sampling Works. Owned by Franklin Ballou; located in Silver Plume, on south Clear Creek, 2 miles west of Georgetown; has a capacity for treating 40 tons of ore in 20 hours; main building is 45 by 102 feet and contains a 40-horse power engine, Blake crusher, Cornish rollers, and a complete set of sizing apparatus.

Silver Queen Milling Company. Incorporated May 26, 1875; capital stock, $250,000, in 2,500 shares, $100 each; office Georgetown; organized by W. Willet Rose, Julius G. Pohle, William W. Rose, Jr., William V. Simpson, Charles S. Reinhart.

Specie Payment Mill. Owned by the Sunshine Mining Co., of Troy, New York; John Newman, president and general manager; located at Idaho Springs, on Clear Creek, 39 miles from Denver; contains 25 stamps and other machinery.

Stephens Concentration Mill. Owned by Amos P. Stephens; located in town of Lawson; has a capacity for treating 25 tons of ore in 24 hours.

Stewart's Works. Located in the lower part of Georgetown; contains 20 stamps, furnaces, amalgamating pans, 8 tanks with stirrers, vats, etc.; has a capacity for treating 24 tons of ore daily.

Tennessee Mill. Owned by the Tennessee Milling Co.; located in Lyons Gulch, 2 miles from railroad; has a capacity for treating 10 tons of ore in 24 hours.

Terrible Mill. Owned by the Colorado United Mining Co., England; Hon. William A. Hamill, manager; located on South Clear Creek, in town of Brownville, 3 miles from Georgetown; main building is 85 by 98 feet, contains a 60-horse power engine and 2 large boilers, 2 crushers, 2 rollers manufactured by Fraser & Chalmers of Chicago, 6 Hartz jigs, screens, etc; has a capacity for treating 50 tons of ore in 24 hours.

Washington Crushing and Sampling Mill. Owned by L.F. Olmsted and Franklin Ballou; located on Main Street, Georgetown; contains a 20-horse power engine, Dodge crusher, Cornish rollers, etc.

Whale Concentration Mill. Owned by the Hukill Gold and Silver Mining Co.; located on the Colorado Central Railroad, 1 mile from Idaho Springs; has a capacity for treating 20 tons of ore daily; main building 75 by 140 feet, constructed of brick.

White's Concentration Mill. Owned by S.H. White; located on Clear Creek, in the town of Idaho Springs; has a capacity for treating 50 tons of ore daily; main building is 40 by 80 feet; contains a 70-horse power turbine water wheel, Blake crusher, Cornish rollers, jugs, screens, etc.

Along the narrow mountain roads around Idaho Springs are numerous abandoned mines, lrospect holes, sluices, and other remnants of the gold rush.

Photo taken June 1908, The Argentine Central Railway en route to Mt. McClellan, with passengers posing along the snow drifts.

Recent view from "Oh, My God Highway" between Russell Gulch and Idaho Springs. Today, tailings piles from these long-abandoned mines pose an environmental threat from the cyanide used to extract gold.

CUSTER COUNTY

Thomas B. Corbett reported that Custer County had a population of 3,500 in 1878 and produced bullion valued at $600,000 that year. In 1878, the mining districts were Hardscrabble [alternate spelling, Hardscrable], Sangre de Cristo, and Verde.

Less than ten years before Thomas Corbett compiled the first edition of his classic mining directory, Custer County was cattle country. In 1870, a cowboy, Daniel Baker, found some galena, and two prospectors, Dick Irwin and Jasper Brown, found some low-grade ore. When more prospectors moved into the area in 1872, the Hardscrabble mining district was formed. By the following spring, a small camp grew into what was later named Rosita, and the first mines were found. In 1874, the Pocahontas Mine was discovered in Rosita, and the town mushroomed into one of the largest cities in Colorado, with hotels, saloons, restaurants, and stores to meet the needs of the flood of gold seekers. By the eve of the 20th century, the mines dried up, and today Rosita is a ghost town.

Albion. Owned by John J. Edwards, Henry S. Webb, and Robert Foster; located near town of Silver Cliff, 8 miles from Rosita; development, 3 shafts, 12, 15, and 40 feet respectively;

Alma. Owned by the Ula Mining and Reduction Co.; Robert Foster, superintendent; located near town of Silver Cliff, 8 miles from Rosita; development, 1 shaft 53 feet deep; the ore ranges in value from 30 to 600 ounces silver per ton.

Alta. Owned by the Alta and Verde Mining Co.; incorporated October 30, 1876; capital stock, $30,000, in 600 shares, $50 each; office Rosita, Colorado; George Boyce, president, Alexander Thornton, vice president and manager, R.N. Clark, secretary and treasurer; claim 300 by 1,500 feet; discovered 1876; located in Verde mining district, 5 miles from Ula and 5 miles from Rosita; course of vein, northeast and southwest, pay vein 1 foot; nature of ore, gray copper and copper pyrites; average value,

50 ounces silver per ton; 1 drift 75 feet in length; the country surrounding this property is heavily timbered and well watered.

Arminius. Owned by Carl Wulsten of Rosita; claim 300 by 1,500 feet; discovered 1877; located on Game Ridge, Hardscrabble mining district, half mile from town of Rosita; course of vein, northwest and southwest, width, 3 feet, pitch 20 degrees northeast, pay vein 2 inches; nature of ore, gray copper, iron and copper pyrites; main shaft 20 feet deep, 1 tunnel 92 feet in length.

Belfast. Owned by Powell, Edwards, Hafford & McComb; claim 300 by 1,500 feet; discovered August 1878; located in town of Silver Cliff, 6 miles from Rosita; developed by a surface opening 30 feet in length, and a shaft 25 feet deep.

Ben Franklin. Owned by Prescott, Maxwell & Thurmond, of Rosita; claim 300 by 1,500 feet; discovered 1877; located on Ben Franklin Hill, Hardscrabble mining district, 2 miles fro town of Rosita; course of vein, northwest and southeast, width 15 feet, pitch 25 degrees south, pay vein 1 foot; nature of ore, gray copper and galena; average value, 125 ounces silver per ton; main shaft 70 feet, and 1 other shaft 53 feet deep, aggregate length of levels 92 feet, being 45 and 67 feet from surface respectively.

Bunker Hill. Owned by Carl Wulsten, of Rosita; claim 300 by 1,500 feet; discovered 1878; located on Bunker Hill, Hardscrabble mining district; 2 miles from town of Rosita; course of vein, northwest and southeast, width 12 feet, pitch 50 degrees southwest, pay vein 4 feet; nature of ore, carbonate of lead and silver; developed by 2 tunnels, 52 and 40 feet in length, respectively.

Cherokee. Owned by Theodore M. Harding, D.J. Hutchinson, Robert Foster, and others; located at foot of the Blue Range, 2 miles from town of Silver Cliff and 8 miles from Rosita.

Columbus. Owned by Carl Wulsten, of Rosita; claim 300 by 1,500 feet; discovered 1875; located on Game Ridge, Hardscrabble mining district, half mile from town of Rosita; course of vein, northwest and southeast, width 3 feet, pitch 10 degrees southwest, pay vein, 3 inches; nature of ore, gray copper and copper pyrites; average value, 150 ounces silver per ton; main shaft 75 feet and one other shaft 35 feet deep; shaft house, 14 by 18 feet.

Custer County Tunnel Co., Colorado. Incorporated February 13, 1877; capital stock, $40,000, in 400 shares, $100 each; office Rosita; trustees: Charles F. Blossom, Charles Baker, W.L. Knight, James A. Melvin, and W.A. Offenbacher.

Cymbeline. Owned by Richard Irwin and Thomas C. Parrish, of Rosita; claim 300 by 1,500 feet; discovered 1873; located on Game Ridge, Hardscrabble mining district, within the limits of Rosita; course of vein, northwest and southwest, width 3 feet, pitch 15 degrees east, pay vein 4 inches; nature of ore, sulphurets and galena; average value, 50 ounces silver per ton; main shaft 80 feet deep.

Del Monte. Owned by Sargeant, Combs & Co., of Rosita; claim 300 by 1,500 feet; discovered 1876; located near Bunker Hill, Hardscrabble mining district, 3 miles from town of Rosita; course of vein, northwest and southeast, width 8 feet, pitch 30 degrees west, pay vein 9 inches; nature of ore, carbonates and galena, average value, 100 ounces silver per ton; main shaft 80 feet and 1 other shaft 60 feet deep.

Denver. Owned by B.D. Spencer and others; claim 300 by 1,500 feet; discovered September 1878; located 2 miles north of Silver Cliff and 10 miles west of Rosita; course of vein, northeast and southwest, width 4 feet; assay value of ore, 47 ounces silver per ton; developed by a shaft 40 feet deep.

Elizabeth. Owned by Daniels Bros. & Co., of Rosita; claim 300 by 1,500 feet; discovered 1876; located on Plymouth Hill, Hardscrabble mining district, within the limits of Rosita; course of vein, northwest and southeast, width 5 feet, pitch 20 degrees north, pay vein 5 inches; nature of ore, gray copper, iron and copper pyrites; average value, 60 ounces silver per ton; main shaft 133 feet and 1 other shaft 65 feet deep; 2 shaft houses, 18 by 20 feet, and 40 by 40 feet, respectively.

Eureka. Owned by H.N. and J.L. Webb, of Rosita; claim 1,200 by 300 feet; discovered 1875; located on Eureka Hill, Hardscrabble mining district, 1 mile from town of Rosita; course of vein, northwest and southeast, width 5 feet, pitch 30 degrees, pay vein 1 foot; nature of ore, gray copper and copper pyrites, average value, 60 ounces silver per ton; main shaft 103 feet, and 1 other shaft, 35 feet deep, 1 level in length, and 60 feet from surface.

Fleetwood. Operated by C.C. Davidson and others; office Denver; claim 100 by 1,500 feet; discovered September 1878; located on the north side of Wet Mountain Valley, Hardscrabble mining district, 3 miles from Ula and 8 miles from Rosita; course of vein, east and west; nature of ore, chloride of silver; development, 2 shafts, 45 and 70 feet respectively; at a depth of 40 feet, 2 assays were made which assayed 339 ounces, and at a depth of 60 feet the ore assayed 180 ounces silver per ton.

G.V.F. Owned by A.M. McElhenney and W. Emigh, of Rosita; claim 300 by 300 feet; discovered 1874; located in Nebraska Gulch, Hardscrabble mining district, 1 mile from town of Rosita; vertical vein, running northwest and southeast, width 4 feet, pay vein 4 inches; nature of ore, gray copper; average value, 120 ounces silver per ton; main shaft 105 feet deep, 1 level 25 feet in length and 100 feet from surface; shaft house 12 by 16 feet.

General Custer. Owned by Duncan, Adams & Webb, of Rosita; claim 300 by 1,500 feet; discovered 1875; located on Robinson Hill, Hardscrabble mining district, half mile from town of Rosita; vertical vein, running east and west, width 2 feet, pay vein 6 inches; nature of ore, gray copper and iron pyrites; average value, 50 ounces silver per ton; main shaft 60 feet, and 1 other shaft, 40 feet deep, 1 level 20 feet in length and 50 feet from surface.

Golden Era. Owned by Edmund C. Bassick, of Rosita; this property comprises 20 acres of mineral land, located on Tyndall Hill, Hardscrabble mining district, 3 miles from town of Rosita; the nature of the ore is galena, zinc blende, sylvanite, and free gold, and averages $500 per ton. The mine has already produced over $500,000, besides a pile of ore remaining at the mine estimated to be worth $300,000. It is developed by a shaft 18 feet in width by 20 feet in length, and 300 feet deep, 2 tunnels 175 and 450 feet in length, respectively, and 4 levels, ranging in depth from 40 to 200 feet. A tunnel intersects the shaft at a depth of 200 feet, at which point a chamber has been excavated 30 by 30 feet for machinery. It is estimated by experts that there is fully $1,000,000 worth of ore in sight in this mine. On the surface is a shaft house 20 by 20 feet, and an ore house 16 by 40 feet.

Golden Fleece. Owned by Ely Gill and H.C. Nichols; office Ula; claim 300 by 1,500 feet; discovered December 24, 1878; located 2 miles from town of Silver Cliff and 8 miles from Rosita; value of ore, 34 ounces silver per ton; 1 shaft 25 feet deep.

Governor Pitkin. Owned by Charles F. Burrell; office Denver; claim 300 by 1,500 feet; discovered September 1878; located 3 miles from town of Silver Cliff and 7 miles from Rosita; 1 shaft 12 feet deep.

Horn and Chloro Silver. Owned by R.J. Edwards, Robert Powell, George H. Hafford, and others; office Silver Cliff and Denver, Colorado; claim 300 by 2,300 feet; discovered August 1878; located south of Round Mountain, Hardscrabble mining district, 4 miles from Ula and 7 miles from Rosita; course of vein, north and south; nature of ore, chloride of silver; average value, 100 ounces silver per ton; developed by an open cut 50 feet in length and a shaft 50 feet deep; yield $5,000.

Humboldt [alternate spelling, Humbolt]. Owned by the Humboldt Silver Mining Co.; incorporated October 20, 1875; capital stock, $74,000, in 74 shares, $1,000 each; office Rosita, Colorado; George Boyce, president, Alexander Thornton, vice president and manager, R.N. Clark, secretary; claim 300 by 1,900 feet, patented; discovered 1874; located on Lucille Hill, Hardscrabble mining district, within limits of Rosita; course of vein, northwest and southeast; width 2 ½ feet pitch 10 degrees southwest, pay vein 6 inches; nature of ore, gray copper and copper pyrites; average value, 125 ounces silver per ton; main shaft 420 feet deep, and other shafting aggregating 895 feet, aggregate length of levels, 2,150 feet; shaft house 40 by 80 feet, containing an engine and a steam pump; total yield to January 1878, $225,604.

Indian Chief. Owned by Alfred D. Avery, Charles R. and William S. Hurd; office Denver and Silver Cliff, Colorado; claim 300 by 1,500 feet; discovered 1878; located on Round Mountain, 1 mile from town of Silver Cliff and 6 miles from Rosita; course of vein, northwest and southeast; nature of ore, chloride of silver; average value, 125 ounces silver per ton; developed by a shaft 60 feet deep.

Inter-Ocean Tunnel Co. Incorporated December 9, 1874; capital stock, $50,000, in 14 shares; office Rosita, Colorado; organized by G.S. Barlow, Z. Swarenger, William A. Lyons, Ed. P. Smith, Frank A. Smith, John C. Myers, and H.N. Kelly.

Jersey. Owned by H.N. and J.L. Webb, of Rosita; claim 300 by 1,500 feet; discovered 1875; located on Jersey Hill, Hardscrabble mining district, half mile from town of Rosita; vertical vein running northwest and southeast, width 2 feet, pay vein 4 inches; nature of ore, gray copper and copper pyrites; average value, 50 ounces silver per ton; main shaft, 50 feet deep, 1 level 30 feet in length and 50 feet from surface.

King of the Valley. Owned by August Rische, D.R. and Alexander M. Cassiday, J.S. Smith, and A.J. Woodside; office Denver; claim 300 by 1,500 feet; discovered August 1878; located half mile from town of Silver Cliff, 3 miles from town of Ula and 6 miles from Rosita; course of vein, east and west, width 80 feet; nature of ore, chloride of silver; average value, 140 ounces silver per ton; developed by 4 shafts 40, 50, 65, and 90 feet, respectively, and a surface opening 80 feet in length; yield $6,000.

Leavenworth. Owned by Paul Goerke and Theodore F. Braun, of Rosita; claim 300 by 1,500 feet; discovered 1874; located on Robinson Hill, Hardscrabble mining district, half mile from town of Rosita; course of vein, northwest and southeast, width 12 feet, pitch 10 degrees south, pay vein 18 inches; nature of ore, copper pyrites and gray copper; average value, 125 ounces silver per ton; main shaft 107 feet, and 1 other shaft 80 feet deep, aggregate length of levels 150 feet, being 50 and 90 feet from surface, respectively; shaft house 80 by 20 feet.

Leftwick. Owned by Charles F. Burrell and others; office Denver; claim 300 by 1,500 feet; discovered July 1878; located on Horn Silver Hill, 1 mile from town of Silver Cliff and 7 miles from Rosita; course of vein, northeast and southwest; nature of ore, chloride and horn silver; average assay value, 65 ounces silver per ton; developed by a shaft 25 feet deep.

Leviathan. Owned by Benjamin Mattice and F.A. Raynolds; office Pueblo; claim 350 by 750 feet; discovered 1873; located on Lucille Hill, half mile from Rosita and 30 miles from Canon City; course of vein, northwest and southeast width 23 feet, pay vein 4 inches; nature of ore, gray copper; value, 90 ounces silver per ton; main shaft 125 feet deep, 2 levels 50 and 100 feet in length, respectively; shaft house 20 by 30 feet; yield to date $4,000.

Lone Star. William H. Mills, proprietor; office Ula; claim 300 by 1,500 feet; discovered 1876; located on Texas Creek, Sangre de Cristo mining district, 6 miles from Ula and 15 miles from Rosita; course of vein, east and west, width 11 feet, pitch 15 degrees; nature of ore, copper pyrites; assay value, 4 ounces gold per ton; 1 tunnel 30 feet and 1 level 20 feet in length.

Lucille. Owned by Hoyt, Bangs, and others, of Rosita; claim 300 by 1,500 feet; discovered 1873; located on Lucille Hill, Hardscrabble mining district, within limits of Rosita; course of vein, northwest and southeast, width 6 feet, pitch 20 degrees north, pay vein 4 inches; nature of ore, galena and gray copper; average value, 300 ounces silver per ton; main shaft 55 feet deep, aggregated length of other shafting 165 feet, aggregate length of levels 100 feet; shaft house 16 by 24 feet.

Michigan Gold and Silver Mining Co. Organized under the laws of the State of Michigan, August 28, 1878; capital stock, $500,000, in 50,000 shares, $10 each; offices Rosita, Colorado, and Hudson, Michigan; trustees: Stephen A. Edon, Jacob J. Daniels, Harry A. Branch, Charles E. Lawrence, and J.K. Boies.

Nevada. Owned by the Ula Mining and Reduction Co.; Robert Foster, superintendent; located near town of Silver Cliff; 2 miles from town of Ula [now a ghost town] and 8 miles from Rosita.

Old Colony. Owned by the Ula Mining and Smelting Co.; Robert Foster, superintendent; located near town of Silver Cliff, Hardscrabble mining district, 8 miles from Rosita; developed by 2 shafts, 30 to 40 feet respectively.

Plata Verde. Owned by Robinson & Junkins; office, Silver Cliff; claim 300 by 1,500 feet; discovered July 1878; located on Round Mountain, 1 mile east of Silver Cliff and 7 miles from Rosita; nature of ore, chloride of silver, which ranges in value from 42 to 563 ounces silver per ton; developed by an open cut, 15 feet in width by 40 feet in length; yield to date $5,000.

Pocahontas. Owned by Theodore W., Hiero B., and A.J. Herr, H.C. Lehman, and A.J. Ballard; offices Denver and Rosita, Colorado; claim 300 by 1,385 feet; discovered 1874; located on Lucille Hill, within the limits of the town of Rosita; course of vein, northeast and southwest, average width, 25 feet, pitch 18 degrees southwest, pay vein 12 inches; nature of ore, baryta, silver, copper, iron, antimony, and sulphur, with traces of gold; average value, 150 ounces silver per ton; developed by a main shaft, 210 feet deep, and 3 other shafts, which aggregate 500 feet, a tunnel 250 feet, and levels aggregating 2,000 feet in length; the main shaft is 8 feet in width by 16 feet in length, is not on the vein and extends 71 feet below the lower level, which is 300 feet from the surface. It is the intention of the owners to run adits from the shaft to the vein at each 100 feet. On the surface is a shaft house, 30 by 40 feet, containing a 30-horse power engine, an air compressor and steam pump, 2 large ore assorting houses, 2 blacksmith shops, a boarding house, and 3 dwelling houses; railroad track and iron cars. The standard coin value of ore sold from this mine is $280,000, and ore now at the mine is estimated to be worth $300,000. [Soon after the Pocahontas began producing, two men from Denver, a self-styled Colonel named Boyd and a banker named Stewart, forcibly jumped the claim and installed their own men. For several days, miners were terrorized, the mine was besieged, and the town of Rosita was shuttered. The citizens of Rosita joined local law enforcement, and violently took back the mine.]

Polonia. Owned by Melvin, Bothsell & Goerke, of Rosita; claim 300 by 1,200 feet; discovered 1874; located in Nebraska Gulch, Hardscrabble mining district, 1 mile from the town of Rosita; course of vein, east and west, width 5 feet, pitch 10 degrees north, pay vein 8 inches; nature of ore, gray copper and iron pyrites; average value, 90 ounces silver per ton; main shaft 118 feet, and 1 other shaft 55 feet deep, being 75 and 95 feet from surface respectively.

Racine Boy. Owned by John W. Bailey, Isaac T. Beck, Ferdinand C. Taylor, John J. Edwards, Rodney Curtis, and George E. Snider; office Silver Cliff and Denver, Colorado; claim 300 by 1,500 feet; discovered July 1878; located on the north side of Wet Mountain Valley, Hardscrabble mining district, 3 miles from Ula and 8 miles from Rosita; course of vein, northeast and southwest; nature of ore, chloride of silver; average value, 150 ounces silver per ton; developed by a tunnel 60 feet in length, a pit 30 feet deep and 30 feet square, 1 shaft 35 feet deep and 3 surface openings, aggregating 60 feet in length; yield to date $75,000.

Rosita Mining Co. Incorporated January 15, 1876; capital stock, $100,000, in 1,000 shares, $100 each; office Rosita; organized by Tower Thomasson, Henry T. Blake, Carl Wulsten, William Schaefer, Frank Canins, Robert Powell, and August Strehlong.

Royal Gem. Owned by M.A. Austin & Co., of Rosita; claim 300 by 800 feet, discovered 1875; located on Lucille Hill, Hardscrabble mining district, within limits of Rosita; course of vein, northwest and southeast, width 6 feet, pitch 20 degrees north, pay vein 5 inches; nature of ore, galena and gray copper; average value, 200 ounces silver per ton; main shaft 150 feet deep, 2 levels, aggregate length 110 feet, being 50 and 125 feet from surface respectively.

Ruby Cliff. Owned by Robert Powell, R.J. Edwards, George H. Hafford, and others; located 1 mile west of Silver Cliff and 8 miles from Rosita; developed by an open cut 15 feet in length and a shaft 15 feet deep.

San Domingo. Owned by Thyng and Co.; located on Blue Ridge, 3 miles northeast of Silver Cliff, and 10 miles from Rosita; nature of ore, galena; average value. 100 ounces silver per ton; developed by a surface opening 250 feet in length, 4 shafts ranging in depth from 10 to 25 feet.

Senator. Owned by the Hoyt Mining Co., of Colorado; claim 300 by 1,500 feet; discovered 1873; located on Senator Hill, Hardscrabble mining district, within the limits of Rosita; course of vein, northwest and southeast, width 6 feet, pitch 10 degrees north; nature of ore, galena and silver glance; average value, 150 ounces silver per ton; main shaft 65 feet deep, aggregate depth of other shafting 70 feet.

Silver Cliff. Owned by R.J. Edwards, Robert Powell, George H. Hafford, and others; offices Silver Cliff and Denver, Colorado; claim 300 by 1,500 feet; discovered June 29, 1878; located north side of Wet Mountain Valley, Hardscrabble mining district, 3 miles from Ula and 8 miles from Rosita; course of vein, northeast and southwest; nature of ore, chloride of silver; average value, 40 ounces silver per ton; developed by a shaft 40 feet deep and surface openings, aggregating 100 feet in length.

Stephens. Owned by Paul Goerke and S. Smith, of Rosita; claim 300 by 1,500 feet; discovered 1874; located on Robinson Hill, Hardscrabble mining district, near limits of Rosita; course of vein, northwest and southeast, width 12 feet, pitch 10 degrees south, pay vein 8 inches; nature of ore, gray copper, iron and copper pyrites; average value, 125 ounces silver per ton; main shaft 70 feet and 1 other shaft 40 feet deep.

Tucson. George W. Monk, proprietor; office Pueblo; claim 300 by 1,500 feet; discovered 1874; located on Texan Peak, Sangre de Cristo mining district, 6 miles from Ula; course of vein, east and west, width 30 feet, pay vein 4 feet, pitch 25 degrees; nature of ore, copper pyrites and free gold; average value, $156 per ton; opened on the surface for a distance of 500 feet.

Twenty-Six. Owned by Paul Goerke and F. Schmit, of Rosita; claim 300 by 1,100 feet; discovered 1874; located in Nebraska Gulch, Hardscrabble mining district, 1 mile from town of Rosita; course of vein, east and west, width 5 feet, pitch 10 degrees south, pay vein 10 inches;

nature of ore, gray copper, copper and iron pyrites; average value, 120 ounces silver per ton; main shaft 100 feet deep, 1 level 100 feet in length and 95 feet from surface; shaft house 16 by 24 feet.

Ula Mining and Reduction Company. Incorporated January 13, 1879; capital stock, $300,000, in 3,000 shares, $100 each; offices Denver and Silver Cliff, Colorado; John W. Smith, president, J.H. Jones, vice president, A.J. Williams, treasurer, S.A. Shepherd, secretary, Robert Foster, superintendent. The property of this company is located near town of Silver Cliff, Hardscrabble mining district, 2 miles from town of Ula and 8 miles from Rosita, the county seat. It comprises the Alma, Old Colony, Norwich, Crystal Quartz, Haxtun, New York, Olive Branch, Hahns Peak, Sayre, Denver City, and Nevada lodes, which cover over 90 acres of surface ground; the claims are 300 feet in width by 1,500 feet in length, were discovered since July 1878, and have been discovered, surveyed, recorded an developed according to law. Shafts have been sunk from 10 to over 60 feet in depth, and mineral has been found in all of them, varying in richness and quantity. The close proximity of these lodes to the well-known rich mines of the camp gives sufficient assurance that they will, when properly developed, be of great value. Good timber and waterpower can be obtained within a convenient distance of this property.

Verde. Owned by the Alta and Verde Mining Co.; incorporated October 30, 1876; capital stock, $30,000, in 600 shares $50 each; office Rosita, Colorado; George Boyce, president, Alexander Thornton, vice president and manager, R.N. Clark, secretary and treasurer; claim 300 by 1,500 feet; discovered 1876; located in Verde mining district, 5 miles from Ula and 5 miles from Rosita; course of vein, northwest and southeast, width 4 feet, pitch 45 degrees southwest, pay vein, 1 foot; nature of ore, gray copper and copper pyrites; average value, 50 ounces silver per ton; 1 drift 75 feet in length. The country surrounding this property is heavily timbered and well watered.

Victor. Owned by Carl Wulsten, of Rosita; claim 300 by 1,240 feet; discovered 1878; located on Lucille Hill, Hardscrabble mining district; 1 mile from town of Rosita; course of vein, east and west, width 2 feet, pitch 15 degrees south, pay vein 2 inches; nature of ore, gray copper and carbonates; 2 tunnels 110 and 80 feet in length, respectively.

Victoria. Owned by J.A. Melvin, of Rosita; claim 300 by 1,500 feet; discovered 1874; located on Lucille Hill, Hardscrabble mining district, near limits of Rosita; course of vein, northeast and southwest, width 15 feet, pitch 10 degrees north, pay vein 8 feet; nature of ore, gray copper and native silver, average value, 40 ounces silver per ton; main shaft 60 feet and 1 other shaft 30 feet deep, 1 level 60 feet in length and 25 feet from surface.

White Pine. Owned by B.D. Spencer and others; office, Denver; claim 300 by 1,500 feet; discovered August 1878; located on Round Mountain, Hardscrabble mining district, 2 miles northwest of Silver Cliff and 8 miles from Rosita.

Woodburn. Owned by Theodore M. Harding, D.J. Hutchinson, Lafayette Cornwell, and Robert Foster; located one mile from town of Silver Cliff and 8 miles from Rosita; developed by a shaft 25 feet deep.

Virginia. Owned by the Virginia Mining Co., of Rosita; incorporated March 30, 1876; capital stock, $60,000, in 6,000 shares, $10 each; office Rosita; Mayland Cuthbert president, R.N. Clark, vice president, Thomas C. Parrish, secretary and treasurer; claim 300 by 1,300 feet, patented; discovered 1873; located in Hardscrabble mining district, within limits of Rosita; course of vein, northwest and southeast, width 2 ½ feet, pitch 10 degrees southwest, pay vein 10 inches; nature of ore, gray copper and copper pyrites; average value, 115 ounces silver per ton; main shaft 355 feet deep and 5 other shafts, aggregating 161 feet, 1 tunnel, 140 feet in length, which intersects the vein at a depth of 184 feet, aggregate length of levels 756 feet, being 22, 25, 122, 184, and 235 feet from surface respectively; shaft house 40 by 85 feet, containing an engine, boiler, and 2 pumps, total yield to January 1878, $18,547.

Custer County Coal Mines

Canon City Coal Co. Incorporated October 16, 1878; capital stock, $250,000, in 5,000 shares, $50 each; office Canon City; organized by James W. Gaff, Jr., James D. Parker, and William H. May.

Custer County Ore Mills

Crushing and Sampling Mill. Scheyer, Dillingham & Co.; located in town of Silver Cliff, 3 miles from Ula and 7 miles from Rosita; contains a 25-horse power engine and other machinery, has a capacity for treating 25 tons of ore daily.

Mallett Reduction Works. Owned by the Mallet Reduction and Sampling Co.; incorporated October 22, 1877; capital stock, $50,000, in 2,000 shares, $25 each; located in Canon City; has a capacity for treating 14 tons of ore daily. The main building is 75 by 200 feet and contains an engine, a turbine wheel, Blake crusher, Ball pulverizer, 5 stamps, vats, pans, etc.

Pennsylvania Reduction Works. Owned by the Pennsylvania Reduction Co., P.H. Van Diest, superintendent; located in town of Rosita; has a capacity for treating 10 tons of ore daily; the main building is 80 by 120 feet, and contains all necessary machinery.

Rosita Reduction Works. Owned by John Q. Watkins, of Kansas City; located in town of Rosita; has a capacity for treating 5 tons of ore daily. The main building is 40 by 130 feet, and contains an engine, boilers, Blake crusher, Ball pulverizer, vats, pans, etc.

GILPIN COUNTY

Thomas B. Corbett reported that Gilpin County had a population of 10,000 in 1878 and produced bullion valued at $2,200,000 for that year. In 1878, the mining districts included: Central City, Enterprise, Eureka, Gregory, Hawkeye, Illinois, Illinois Central, Lake, Mountain City, Nevada, Russell, Silver Lake, and Wide Awake.

Adeline. Owned by Henry M. Teller, of Central City; claim 50 by 200 feet, patented; located on Gunnell Hill, Nevada mining district, within the limits of Central City.

Aduddel. Owned by Robert G. Aduddel, M.D., and heirs of Joab Jones; office Central City; claim 150 by 3,000 feet, patented; discovered 1868; located at the head of Elkhorn Gulch, Russell mining district, 2 miles from Central City; course of vein, northeast and southwest, width 4 feet, pay vein 3 to 8 inches, pitch 35 degrees north, until at depth of 70 feet, when the vein becomes vertical; nature of ore, iron pyrites and zinc blende; average value, $100 per ton; main shaft 120 feet deep; shaft house 16 by 24 feet, containing a 15-horse power engine and hoisting apparatus.

Aetna. Horace M. Hale, proprietor; office Central City; claim 50 by 900 feet, patented; discovered 1866; located on the eastern slope of Casto Mountain, Gregory mining district, within the limits of Central City; course of vein, northeast and southwest, width 4 feet, pay vein 18 inches, pitch 23 degrees south; nature of ore, galena, zinc blende, iron and copper pyrites; average value, 3 ounces gold per cord; main shaft 100 feet deep, 1 tunnel 450 feet in length.

Aetna. Owned by the Aetna Gold Mining Co.; George E. Randolph, agent; office Central City; claim 50 by 900 feet; discovered 1859; located on Quartz Hill, Nevada mining district, 1 mile from Central City; course of vein, northeast and southwest, width 3 feet; nature of ore, iron and copper pyrites; average value of mill ore, 4 ounces gold per cord, smelting ore $30 to $120 per ton; main shaft 500 feet deep, 1 level 300 feet in length.

Alabama. Kelty & Tomlinson, proprietors; office Black Hawk; claim 150 by 1,500 feet; discovered 1876; located between Chase Gulch and Clear Creek, within the limits of Black Hawk; course of vein, northeast and southwest, width 2 feet; nature of ore, galena, iron and copper pyrites; average value, 4 ounces gold per cord; 1 tunnel 100 feet in length, 4 shafts aggregate depth 100 feet.

Alaska. Owned by William H. Bush & Co.; discovered in Enterprise mining district, 1 mile from Central City; average value of ore, $40 per ton; main shaft 150 feet deep.

Alger-Kansas. Owned by Jacob Tascher, of Central City; discovered 1859; located on Quartz Hill, 1 mile from Central City; value of ore, 3 to 7 ounces gold per cord; will be intersected by the Central City Tunnel at a depth of 75 feet.

Alps and Mackie. Owned by the Cleveland Gold Mining Co.; George E. Hutchinson, of Central City, manager; claim comprised 750 linear feet on the Alps lode and 800 linear feet on the Mackie lode; discovered 1863; located on Quartz Hill, Nevada mining district, 2 miles from Central City; course of vein, northeast and southwest, width 2 feet, pitch south; nature of ore, gray copper, iron and copper pyrites; average value of mill ore, 6 ounces gold per cord, smelting ore, $80 per ton; main shaft 340 feet deep, aggregate depth of levels 1,850 feet; building 25 by 100 feet, containing a 40-horse power engine and 1 steam pump.

Amazon. Owned by Updegraf, Kimber & Co.; office Central City; claim 50 by 500 feet; discovered 1865; located on Gunnell Hill, Eureka mining district, within the limits of Central City; course of vein, northeast and southwest, width 5 feet; nature of ore, iron and copper pyrites; average value, 4 ounces gold per cord; main shaft 125 feet deep, 1 level 100 feet from surface and 50 feet in length.

American Flag. Owned by the American Flag Gold Mining Co; John Van Nest, president, A.J. Van Deventer, secretary; Andrews N. Rogers, agent; office Central City, Colorado, and New York City; claim 50 by 550 feet; discovered 1860; located on Quartz Hill, Nevada mining district, 1 mile from Central City; course of vein, northeast and southwest, width 5 feet; nature of ore, galena, iron and copper pyrites; average value, 6 ounces gold per cord; main shaft 440 feet deep, 1 level 300 feet in length; yield, $250,000.

Americus. James Fraser, proprietor; office Central City; claim 150 by 1,000 feet; discovered 1877; located on Bobtail Hill, Gregory mining district, within the limits of Black Hawk; course of vein, northeast and southwest, pitch 10 degrees north, width 3 feet; nature of ore, iron and copper pyrites; average value, 3 ounces gold per cord; main shaft 40 feet deep.

Arapahoe. Owned by the Belden & Tennal Mining Co.; office Central City; claim 150 by 1,500 feet, patented; discovered 1877; located on Casto Mountain, within the limits of Central City; width of vein 3 feet, pay streak 2 feet; average value of ore, 5 ounces gold per cord; reached by the Belden Tunnel at a distance of 530 feet from the mouth, and intersected at a depth of 450 feet.

Arlington. Owned by Sidney W. Tyler and George H. Gray; office Central City; claim 150 by 1,100 feet; discovered 1875; located on Eureka Hill, Eureka mining district, within the limits of Central City; width of vein 3 feet, pay streak 18 inches; average value of ore, 5 ounces gold per cord; main shaft 100 feet deep, 1 level 40 feet in length.

Ashtabula. Owned by the La Cross Gold Mining Co., New York; claim 50 by 800 feet; discovered 1860; located on Gunnell Hill, Nevada mining district, 2 miles from Central City; vertical vein, running northeast and southwest, width 1 feet, pay vein 4 foot; nature of ore, iron and copper pyrites; average value, 6 ounces gold per cord; main shaft 150 feet deep, and 4 other shafts, aggregating 200 feet in depth, aggregate length of levels, 200 feet; total yield, $20,000.

Aurum and Argentum Mining Co. Incorporated April 1, 1878; capital stock, $100,000, in 10,000 shares. $10 each; offices Central City, Colorado, St. Louis, Missouri, and New York City; trustees: James T. Cooper, Hugh E. Bayle, William G. Miller, J.W. Holman, and W.C. Bragg.

Baker. Owned by Joseph Hafer and William Taylor; office Central City; claim 50 by 400 feet; discovered 1860; located in Central City mining district; width of vein, 2 feet; average value of ore, 5 ounces gold per cord; 2 shafts aggregate depth 250 feet, 1 tunnel 150 feet in length, 2 levels 175 feet in length each, being 100 and 200 feet from surface, respectively.

Baker, Extension of Winnebago. A.H. Baker, proprietor; office Central City; claim 150 by 200 feet; discovered 1862; located on Casto Mountain, within the limits of Central City; course of vein, east and west, width 5 feet, pitch 5 degrees north; nature of ore, iron and copper pyrites; average value, 6 ounces gold per cord; 1 shaft 50 feet deep.

Bates. Owned by J. Baker; office San Francisco, California; claim 50 by 360 feet; discovered 1859; located on Bates Hill, within the limits of Black Hawk; course of vein, northeast and southwest, width 5 feet, pitch 15 degrees south; nature of ore, galena, iron and copper pyrites; average value of mill ore, 9 ounces gold per cord, smelting ore, $60 per ton; 3 shafts aggregate depth 400 feet.

Bates, Nos. 1 and 2 West. Owned by the Union Gold Co.; located on Bates Hill, within the limits of Black Hawk; main shaft, 410 feet deep; these claims produced in 16 months $205,000.

Bates and Baxter, Nos. 1, 2, and 3, East. Owned by the Bates and Baxter Co.; located on Bates Hill, within the limits of Black Hawk; main shaft 388 feet deep, 81 cords of ore from this property yielded 1,092 ounces of gold.

Bates-Hunter. William J. Barker, Frank A. Brunell, and Jeremiah Borham, proprietors; office Denver; claim 50 by 300 feet, patented; discovered 1859; located on Gregory Hill, Gregory mining district, within the limits of Central City; course of vein, northeast and southwest, pitch 20 degrees south, width 6 feet; nature of ore, iron and copper pyrites; average value 8 ounces gold per cord, smelting ore, $75 per ton; main shaft 400 feet and 1 other 280 feet in depth, 4 levels, aggregate length 700 feet; shaft and engine house 40 by 60 feet, containing a 15-horse power engine and hoisting apparatus; yield to date, $200,000.

Bazee. Owned by Henry Chatillon, of Central City; claim 150 by 1,500 feet; discovered 1877; located on Winnebago Hill, near the limits of Central City; course of vein, northeast and southwest; width 3 feet, pitch 25 degrees south, pay vein 8 inches; nature of ore, iron and copper pyrites; average value of mill ore, 4 ½ ounces gold per cord; main shaft 25 feet deep; total yield, $400.

Bedford County. Owned by James B. and Lemuel M. Bradley and Drury R. Moss; office Black Hawk; claim 50 by 1,000 feet; discovered 1866; located on Bobtail Hill, Enterprise mining district, within the limits of Black Hawk; course of vein, northeast and southwest, width 2 feet, pitch north; nature of ore, iron and copper pyrites; average value, 50 ounces gold per cord, 4 shafts, aggregate depth 250 feet, 1 level 50 feet from surface and 25 feet in length.

Belden Tunnel. Owned by the Belden and Tennal Mining Co., of Colorado; incorporated August 19, 1875; capital stock, $500,000, in 100,000 shares, $5 each; offices Central City and Denver, Colorado; David D. Belden, president, Frederick Kruse, vice president, Alonzo Furnald, secretary; claim 150 by 3,000 feet, patented; located in Chase Gulch, at the foot of CastoMountain, half mile from Central City and 1 mile above Black Hawk; length of tunnel 550 feet, course southeasterly, when completed, will be 3,000, inlength commencing in Chase Gulch and terminating in Gregory Gulch, completely undercutting Casto Mountain.

Big Thunder. Willard Teller, proprietor; office Denver; claim 50 by 1,600 feet, patented; located on Quartz Hill, Illinois mining district, half mile from Central City; course of vein, northeast and southwest, width 2 ½ feet; nature of ore, iron and copper pyrites; main shaft 50 feet deep, 1 drift 60 feet in length.

Billings. Owned by Elias Goldman and Jacob Mack, of Central City; claim 50 by 600 feet, patented; discovered 1861; located on Casey Road, within the limits of Central City; course of vein, northeast and southwest, width 4 feet, pitch 30 degrees south, pay vein 20 inches; nature of ore, iron and copper pyrites; average value, 4 ½ ounces gold per cord; main shaft 70 feet deep, 1 drift 25 feet in length.

Billings. Owned by Henry Sweeder; office Central City; claim 150 by 1,500 feet, patented; discovered 1862; located on Casto Mountain, near Central City; course of vein, northeast and southwest, width 4 feet, pay vein 18 inches; nature of ore, galena, iron and copper pyrites; average value of mill ore, 5 ounces gold per cord, smelting ore, $40 per ton; main shaft 115 feet deep.

Black Cloud. Owned by Robert B. Smock and John Chackfield; office Central City; claim 50 by 1,500 feet; discovered 1876; located in Gregory Gulch, within the limits of Central City; course of vein, northeast and southwest, pitch 10 degrees south, width 3 ½ feet; nature of ore, iron and copper pyrites; average value of mill ore, 2 ½ ounces gold per cord, smelting ore, $40 per ton; main shaft 50 feet deep.

Black Hawk Tunnel. Owned by the Black Hawk Tunnel and Mining Co.; incorporated 1876; capital stock, $250,000, in 25,000 shares, $10 each; office Black Hawk; L.C. Snyder, president, S.H. Bradley, vice president, T.D. Sears, secretary; claim 50 by 3,000 feet; discovered 1876;

located at the base of Casto Mountain, in Chase Gulch within the limits of Black Hawk.

Black Quartz. Owned by the Belden and Tennal Mining Co.; David D. Belden, president and superintendent; office Central City; claim 150 by 1,500 feet, patented; discovered 1877; located on Casto Mountain, Central City mining district; width of vein, 2 ½ feet; average value of ore, 5 ounces gold per cord; reached by the Belden Tunnel at a distance of 300 feet from the mouth, and intersected at a depth of 150 feet.

Bobtail. Owned by the Consolidated Bobtail Gold Mining Co.; incorporated under the laws of the State of New York, November 19, 1869; capital stock $1,136,630; George A. Hoyt, president, John Stanton, Jr., secretary and treasurer, Andrews N. Rogers, agent, Frank H. Messinger, assistant superintendent; offices Black Hawk, Colorado, and New York City; claim 150 by 866 2/3 feet, patented discovered 1859; located on Bobtail Hill, Gregory mining district, in town of Black Hawk; course of vein, northeast and southwest, width 3 feet, pitch 10 degrees north; reached by the Bobtail tunnel at a distance of 1,150 feet from the mouth, and intersected at a depth of 475 feet; nature of ore, iron and copper pyrites, average value of mill ore, 8 ounces gold per cord, smelting ore, $40 to $200 per ton; main shaft 800 feet deep, and 8 other shafts, aggregating 3,300 feet, 4 levels aggregating 3,467 feet in length; 2 double engines for hoisting ore, 80-horse power each, 1 duplex engine, 40 horse power, and 1 engine for compressing air for drilling purposes.

Bobtail. Owned by the New York and Colorado Co.; B.T. Wells, manager; offices Black Hawk, Colorado, and 52 William Street, New York City; claim 225 linear feet; located in Gregory mining district, within the limits of Black Hawk; course of vein, northeast and southwest, width 3 feet, pitch 10 degrees north; nature of ore, iron and copper pyrites; average value of mill ore, 4 ounces gold per cord; main shaft 100 feet deep.

Bobtail. Whipple Gold Mining Co.; claim 50 by 100 feet; discovered 1862; located on Gregory Hill; main shaft 130 feet deep; average value of ore, 4 ounces gold per cord.

Bobtail, Nos. 3 and 4 West, and No. 5 East. Mrs. Francis Field, proprietor; Henry W. Lake, manager; office Black Hawk; 2 patented claims 50 by 260 and 50 by 33 1/3 feet respectively; discovered 1859; located on Bobtail Hill, Gregory mining district, within the limits of Black Hawk; course of vein, northeast and southwest, width 3 feet, pay vein 15 inches; nature of ore, iron and copper pyrites; average value of mill ore, 4 ounces gold per cord, smelting ore, $50 per ton; claim No. 5 east has a shaft 600 feet deep and is reached by the Bobtail Tunnel at a distance of 50 feet from the mouth, and intersected at a depth of 450 feet; the shaft house if 24 by 36 feet, and the total yield of the mine $500,000; claim 3 and 4 west have a shaft 500 feet deep and a level 150 feet in length, and a shaft house 34 by 36 feet containing a 12-horse power engine and hoisting apparatus; yield to date, $12,000.

Bobtail, No. 6 West. Henry W. Lake, proprietor; office Black Hawk; claim 50 by 76 feet, patented; discovered 1859; located on Bobtail Hill, Gregory mining district, within the limits of Black Hawk; course of vein, northeast and southwest, pitch 5 degrees north, width 3 ½ feet, pay vein 15 inches; nature of ore, iron and copper pyrites, average value of mill ore, 4 ounces gold per cord, smelting ore, $50 per ton; main shaft 50 feet deep; total yield, $500.

Bobtail, No. 8 East. Owned by John H. Lafrenz, John Thaller, and Joseph Meugg; office Central City; claim 150 by 110 feet, patented; discovered 1860; located on Bobtail Hill, Gregory mining district, within the limits of Black Hawk; course of vein, northeast and southwest, width 4 feet, pay vein 12 inches; nature of ore, iron and copper pyrites; average value of mill ore, 5 ounces gold per cord, smelting ore, $100 per ton; main shaft 640 feet deep, 2 levels aggregating 160 feet in length; shaft house 30 by 68 feet containing a hoisting apparatus and a 50-horse power engine.

Bobtail, Nos. 9, 10, 11, and 12. D. Sullivan, proprietor; office Central City; claim 150 by 350 feet, patented; discovered 1859; located on Bobtail Hill, Gregory mining district, within the limits of Black Hawk, course of vein, northeast and southwest, pitch north, width 5 feet, pay vein 12 inches; nature of ore, iron and copper pyrites; average value of mill ore, 6 ounces gold per cord, smelting ore, $100 per ton; 1 level 100 feet in length, and 600 feet from surface; operated through a tunnel from an adjoining claim.

Bobtail Tunnel. Owned by the Bobtail Tunnel Co.; incorporated January 1863; Andrews N. Rogers, president, John Stanton, Jr., treasurer, Frank H. Messinger, secretary; offices Black Hawk, Colorado, and 25 Nassau Street, New York City; claim 1,500 by 3,000 feet; discovered 1863; located at the base of Bobtail Hill, Gregory mining district, in town of Black Hawk; length of tunnel 2,200 feet, intersects the Fiske lode at a distance of 550 feet and the Bobtail lode at a distance of 1,150 feet from the mouth.

Booth. Owned by Ed. B. Warner; office Central City; claim 150 by 1,500 feet; discovered 1862; located on Saratoga Hill, Russell mining district, 1 mile from Central City; course of vein, northeast and southwest, width 3 feet, pay vein 18 inches; nature of ore, iron and copper pyrites; average value, 5 ounces gold per cord; shafts aggregating 200 feet in depth.

Borton. Located on Quartz Hill, Nevada mining district, 1 mile from Central City; Professor Nathaniel P. Hill owns 500 feet on this vein and James Clark owns 400 feet; will be intersected by the Quartz Hill Tunnel at a depth of 550 feet.

Bowker. Owned by L.A. Johnson; located in Nevada mining district; nature of ore, galena and copper pyrites; value, 1 ounce gold and 28 ounces silver per cord, and 40 percent lead.

Briggs Property. Owned by J. Smith, and Charles H. and George W. Briggs; office Black Hawk; claims consist of 540 linear feet on the Briggs lode, and 540 linear feet on the Gregory lode, all of which is patented; discovered 1859; located on Colorado Central Railroad, Main Street, Black Hawk; course of veins, northeast and southwest, width 8 to 12 feet, pitch 7 to 100 feet south; nature of ore, iron and copper pyrites; main shaft 915 feet, and a pump shaft 575 feet deep; over the mine is erected a 50-stamp mill with a capacity for treating 40 tons of ore daily. It contains a 100-horse power engine for hoisting and crushing ore, an 80-horse power pumping engine, with a 14-inch Cornish pump, and a reserve engine of 30-horse power.

Buell Property. Owned by Bela S. Buell, of Central City; claim 50 by 1,000 feet, patented; discovered 1862; located in Gregory Gulch, in town of Central City; course of vein, northeast and southwest, width 8 feet, pitch 5 degrees south, pay vein 6 inches to 8 feet; nature of ore, iron and copper pyrites; average value of mill ore, 5 ounces gold per cord, smelting ore, $100 per ton; main shaft 600 feet deep, aggregate depth of five other shafts, 1,100 feet, 7 levels, 25, 90, 180, 250, 300, 400, and 500 feet from surface, respectively, aggregating 5,000 feet in length; mill site, comprising 67,500 square feet of ground, on which is erected a fire-proof stone building 50 by 135 feet, containing 60 stamps, 1 engine, 100-horse power, double upright hoisting engine, 75-horse power, 2 steam pumps, hoisting apparatus, and 6 Freiberg pans; total yield, $1,000,000,

Bugher. Owned by Joseph B. Hafer and William Taylor, of Central City; claim 50 by 75 feet; discovered 1875; located within the limits of Central City; course of vein, northeast and southwest, width 4 feet, pitch 10 degrees south, pay vein 2 feet, nature of ore, iron and copper pyrites; average value of mill ore, 5 ounces gold per cord; main shaft 140 feet deep, 1 tunnel 140 feet in length; total yield, $8,000.

Bugle Tunnel. Owned by Bela S. Buell; located on Gregory Street, Central City; course south; will intersect 27 lodes at various depths.

Bull of the Woods. Owned by David Evans and Joseph B. Hafer, of Central City; claim 50 by 585 feet; discovered 1867; located within the limits of Central City; course of vein, northeast and southwest, width, 3 feet, pitch 25 degrees north, pay vein 2 feet; nature of ore, iron and copper pyrites; average value of mill ore, 6 ounces gold per cord, smelting ore, $80 per ton; main shaft 200 feet deep, 3 levels aggregating 200 feet in length; shaft house 25 by 50 feet; total yield, $15,000.

Burroughs. Owned by the La Crosse Gold Mining Co.; incorporated 1864; capital stock, $2,000,000, in 200,000 shares, $10 each; offices Central City, Colorado, and New York City; John Van Nest, president, Henry Smith, secretary, Andrews N. Rogers, superintendent; claim 50 by 305 feet, patented; located on Quartz Hill, Nevada mining district, 2 miles from Central City; course of vein, northeast and southwest, width 3 feet, pay vein 18 inches, pitch 30 degrees south; nature of ore, iron and copper pyrites; average value of mill ore, 6 ounces gold per cord, smelting ore, $80 per ton; main shaft 400 feet deep, 20-horse power engine and a hoisting apparatus; yield to date, $60,000.

Burroughs. Owned by the First National Gold Mining Co., of Boston, Massachusetts; claim, 50 by 1,002 feet; operated by the Monmouth Consolidated Gold Mining Co.; S. Sullivan, superintendent. This property is patented, is well developed, and has paid largely.

Burroughs. Owned by Henry M. and Willard Teller, of Central City; claim 50 by 22 feet; located on Quartz Hill, 1 mile from Central City.

Burroughs, Extension. Owned by Richard Mackey; office Bald Mountain; claim 50 by 400 feet, patented; discovered 1864; located on Quartz Hill, Nevada mining district, 2 miles from Central City; course of vein, nearly east and west, width 3 feet, pitch north; nature of ore, iron and copper pyrites; average value, 5 ounces gold per cord; 3 shafts aggregating 690 feet, and 4 levels, aggregating 425 feet in length, shaft house, engine, and hoisting works.

Butler. Owned by Hon. Jerome B. Chaffee and James D. Wood; office Central City; claim 150 by 1,500 feet; discovered 1859; located on Gunnell Hill, within the limits of Central City; course of vein, northeast and southwest, width 4 feet, pitch 20 degrees south, pay vein 20 inches; nature of ore, iron and copper pyrites; average value of mill ore, 4 ½ ounces gold per cord, smelting ore $85 per ton; main shaft 185 feet deep, aggregate length of levels, 125 feet; shaft house 16 by 20 feet.

Caledonia. Owned by Samuel Cushman, of Deadwood, Wyoming; claim 150 by 1,500 feet; discovered 1860; located in Hawkeye mining district, near Wide Awake [now a ghost town], 6 miles from Central City; course of vein, northeast and southwest; nature of ore, iron pyrites; average value, 10 ounces gold per cord; main shaft 267 feet, 5 other shafts 40 to 140 feet deep.

Calhoun. Owned by Horatio E. and Morris Hazard; office Central City; claim 150 by 1,500 feet; discovered 1877; located on Alps Hill, Russell mining district, 1 mile from Central City; width of vein, 2 feet; average value of mill ore, 6 ounces gold per cord, smelting ore $62 per ton; main shaft, 175 feet deep, 1 level 75 feet from surface and 77 feet in length.

Calhoun. Owned by the Rockdale Gold Mining Co., New York; Edward W. Henderson, agent; office Central City; claim 50 by 650 feet, patented; located in Leavenworth Gulch, Russell mining district, 2 miles from Central City; vertical vein, running northeast and southwest, width 4 feet; nature of ore, iron and copper pyrites; main shaft 125 feet deep.

California. Joseph Standley, proprietor; office Bald Mountain; claim 50 by 650 feet, patented; discovered 1859; located on Quartz Hill, Nevada mining district, 2 miles from Central City; course of vein, early east and west, width 3 feet, pitch south; nature of ore, iron and copper pyrites; average value of mill ore, 7 ounces gold per cord, smelting ore, $70 per ton; main shaft 728 feet deep, aggregate depth of other shafting 1,800 feet, 3 levels aggregating 1,000 feet in length; stone building with iron roof, 34 by 64 , with a wind 18 by 28 feet, containing a 50-horse power engine, hoisting apparatus, etc; total yield, $500,000.

California, No. 4 East. Wm. M. Roworth, proprietor; office Central City; claim 50 by 100 feet, patented; discovered 1860; located on Quartz Hill, Nevada mining district, 2 miles from Central City; course of vein, nearly east and west, width 3 feet; nature of ore, iron and copper pyrites; average value of mill ore, 6 ounces gold per cord, smelting ore, $100 per ton; main shaft 375 feet deep; yield, $80,000.

Camp Grove. Owned by the Monmouth consolidated Gold Mining Co.; D. Sullivan superintendent; claim 100 by 340 feet; located on Quartz Hill, Nevada mining district, 2 miles from Central City.

Canandaigua. Owned by Jason E. Scobey; office Black Hawk; claim 50 by 357 feet, patented; discovered 1860; located in Spring Gulch, within the limits of Central City; course of vein, northeast and southwest, width 3 feet, pitch south; nature of ore, galena, iron and copper pyrites; average value of mill ore, 7 ounces gold per cord, smelting ore, $125 per ton; main shaft 100 feet deep, 1 level 95 feet from surface and 70 feet in length.

Carr. Owned by L.C. Snyder, Horace H. Atkins, and B.O. Russell; office Black Hawk; claim 150 by 1,300 feet, patented; discovered 1860; located on Bobtail Hill, Gregory mining district, within the limits of Black Hawk; course of vein, east and west, width 6 feet; nature of ore, copper pyrites; average value of mill ore, 12 ounces gold per cord, smelting ore $140 per ton; main shaft 225 feet deep.

Cashier. Owned by the Cashier Mining Co., of New York; incorporated 1878; capital stock, $500,000, in 250,000 shares, $2 each; office Central City, Colorado, and 29 Broad Street, New York City; Stephen B. French, president, John Sherry, secretary and treasurer, Joseph W. Holman, general superintendent and agent; claim 150 by 1,500 feet, patented; discovered 1874; located on Mammoth Hill, Gregory mining district, 400 feet from limits of Central City; course of vein, northeast and southwest, pitch 5 degrees south, width 4 feet, pay vein 18 inches; nature of ore, iron and copper pyrites; average value of mill ore, 3 ½ ounces gold per cord, smelting or $100 per ton; main shaft, 200 feet deep and 14 other shafts aggregating 280 feet, 1 level 60 feet in length, and 30 feet from surface; yield to date, $12,000.

Cashier, North. Owned by the Cashier Mining Co., of New York; Joseph W. Holman, general superintendent and agent; offices Central City, Colorado, and 29 Broad Street, New York City; claim 150 by 1,500 feet; discovered 1875; located on Mammoth Hill, Gregory mining district, 400 feet from limits of Central City; course of vein, northeast and southwest, pitch 5 degrees north, width 3 feet, pay vein 14 inches; nature of ore, iron and copper pyrites; average value, 3 ounces gold per cord; 1 shaft 40 feet deep.

Cashier, South. Owned by the Cashier Mining Co., of New York; Joseph W. Holman, general superintendent and agent; offices Central City, Colorado, and 29 Broad Street, New York City; claim 150 by 1,500 feet; discovered 1875; located on Mammoth Hill, Gregory mining district, 400 feet from the limits of Central City; course of vein, northeast and southwest, pitch 5 degrees north, width 4 feet, pay vein 12 inches; nature of ore, iron and copper pyrites; average value, 3 ½ ounces gold per cord; 1 shaft 110 feet deep; yield to date, $2,000.

Casto. Located on Casto Mountain, Central City mining district, near town of Central City.

Centennial Tunnel Mining Co. Incorporated October 23, 1876; capital stock, $500,000, in 500,000 shares, $1 each; office Central City; organized by John Tierney, William Burke, Charles Peterson, and Theodore H. Becker.

Central City. Owned by Azor A. Smith and others, of Black Hawk; claim 50 by 600 feet; discovered 1860; located at the junction of Spring and Gregory streets, Central City; vertical vein, trending northeast and southwest, width 7 feet, nature of ore, iron and copper pyrites; value, 20 ounces gold per cord, main shaft 140 feet deep, 1 level 60 feet in length; yield to date, $4,000.

Central City Tunnel. Owned by the Central City Tunnel Co.; organized under the laws of the State of New York in the year 1876; capital stock, $1,200,000, in 12,000 shares, $100 each; offices, Central City, Colorado, and 161 Broadway, New York City; J.W. Alder, president, J.C. Cockey, secretary, David G. Wilson, superintendent; claim 1,500 by 3,000 feet, located at the base of the eastern slope of Quartz Hill, within the limits of Central City; length of tunnel, 450 feet, course southwest; has reached 3 lodes: lode No. 1 at a distance of 100 feet, No. 2 at a distance of 180 feet, and No. 3 at a distance of 200 feet from the mouth; will intersect 30 lodes or mineral veins, among which are the Rice, Josephine, Lewis, Fortune, Fourth of July, Burroughs, Columbia, Gardner, Illinois, Sap, and Register; shaft house 25 by 50 feet, constructed of stone with iron roof, containing a drilling machine and apparatus, 2 engines, 10- and 25-horse power, respectively; the tunnel is 5 ½ feet wide by 6 high, is well timbered and supplied with iron track and cars.

Champion. Owned by Joseph Standley; office Bald Mountain; claim 50 by 1,000 feet, patented; discovered 1865; located on Nevada Flats, Eureka mining district, 1 mile from Central City; width of vein, 2 ½ feet; 1 shaft 130 feet deep.

Champion Mining and Reducing Co. Incorporated October 9, 1878; capital stock, $300,000, in 3,000 shares, $100 each; office Central City; trustees: Alfred Cowles, Charles T. Tugo, Charles A. Mair, Frank E. Morse, Edson Keith, Thomas I. Richman, and Francis G. Staltonstall.

Chihuahua. Owned by William H. Bush & Co.; discovered in Wide Awake mining district; width of vein 20 feet; value of ore, $50 per ton; developed by a tunnel 300 feet in length.

Clark Gardner. Owned by the Clark Gardner Mining Co.; organized under the laws of the State of New York, September 4, 1865; capital stock, $300,000, in 3,000 shares, $100 each; offices Central City, Colorado, Rome and New York City in the State of New York; John Striker, president, Thomas Striker, vice president, Samuel Wardwell, secretary, A.B. Clark, superintendent; claim 200 by 300 feet, patented; located on Quarts Hill, Nevada mining district, 1 mile from Central City; course of vein, east and west, width 10 feet, pitch south; nature of ore, sulphurets of copper and iron; average value of mill ore, 6 ounces gold per cord, smelting ore, $100 per ton; 2 shafts, aggregating 400 feet in depth, 2 levels, 200 and 300 feet from surface, respectively, aggregating 500 feet in length; shaft and engine house 40 by 80 feet, constructed of stone, containing a hoisting apparatus and a 30-horse power engine.

Clay County. Hon. Jerome B. Chaffee, proprietor; office Denver; claim 50 by 1,500 feet, patented; discovered 1859; located in Lake Gulch, Lake mining district, 1 mile from Central City; course of vein, northeast and southwest, width 3 feet, pitch 10 degrees north; nature of ore, iron and copper pyrites, average value 12 ounces gold per cord; main shaft 250 feet deep, and 2 other shafts, aggregating 600 feet, 2 levels 120 and 225 feet from surface, respectively, aggregating 500 feet in length; shaft house, 20 by 40 feet, blacksmith shop 16 by 20 feet, mill building 60 by 60 feet, 2 dwelling houses 20 by 60 and 35 by 40 feet, respectively, 12 acres of ground and a barn; yield to date, $400,000.

Cleveland Gold Mining Co. Organized under the laws of the State of New York; capital stock, $250,000, in 25,000 shares, $10 each; offices Central City, Colorado, and 54 William Street, New York City; H.A. Johnson, president, R.A. Johnson, secretary.

Clifton. Owned by the Clifton Mining Co., London, England; claim 150 by 1,400 feet; located on Virginia Hill, 3 miles from Central City.

Clipper. Owned by Ezra D. Frits, Joseph A. Thatcher, and others; claim 150 by 1,500 feet; discovered 1876; located on Casto Mountain, within the limits of Central City; course of vein, east and west, width 4 feet, pay vein 12 inches, pitch south; nature of ore, iron and copper pyrites, average value of mill ore 5 ounces gold per cord; 2 shafts, aggregate depth 135 feet.

Coaley. Owned by the Coaley and Gilpin Mining Co., of Black Hawk; claim 50 by 3,000 feet, patented; discovered 1867; located in Silver Gulch, Enterprise mining district, near the limits of Black Hawk; course of vein, northeast and southwest, width 3 ½ feet, pitch 20 degrees north, pay vein 1 to 2 feet; nature of ore, sulphurets of silver and lead; average value, 500 ounces silver per cord; main shaft 200 feet deep, 1 tunnel 140 feet, and 3 levels, aggregating 300 feet in length; shaft house 30 by 40 feet, hoisting apparatus and a 16-horse power engine; total yield, $65,000.

Coleman Gold Mining Co., New York. Capital stock, $1,000,000, in 100,000 shares, $10 each; office 80 Wall Street, New York city; Thomas I. Minford, president, Henry W. Lake, manager.

Colorado. Owned by Thomas Seary and Sylvester Nichols; office Central City; claim 150 by 1,500 feet; discovered 1864; located on Quartz Hill, Nevada mining district, within the limits of Central City; course of vein, early east and west, pitch south, width 3 ½ feet; nature of ore, iron and copper pyrites; average value of mill ore, 4 ounces gold per cord; main shaft 100 feet deep, 1 tunnel 225 feet in length; yield to date, $5,000.

Columbia. Thomas Seary, proprietor; office Central City; claim 50 by 100 feet; discovered 1862; located on Quartz Hill, Nevada mining district, within the limits of Central City; course of vein, nearly east and west, pitch south, width 4 feet; nature of ore, iron and copper pyrites; average value, 6 ounces gold per cord; main shaft 75 feet deep, 1 drift 75 feet in length.

Columbia. Owned by Hon. Henry M. Teller of Central City; claim 50 by 1,000 feet, patented; located on Quartz Hill, within the limits of Central City; course of vein, northeast and southwest; 2 shafts, 60 and 100 feet, respectively.

Columbus. William J. Barker, proprietor; office Denver; claim 150 by 1,000 feet, patented; discovered 1859; located in Saw-pit Gulch, Russell mining district, 1 mile from Central City; vertical vein, running northeast and southwest, width 5 feet, pay vein 3 ½ feet; nature of ore, iron and copper pyrites; average value, $45 per ton; main shaft 110 feet deep, 1 level 80 feet in length.

Columbus. James Frazer and Mary Hepburn, proprietors; office Central City; claim 50 by 300 feet; discovered 1862; located on Bobtail Hill, Gregory mining district, within the limits of Black Hawk; course of vein, northeast and southwest, pitch 15 degrees north, width 4 feet, pay vein 15 inches; nature of ore, iron and copper pyrites; average value, of mill ore, 3 ounces gold per cord, smelting ore, $50 per ton; main shaft 270 feet deep, aggregate length of levels, 250 feet.

Comet. James H. Reed, proprietor; office Black Hawk; claim 50 by 800 feet; discovered 1865; located in Illinois Gulch, half mile from Central City; course of vein, northeast and southwest, pitch 10 degrees north, width 3 ½ feet; nature of ore, sulphurets of iron and copper; average value, 3 ounces gold per cord; main shaft 40 feet deep; yield to date, $500.

Comstock. Owned by Ezra D. Frits, Joseph A. Thatcher, and others, of Central City; claim 150 by 1,500 feet; discovered 1876; located on Casto Mountain, within the limits of Central City; course of vein, northeast and southwest, width 5 feet, pitch 25 degrees south; nature of ore, iron and copper pyrites; average value of mill ore, 5 ounces gold per cord, smelting ore, $100 per ton; main shaft 450 feet deep; 3 levels, 130, 170, and 360 feet from surface, respectively, aggregating 400 feet in length; good shaft house, and a 10-horse power engine; yield to date, $50,000.

Confidence. Owned by Thomas H. Potter, of Central City; claim 50 by 1,400 feet, patented; located on Quartz Hill, Nevada mining district, near limits of Central City.

Consolidated Bobtail Gold Mining Co. Incorporated under the laws of the State of New York, November 19, 1869; capital stock, $1,053,300, in 210,660 shares. $5 each; offices Black Hawk, Colorado, and New York

City; George A. Hoyt, president, John Stanton, Jr., secretary and treasurer, Andrews N. Rogers, agent, Frank H. Messinger, assistant superintendent.

Continental Mill and Mining Co.'s Property. Located in Quartz Valley, 1 mile from Central City; claim 150 by 2,000 feet; main shaft 120 feet deep.

Cork. Owned by Hon. Henry M. Teller, of Central City; claim 50 by 600 feet, patented; discovered at Gunnell Hill, Nevada mining district, within the limits of Central City.

Corydon. Owned by Hon. Henry M. Teller; office Central City; claim 50 by 900 feet, patented; located on Gunnell Hill, Nevada mining district, within limits of Central City; vertical vein, running northeast and southwest, width 3 feet; nature of ore, iron and copper pyrites; average value, 4 ounces gold per ton; main shaft 270 feet deep, aggregate length of levels, 450 feet, 1 tunnel 170 feet in length; total yield, $30,000.

Cotton, East Half of Nos. 1, 2, and 3. Mrs. Francis Field, proprietor, Henry W. Lake, manager; office Black Hawk; claim 50 by 350 feet; discovered 1859; located on Bobtail Hill, Gregory mining district, within the limits of Black Hawk; vertical vein, trending northeast and southwest, width 3 feet, pay vein 12 inches; nature of ore, iron and copper pyrites; average value, $200 per cord; main shaft 130 feet deep, and 3 other shafts aggregating 140 feet, 1 level 100 feet in length; yield to date, $10,000.

Cotton, 50 feet West of Discovery. Henry W. Lake, proprietor; office Black Hawk; claim 50 by 50 feet; discovered 1859; located on Bobtail Hill, Gregory mining district, within the limits of Black Hawk; vertical vein, trending northeast and southwest, width 3 feet, pay vein 20 inches; nature of ore, iron and copper pyrites; average value of mill ore $75 per cord, smelting ore $40 per ton; main shaft 70 feet deep; yield to date, $3,000.

Crispin. Enos K. Baxter, proprietor; office Central City; claim 50 by 1,600 feet, patented; discovered 1870; located on Quartz Hill, Nevada mining district, within the limits of Central City; course of vein, northeast and southwest, width 3 feet, pay vein 18 inches, pitch 15 degrees north; nature of ore, sulphurets of copper and iron; average value 5 ounces gold per cord; main shaft 170 feet deep.

Crown Point and Virginius. Job V. Kimber & Co., proprietors; office Central City; claim 150 by 1,500 feet, patented; discovered 1859; located at the head of Virginia Canyon, Russell mining district, 2 miles from Central City; course of vein, northeast and southwest, width 3 feet, pay vein 1 foot, pitch 20 degrees north; nature of ore, copper pyrites; average value of mill ore, 14 ounces.

Curlew. Kelty & Tomlinson, proprietors; office Black Hawk; claim 150 by 1,500 feet; located between Chase Gulch and Clear Creek, within the limits of Black Hawk; width of vein, 1 foot; nature of ore, galena, iron and copper pyrites; 1 shaft 25 feet deep, 1 tunnel 125 feet in length; gold per cord, smelting ore, $80 per ton; main shaft, 125 feet deep, aggregated length of levels 300 feet; yield to date, $60,000.

Dallas. Owned by the Dallas Gold and Silver Mining Co., incorporated June 8, 1878; capital stock, $1,000,000, in 100,000 shares, $10 each; offices Black Hawk, Colorado, and Chicago, Illinois; organized by Samuel Mishler, William Germain, Mark F. Bebee, and James B. Norton; claim 150 by 2,642 feet, patented; discovered 1860; located on Dallas Hill, Enterprise mining district, within the limits of Black Hawk; vertical vein, trending northeast and southwest, width 5 feet; nature of ore, galena, iron and copper pyrites; average value of mill ore, 7 ounces gold per cord, smelting ore, $80 per ton; 6 shafts, aggregating 300 feet in depth, 3 levels, 60, 40, and 100 feet from surface, respectively, aggregating 500 feet in length. This property has paid in full the expense of development, and has on hand a surplus working capital.

Dallas, Extension. Owned by the Dallas Gold and Silver Mining Co., office Black Hawk; claim 150 by 1,300 feet, patented; discovered 1860; located on North Clear Creek, within the limits of Black Hawk; course of vein, nearly east and west, width 4 feet, pitch south; nature of ore, galena, iron and copper pyrites; average value 4 ounces gold per cord; 1 shaft 25 feet deep, 1 tunnel 165 feet in length.

Delaware. Owned by the Delaware Mining Association, of Boston; claim 150 by 500 feet; located in Russell Gulch, Russell mining district, 2 miles from Central City; course of vein, northeast and southwest, width 4 feet, pitch 10 degrees north, pay vein 18 inches; nature of ore, iron pyrites; average value, 5 ounces gold per cord; main shaft 375 feet, and 1 other shaft 150 feet deep, 1 drift 150 feet in length and 150 feet from surface; shaft house and mill building 40 by 60 feet each. ["Oh, My God Road" is the shortcut linking Russell Gulch and Idaho Springs. Most travelers on the road agree that it is well named.]

Delaware, Nos. 5 and 6 East. Elias Goldman and Jacob Mack, proprietors; office Central City; claim 50 by 200 feet; located on Alps Hill, Russell mining district, 2 miles from Central City; course of vein, northeast and southwest, width 4 feet, pay vein 18 inches, pitch 10 degrees north; nature of ore, galena, iron and copper pyrites; average value, $40 per ton.

Del Norte. Andrew M. Emery, proprietor; office Central City; claim 50 by 1,500 feet; discovered 1877; located within the limits of Black Hawk; course of vein, northeast and southwest, pitch 30 degrees west, width 4 feet, pay vein 2 feet; nature of ore, sulphurets of iron and copper; average value, $50 per ton; main shaft 80 feet deep.

Denmark. Owned by Joseph W. Holman; office Central City; claim 150 by 1,425 feet, patented; located on the Bobtail lode, Gregory mining district, within the limits of Black Hawk; course of vein, northeast and southwest, pitch 10 degrees south, width, 5 feet, pay vein 18 inches; nature of ore, iron and copper pyrites; average value of mill ore, 12 ounces gold per cord, smelting ore, $150 per ton; main shaft 134 feet, and 1 other shaft 30 feet in depth; 1 level 40 feet in length and 50 feet from surface; yield to date, $5,000.

Dillon. Owned by William Dillon, of Central City; claim 150 by 1,500 feet; located on Casto Mountain, within the limits of Central City; course of vein, northeast and southwest, pitch south, width 3 feet, pay vein 2 feet; nature of ore, sulphurets of copper and iron; average value, 4 ounces gold per cord; 2 shafts, aggregate depth 105 feet, 1 tunnel 14 feet in length.

Dorchester. William J. Barker, proprietor; office Denver; claim 50 by 300 feet; discovered 1859; located on Nottaway Hill, Russell mining district, 1 mile from Central City; course of vein, northeast and southwest, width 2 feet, pitch 3 degrees west; nature of ore, iron and copper pyrites; average value of mill ore, 5 ounces gold per cord, smelting ore, $85 per ton; main shaft 110 feet deep, 2 levels, 75 and 100 feet in length, respectively.

East Leavenworth. John F. Cheatley, George and R. H. Martin, proprietors; office Central City; claim 50 by 1,200 feet; discovered 1876; located on Leavenworth Hill, Russell mining district, 2 miles from Central City; width of vein, 3 feet, pitch north; nature of ore, iron and copper pyrites; average value, 6 ounces gold per cord; aggregated depth of shafting, 600 feet.

Ellery. Owned by the Belden & Tennal Mining Co.; David D. Belden, superintendent; office Central City; claim 150 by 2,000 feet, patented; discovered 1875; located on Casto Mountain, within the limits of Central City; width of vein 3 feet; average value of ore, 6 ounces gold per cord; reached by the Belden Tunnel at a distance of 100 feet from the mouth, and intersected at t depth of 300 feet.

Elvaretta. Thomas Davidson & Co., proprietors; office Bald Mountain; claim 100 by 1,500 feet; discovered 1860; located on Quartz Hill, Nevada mining district, 2 miles from Central City; width of vein 18 inches; nature of ore, galena; average value, 5 ounces gold per cord; 3 shafts, aggregating 280 feet in length.

English-Kansas. James C. Fagan, proprietor; offices Central City, Colorado, and New York City; claim 50 by 700 feet, patented; located on Quartz Hill, Nevada mining district, near limits of Central City; course of vein, northeast and southwest, width 4 feet, pitch 15 degrees south; nature of ore, iron and copper pyrites; average value, 8 ounces gold per cord; main shaft 450 feet deep; good shaft house and machinery.

Enterprise. Owned by J.A. Perley, Kelly & Co., of Black Hawk; claim 50 by 200 feet; discovered 1867; located in Silver Gulch, Enterprise mining district, within the limits of Black Hawk; vertical vein, running northeast and southwest, width 4 ½ feet; nature of ore, sulphurets of silver; average value, 150 ounces silver per cord; main shaft 120 feet deep, 1 level 75 feet in length; shaft house 25 by 30 feet.

Essex. Owned by Jason E. Scobey and J. Bohram; office Central City; claim 50 by 1,400 feet, patented; discovered 1862; located in Prosser Gulch, Central City mining district; course of vein, northeast and southwest, width 6 feet, pitch north; nature of ore, decomposed quartz; average value, 4 ounces gold per cord; 2 shafts aggregating 75 feet in depth.

Essex Tunnel. Owned by the Essex Gold and Silver Mining and Tunneling Co., of Newark, New Jersey; capital stock, $500,000; claim 1,500 by 2,000 feet; located on Pewabic Mountain, Russell mining district, 50 miles from Denver; length of tunnel 875 feet.

Eureka. Alexander Taylor, proprietor; office Black Hawk; claim 50 by 1,700 feet; discovered 1864; located on Eureka Hill, Eureka mining district, near limits of Central City; course of vein, northeast and southwest, width 6 feet, pitch 35 degrees north; nature of ore, sulphurets of copper and iron; average value, 9 ounces gold per cord; 10 shafts, aggregating 1,000 feet, and 4 levels, aggregating 500 feet in length; yield to date, $60,000.

Express. Owned by Harley B. Morse, of Central City; claim 150 by 1,500 feet, patented; located on Quartz Hill, Nevada mining district, 2 miles from Central City; course of vein, northeast and southwest, pitch north, width 5 feet, pay vein 3 feet; nature of ore, galena and iron pyrites; main shaft 30 feet deep.

Extension. Owned by Hon. Henry M. Teller, of Central City, claim 50 by 1,700 feet, patented; discovered 1865; located on Quartz Hill, Illinois Central mining district, near limits of Central City.

Fagan-Gunnell. Operated by the Monmouth Consolidated Gold Mining Co., of New York; D. Sullivan, superintendent; claim 100 by 800 feet, patented; located on Gunnell Hill, near Central City; main shaft 500 feet deep.

Fairfield. Owned by the Hill Gold Mining Co., of Providence, Rhode Island; Prof. Nathaniel P. Hill, manager; office Denver; claim 50 by 600 feet; discovered 1861; located at head of Russell Gulch, 2 miles from Central City; course of vein, north and south, width 3 feet, pitch west, pay vein 18 inches; nature of ore, iron and copper pyrites; average value of mill ore, 16 ounces gold per cord, smelting ore, $100 per ton; main shaft 150 feet deep, 1 level 120 feet in length and 60 feet from surface.

Fennelly. Owned by Thomas H. Potter, Sylvester Nichols, William Fennelly, and Robert Kerwan; office Central City; claim 150 by 1,500 feet, patented; discovered 1863; located on Caledonia Hill, Hawkeye mining district, 3 miles from Central City; course of vein, nearly east and west, pitch north, width 4 ½ feet; nature of ore, iron and copper pyrites; average value of mill ore, 3 ½ ounces gold per cord; main shaft 125 feet deep, and 6 other shafts aggregating 125 feet in depth, 1 tunnel 100 feet and 1 level, 60 feet in length; shaft house, whim and hoisting apparatus; total yield, $10,000.

First Centennial. Owned by L.C. Snyder and R.S. Haight; office Central City; claim 150 by 1,500 feet, patented; discovered 1875; located on Casto Mountain, within the limits of Central City; course of vein, northeast and southwest, width 5 feet, pay vein 3 feet, pitch 27 degrees south; nature of ore, galena, iron and copper pyrites; average value of mill ore, 10 ounces gold per cord, smelting ore, $125 per ton; main shaft 140 feet, 1 tunnel 200 feet in length and 60 feet from surface, 1 level 260 feet in length.

First National. Owned by the First National Mining Co., incorporated October 4, 1871; capital stock, $1,000,000, in 50,000 shares, $20 each; offices Nevada, Colorado, and Boston, Massachusetts; organized by Benjamin S. and William J. Rotch, Samuel H. Gookin, Thomas J. and W. Tracy Custis; patented claim; located on Quartz Hill, Nevada mining district, 1 mile from Central City; course of vein, east and west, width 3 feet, pitch south; nature of ore, iron and copper pyrites; average value of mill ore, 7 ounces gold per cord, smelting ore $75 per ton; main shaft 650 feet, and 1 other 300 feet deep, aggregate length of levels 1,800 feet; shaft house and machinery.

Fisk. Operated by George W. Mabee, of Central City; claim 50 by 190 feet; discovered 1860; located on Bobtail Hill, Gregory mining district, within the limits of Black Hawk; course of vein, nearly east and west, width 4 feet, pay vein, 3 feet, pitch north and south; nature of ore, iron and copper pyrites; average value, 6 ounces gold and 24 ounces silver per cord; main shaft 615 feet deep, 7 levels, aggregating 2,500 feet in length; reached by the Bobtail Tunnel at a distance of 700 feet from the mouth and intersected at a depth of 280 feet. On the surface is a shaft house, 16 by 20, an engine house, 30 by 40, sorting room, 20 by 25, and a cobbing house, 10 by 20 feet, a 30-horse power engine and hoisting apparatus. The yield of this mine for the 6 years previous to 1876 was $300,000.

Fisk. New York and Colorado Co.; B.T. Wells, manager; owns 250 linear feet on this lode, which is located in Gregory mining district, in town of Black Hawk; course of vein, northeast and southwest, width 4 feet, pitch 10 degrees south; nature of ore, iron and copper pyrites; average value of mill ore, 2 ½ ounces gold per cord, smelting ore, $100 per ton; main shaft 420 feet deep, 1 tunnel 150 feet in length, connected with the Bobtail Tunnel.

Fisk. Whipple Gold Mining Co.; owns 50 by 200 feet; located in Gregory mining district, within the limits of Black Hawk; course of vein, northeast and southwest, width 4 feet, pitch 10 degrees south; nature of ore, iron and copper pyrites; average value 5 ounces gold per cord; main shaft 190 feet deep.

Fisk. Owned by the Fisk Gold Mining Co., of New York; claim, 50 by 200 feet, patented; located on Bobtail Hill, Gregory mining district, within the limits of Black Hawk.

Flack. Located on the northwestern slope of Quartz Hill, Nevada mining district, 2 miles from Central City; total yield, $500,000.

Foot and Simmons. Located on the northeastern slope of Gregory Hill, in the town of Black Hawk.

Forks. Located on the west end of Quartz Hill, Nevada mining district, 2 miles from Central City; course of vein, northeast and southwest; average width 8 feet; nature of ore, galena iron and copper pyrites; yield, $500,00.

Forrester. Owned by Nicholas C. Forrester and Elijah C. Sherman; office Denver; claim 150 by 1,500 feet; discovered 1876; located in Silver Lake mining district, 5 miles from Central City; course of vein, northeast and southwest, width 2 ½ feet, pitch 5 degrees west, pay vein 4 to 12 inches; nature of ore, gray copper and galena; average value, 200 ounces silver per ton, 1 tunnel 150 feet in length; aggregate length of levels, 175 feet; total yield, $8,000.

Fortune. Owned by the Central City Tunnel Co.; David G. Wilson, superintendent; offices Central City, Colorado, and 161 Broadway, New York City; claim 1860; located on Quartz Hill, within the limits of Central City; course of vein, northeast and southwest, pitch 5 degrees south, width 4 feet; nature of ore, sulphurets of copper and iron; average value, 6 ounces gold per cord; main shaft 130 feet deep; operated through the Central City Tunnel; yield to date, $2,000.

Fourth of July. William H. Doe, Jr. and Herbert Waterman, proprietors; office Bald Mountain; claim 150 by 1,500 feet; discovered 1862; located on Quartz Hill, Nevada mining district, 1 mile form Central City; course of vein, east and west, width 3 feet, pay vein 18 inches, pitch 3 degrees south; nature of ore, galena, iron and copper pyrites; average value, 10 ounces gold per cord; 6 shafts aggregating 200 feet in depth, and

3 levels, 80, 90, and 120 feet from surface, respectively, aggregating 180 feet in length.

Freeman. Owned by Thomas Seary, of Central City; claim 150 by 1,500 feet; discovered 1865; located on Quartz Hill, Nevada mining district, within the limits of Central City; width of vein 4 feet; average value of ore, 6 ½ ounces gold per cord; 2 shafts aggregating 200 feet in depth, and 3 drifts aggregating 200 feet in length.

Fremont. Owned by Harry A. La Paugh, of Denver; claim 150 by 1,500 feet; discovered 1876; located in Silver Lake mining district, 5 miles from Central City; course of vein, east and west, width 15 feet; nature of ore, iron and copper pyrites and bismuth; average value of mill ore, 6 ounces gold per cord, smelting ore, $250 per ton; main shaft 150 feet deep, 1 tunnel 250 feet deep, and 3 levels aggregating 400 feet in length; shaft house and 2 dwelling houses.

Furnald. Owned by the Belden & Tennal Mining Co., David D. Belden, superintendent; office Central City; claim 150 by 2,000 feet, patented; discovered 1877; located on Casto Mountain, Central City mining district; width of vein, 2 feet; average value of ore, 5 ounces gold per cord; reached by the Belden Tunnel [see separate listing] at a distance of 200 feet from the mouth, and intersected at a depth of 300 feet.

Gardner. Located on the northeastern slope of Quartz Hill, between the Roderick Dhu and California lodes; Nevada mining district, 1 mile from Central City; has produced over $500,000.

Garrison. Located on the Kansas lode on Quartz Hill, Nevada mining district, 2 miles from Central City.

Gaston. Located on Bates Hill, Gregory mining district, within the limits of Black Hawk.

Gauntlet. J.C. Franks, proprietor; office Central City; claim 50 by 750 feet; discovered 1873; located on Quartz Hill, Illinois Central mining district, within the limits of Central City; course of vein, northeast and southwest, width 3 feet, pay vein 1 foot, pitch 5 degrees north; nature of ore, galena and iron pyrites; average value of mill ore, 5 ounces gold per cord, smelting ore, $75 per ton; 3 shafts, aggregating 200 feet in depth, 1 level 75 feet in length and 90 feet from surface.

German. Owned by J.F. and George J. Kline; office Central City; claim 50 by 1,850 feet, patented; discovered 1859; located on German Hill, Mountain City mining district, within the limits of Central City; course of vein, east and west, width 4 feet, pitch north; nature of ore, iron and copper pyrites; average value of smelting ore, $50 per ton; 2 shafts aggregating 490 feet in depth, 1 level 350 feet in length and 200 feet from surface.

German. Owned by the Susquehanna Gold Mining Co.; Frank H. Messinger, agent; offices Black Hawk, Colorado, and 33 North Water Street, Philadelphia; Albert G. Buzby, president, Samuel Broadbent, treasurer; claim 50 by 815 feet, patented; discovered 1860 located on Mammoth Hill, Gregory mining district, within the limits of Central City; course of vein, northeast and southwest, width 3 feet, pitch north; nature of ore, iron and copper pyrites; average value of mill ore, 5 ounces gold per cord, smelting ore, $40 per ton; 2 shafts aggregating 400 feet in depth.

German Branch. J.F. and George J. Kline, proprietors; office Central City; claim 50 by 2,000 feet; discovered 1868; located on German Hill, Mountain City mining district, within the limits of Central City; course of vein, east and west, width 3 feet, pitch 45 degrees north; nature of ore, iron and copper pyrites; average value of mill ore, 10 ounces gold per cord, smelting ore, $100 per ton; 2 shafts, aggregating 380 feet in depth, 1 level 80 feet in length and 175 feet from surface.

German Tunnel. Owned by German Tunnel Co., incorporated January 1875; capital stock, $15,000, in 30 shares, $500 each; office Central City; organized by Alexander Carstens, Gustave Kruse, Charles Bamberry, Deitmar Schneider, Gotfried Krug, Henry Becker, Joseph Kramer, John Kruse, Jacob Pressler, Philip Pressler, Fidel Pflum, Michael Thies, and Jacob Tosser. The tunnel is located on Gregory Street, Central City, is 500 feet in length, and has cut 5 lodes.

Gerrione. Owned by Robert G. Aduddel, M.D., of Central City; claim 50 by 750 feet; discovered 1868; located on Quartz Hill, Nevada mining district, 2 miles from Central City; vertical vein, running northeast and southwest, width 4 feet; nature of ore, iron pyrites, average value of mill ore, 3 ½ ounces gold per cord, smelting ore, $60 per ton; main shaft 125 feet deep; shaft house, 15 by 18 feet.

Gettysburg. Owned by Joseph B. Hafer, of Central City; claim 50 by 725 feet; located on Bates Hill, within the limits of Central City, vertical vein, running northeast and southwest, width 3 feet, pay vein 18 inches;

nature of ore, iron and copper pyrites; average value of mill ore, 8 ounces gold per cord, smelting ore, $40 per ton; main shaft 120 feet deep, 1 tunnel 60 feet in length; total yield, $6,000.

Gibson. Owned by Henry M. Teller and Alva Mansur; office Central City; claim 50 by 300 feet, patented; located on Quartz Hill, Illinois Central mining district, near the limits of Central City.

Gilpin. Owned by the Gilpin Gold Mining Co.; John G. Tappan, president; office Boston, Massachusetts; claim 50 by 262 feet, patented; discovered 1859; located on Quartz Hill, Nevada mining district, 1 mile from Central City; course of vein, northeast and southwest, width 3 feet, pitch south; nature of ore, iron and copper pyrites; average value of mill ore, 6 ounces gold per cord, smelting ore, $100 per ton; main shaft 600 feet, and 1 other 150 feet deep.

Gilpin. Owned by the Coaley and Gilpin Mining Co., of Black Hawk; claim 50 by 2,000 feet, patented; discovered 1867; located on Silver Gulch, Enterprise mining district, within the limits of Black Hawk; course of vein, northeast and southwest, width 4 ½ feet, pitch 5 degrees north; nature of ore, sulphurets of silver and lead; average value, 150 ounces silver per cord; shaft 200 feet deep, 1 tunnel 550 feet and 3 levels aggregating 125 feet in length; total yield, $65,000.

Glennan, 1 to 5 West. James H. Ried, proprietor; office Black Hawk; claim 50 by 400 feet; discovered 1865; located on Bullion Mountain, Enterprise mining district, within the limits of Black Hawk; course of vein, northeast and southwest, pitch 5 degrees north, width 3 feet, pay vein 2 feet; nature of ore, galena, iron and copper pyrites; average value 5 ½ ounces gold per cord and 20 ounces silver per ton; main shaft 70 feet deep, 1 level 20 feet in length.

Glennan, From 1 to 9 East. James Kelly, proprietor; office Black Hawk; claim 50 by 872 feet, patented; discovered 1865; located on Bullion Mountain, Enterprise mining district, with the limits of Black Hawk; course of vein, northeast and southwest, pitch 5 degrees north, width 3 feet, pay vein 2 feet; nature of ore, iron and copper pyrites, galena and zinc blende; average value, 5 ½ ounces silver per ton; main shaft 70 feet, and 1 other shaft 65 feet deep, 1 tunnel 100 feet in length.

Gold Dirt. Owned by John Q.A. Rollins, of Rollinsville; located on Gold Dirt Hill, 8 miles from Central City; main shaft 500 feet deep; yield, $400,000.

Gold Ring. Owned by Horatio E. and Morris Hazard; office Central City; claim 50 by 1,500 feet; discovered 1873; located on Leavenworth Hill, Russell mining district, 1 mile from Central City; course of vein, northeast and southwest, width 2 feet, pitch north; nature of ore, iron and copper pyrites; average value, 9 ounces gold per cord; main shaft 100 feet deep.

Gold Rock Mining Co., New York. Edmund Driggs, president; capital stock, $500,000, in 50 shares of $100 each.

Golden Anchor. Owned by William Nicholson and O.M Williams; office Central City; claim 150 by 1,500 feet; discovered 1872; located on Wide Awake Hill, Hawkeye mining district, 6 miles from Central City; course of vein, nearly east and west, pitch 35 degrees south, width 3 ½ feet, pay vein 2 feet; nature of ore, iron pyrites; average value, 5 ounces gold per cord; main shaft 45 feet deep, 2 tunnels on vein 175 feet apart, 30 and 100 feet in length, respectively; total yield, $5,000.

Golden Eagle. Owned by A.C. Johnson, of Black Hawk; claim 50 by 600 feet; discovered 1863; located on Bobtail Hill, Gregory mining district, near limits of Central City; course of vein, northeast and southwest, width 4 feet, pitch 3 degrees south; nature of ore, iron pyrites; average value of mill ore, 10 ounces gold per cord, smelting ore, $100 per ton; 3 shafts, aggregating 250 feet in depth, 1 level 30 feet in length and 110 feet from surface.

Golden Eagle, Nos. 3, 4, 5, and 6 East. Owned by James H. Ried; office Black Hawk; claim 50 by 400 feet; discovered 1863; located on Bobtail Hill, half mile from Central City; course of vein, northeast and southwest, pitch 10 degrees north, width 3 feet, pay vein 2 to 2 feet; nature of ore, iron and copper pyrites; average value 7 ½ ounces gold per cord; main shaft 100 feet deep; yield to date, $5,000.

Gomer. Owned by Edward W. Williams, William and Edward Jones, and Owen Hughes; office Central City; claim 150 by 1,500 feet; discovered January 1878; located in Russell Gulch, Russell mining district, 3 miles from Central City; course of vein, northeast and southwest, pitch 5 degrees north, width 3 feet, pay vein 18 inches; nature of ore; iron and copper pyrites; average value of mill ore, 5 ounces gold per cord; main shaft 100 feet deep, 1 drift 300 feet in length. [William and Edward Jones were both born in Wales and lived with their families next door to each other in Russell Gulch. In the 1880 Federal Census, both reported that they were miners. Richard Owen Hughes, also born in Wales, was a boarder in the William Jones household. Their partner Edward W. Williams also lived in Russell Gulch with his wife Annie and younger sister Elizabeth; all three were born in Wales. He reported his occupation as grocer.]

Granby Gold and Silver Mining Co., of Colorado. Incorporated June 25, 1875; capital stock, $50,000, in 5,000 shares, $10 each; office Denver; organized by J.H. Jones, A.J. Williams, E. Winsrow Cobb, J. George Hoffer, and Charles F. Leimer.

Grand Army-Gunnell. Job V. Kimber & Co., proprietors; office Black Hawk; claim 50 by 1,600 feet, patented; discovered 1871; located on the Gunnell lode, within the limits of Central City; course of vein, east and west, width 4 feet, pitch 15 degrees south; nature of ore, iron and copper pyrites; average value of mill ore, 5 ounces gold per cord, smelting ore, $50 per ton; main shaft 600 feet, and 1 other 95 feet deep, 4 levels, 200, 357, 300, and 450 feet from surface, respectively, aggregating 500 feet in length.

Granite. Owned by Hon. Henry M. Teller, of Central City; claim 50 by 700 feet, patented; discovered in Gregory mining district, near Black Hawk.

Grant County. Owned by William M. Rule of Central City; claim 100 by 1,600 feet, discovered in Chase Gulch, near Black Hawk.

Gregory. Discovered in Gregory mining district, in town of Black Hawk; discovered 1859; yield to date, $9,000,000; the Briggs Bros. own 540 linear feet on this lode, and the New York and Colorado Co., 800 feet.

Gregory-Second. Located on Bates Hill, Enterprise mining district, near Black Hawk, runs parallel to the Gregory, and has a shaft 400 feet deep.

Grizzly. Owned by Charles McKee of Central City; claim 50 by 1,400 feet, patented; discovered 1876; discovered in Davenport Gulch, Russell mining district, 2 miles from Central City; width of vein 15 inches;

average value of ore, 8 ounces gold per cord; main shaft 150 feet deep, 1 level 60 feet in length and 40 feet from surface.

Gunnell. Owned by William Fullerton, Job V. Kimber, Richard Mackey, Mosley & Ballard, and Edward F. Clinton; office Central City; claim 50 by 454 feet, patented; discovered 1859; located on Gunnell Hill, Eureka mining district, within the limits of Central City; course of vein, east and west, width 3 feet, pitch south; nature of ore, iron pyrites; main shaft 700 feet deep. On the surface is a large shaft house, constructed of stone, containing engines and machinery for hoisting and handling ore. The yield of this mine for one year, under the present management, was $300,000.

Gunnell. Owned by the University Gold Mining Co., Central City; Henry W. Lake, manager; claim 50 by 98 feet; located on Gunnell Hill, within the limits of Central City; course of vein, east and west, average width 3 feet, pitch 40 degrees south, pay vein 1 foot; nature of ore, copper pyrites, average value of mill ore, 8 ounces gold per cord, smelting ore, 8 ounces gold per cord, smelting ore, $80 per ton; main shaft 550 feet deep, aggregate length of levels 500 feet; shaft house, 40 by 60 feet, hoisting apparatus and a 20-horse power engine; total yield, $120,000.

Gunnell. Owned by Miss Mary T. Scudder; John Scudder, manager; office Central City; claim 100 by 600 feet; located on Gunnell Hill, within the limits of Central City; course of vein, nearly east and west, average width 3 feet pitch 40 degrees south, pay vein 1 foot; nature of ore, copper pyrites; average value of mill ore, 9 ounces gold per cord, smelting ore, $70 per ton; main shaft 350 feet, and 3 other shafts, aggregating 100 feet in depth; aggregated length of levels 600 feet; total yield, $230,000.

Gunnell. Owed by the Coleman Gold Mining Co., of New York; claim 50 by 155 feet; located on Gunnell Hill, within the limits of Central City; course of vein, east and west, average width 3 feet, pitch 40 degrees south, pay vein 1 foot; nature of ore, copper pyrites, average value of mill ore, 7 ounces gold per cord, smelting ore, $80 per ton; main shaft 450 feet deep; aggregate length of levels, 600 feet; shaft house, 30 by 40 feet, hoisting apparatus and a 20-hourse power engine; total yield, $200,000.

Gunnell. Owned by Hal. Sayre and Horace H., Atkins, of Central City; claim 50 by 449 feet; located on Gunnell Hill, within the limits of Central City; course of vein, northeast and southwest, width 3 feet, pitch 30 degrees south, pay vein 2 feet; nature of ore, copper pyrites; average value of mill ore, 4 ounces gold per cord, smelting ore, $60 per ton; main shaft 230 feet; shaft house, 20 by 40 feet; total yield, $4,000.

H.M. Woods. Owned by John S. Hough, Thomas Shanley, and others; office Lake City; claim 300 by 1,500 feet; located one mile from Capitol City; Galena mining district, 10 miles from Lake City; course of vein, northeast and southwest, width 4 feet, pay vein 8 inches; nature of ore, iron and copper pyrites; value, 20 ounces silver per ton; 1 drift 15 feet in length.

Harper-California. Owned by William M. Roworth; offices Central City and Denver, Colorado; claim 50 by 100 feet, patented; discovered 1860; located on Quartz Hill, 2 miles from Central City; course of vein, east and west, width 3 feet; nature of ore, iron and copper pyrites; average value of mill ore, 6 ounces gold per cord, smelting ore, $100 per ton; main shaft 375 feet deep; shaft house and hoisting apparatus; yield, $80,000.

Harry. Owned by Henry Chatillon, of Central City; claim 150 by 1,500 feet; discovered 1877; located on Winnebago Hill, Quartz Valley mining district, near limits of Central City; vertical vein, running northeast and southwest, width 3 ½ feet, pay vein 6 inches; nature of mill ore, galena, iron and copper pyrites; average value of mill ore, 6 ounces gold per cord, smelting ore, $40 per ton; main shaft 40 feet deep.

Harsh. Owned by the Rockdale Gold Mining Co., New York; Edward W. Henderson of Central City, agent; claim 50 by 800 feet, patented; discovered 1861; located in Leavenworth Gulch, Russell mining district, 2 miles from Central City; course of vein, northeast and southwest, width 4 feet; nature of ore, iron and copper pyrites; average value of mill ore, 5 ounces gold per cord; main shaft 180 feet deep; mill building, 50 by 100 feet.

Hawley-Gardner. Orrin Thurber, proprietor, Sidney B. Hawley, manager; office Bald Mountain; claim 50 by 354 ¼ feet, patented; discovered 1860; located on Quartz Hill, 1 mile from Central City; course of vein, east and west; width, 4 feet, pitch south; nature of ore, iron and copper pyrites; main shaft, 600 feet deep, 4 levels aggregating 600 feet in length.

Herbert Spencer. Owned by the Belden and Tennal Mining Co.; David D. Belden, superintendent; office Central City; claim 150 by 1,500, patented; discovered 1877; located on Casto Mountain, within the limits of Central City; width of vein 3 feet; average value of ore, 5 ounces gold per cord; reached by the Belden Tunnel, at a distance of 450 feet from the mouth.

Hercules. Henry J. Hawley and Barney Cochrane, proprietors; office Central City; claim 150 by 1,500 feet; discovered 1877; located near the limits of Central City; course of vein, northeast and southwest, width 5 feet, pay vein 10 inches, pitch 45 degrees south; nature of ore, iron and copper pyrites; average value, 3 ounces gold per cord; main shaft 65 feet deep.

Hidden Treasure – California. Owned by Aaron M. Jones, John Johnson, and Samuel V. Newell; offices Bald Mountain and Denver, Colorado; claim, 50 by 400 feet, patented; discovered 1859; located on Quartz Hill, in town of Nevadaville, 2 miles from Central City; course of vein, nearly east and west, width 4 feet, pay vein 2 feet, pitch 4 degrees south; nature of ore, iron and copper pyrites; average value of mill ore, 8 ounces gold per cord, smelting ore, $90 per ton; main shaft, 620 feet deep, and 2 other shafts 200 feet each, 2 levels 500 and 550 feet from surface, respectively, aggregating 800 feet in length; on the surface is a shaft house, 15 by 16 feet, and an ore house, 20 by 60 feet, hoisting apparatus and a 40-horse power engine; the average monthly yield of this mine for the last 5 months has been $15,000, making a total to date of $75,000.

Hill House. Owned by Robert G. Aduddel, M.D., and David Barnes, of Central City; claim 150 by 1,033 feet; located on the southern slope of Quartz Hill, Illinois mining district, 1 mile from Central City; vertical vein, running northeast and southwest, width 3 feet; nature of ore, iron pyrites; average value of mill ore, 6 ounces gold per cord, smelting ore, $80 per ton; main shaft 100 feet deep.

Holman. Owned by William H. Bush; office Central City; claim 150 by 1,000 feet, patented; located on Quartz Hill, Nevada mining district, within the limits of Central City.

Home Prospecting Association. Incorporated July 28, 1875; capital stock, $1,000, in 1,000 shares, $1 each; office Central City; organized by Frank C. Young, George E. Randolph, Joseph H. Bostwick, Dennis Sullivan, W. Edmundson, George W. Mabee, Sylvester Nichols, Robert A. Campbell, and Foster Nichols.

Howe. James McNasser, proprietor; office Denver; claim 150 by 1,500 feet; discovered 1863; located on Quartz Hill, Nevada mining district, 1 mile from Central City; course of vein, northeast and southwest, width 2 feet, pitch north; nature of ore, iron and copper pyrites; average value, 6 ounces gold per cord; 3 shafts aggregating 100 feet in depth.

Hubert. Owned by the Hense Estate, Barney Koch, administrator; office Central City; claim 50 by 1,500; discovered 1862; located on Prize Hill, Nevada mining district, 2 miles from Central City; course of vein, east and west, width 3 feet, pay vein 18 inches, pitch 20 degrees south; nature of ore, iron and copper pyrites; average value of mill ore, 6 ounces gold per cord, smelting ore, $75 per ton; 2 shafts aggregating 150 feet in depth, 2 levels, 75 and 100 feet from surface, respectively, aggregating 100 feet in length.

Hunter. Owned by the Susquehanna Gold Mining Co.; Frank H. Messinger, agent; offices Black Hawk, Colorado, and 33 North Water Street, Philadelphia, Pennsylvania; claim 50 by 200 feet; discovered 1864; located on Mammoth Hill, Gregory mining district, within the limits of Central City; vertical vein, trending, northeast and southwest, width 2 feet; nature of ore, iron and copper pyrites; 3 shafts, aggregating 300 feet in depth.

Illinois. Owned by Thomas H. Potter and George W. Jacobs; office Central City; claim 50 by 1,000 feet, patented; discovered 1859; located on the eastern slope of Quartz Hill, Nevada mining district, half mile from Central City; course of vein, northeast and southwest; width 4 feet, pitch 5 degrees south; nature of ore, iron and copper pyrites; average value, $40 per ton; main shaft 230 feet, and 1 other shaft 187 feet deep, 2 levels 555 and 350 feet in length, respectively; over the mine is a large and well constructed building containing a 22-stamp mill, an engine, tubular boiler, hoisting apparatus, etc.; yield, $350,000.

Indiana. Owned by Harley B. Morse, of Central City; claim 100 by 1,300 feet; discovered 1859; located on the west end of Quartz Hill, 2 miles from Central City; course of vein, east and west, pitch 20 degrees south, width 4 feet, pay vein 2 feet; nature of ore, galena, iron and copper pyrites; average value of mill ore, 4 ounces gold per cord, smelting ore, $50 per ton; main shaft 260 feet deep.

Irish Flag. Owned by James C. Fagan; located on the northwest slope of Quartz Hill, 2 miles from Central City; course of vein, east and west.

Irene. Owned by Henry Chatillon, of Central City; claim 150 by 1,500 feet; discovered 1877; located on Winnebago Hill, near the limits of Central City; vertical vein, running northeast and southwest, width 3 feet, pay vein 14 inches; nature of ore, iron and copper pyrites; average value of mill ore, 7 ounces gold per cord, smelting ore, $80 per ton; main shaft 125 feet deep; total yield, $3,000.

Irving. Owned by Harley B. Morse, of Central City; claim 50 by 400 feet, patented; located on Nevada Flats [now a ghost town], 2 miles from Central City; course of vein, northeast and southwest, pitch north, width 5 feet, pay vein 3 feet; nature of ore, iron pyrites and galena; average value, $30 per ton; main shaft 40 feet deep.

Jackson. Located on the north slope of Gunnell Hill, 1 mile from Central City.

James Henry. Henry Bolthoff, Sidney W. Tyler, and Jacob Pressler, proprietors; office Central City; claim 150 by 1,400 feet; discovered 1875; located on Gunnell Hill, Eureka mining district, within the limits of Central City; course of vein, northeast and southwest, width 4 feet, pay vein 18 inches; nature of ore, iron and copper pyrites; average value of mill ore, 5 ounces gold per cord, smelting ore, $40 per ton; main shaft 150 feet, and 1 other 45 feet deep, 4 levels, aggregating 160 feet in length.

Jessie. Owned by the Victory Mining Co., of Wisconsin; O.G. Raynor, superintendent; offices Central City, Colorado, and Fon du Lac, Wisconsin; claim 150 by 750 feet, patented; discovered 1874; located on Quartz Hill, Illinois Central mining district, within the limits of Central; course of vein, southwest and northeast, width 4 feet, pitch south; nature of ore, iron and copper pyrites; average value of mill ore, 6 ounces gold per cord, smelting ore, $100 per ton; 1 shaft 50 feet deep, 1 level 25 feet in length.

Jessie, Westerly Half. Owned by L.E. Hoskins, of Oshkosh, Wisconsin; claim 150 by 750 feet; located on Quartz hill, Illinois Central mining district, within the limits of Central City; 1 shaft 70 feet deep, 1 level 30 feet in length.

John L Emerson. Owned by John L. and Charles H. Emerson and Wilbur W. Flagg; office Central City; claim 150 by 4500 feet; discovered 1867; located on Belleview Mountain, Virginia mining district, 3 miles from Central City; course of vein, northeast and southwest, pitch 5 degrees north, width 5 feet; nature of ore, galena, iron and copper pyrites; main shaft 187 feet deep, 3 drifts aggregating 185 feet in length.

Jones. Owned by Daniel McGonigal, of Bald Mountain; claim 50 by 200 feet; discovered 1862; located on Prize Hill, Nevada mining district, 2 miles from Central City; course of vein, east and west, width 2 feet, pitch 5 degrees south, pay vein 18 inches; nature of ore, iron and copper pyrites; average value of mill ore, 8 ½ ounces gold per cord, smelting ore, $110 per ton; main shaft 300 feet deep, 3 levels 230, 240, and 300 feet from surface, respectively, aggregating 450 feet in length; good shaft house, and machinery.

Justice. Owned by William M. Roworth, M.H. Root, Hiram Hill, and others; office Central City; claim 150 by 1,600 feet, patented; discovered 1860; located in Lake Gulch, near the limits of Black Hawk; vertical vein, trending northeast and southwest, width 4 feet; nature of ore, iron and copper pyrites; average value of mill ore, 15 ounces gold per cord; main shaft 280 feet deep; yield to date, $50,000.

Justice. William M. Roworth, proprietor; office Central City; claim 50 by 1,100 feet; discovered 1860; located near Lake Gulch, Lake mining district, 1 mile from Central City; vertical vein, trending northeast and southwest, width 3 feet; nature of ore, iron and copper pyrites; average value, 6 ounces gold per cord; 1 shaft, 125 feet deep.

Kansas. Owned by the La Crosse Gold Mining Co., New York; Andrews N. Rogers, superintendent; offices Central City, Colorado, and 59 William Street, New York City; claim 50 by 120 feet, patented; discovered 1859; located on Quartz Hill, Nevada mining district, 2 miles from Central City; course of vein, northeast and southwest; width 3 feet, pitch 30 degrees south, pay vein 1 foot; nature of ore, iron and copper pyrites; average value of mill ore, 5 ounces gold per cord, smelting ore, $100 per ton; main shaft 60 feet deep; reached by the La Crosse Tunnel at a distance of 150 feet from the mouth and intersected at a depth of 60 feet.

Kansas. Owned by the Boston and Colorado Gold Mining Co., of Boston, Massachusetts; H.S. Chase, president; claim 50 by 190 feet; located on Quartz Hill, 2 miles from Central City.

Kansas. Owned by the University Gold Mining Co.; claim 50 by 200 feet; located on Quartz Hill, 2 miles from Central City; main shaft 600 feet deep; yield for 18 months $120,000.

Kent County. Richard Mackey, proprietor; office Bald Mountain; claim 50 by 3.256 feet, patented; discovered 1860; located on the northwestern slope of Quartz Hill, Nevada mining district, 2 miles from Central City; course of vein, east and west, width 3 ½ feet, pay vein 2 feet; pitch 15 degrees south; nature of ore, sulphurets of copper and iron, average value of mill ore, 4 ½ ounces gold per cord; 3 shafts aggregating 600 feet in depth, 5 levels 250, 300, 400, 475, and 550 feet from surface, respectively, aggregating 1,100 feet in length; good shaft house, containing a 60-horse power engine, for hoisting and pumping. [This mine was intersected by the Argo Tunnel in Idaho Springs.]

Keystone, and Extension. James McNasser, proprietor; office Denver; claim 150 by 2,200 feet, patented; discovered 1861, located on Quartz Hill, Nevada mining district, 1 mile from Central City; course of vein, northeast and southwest, width 2 feet, pitch 25 degrees north; nature of ore, galena, iron and copper pyrites; 3 shafts, aggregating 230 feet in depth.

Kilbourn Mining Co. Incorporated October 4, 1873; capital stock, $50,000, in shares of $100; office Central City; organized by F. Kilbourn, H.C. Lett, H.M. Atkinson, and A. Kilbourn.

King. Owned by John M. Ross, Alexander McLeod, John B. and Daniel J. McKay; office Central City; claim 150 by 1,500 feet; discovered 1860; located on Nevada Flats [now a ghost town], Nevada mining district, 1 mile from Central City; width of vein 4 feet; average value of ore, 6 ounces gold per cord; 2 shafts, aggregated depth 240 feet, 2 levels aggregating 500 feet in length.

Kinney-Gardner. Owned by D. Sullivan & Co., of Central City; claim 100 by 650 feet, patented; discovered 1862; located on Quartz Hill, Nevada mining district, 2 miles from Central City; course of vein, nearly east and west, width 3 feet, pitch 10 degrees south, pay vein 2 feet; nature of ore, copper pyrites; average value of mill ore, 9 ounces gold per cord; smelting ore, $125 per ton; main shaft 300 feet deep; shaft house, 15 by 15 feet, with hoisting apparatus; total yield, $30,000.

Kip. Owned by Bela S. Buell, of Central City; claim 50 by 1,200 feet, patented; located on Mammoth Hill, Gregory mining district, within the limits of Central City; course of vein, northeast and southwest, width 6 feet, pitch 5 degrees north; nature of ore, iron and copper pyrites; average value of mill ore, 3 ounces gold per cord, smelting ore, $100 per ton; main shaft 300 feet, and 1 other, 260 feet deep, aggregate length of levels 1,200 feet.

Kirk. Located in the Illinois Central mining district, 1 mile from Central City; claim 100 by 1,500 feet, patented.

Lacrelus. William H. Doe, Sr., proprietor; office Central City; claim 150 by 1,500 feet; discovered 1878, located on Eureka Hill, Eureka mining district, near the limits of Central City; course of vein, east and west, width 3 feet, pay vein 2 feet, pitch 5 degrees north; nature of ore, iron and copper pyrites; average value of mill ore, 4 ounces gold per cord, smelting ore, $50 per ton; main shaft, 25 feet deep.

La Crosse Tunnel. Owned by the La Crosse Gold Mining Co., organized under the laws of the State of New York in the year 1864; capital stock, $2,000,000, in 200,000 shares, $10 each; John Van Nest, president, Henry Smith, secretary, Andrews N. Rogers, superintendent; offices Central City, Colorado, and 59 William Street, New York City; length of tunnel, 956 feet; located in Nevada Gulch at the base of Quartz Hill, 1 mile from Central City.

Lake View. Owned by Michael and Patrick Flynn, Thomas Kennedy, Alfred McGruder, and Patrick Maher; office Central City; claim 150 by 3,000 feet; discovered 1860; located on Mammoth Hill, Nevada mining district, 1 mile from Central City; width of vein 2 ½ feet; average value of ore, 5 ounces gold per cord; 2 shafts, aggregating 125 feet in depth.

Lamberson & Warren. Owned by Harley B. Morse, of Central City; claim 50 by 1,600 feet, patented; discovered 1867; located on Nevada Flats, Nevada mining district, 2 miles from Central City; course of vein, east and west, pitch 20 degrees north, width 4 feet, pay vein, 2 feet; nature of ore, pyrites of iron, galena and zinc blende; average value, $50 per ton; main shaft 50 feet deep.

Leavenworth. Horatio E. and Morris Hazard, proprietors; office Central City; claim 50 by 442 feet; discovered 1878; located on Leavenworth Hill, Russell mining district, 1 mile from Central City; course of vein, northeast and southwest, width 2 feet, pitch south; nature of ore, iron and copper pyrites; average value of mill ore, 7 ounces gold per cord; main shaft 80 feet deep.

Leavenworth. Alexander Taylor, proprietor; office Black Hawk; claim 50 by 1,250 feet; discovered 1859; located on Leavenworth Hill, Russell mining district, 2 miles from Central City; vertical vein, trending northeast and southwest, width 4 feet, pay vein 3 feet; nature of ore, sulphurets of iron and copper; average value, 6 ounces gold per cord; 2 shafts aggregating 300 feet in depth, 2 levels, aggregate length 250 feet.

Leavitt. Bela S. Buell, proprietor; office Central City; claim 50 by 1,000 feet; discovered 1862; located in Gregory Gulch, within the limits of Central City; course of vein, northeast and southwest, width 8 feet, pay vein 6 inches to 8 feet, pitch 5 degrees south; nature of ore, iron and copper pyrites; average value of mill ore, 4 ounces gold per cord, smelting ore, $100 per ton; main shaft 550 feet deep and 5 other shafts aggregating 1,100 feet, 7 levels, 25, 90, 180, 250, 300, 400, and 500 feet from surface respectively, aggregating 5,000 feet in length; has produced largely.

Legal Tender Mining Co. Incorporated under the laws of the State of Iowa, July 1878; capital stock, $60,000, in 6,000 shares, $10 each; offices Central City, Colorado, and Marshaltown, Iowa; John Turner, president, Frank L. Downend agent; the property of this company comprises the Taylor Tunnel in Clear Creek County and the Clay County lode [see separate listing] in Gilpin County.

Lewis. William H. Bush, proprietor; office Central City; claim 150 by 1,300 feet, patented; located on Quartz Hill, Nevada mining district, within the limits of Central City; course of vein, northeast and southwest, width, 3 ½ feet, pitch south; nature of ore, iron and copper pyrites, average value of mill ore, 9 ounces gold per cord, smelting ore, $150 per ton; 2 shafts aggregating 200 feet in depth, 2 levels aggregating 130 feet in length.

Linden Castle. Owned by James H. Ried, Thomas Clery, and Thomas Skelley; office Black Hawk; claim 50 by 3,000 feet; discovered 1866; located in Slaughter House Gulch, half mile from Black Hawk; course of vein, northeast and southwest, pitch 12 degrees north, width 3 feet, pay vein 1 to 10 inches; nature of ore, sulphurets of iron and copper; average value, $100 per ton; main shaft 100 feet deep, 1 drift 20 feet in length; yield to date, $1,200.

Little Bessie. Owned by John Palmer and William McBreen; office Bald Mountain; claim 100 by 1,500 feet; discovered 1860; located on Quartz Hill, Nevada mining district, 2 miles from Central City; width of vein, 1 foot; nature of ore, galena and iron pyrites; average value, 5 ounces gold per cord; main shaft 100 feet deep.

Little Comstock. Taylor & Chackfield, proprietors; office Black Hawk; claim 50 by 1,000 feet; discovered 1874; located on U.P.R. Hill, within the limits of Central City; vertical vein, trending northeast and southwest, width 3 feet, pay vein 2 feet; nature of ore, sulphurets of iron and copper; average value, 8 ounces gold per cord; 1 shaft 60 feet deep, 1 level 70 feet in length.

Little Gunnell Gold Mining Co. Organized under the laws of the State of New York; offices Central City, Colorado, and New York City; Lewis C. Rockwell, agent.

Los Angelos. George W. Estabrook, proprietor; office Bald Mountain; claim 150 by 800 feet, discovered 1865; located on Russell Hill, Russell mining district, 2 miles from Central City; vertical vein, trending east and west, width 2 feet, nature of ore, iron and copper pyrites; average value of mill ore, 6 ounces gold per cord; 3 shafts, aggregating 100 feet in depth, 1 level 350 feet in length.

Louisiana. Owned by Jacob Mack, of Denver; claim 50 by 1,200 feet; discovered 1861; located in Russell Gulch, Russell mining district, 1 mile from Central City; course of vein, northeast and southwest, width 4 feet, pitch 10 degrees south, pay vein 6 inches; nature of ore, iron and copper pyrites; average value, $40 per ton; main shaft 70 feet deep, 1 drift 25 feet in length.

Lyman. Owned by Hon. Henry M. Teller, of Central City; claim 50 by 500 feet, patented; located on the southern slope of Gunnell Hill, Nevada mining district, within the limits of Central City.

M.B.H. Owned by M.B. Hyndman and Abel Knight; office Central City; claim 150 by 1,500 feet, patented; located on Quartz Hill, Nevada mining district, 1 mile from Central City; course of vein, northeast and southwest, width 3 feet, pitch north; nature of ore, iron and copper pyrites; average value, 7 ounces gold per cord; main shaft 65 feet deep.

Mammoth. Owned by the Gold Rock Mining Co., New York; claim 50 by 232 feet, patented; located on Mammoth Hill, Gregory mining district, within the limits of Central City; vertical vein running east and west, width 6 feet; nature of ore, pyrites of iron; average value of mill ore,

5 ounces gold per cord, smelting ore, $25 per ton; main shaft 400 feet deep, 1 drift 100 feet in length; shaft house, 40 by 60 feet.

Mammoth. Owned by Mrs. Louisa A. Willard; claim 50 by 1,582 feet, patented; discovered 1859; located on Mammoth Hill, Gregory mining district, within the limits of Central City; vertical vein, running east and west, width 7 feet, pay vein 5 feet; nature of ore, pyrites of iron; average value of mill ore, 3 ounces gold per cord, smelting ore, $22 per ton, main shaft 140 feet deep, 1 level 150 feet in length; yield for the past 2 years from surface ore, $10,000.

Mammoth Side. Owned by Harley B. Morse, of Central City; claim 150 by 1,300 feet; discovered 1876; located on Mammoth Hill, within the limits of Central City; course of vein, northeast and southwest, pitch 5 degrees south, width 6 feet; nature of ore, pyrites of iron; average value of mill ore, 2 ½ ounces gold per cord, smelting ore, $25 per ton; main shaft 100 feet deep.

Mammoth, 6, 7, and 8, West. Owned by Edward L. Salisbury, of Central City; claim 150 by 300 feet; discovered 1859; located on Gregory Hill, Gregory mining district, within the limits of Central City; course of vein, nearly east and west, pitch 5 degrees north, width 20 feet; nature of ore, iron and copper pyrites; average value of mill ore, 5 ounces gold per cord, smelting ore, $12 per ton; main shaft 280 feet deep; total yield, $10,000; the Bobtail Tunnel is now within 150 feet of this property.

Mammoth Gulch Placer Mining Co. Incorporated July 1, 1878; capital paid up, stock $500,000, in 50,000 shares, $10 each; offices Georgetown, Colorado, and New York City; trustees: Timothy G. Negus, Charles A. Martine, Frank M. Taylor, Elisha S. Weaver, and Jacob H. Abeel.

Manhattan, on Fisk Lode. Owned by the Manhattan Gold Mining Co.; George E. Randolph, agent; office Central City; claim 50 by 190 feet, patented; discovered 1859; located on Bobtail Hill, Gregory mining district, within the limits of Black Hawk; width of vein 3 feet; nature of ore, iron and copper pyrites; average value of mill ore, 5 ounces gold per cord, smelting ore, $50 per ton; 1 shaft 600 feet deep; intersected by the Bobtail Tunnel.

Marine. Taylor Lindsley, proprietor; office Central City; claim 150 by 900 feet, patented; discovered 1860; located on Gunnell Hill, Eureka mining district, within the limits of Central City; course of vein, east and west, width 4 feet, pitch south; nature of ore, galena and sulphurets of iron; average value of mill ore, 4 ½ ounces gold per cord, smelting ore, $75 per ton; 2 shafts aggregating 140 feet in depth, 1 tunnel 200 feet in length; yield to date, $4000.

Matt France. Willard and Henry M. Teller and Matt. France, proprietors; office Central City; claim 50 by 1,600 feet, patented; located on Quartz Hill, Nevada mining district, half mile from Central City; course of vein, northeast and southwest, width 2 ½ feet; nature of ore, iron and copper pyrites; average value, $60 per ton; main shaft 100 feet, and 1 other shaft 75 feet deep, aggregate length of levels 100 feet.

Maryland. Theodore H. Becker, proprietor; office Central City; claim 50 by 1,500 feet; discovered 1859; located in Chase Gulch, within the limits of Central City; course of vein, northeast and southwest, width 2 feet, pitch 15 degrees south; nature of ore, galena, zinc blende, and sulphurets; average value of mill ore, 7 ounces gold per cord, smelting ore, $75 per ton; main shaft 125 feet deep, 2 levels 75 and 100 feet from surface, respectively, aggregating 225 feet in length.

Maurer. Located on Gold Dirt Hill, Independent mining district, 8 miles from Central City.

Mazeppa. Discovered in Russell mining district, 3 miles from Central City; 1 shaft 125 feet deep; yield, $6,000.

McAdams. Owned by Harley B. Morse, of Central City; claim 100 by 1,500 feet, patented; located on Mammoth Hill, within the limits of Central City; course of vein, northeast and southwest; pitch 5 degrees south, width 3 feet, pay vein 1 foot; nature of ore, pyrites of iron; average value of mill ore, 2 ½ ounces gold per cord; main shaft 50 feet deep.

McAllister. Owned by Thomas McAllister and Simon Ewing; office Central City; claim 50 by 1,500 feet, patented; discovered 1872; located on Mammoth Hill, Gregory mining district, within the limits of Central City; course of vein, northeast and southwest, pitch 5 degrees south, width 3 feet, pay vein 18 inches; nature of ore, iron and copper pyrites; average value, 4 ounces gold per cord; main shaft 150 feet deep; yield to date, $10,000.

Mechanics. Owned by Charles F. Hendrie and Henry Boltoff, of Central City; claim 150 by 1,500 feet; discovered 1875; located on Gunnell Hill, Central City mining district, within the limits of Central City; course of vein, northeast and southwest, width 2 feet, pay vein 3 inches to 1 foot, pitch 10 degrees south; nature of ore, iron pyrites; average value, 4 ounces gold per cord; main shaft 50 feet deep operated through the Mechanics Tunnel.

Mechanics Tunnel. Owned by Charles F. Hendrie and Henry Boltoff, of Central City; claim 150 by 3,750 feet; discovered 1875; located at the base of Gunnell Hill, within the limits of Central City; course southwest; length 200 feet.

Mendota. William Hannah, proprietor; office Central City; claim 150 by 1,500 feet; discovered 1876; located at the head of Virginia Canyon, Virginia mining district, 3 miles from Central City; course of vein, northeast and southwest, width 3 feet, pitch north; nature of ore, iron and copper pyrites; average value 4 ½ ounces gold per cord; 1 shaft 100 feet deep.

Mercer County. Mercer Gold Mining Co., of New York, proprietors; Sidney Davis, of Bald Mountain, agent; claim 50 by 500 feet; located on Quartz Hill, Nevada mining district, 1 mile from Central City; vertical vein, running northeast and southwest, width 5 feet; nature of ore, iron and copper pyrites; average value of mill ore, 5 ounces gold per cord, smelting ore, $80 per ton; main shaft 200 feet deep 2 drifts, 50 and 150 feet from surface, respectively, aggregating 160 feet in length.

Mineral and Extension. James McNasser, proprietor; office Denver; claim 75 by 2,400 feet, patented; discovered 1862; located on Quartz Hill, Nevada mining district, 1 mile from Central City; course of vein, north and south, width 3 feet, pitch 25 degrees west; nature of ore, galena, iron and copper pyrites; 2 shafts aggregating 100 feet in depth.

Minnie. A.P. Baker and E.W. Tyler, proprietors; office Central City; claim 150 by 1,500 feet; discovered 1876; located on Gregory Hill, Gregory mining district, near the limits of Central City; vertical vein, running northeast and southwest, width 3 feet, pay vein 18 inches; nature of ore, iron and copper pyrites; average value, 8 ounces gold per cord; main shaft 150 feet deep shaft house, 14 by 20 feet; yield, $7,000.

Missouri. Mercer Gold Mining Co., of New York, proprietors; Sidney Davis, of Bald Mountain, agent; claim 50 by 500 feet, patented; located on Quartz Hill, Nevada Mining District, half mile from Central City; course of vein, northeast and southwest, width 4 feet, pitch 10 degrees north; nature of ore, iron and copper pyrites; average value, 7 ounces gold per cord, smelting ore, $80 per ton; main shaft 200 feet, and 1 other shaft 200 feet deep; 3 levels, 50, 75, and 160 feet from surface, respectively, aggregating 300 feet in length.

Monmouth-Kansas. Owned by the Monmouth Consolidated Gold Mining Co., New Jersey; incorporated 1875; capital stock, $450,000; Asburton Fountain, president, James C. Fagan, vice president, Charles Fountain, treasurer, D. Sullivan, superintendent; offices Central City, Colorado, and Matawan, New Jersey; claim 50 by 600 feet, patented; discovered 1859; located on Quartz Hill, Nevada mining district, 1 mile from Central City; course of vein, northeast and southwest, width 4 feet, pitch 15 degrees south; nature of ore, mispickle, zinc blende, iron and copper pyrites; average value of mill ore, 7 ounces gold per cord, smelting ore, $100 per ton; main shaft 775 feet deep, aggregate depth of other shafting, 500 feet; aggregate length of levels, 6,000 feet; shaft and engine house, 50 by 60 feet with a wing 30 by 40 feet, containing 2 engines, 40- and 70-horse power, respectively, 4 boilers and a 6-inch Cornish pump; mill building 80 by 120 feet containing 60 stamps, weighting 525 pounds each, 1 steam pump and a 75-horse power engine with boilers; capacity for treating 48 tons of ore in 24 hours; yield in one year under present management, $100,000.

Monroe. Owned by the La Crosse Gold Mining Co., New York; Andrews N. Rogers, superintendent; offices Central City, Colorado, and New York City; claim 50 by 400 feet, patented; discovered 1860; located on Quartz Hill, Nevada mining district, 2 miles from Central City; course of vein, northeast and southwest, width 3 feet, pitch 20 degrees south, pay vein 1 foot; nature of ore, iron and copper pyrites; average value, 4 ounces gold per cord; reached by the La Crosse Tunnel at a distance of 160 feet from the mouth and intersected at a depth of 70 feet.

Montie, Extension. Owned by O.M. Albro and H.E. Hyatt; office Bald Mountain; claim 150 by 1,000 feet; discovered 1863; located on Quartz Hill, Illinois Central mining district, 1 mile from Central City; vertical vein, trending northeast and southwest, width 4 feet; nature of ore, iron and copper pyrites; average value, 4 ounces gold per cord; main shaft 50 feet deep.

Montrose. Owned by Peter Thompson & Co., Chicago, Illinois; claim 50 by 1,000 feet, patented; discovered 1869; located on Eureka Hill, Eureka mining district, within the limits of Central City; width of vein, 6 feet; average value of ore, 7 ounces gold per cord, smelting ore, $65 per ton; main shaft 65 feet deep.

Mount Desert. Owned by James C. Fagan; claim 50 by 831 feet; located on the northwestern slope of Quartz Hill, Nevada mining district, 2 miles from Central City; course of vein, east and west; main shaft 300 feet deep.

Mountain City. Owned by Christopher C. Miller, Frederick Kuhster, Joseph Case, and Franz Hatch, of Central City; claim 50 by 1,100 feet, patented; discovered 1865; located on Mammoth Hill, Gregory mining district, within the limits of Central City; course of vein, northeast and southwest, width 3 feet, pitch 25 degrees north, pay vein 2 inches to 3 feet; nature of ore, iron and copper pyrites; average value of mill ore, 5 ounces gold per cord, smelting ore, $100 per ton; main shaft 225 feet, and 3 other shafts, aggregating 415 feet in depth, 4 levels aggregating 200 feet in length; reached by the Centennial Tunnel; yield to date, $1,800.

Mountain Rose. Owned by Thomas McAllister and Simon Ewing; office Central City; claim, 50 by 1,400 feet; discovered 1869; located at head of Packard Gulch, 1 mile from Central City; course of vein, northeast and southwest, pitch 5 degrees north, width 1 foot; nature of ore, iron and copper pyrites; average value, 7 ounces gold per cord; main shaft 50 feet deep.

Napoleon. Belden and Tennal Mining Co., proprietors; David D. Belden, superintendent; office Central City; claim 150 by 1,500 feet, patented; discovered 1877; located on Casto Mountain, within the limits of Central City; course of vein, east and west, width 5 feet, pay vein 2 ½ feet; nature of ore, galena, iron and copper pyrites; average value, 4 ½ ounces gold per cord; reached by the Belden Tunnel at a distance of 500 feet from the mouth.

Narragansett. Owned by the Narragansett Mining Co.; Frank H. Messinger, agent; claim 50 by 400 feet, patented; located within the limits of Central City; width of vein 5 feet; average value of ore, 4 ounces gold per cord; 2 shafts, aggregating 800 feet in depth.

National. Owned by Robert B. Smock; office Central City; claim 150 by 1,500 feet; discovered 1876; located on the eastern slope of Quartz Hill, within the limits of Central City; course of vein, east and west, pitch 15 degrees south, width 3 feet, pay vein 18 inches; nature of ore, sulphurets of copper and iron; average value of mill ore, 3 ounces gold per cord, smelting ore, $21 per ton; main shaft 105 feet, and 1 other shaft 90 feet deep, 2 tunnels aggregating 210 feet, and 2 drifts aggregating 140 feet in length; shaft house, 26 by 18 feet.

National Mining Co. Incorporated November 20, 1869; capital stock, $5,000,000, in 5,000 shares, $1,000 each; office Central City; Frank C. Messinger, president, Mary Johnson, vice president, Peter C. Johnson, treasurer, Charles M. Leland, secretary, George H. Hill, chief engineer.

Nemaha, Nos. 1 and 2 East. Owned by Mrs. Francis Field; Henry W. Lake, of Black Hawk, manager; claim 50 by 100 feet; located on Bobtail Hill, within the limits of Black Hawk, vertical vein, running northeast and southwest, width 3 feet, pay vein 6 to 12 inches; nature of ore, iron and copper pyrites; average value of mill ore, 4 ounces gold per cord; main shaft 80 feet deep, tunnel 40 feet in length; yield, $1,000.

Nehama, No. 1, West. Owned by Henry W. Lake, of Black Hawk; claim 50 by 100 feet; discovered 1860; located on Bobtail Hill, within the limits of Black Hawk; vertical vein, running northeast and southwest, width 3 feet, pay vein 6 to 12 inches; nature of ore, iron and copper pyrites; average value of mill ore, 4 ounces gold per cord; main shaft 25 feet deep; yield, $200.

New Foundland [sic]. Discovered in Nevada mining district, 1 mile from Central City; width of vein, 4 feet; main shaft 400 feet deep; yield, $50,000.

New York and Colorado Co. Organized under the laws of the State of New York, in the year 1872; capital stock, $1,000,000, in 50,000 shares, $20 each; A.L. Pritchard president, W.C. Sheldon, vice president, H. Groenmayer, secretary, B.T. Wells, superintendent and general manager; offices Black Hawk, Colorado, and 52 William Street, New York City; claims consist of 800 linear feet on the Gregory lode and 550 linear feet on the Briggs lode, all of which is patented and located on the Colorado Central Railroad, Main Street, Black Hawk; course of vein, northeast and southwest, width 4 ½ feet, pitch 10 degrees south; nature of ore, iron and copper pyrites; average value of mill ore, 3 ½ ounces gold per cord, smelting ore, $100 per ton; main shaft 725 feet, air shaft 900 feet, and 1 other shaft 500 feet deep, 1 tunnel 200 feet in length, 4 levels 500, 570, 640, and 700 feet from surface, respectively, aggregating 2,000 feet in length; on the surface is a mill building, 50 by 100 feet, containing 40 stamps, weighing 570 pounds each, 1 engine, 100-horse power, 4 Bartola pans and a 6-inch Cornish pump; the mill has a capacity for treating 40 tons of ore in 24 hours.

Nimrod. Owned by M. Halm; offices Central City, Colorado, and Columbus, Ohio; claim 50 by 1,200 feet, patented; discovered 1862; located on Quartz Hill, Illinois mining district, 1 mile form Central City; course of vein, northeast and southwest, width 3 feet, pitch 5 degrees north, pay vein 6 to 12 inches; nature of ore, galena and iron pyrites; average value of mill ore, 5 ounces gold per cord, smelting ore, $50 per ton; main shaft 315 feet, and 4 other shafts aggregating 240 feet in depth, 2 levels, 170 and 200 feet from surface, respectively, aggregating 500 feet in length; shaft house, 20 by 75 feet, containing a hoisting apparatus, and a 10-horse power engine.

Nothrup. Owned by the Colorado Central Gold and Silver Mining Co.; offices Central City, Colorado, and Morrison, Illinois; claim 150 by 1,500 feet; discovered 1862 located on Quartz Hill, Nevada mining district, 2 miles from Central City; width of vein 2 ½ feet; average value of ore, 4 ounces gold per cord, smelting ore, $100 per ton; main shaft, 80 feet deep, 1 level 50 feet from surface, and 50 feet in length.

Nottaway, Half of Nos. 6 and 7, West. Owned by Robert M.D. Morrison; office Central City; claim 150 by 150 feet; located in Lake Gulch, Lake mining district., 2 miles from Central City; course of vein, northeast and southwest, pitch 10 degrees north, width 3 feet; nature of ore, iron and copper pyrites; average value of mill ore, 6 ounces gold per cord; main shaft 50 feet deep; yield, $600.

Nottaway. Owned by Hon. Jerome B. Chaffee; office Denver; claim 50 by 1,500 feet, patented; discovered 1862; located in Lake Gulch, Lake mining district, 1 mile from Central City; course of vein, nearly east and west, pitch 10 degrees north, width 3 feet, pay vein 12 inches to 3 feet; nature of ore, iron and copper pyrites; average value of mill ore, 16 ounces gold per cord, smelting ore, $350 per ton; main shaft, 350 feet, and 4 other shafts aggregating 400 feet in depth, 2 levels, 100 and 200 feet from surface, respectively, aggregating 650 feet in length; yield $200,000.

O.K. James W. Hannah, proprietor; office Denver; claim 150 by 1,500 feet, patented; discovered 1860; located on Mammoth Hill, Gregory mining district, within the limits of Central City; course of vein, northeast and southwest, width 42 inches, pitch 20 degrees south; nature of ore, iron and copper pyrites; average value of mill ore, 6 ounces gold per cord, smelting ore, $25 to $125 per ton; main shaft 325 feet deep and other shafting aggregating 300 feet, 4 levels 40, 115, 225, and 285 feet from surface, respectively, aggregating 700 feet in length; shaft and engine house, 45 by 60 feet, 20-horse power engine, hoisting apparatus and 60 feet of iron tramway; yield to date, $21,000.

Old Bullion. Owned by Hon. Henry M. Teller, of Central City; claim 50 by 100 feet, patented.

Omaha. Henry W. Lake, proprietor; office Black Hawk; claim 50 by 200 feet; discovered 1860; located at the head of Packard Gulch, Gregory mining district, 2 miles from Black Hawk; course of vein, northeast and southwest, pitch north, pay vein 6 inches; 3 shafts aggregating 100 feet in length.

Omaha. Owned by Mrs. Francis Field, Henry W. Lake, and C.W. Mathers; office Black Hawk; claim 150 by 1,200 feet, patented; discovered 1859; located at the head of Packard Gulch, Gregory mining district, 1 mile form Central City; course of vein, northeast and southwest, pitch 10 degrees north, width 3 feet, pay vein 6 inches to 1 foot; nature of ore, iron and copper pyrites; average value, $100 per ton; main shaft 130 feet, and 3 other shafts aggregating 200 feet in depth, 2 open cuts, 40 feet in length each; yield, $2,000.

Omaha, Nos. 2 and 3 East. Owned by the New York and Colorado Co.; B.T. Wells, manager; offices Black Hawk, Colorado, and 52 William Street, New York City; claim 200 linear feet; discovered at the head of Packard Gulch, 1 mile from Central City; course of vein, northeast and southwest, width 3 feet, pitch 10 degrees north.

Ophir-Burroughs. Owned by the Ophir Gold Mining Co.; George E. Randolph, agent; office Central City; claim 50 by 462 feet, patented; discovered 1859; located on Quartz Hill, Nevada mining district, 1 mile from Central City; course of vein, northeast and southwest, width 3 feet; pitch south; nature of ore, iron and copper pyrites; average value of mill ore, 8 ounces gold per cord, smelting ore, $50 to $150 per ton; main shaft

1,000 feet, and 3 other shafts aggregating 1,000 feet in depth; has produced largely.

Ophir-Kansas. Owned by the Ophir Gold Mining Co.; George E. Randolph, agent; office Central City; claim 50 by 200 feet, patented; discovered 1859; located on Quartz Hill, Nevada mining district, 1 mile from Central City; course of vein, northeast and southwest, width 3 feet; nature of ore, iron and copper pyrites; average value of mill ore, 6 ounces gold per cord, smelting ore, $75 to $100 per ton; main shaft 450 feet deep.

Parent. Owned by Horace H. Atkins, of Georgetown; located on the summit of the hill south of the Gregory; discovered 1864; main shaft 225 feet.

Peck. Owned by A.C. Johnson, Black Hawk; claim 50 by 600 feet; discovered 1859; located in Leavenworth Gulch, Illinois mining district; 2 miles from Central City; width of vein 4 feet; average value of ore, 3 ½ ounces gold per cord, smelting ore, $100 to $150 per ton; main shaft 84 feet deep.

Peck. Owned by William J. Barker and Henry J. Hawley; office Central City; claim 150 by 400 feet; discovered 1868; located in Leavenworth Gulch, Illinois mining district, 1 mile from Central City; vertical vein, trending northeast and southwest, width 5 feet; nature of ore, iron and copper pyrites; average value, 3 ½ ounces gold per cord; 3 shafts aggregating 300 feet in depth.

Peck & Thomas. Owned by Hon. Henry M. Teller, of Central City; claim 50 by 600 feet, patented; discovered 1860; located on Mammoth Hill, Gregory mining district, within the limits of Central City.

Pendleton. Owned by the Gold Rock Mining Co., New York; claim 50 by 1,600 feet, patented; discovered 1859; located at the head of Virginia Canyon, Russell mining district, 2 miles from Central City; course of vein, northeast and southwest, pitch 5 degrees south, width 4 feet, pay vein 18 inches; nature of ore, iron pyrites and galena; average value of mill ore, 3 ounces gold per cord, smelting ore, $60 per ton; main shaft, 200 feet deep; mill building 40 by 80 feet.

Perigo. Owned by John Q. A. Rollins and H.P. Cowenhoven; office Central City, and Boulder, Colorado; claim 150 by 200 feet, patented; discovered 1861; located on Dust Mountain, Independent mining district, 8 miles from Central City; width of vein, 2 ½ feet; average value of ore, 8 ounces gold per cord; 1 tunnel 50 feet in length, 1 level 100 feet from surface and 350 in length.

Perrin. Owned by Thomas H. Potter, of Central City; claim 50 by 500 feet, patented; located on Leavenworth Hill, Russell mining district, 2 miles from Central City.

Perrin. Owned by the Perrin Gold mining Co.; office Central City; claim 50 by 400 feet, patented; located on Leavenworth Hill, Russell mining district, 2 miles from Central City; main shaft 260 feet deep.

Pewabic. Owned by Edward S. Perrin; office Central City; claim 50 by 1,000 feet, patented; discovered 1865; located on Pewabic Mountain, Russell mining district, 2 miles from Central City; course of vein, northeast and southwest, pitch 35 degrees north, width 4 feet; nature of ore, iron and copper pyrites; average value, $100 per ton; main shaft 270 feet, and 1 other shaft 125 feet deep, 1 drift 125 feet in length; shaft house, 28 by 48 feet, containing a 20-horse power engine and hoisting apparatus; yield to date, $75,000.

Pewabic. Owned by Henry E. Lyon; office Georgetown; claim 50 by 400 feet, patented; discovered 1865; located in Graham Gulch, Russell mining district, 2 miles from Central City; course of vein, northeast and southwest, pitch 35 degrees north, width 4 feet; nature of ore, iron and copper pyrites; average value, $100 per ton; main shaft 225 feet deep, 1 drift 100 feet in length; shaft house, 40 by 45 feet, containing a 40-horse power engine and hoisting apparatus; total yield, $75,000.

Pewabic, No. 3. Owned by Hon. Henry M. Teller, of Central City; claim 50 by 1,200 feet, patented; discovered in Graham Gulch, on Pewabic Mountain, Russell mining district, 2 miles from Central City.

Phenix Mining and Mineral Land Co., of Colorado. Incorporated June 3, 1876; capital stock, $200,000, in 20,000 shares, $10 each; office Denver; organized by John W. and William B. Russell, Clarence P. Elder, Eli M. Ashley, and Wilbur C. Lothrop.

Philadelphia-Gardner. Owned by the Monnier Mining Co., of Philadelphia; claim 50 by 200 feet, patented; discovered 1860; located on Quartz Hill, Nevada mining district, 2 miles from Central City; course of vein, nearly east and west, width 4 feet, pitch 10 degrees south, pay vein 3 feet; nature of ore, copper pyrites; average value of mill ore, 8 ounces gold per cord, smelting ore, $125 per ton; main shaft 500 feet deep, aggregate length of levels 600 feet; shaft house, 40 by 60 feet, containing a hoisting apparatus and a 25-horse power engine; total yield, $260,000.

Philadelphia-Gardner. D. Sullivan, proprietor; office Central City; claim 60 by 500 feet, patented; discovered 1871; located on Quartz Hill, Nevada mining district, 1 mile from Central City; width of vein 4 feet; average value of ore, 5 ounces gold per cord; main shaft 230 feet deep, 3 levels 75, 85, and 230 feet from surface, respectively, aggregating 165 feet in length.

Phoenix-Burroughs. Owned by the La Crosse Gold Mining Co.; Andrews N. Rogers, agent; office Central City; claim 115 by 305 feet; discovered 1860; located on Quartz Hill, Nevada mining district, 1 mile from Central City; width of vein 3 feet; average value of ore, 8 ounces gold per cord, smelting ore, $80 to $100 per ton; main shaft 340 feet deep; reached by tunnel at a distance of 1,000 feet from the mouth and intersected at a depth of 165 feet, 1 level 300 feet in length.

Phoenix Mining Co., of Colorado. Organized under the laws of the State of New York; Moses S. Moody, president, Robert M. Stratton, secretary, Andrews N. Rogers, of Central City, agent.

Pierce. Willard Teller, proprietor; office Denver; claim 50 by 500 feet, patented; discovered 1861; located at junction of Spring and Nevada Gulches, within limits of Central City; course of vein, northeast and southwest, width 3 feet; nature of ore, iron and copper pyrites; main shaft 75 feet deep.

Pleasant View. Owned by the Pleasant View Mining Co.; incorporated January 3, 1876; capital stock, $200,000, in 20,000 shares, $10 each; offices Central City and Denver, Colorado; organized by William M. Roworth, J. Jay Joslin [a successful Denver merchant], James Hutchinson, Charles H. Briggs, and Sumner T. McKnight; claim 150 by 1,000 [?] feet; discovered 1864; located on the eastern end of Gunnell Hill, within the limits of Central City; main shaft 300 feet deep; on the surface is a shaft house and a 12-horse power engine; yield to date, $10,000.

Pocahontas. Owned by Ed. B. Warner, George Martin, and William H. Cheatley; office Central City; claim 150 by 1,000 feet; discovered 1874; located on Saratoga Hill, Russell mining district, 1 mile from Central City;

width of vein 2 feet; average value of ore, 5 ounces gold per cord; main shaft 50 feet deep.

Portage. A.C. Johnson, proprietor; office Black Hawk; claim 50 by 800 feet; discovered 1863; located in Chase Gulch, within the limits of Central City; course of vein, northeast and southwest, width 2 ½ feet, pitch 5 degrees south; nature of ore, galena, zinc blende, and iron pyrites; average value, 6 ounces gold per cord, smelting ore, $60 to $100 per ton; 2 shafts aggregating 100 feet in depth, 2 levels, 45 and 60 feet from surface, respectively, aggregating 125 feet in length.

Pratt. Willard Teller, proprietor; office Denver; claim 50 by 354 feet, patented; discovered 1862; located on Quartz Hill, Illinois Central mining district, half mile from Central City; vertical vein, running northeast and southwest, width 3 feet, pay vein 18 inches; nature of ore, iron and copper pyrites; average value, 4 ounces gold per cord; main shaft 150 feet deep; yield, $1,000.

Prize. William M. Roworth, proprietor; office Central City; claim 150 by 200 feet; discovered 1862; located on Prize Hill, Nevada mining district, 2 miles from Central City; width of vein 4 feet; average value of mill ore, 5 ounces gold per cord, smelting ore, $50 to $75 per ton; 2 shafts, aggregating 250 feet in depth.

Prospector. J.O. Wheeler, proprietor; office Black Hawk; claim 50 by 1,500 feet; discovered 1867; located on Bates Hill, Enterprise mining district, within the limits of Black Hawk; width of vein 6 feet; average value of ore, 3 ounces gold per cord, smelting ore, $110 per ton; main shaft 35 feet deep; reached by the Prospector Tunnel [see separate listing] at a distance of 200 feet from the mouth and intersected at a depth of 175 feet.

Prospector Tunnel. Owned by J.O. Wheeler, D. Pippin, and E.C. Beach; office Black Hawk; claim 50 by 1,500 feet; discovered 1866; located in Chase Gulch, foot of Bates Hill, Enterprise mining district, within the limits of Black Hawk; length of tunnel, 300 feet.

Prussian. George J. and J.F. Kline, proprietors; office Central City; claim 100 by 1,500 feet; discovered 1868; located on German Hill, Mountain City mining district, half mile from Central City; width of vein 18 inches; average value of ore, 8 ounces gold per cord; main shaft 50 feet deep.

Pyrenees. Owned by George W. Estabrook & Co.; office Bald Mountain; claim 150 by 1,850 feet; discovered 1863; located on Alps Hill, Illinois mining district, 2 miles from Central City, course of vein, east and west, width 2 feet, pitch south; nature of ore, galena, iron and copper pyrites; average value of mill ore, 7 ounces gold per cord, smelting ore, $55 per ton; main shaft 300 feet deep, 3 levels, aggregating 500 feet in length; shaft house, containing a 10-horse power engine and hoisting apparatus.

Quartz Hill Tunnel. Owned by the Quartz Hill Tunnel Co.; incorporated February 1866; Henry Altvater, president, Henry J. Kruse, secretary, Henry Kruse, treasurer, John H. Lafrenz, superintendent; capital stock, $50,000, in 50 shares, $1,000 each; office Central City; claim 150 by 10,000 feet; located in Nevada Gulch, on the eastern slope of Quartz Hill, within the limits of Central City; course, southwest, length, 1,025 feet; will intersect the Burroughs, Borton, Illinois, and other prominent lodes.

Ranney. Owned by the Belden and Tennal Mining Co.; David D. Belden, superintendent; office Central City; claim 150 by 1,500 feet, patented; discovered 1877; located on Casto Mountain, within the limits of Central City; width of vein 2 ½ feet, pay vein 18 inches; average value of ore, 7 ounces gold per cord; reached by the Belden Tunnel at a distance of 475 feet from the mouth, and intersected at a depth of 350 feet.

Rara Avis. Located on Virginia Hill, Idaho mining district, 3 miles from Central City; claim 150 by 1,400 feet; nature of ore, galena and gray copper.

Rattler. Owned by Richard Bond and James H. Thompson; office Central City; claim 150 by 1,500 feet; discovered 1877; located at the forks of the Prosser and Eureka gulches, within the limits of Central City; course of vein, northeast and southwest, width 3 feet, pitch 40 degrees north; pay vein 15 inches; nature of ore, iron and copper pyrites and galena; average value of mill ore, 12 ounces gold per cord, smelting ore, $60 to $160 per ton; main shaft 55 feet deep.

Register. Willard Teller, proprietor; office Denver; claim 50 by 800 feet, patented; discovered 1866; located on Quartz Hill, Illinois Central mining district, half mile from Central City; vertical vein, running northeast and southwest, width 2 ½ feet; nature of ore, iron and copper pyrites; average value 5 ounces gold per ton; main shaft 125 feet deep; 2 levels, 50 feet in length, respectively.

Register. Owned by Hon. Henry M. Teller; office Central City; claim 50 by 1,500 feet, patented; discovered 1860; located on Quartz Hill, Illinois mining district, half mile from Central City; course of vein, northeast and southwest, width 4 feet; pitch south; nature of ore, iron and copper pyrites; average value of mill ore, 9 ounces gold per cord, smelting ore, $150 to $1,140 per ton; main shaft 150 feet deep, 2 levels 50 and 100 feet from surface, respectively

Rialto. Thomas Mullen, proprietor; office Central City; claim 150 by 1,500 feet; located within the limits of Central City; course of vein, north and south; pitch southeast; width 2 ½ feet; nature of ore, iron and copper pyrites; average value, 4 ounces gold per cord; main shaft 200 feet deep.

Rob Roy. Owned by Kelty and Tomlinson; office Black Hawk; claim 150 by 1,500 feet; discovered 1876; located between Chase Gulch and Clear Creek, within the limits of Black Hawk; course of vein, northeast and southwest, width 18 inches, pitch south; nature of ore, galena, iron and copper pyrites; average value 3 ounces gold per cord; main shaft 50 feet deep, 1 tunnel 150 feet in length.

Rockdale Gold Mining Co. Edward W. Henderson, agent; offices Central City, Colorado, and 24 Broadway, New York City; J. Fred Pierson, secretary and treasurer.

Rocky Mountain Mining Bureau of Mines. Incorporated September 27, 1872; capital stock, $25,000, in 250 shares, $100 each; office Central City; organized by Horace H. Atkins, Hal Sayre, and Foster Nichols.

Roderick Dhu. Owned by the Louisville Mining Co.; office Central City; claim 50 by 385 feet, patented; discovered 1860; located on Quartz Hill, Nevada mining district, near the limits of Central City; course of vein, northeast and southwest, pitch 5 degrees south, width 3 feet, pay vein 2 feet; nature of ore, iron and copper pyrites; average value of mill ore, 6 ounces gold per cord; smelting ore, $50 per ton; main shaft 540 feet deep; shaft house, 15 by 20 feet; yield, $300,000.

Roderick Dhu. Owned by the Rochester Gold Mining Co., New York; office Black Hawk, Colorado, and Rochester, New York; claim 50 by 217 feet, patented; capital stock, $100,000; discovered 1860; located on

Quartz Hill, Nevada mining district, near the limits of Central City; course of vein, northeast and southwest, pitch 5 degrees south, width 3 feet, pay vein 2 feet; nature of ore, iron and copper pyrites; average value of mill ore, 5 ounces gold per cord, smelting ore, $50 per ton; main shaft 400 feet deep, 3 levels, aggregate length 500 feet; yield, $400,000.

Roderick Dhu. Owned by Stevens & Deaver Gold Mining Co., New York; incorporated 1864; claim 50 by 500 feet, patented; capital stock, $1,000,000, 100,000 shares, $10 each; J.P. Geraud Foster, president, Edward W. Henderson, of Central City, manager; located on Quartz Hill, Nevada mining district, near the limits of Central City; course of vein, northeast and southwest, pitch 5 degrees south, width 3 feet, pay vein 2 feet; nature of ore, iron and copper pyrites; average value of mill ore, 7 ounces gold per cord, smelting ore, $100 per ton; main shaft 250 feet deep; yield, $200,000.

Running. Horace Humphreys and Rodney French, proprietors; William Fisher, superintendent; claim 50 by 900 feet, patented; discovered 1859; located on Colorado Central Railroad, within the limits of Black Hawk; vertical vein, running northeast and southwest, width 3 ½ feet, pay vein 1 foot; nature of ore, iron and copper pyrites; average value of smelting ore, $35 per ton; main shaft 220 feet, and 4 other shafts, aggregating 300 feet in depth, 1 level 75 feet in length and 220 feet from surface; shaft house, 25 by 50 feet.

San Juan. Owned by E.W. Stevens, James H. Cameron, and Charles W. Anderson; office Bald Mountain; claim 50 by 800 feet; discovered 1877; located on Quartz Hill, Illinois mining district, 1 mile from Central City; width of vein, 4 feet; nature of ore, iron and copper pyrites; 1 shaft, 50 feet deep.

Sapphire. Located on the north slope of Gunnell Hill, near Gunnell lode, 1 mile from Central City.

Saratoga. Owned by Samuel I. Lorah and Corbit Bacon, of Central City; claim 150 by 2,000 feet, patented; discovered 1864; located on Williss Gulch, Russell mining district, 1 mile from Central City; vertical vein, running northeast and southwest, width 4 feet. pay vein 3 feet; nature of ore, iron and copper pyrites; average value of mill ore, 5 ounces gold per cord; main shaft 140 feet and 4 other shafts, aggregating 420 feet in depth, 1 tunnel, 415 feet, and 2 levels aggregating 150 feet in length; has produced largely in former years and prior to present ownership; yield for the last 6 months, $6,000.

Satisfaction. Owned by John Scudder and the heirs of John Hense; office Central City; claim 100 by 1,500 feet, patented; discovered 1860; located near the center of Central City; vertical vein, running nearly east and west, width 5 feet, pay vein 3 feet; nature of ore, iron pyrites; average value, 3 ounces gold per cord; main shaft 225 feet deep, aggregate length of levels 200 feet; yield, $12,000.

Searle. Owned by William Rampage, of Central City; claim 150 by 1,400 feet; located on Virginia Hill, 3 miles from Central City; nature of ore, gray copper and galena; main shaft 140 feet deep.

Shaft. Owned by O.M. Albro and H.E. Hyatt; office Bald Mountain; claim 50 by 2,900 feet, patented; discovered 1865; located on Prize Hill, Nevada mining district, 1 mile from Central City; course of vein, east and west, width 3 feet, pitch south; nature of ore, iron and copper pyrites; average value, 4 ounces gold per cord; main shaft 125 feet deep.

Shanks. Owned by William H. Bush, Central City; claim 150 by 1,500 feet, patented; located on Quartz Hill, Central City mining district.

Shaw. Richard H. Rickard, proprietor; office 19 Nassau Street, New York City; claim 50 by 500 feet, patented; discovered 1861; located on Quartz Hill, Illinois mining district, 1 mile form Central City; vertical vein, trending northeast and southwest, width 8 feet; nature of ore, iron and copper pyrites; 1 shaft 100 feet deep.

Sierra Madre Tunnel Co. Incorporated May 14, 1873; capital stock, $5,000,000, in shares of $100 each; office Denver; organized by George W. Heaton, Peter C. Johnson, and Hiram G. Bond.

Silver King. Owned by Charles F. Hendrie and Henry Boltoff, of Central City; claim 150 by 1,500 feet; discovered 1876; located in Spring Gulch, Independent mining district, 10 miles from Georgetown; course of vein northeast and southwest, pitch 25 degrees southeast, width 4 feet; nature of ore, galena and gray copper; average value $40 per ton; main shaft 40 feet deep, 1 tunnel on vein 100 feet, and 1 cross cut tunnel, 60 feet in length.

Sioux City. Located on Bobtail Hill, Gregory mining district, within the limits of Black Hawk.

Sleepy Hollow. Owned by C.K. Pevey and Mifflin Rasin; office Central City; claim 150 by 1,400 feet; discovered 1873; situated on Bobtail Hill, Gregory mining district, within the limits of Black Hawk; course of vein, northeast and southwest; pitch south; width 18 inches; nature of ore, iron and copper pyrites and galena; average value, 7 ounces gold per cord; main shaft 200 feet deep.

Smith. Owned by Thomas I. Richman; office Central City; claim 50 by 1,000 feet, patented; discovered 1859; located in Chase Gulch, within the limits of Central City; course of vein, northeast and southwest, width 4 feet, pay vein 18 inches, pitch north and south; nature of ore, galena, iron and copper pyrites; average value of mill ore, 5 ounces gold per cord, smelting ore, $100 per ton; 2 shafts, aggregating 300 feet in length.

Smith. Owned by Harper M. Orahood, of Central City; claim 150 by 600 feet, patented; discovered 1859; located in Enterprise mining district, between Chase Gulch and North Clear Creek, within the limits of Black Hawk; course of vein, northeast and southwest, width 3 feet; nature of ore, galena, iron and copper pyrites; average value of mill ore, 5 ounces gold per cord, smelting ore, $40 per ton; main shaft 130 feet deep, 2 levels 50 feet in length each; yield, $12,000.

Sophia. Alexander Taylor, proprietor; office Black Hawk; claim 50 by 1,200 feet; discovered 1864; located on Eureka Hill, Eureka mining district, within the limits of Central City; course of vein, northeast and southwest, width 4 feet, pay vein 2 feet, pitch 20 degrees north; nature of ore, sulphurets of iron and copper; average value of mill ore, 8 ounces gold per cord; main shaft 125 feet deep.

South Lewis. William H. Bush, proprietor; office Central City; claim 150 by 1,500 feet, patented; discovered 1870; located on Quartz Hill, Nevada mining district.

Springdale. Owned by Charles McKee and E.T. Carr; office Central City; claim 150 by 1,500 feet, discovered 1864; located near Russell Gulch, Russell mining district, 2 miles from Central City; width of vein 18 inches; average value of mill ore, 8 ounces gold per cord; 3 shafts, aggregating 160 feet in depth.

Spur. Alexander Taylor, proprietor; office Black Hawk; claim 50 by 700 feet; discovered 1859; located on Eureka Hill, Eureka mining district, within the limits of Central City; course of vein, northeast and southwest, width 4 feet, pay vein 2 feet; nature of ore, sulphurets of iron and copper; average value, 6 ounces gold per cord; 2 shafts, aggregating 125 feet in depth, 1 level 50 feet from surface and 100 feet in length.

Sterling. Owned by Thomas H. Potter and Henry J. Hawley; office Central City; claim 150 by 1,500 feet; locate 1875; located on Bates Hill, within the limits of Central City; course of vein, northeast and southwest, width 2 feet, pitch 45 degrees south; nature of ore, iron and copper pyrites; average value of mill ore, 3 ½ ounces gold per cord; main shaft 150 feet deep, 2 levels aggregating 75 feet in length.

Stevens & Deaver Gold Mining Co., New York. Incorporated 1864; capital stock, $1,000,000, in 100,000 shares of $10 each; J.P. Geraud Foster, president, Edward W. Henderson, of Central City, manager.

St. Louis. Kilbourne, Jenkins & Co., proprietors; office Central City; claim 150 by 400 feet; located on Gunnell Hill, within the limits of Central City; course of vein, northeast and southwest, width 3 feet, pitch 15 degrees south; nature of ore, iron and copper pyrites; average value of mill ore, 10 ounces gold per cord; main shaft 350 feet, and 1 other shaft 200 feet deep; 3 levels 100, 200, and 300 feet from surface, respectively, aggregating 200 feet in length; shaft house and hoisting machinery.

St. Louis No. 2. Owned by George A. Patten, M.F. Bebee, Frank and Asahel Kilbourne; office Central City; claim 150 by 3,000 feet; discovered 1877; located on Gunnell Hill, within the limits of Central City; course of vein, northeast and southwest, width 3 feet, pitch 30 degrees south; nature of ore, iron, copper, and galena; 3 shafts, aggregating 250 feet in depth, 1 level 150 feet in length.

Sucker. Alexander Taylor, proprietor; office Black Hawk; claim 50 by 700 feet; discovered 1859; located on Eureka Hill, Eureka mining district, within the limits of Central City; vertical vein, trending northeast and southwest, width 4 feet, pay vein 2 feet; nature of ore, sulphurets of iron and copper; average value of mill ore, 7 ounces gold per cord; 3 shafts, aggregating 150 feet in depth.

Suderberg. Owned by the First National Bank of Denver and Edward L. Salisbury, of Central City; claim 50 by 1,900 feet, patented; located on

Prize Hill, Nevada mining district, near the limits of Central City; course of vein, early east and west, width 4 feet; nature of ore, zinc blende, iron and copper pyrites; average value of mill ore, 16 ounces gold per cord, smelting ore, $100 to $250 per ton; main shaft 500 feet, and 6 other shafts aggregating 120 feet in depth, 4 levels aggregating 1,200 feet in length; yield, $250,000.

Sugar Tit. Owned by William S. Shaw and William Skelley; office Bald Mountain; claim 50 by 1,200 feet, patented; discovered 1869; located on Quartz Hill, Nevada mining district, 1 mile from Central City; course of vein, east and west, width 3 feet, pay vein 1 foot; nature of ore, galena, iron and copper pyrites; average value, 6 ounces gold per cord; 4 shafts aggregating 312 feet in depth, 3 levels aggregating 400 feet in length.

Sullivan. Owned by D. Sullivan and others; located on Quartz Hill, near Nevada Gulch, 1 mile from Central City; main shaft 180 feet deep.

Summit. Clinton Reed, proprietor; office Central City; claim 150 by 1,500 feet; discovered 1873; located on Casto Mountain, within the limits of Central City; width of vein 3 feet; average value of mill ore, 4 ounces gold per cord; main shaft 50 feet deep.

Susquehanna Gold Mining Co. Organized under the laws of the State of Pennsylvania, Albert G. Buzby, president, Henry E. Lincoln, vice president, Frank H. Messinger, of Black Hawk, agent.

Swansea. John Protheroe and Daniel Harris, proprietors; office Central City; claim 150 by 1,500 feet; located in Davenport Gulch, Russell mining district, 2 ½ miles from Central City; course of vein, northeast and southwest, width 3 feet, pitch 25 degrees north; 1 shaft 30 feet deep.

Teller House. Claim 50 by 400 feet; discovered 1862; located on Casto Mountain, within the limits of Central City; course of vein, northeast and southwest, width 4 feet, pay vein 18 inches, pitch 5 degrees south; nature of ore, iron and copper pyrites; average value of mill ore, 7 ounces gold per cord, smelting ore, $80 per ton; main shaft 65 feet deep, 2 levels, 40 and 60 feet from surface, respectively.

Topeka. Owned by the Great Western Gold Mining Co.; George E. Randolph, agent; office Central City; claim 50 by 200 feet, patented; discovered 1860; located in Russell mining district, 2 miles from Central City.

U.S.M. Owned by L.C. Snyder, Blake & Co; office Black Hawk; claim 150 by 1,050 feet, patented; discovered 1863; located in Gregory Gulch, within the limits of Central City; width of vein 4 feet; average value of mill ore, 3 ½ ounces gold per cord; main shaft 100 feet deep, 1 level 40 feet in length and 45 feet from surface.

Vanderbilt. Owned by Hon. Henry M. Teller, of Central City; claim 50 by 800 feet, patented; located on Gunnell Hill, Nevada Mining district, within the limits of Central City.

Vasa. Owned by Bela S. Buell; patented claim, located on Mammoth Hill, Gregory mining district, within the limits of Central City; course of vein, northeast and southwest, width 6 feet, pitch 5 degrees south, pay vein 2 feet; nature of ore, iron and copper pyrites; average value of mill ore, 4 ounces gold per cord, smelting ore, $100 per ton; main shaft 260 feet deep, and 1 other shaft 250 feet deep; aggregate length of levels, 1,500 feet.

Victor Silver Mining Co. Incorporated May 9, 1877; capital stock, $50,000, in 500 shares, $100 each; office Black Hawk; organized by Thomas J. Oyler, George A. Patten, Frederick Leighton, John Needham, Mark F. Bebee, and Ed. A. Seiwell.

Victory Mining Co. Organized under the laws of the State of Wisconsin; capital stock, $250,000, in 2,500 shares, $100 each; E.C. Gray, president, W.C. Raynor, vice president, J.H. Hauser, secretary and treasurer; offices Central City, Colorado, and Fon du Lac, Wisconsin.

Virginia. Harper M. Orahood, of Central City, proprietor; claim 150 by 1,500 feet, patented; discovered 1859; located on Casto Mountain, within the limits of Central City; course of vein, northeast and southwest, pitch 5 degrees west, width 3 feet, pay vein 14 inches; nature of ore, zinc blende, iron and copper pyrites; average value of mill ore, 3 ½ ounces gold per cord; main shaft 60 feet deep, 1 open cut, 30 feet in length, yield, $10,000.

Wain. Owned by William and Taylor Lindsley and William Wain; office Central City; claim 150 by 1,500 feet, patented; discovered 1877; located in Chase Gulch, within the limits of Black Hawk; course of vein, east and west, width 5 feet, pitch 15 degrees north; nature of ore, galena, iron and copper pyrites; average value of mill ore, 7 ounces gold per cord, smelting ore, $75 to $100 per ton; main shaft 150 feet deep; 1 tunnel 200 feet in length and 75 feet from surface; shaft house, 14 by 36 feet, containing a 15-horse power engine and hoisting apparatus; yield, $2,000.

Washington and Extension. Owned by Lester and Eugene Drake; office Central City; claim 150 by 3,000 feet; discovered 1876; located in Lake Gulch, near the limits of Black Hawk; course of vein, northeast and southwest, pitch 40 degrees north, width 8 feet, pay vein 2 ½ feet; nature of ore, galena, iron and copper pyrites average value of mill ore, 8 ounces gold per cord, smelting ore, $60 per ton; main shaft 85 feet, and 3 other shafts aggregating 160 feet in depth.

Waterman-Kansas. Operated by John R. Berryman and others; located on Kansas lode, Nevada mining district, 1 mile from Central City.

Watermill. Owned by James C. Fagan; claim 50 by 200 feet; located on Quartz Hill, 1 mile from Central City.

West Wyandotte. Owned by William H. Cheatley and George Martin; office Central City; claim 100 by 1,500 feet; located on Leavenworth Mountain, Illinois mining district, 2 miles from Central City; width of vein 3 feet, pay vein 2 feet; average value of mill ore, 4 ounces gold per cord, smelting ore, $70 per ton; 2 shafts, aggregating 200 feet in depth, 2 levels, aggregating 500 feet in length.

Whipple Gold Mining Co. Organized under the laws of Rhode Island; Lyman A. Cook, president, William H. Learned, agent; office Woonsocket, Rhode Island, and Central City, Colorado.

White Cloud. Owned by William Martin, of New York City; claim 50 by 950 feet, patented; discovered 1864; situated at the head of Virginia Canyon, Idaho mining district, 3 miles from Central City; course of vein, northeast and southwest, pitch 35 degrees north, width 4 feet; nature of ore, iron and copper pyrites; average value of mill ore, 15 ounces gold per cord; main shaft 65 feet deep; yield, $5,000.

Whiting. Owned by Kimber Gray & Co.; office Black Hawk; claim 50 by 500 feet, patented; discovered 1859; located on Gunnell Hill, Eureka mining district, within the limits of Central City; course of vein, northeast and southwest, width 2 feet, pitch 35 degrees south; nature of ore, iron and copper pyrites; average value of mill ore, 6 ounces gold per cord; main shaft 125 feet, and 1 other, 60 feet deep.

Whiting. Owned by Edward W. Henderson and Stiles E. Mills, of Central City; claim 50 by 900 feet, patented; discovered 1859; located on Gunnell Hill, Eureka mining district, within the limits of Central City; vertical vein, running east and west, width 5 feet, pay vein 2 feet; nature of ore, iron pyrites; average value of mill ore, 4 ounces gold per cord; main shaft 175 feet deep, 1 level 50 feet in length; shaft house, 16 by 16 feet.

Whitney. Owned by Green Lee, Nathan Sears, and John Hyndman; office Central City; claim 50 by 1,500 feet; discovered 1867; located on Eureka Hill, Eureka mining district, 1 mile from Central City; course of vein, northeast and southwest, pitch 10 degrees south; width 3 feet; nature of ore, galena, iron and copper pyrites; average value of mill ore, $30 per ton; main shaft 100 feet deep, 2 drifts aggregating 150 feet in length.

Williams. Owned by Lester, Eugene, and Alonzo Drake, R. Morrison, M.H. Root, and J.A. Hale; office Central City; claim 150 by 2,000 feet; discovered 1860; located in Lake Gulch, near the limits of Black Hawk; course of vein, northeast and southwest, pitch 40 degrees north, width 5 feet, pay vein 2 feet; nature of ore, gray copper, iron and copper pyrites; average value of mill ore, 8 ounces gold per cord; smelting ore, $80 per ton; main shaft 350 feet, and 8 other shafts aggregating 700 feet in depth, 4 levels aggregating 500 feet in length, being 130, 200, and 290 feet from surface, respectively.

Williams, Nos. 2 and 3 West. Owned by Robert M.D. Morrison; office Central City; claim 50 by 200 feet; located in Lake Gulch, Lake mining district, 2 miles from Central City; course of vein, northeast and southwest, pitch 10 degrees north, width 4 feet, pay vein 18 inches; nature of ore, gray copper, iron and copper pyrites; average value, 5 ounces gold per cord, smelting ore, $80 per ton; main shaft 50 feet deep; yield, $2,000.

Winnebago Property. James W. Hannah, proprietor; office Denver; claim 50 by 1,200 feet; discovered 1859; located on Casto Mountain, within the limits of Central City; course of vein, northeast and southwest, width 45 inches, pitch 10 degrees south; nature of ore, iron pyrites; average value of mill ore, 5 ounces gold per cord, smelting ore, $50 to $150 per ton; main shaft, 450 feet, and other shafting aggregating 600 feet in depth, 4 levels, 100, 190, 330, and 400 feet from surface, respectively, aggregating 1,000 feet in length; over the mine is erected a mill 100 by 120 feet, containing 20 stamps weighing 450 pounds each, 6 Bartola pans, a 40-hourse power engine, and hoisting apparatus; yield, $360,000.

Winnebago. Owned by Ellery, Furnald & Co.; office Central City; claim 150 by 300 feet; discovered 1862; located on Casto Mountain, within the limits of Central City; width of vein 4 ½ feet; average value of mill ore, 8 ounces bold per cord, smelting ore, $70 per ton; main shaft 100 feet deep, 1 level, 250 feet in length, and 100 feet from surface.

Wood. Owned by William H. Doe, Sr., office Central City; claim 150 by 1,600 feet; discovered 1860; located at the forks of the Prosser and Eureka gulches, Eureka mining district, within the limits of Central City; vertical vein, trending northeast and southwest, width 6 feet, pay vein 4 feet; nature of ore, sulphurets of iron and copper; main shaft 160 feet deep.

Wood. Owned by the Rockdale Gold Mining Co., New York; Edward W. Henderson, agent; office Central City; claim 50 by 1,000 feet; located in Leavenworth Gulch, Leavenworth Gulch, Russell mining district, 2 miles from Central City; course of vein, east and west, width 4 feet; nature of ore, iron and copper pyrites; main shaft 125 feet deep.

Wyandotte Consolidated Gold and Silver Mining Co. Incorporated April 29, 1878; office 16 Broad Street, New York City; G.E. Tufts, secretary; this company now owns, by recent consolidation, 8 well defined gold bearing quartz mines; 1 of these alone having already produced from its surface workings over $500,000.

Wyandotte. Owned by Horatio E. and Morris Hazard; office Central City; claim 50 by 375 feet; discovered 1875; located on Leavenworth Hill, Russell mining district, 1 mile from Central City; width of vein, 5 feet; average value of mill ore, 5 ounces gold per cord; main shaft 140 feet deep.

Gilpin County Ore Mills

Black Hawk Mill. Owned by the Consolidated Bobtail Gold Mining Co.; Andrews N. Rogers, agent, Frank H. Messinger, assistant superintendent, located on North Clear Creek, Black Hawk; has a capacity for treating 65 tons of ore daily; the mill contains 75 stamps and 3 engines, 20-horse power each.

Boston and Colorado Smelting Co. Organized under the laws of the State of Massachusetts, May 11, 1867; capital paid up stock, $750,000, in 7,500 shares, $100 each; offices Boston, Massachusetts, Denver and Black Hawk, Colorado; J.W. Converse, president, J. Warren Merrill, treasurer,

Prof. Nathaniel P. Hill, general manager, Henry R. Wolcott, assistant manager, R. Pearce, metallurgist, A. Von Schulz, assayer in charge.

Black Hawk. The Black Hawk Works were commenced in 1867, and consist of a group of buildings covering nearly 5 acres of ground. They are situated on the Colorado Central Railroad and on North Clear Creek, within the limits of Black Hawk, 1 mile from Central City. They have a capacity for treating 60 tons of ore daily, and employ 100 men.

Argo. Argo, the Denver works, located 2 miles north of Denver on the Colorado Central Railroad, was commenced in June 1878, and erected under the supervision of the general manager, Prof. Nathaniel P. Hill. This property comprises 80 acres of land, on which are built the main works of the company, consisting of a refinery 64 feet in width by 293 feet in length, with a wing 40 by 77 feet, a smelting house 283 feet in length by 38 feet in width; an ore house 119 feet in width by 224 feet in length with a wing 33 by 91 feet; a calcining house 119 feet in width by 224 feet in length; an office, counting room and assay laboratory 42 by 81 feet; and a coal shed 51 by 178 feet. These works are constructed of stone, and covered with a superior quality of corrugated iron roofing, and every department is equipped with the latest and most complete machinery selected with especial reference to the extraction of the precious metals in a manner profitable alike to the miner and the mine management. In addition to the foregoing buildings, there are 5 2-story brick dwelling houses of 6 tenements each, a large hotel with accommodations for 100 persons, together with stables, sheds, and other houses of minor importance. The works are surrounded by a substantial stone wall have a capacity for treating 100 tons of ore daily and give employment to 150 men.

Alma, Branch Works. These works were erected in 1874; Henry Williams, superintendent; located in the town of Alma in the county of Park, 63 miles from Black Hawk, and 100 miles from Denver. These works have a capacity for treating 25 tons of ore daily, and employ 10 men.

Boulder Sampling and Crushing Works. These works were erected in 1876; C.G. Duncan, agent; situated corner Pearl and Eighth streets, in the city of Boulder, 43 miles from Denver, and 28 miles from Black Hawk have a capacity for treating 25 tons of ore daily, and employ 6 men.

Bostwick Mill. Owned by Theodore W. Wheeler and Joseph W. Bostwick; located on North Clear Creek, within the limits of Black Hawk; has a capacity for treating 75 cords of ore per month, and contains 25 stamps, a 30-horse power engine, tables, pans, etc. This mill was recently enlarged and improved and is now in good condition.

Buell Mill. Owned by Bela S. Buell; located in Gregory Gulch, within the limits of Central City; has a capacity for treating 50 tons of ore daily, and contains 60 stamps and an 80-horse power engine.

Cleveland Gold Mining Co.'s Mill. Owned by the Cleveland Gold Mining Co.; office 54 William Street, New York City; H.A. Johnson, president; located in Leavenworth Gulch, 2 miles from Central City; contains 16 stamps.

Clayton Mill. Operated by Richard Mackey; located in Nevada Gulch, 2 miles from Central City; has a capacity for treating 3 cords of ore daily, and contains 25 stamps, with other machinery.

Collom's Black Hawk Dressing Works Co. Incorporated August 1, 1873; capital stock, $40,000, in 400 shares, $100 each; offices Black Hawk, Colorado, and Trenton, New Jersey; Bradford H. Lock, of Central City, manager. The works of this company are located on North Clear Creek, within the limits of Black Hawk.

Douglas Mill. Located in Chase Gulch, 1 mile from Black Hawk; contains 25 stamps and has a capacity for treating 2 ½ cords of ore per day.

Eagle Mill. Owned by the consolidated Bobtail Gold Mining Co.; Andrews N. Rogers, agent; located on North Clear Creek, near Black Hawk, contains 20 stamps with other machinery.

Empire Mill. Owned by John and Samuel Mellor; situated in the town of Black Hawk, at the mouth of Chase Gulch; has a capacity for treating 80 cords of ore per month, and contains a 20-horse power engine and 25 stamps.

Fullerton Mill. Owned by W. Fullerton and Job V. Kimber; located on Clear Creek, half mile from Black Hawk; has a capacity for treating 100 cords of ore per month, is run by steam and water power and contains 33 stamps with other machinery.

Field's Mill. Owned by Mrs. Francis Field; located on North Clear Creek, near Boston and Colorado Smelting Works; has a capacity for treating 45 cords of ore per month.

First National Mill. Owned by the First National gold Mining Co.; located in Nevada Gulch, 1 mile from Central City; has a capacity for treating 2 ½ cords of ore daily, and contains 25 stamps.

Golden Flint Mill. Owned by the Golden Flint Mining Co.; William McFarlane, superintendent; located in Gamble Gulch, 8 miles from Central City, and 3 miles from Rollinsville; contains 15 stamps, 20-horse power engine and has a capacity for treating 2 cords of ore daily.

Golden Sampling Works. Represented by H.W. Forman; office Black Hawk.

Gregory Mill. Owned by Briggs Bros.; located on Colorado Central Railroad, Main Street, Black Hawk; contains 50 stamps weighing 700 pounds each, a 50-horse power engine, 11 Bartola pans, tables, etc.; has a capacity for treating 6 cords of ore daily.

Holdbrook Mill. Owned by William Fullerton and Job V. Kimber; located on North Clear Creek; half mile from limits of Black Hawk; has a capacity for treating 50 cords of ore per month, and contains 15 stamps.

Illinois Mill. Owned by Thomas H. Potter and George W. Jacobs; located in the eastern slope of Quartz Hill, half mile from Central City; contains 22 stamps, an engine, boiler, etc.

Kimber & Fullerton Mill. Owned by William Fullerton and Job V. Kimber; located on Clear Creek, half mile from limits of Black Hawk; has a capacity for treating 3 cords of ore daily and contains a 20-horse power engine and a 30-horse power water wheel.

Mackey Mill. Owned by Richard Mackey; located in town of Nevadaville, 2 miles from Central City; has a capacity for treating 4 cords of ore daily and contains 40 stamps.

Mead Mill. Owned by William Fullerton and Job V. Kimber; located on North Clear Creek; has a capacity for treating 75 cords of ore per month, and contains 20 stamps.

Monmouth-Kansas Mill. Owned by the Monmouth Consolidated gold Mining Co.; D. Sullivan, superintendent, located in Nevada Gulch, 1 mile from Central City; has a capacity for treating 48 tons of ore daily; the building is 80 by 120 feet; contains 60 stamps, a 75-horse power engine, vats, pans, etc.

New York Mill. Owned by Thomas H. Potter, George E. Randolph, Henry B. Williams, Samuel Meller, and others; located within the limits of Black Hawk; contains 50 stamps, an engine, boiler, tables, pans, etc.

New York and Colorado Mill. Owned by the New York and Colorado Company; B.T. Wells, manager; located on Clear Creek, within the limits of Black Hawk; contains 40 stamps, a 50-horse power engine, 7 Bartola pans, and an overshot wheel 33 feet in diameter.

Polar Star Mill. Owned by Job V. Kimber and Frank C. Young; located on North Clear Creek, at the intersection of Chase Gulch, Black Hawk; the building is a stone structure 65 by 96 feet, with iron roof, contains 35 stamps, 30-horse power engine, tables, pans, etc.; this mill can be run by steam or water power; the capacity is 3 cords of ore daily.

Randolph Mill. Owned by George E. Randolph, George W. Mabee, and John O. Raynolds; located on North Clear Creek; within the limits of Black Hawk; has a capacity for treating 40 tons of ore daily and contains 50 stamps, a 50-horse power engine, tables, pans, etc.

Richman Mill. Owned by Thomas I. Richman & Co., located on North Clear Creek, within the limits of Black Hawk; has a capacity for treating 35 tons of ore daily, and contains 50 stamps, a 50-horse power engine, tables, pans, etc.

Rollins Mill. Owned by John A.Q. Rollins, of Rollinsville; located 8 miles from Central City.

Sensenderfer Mill. Owned by the Consolidated Bobtail Gold Mining Co.; Andrews N. Rogers, agent; located on North Clear Creek, 1 mile from Black Hawk; has a capacity for treating 50 cords of ore per month and contains 20 stamps.

Tascher Mill. Owned by Jacob Tascher; located in Eureka Gulch, near Central City; contains 20 stamps; has a capacity for treating 50 cords of ore per month.

Tomlinson Mill. Owned by Joseph B. Tomlinson; located on North Clear Creek, within the limits of Black Hawk; ahs a capacity for treating 2 ½ cords of ore per day, and contains 38 stamps, a 30-horse power engine, tables, pans, etc.

Wheeler Mill. Owned by Theodore W. Wheeler and D. Sullivan; located on North Clear Creek, near Black Hawk; has a capacity for treating 75 cords of ore per month, and contains 25 stamps.

Whitcomb Mill. Operated by Lewis & Aulsebrook; located in Nevada Gulch, 2 miles from Central City.

Winnebago Mill. Owned by James W. Hannah, of Denver; located on Casto Hill, within the limits of Central City; the main building is 100 by 120 feet, and contains 20 stamps, 6 Bartola pans, a 40-horse power engine, etc.

Crumbling ruins of the Niagara Mill in Russell Gulch.

In this 2001 Photo, the slowly disintegrating Druid Mine still stands, a monument to Gilpin County's glorious gold rush past.

Today, little remains of Wide Awake, a bustling gold rush town founded in 1860. It exploded from a small tent city into a town with over 500 people.

GRAND COUNTY

Thomas B. Corbett reported that Grand County had a population of 12,000 in 1878. Mining districts included Campbell, Gore Range, and Willow Creek.

Buffalo Patch Placer Diggings. Owned by George W. Day, of Denver; claim comprises 100 acres of placer ground, located on Antelope Creek, Gore Range mining district, 40 miles from Hot Sulphur Springs and 140 miles from Denver.

Crown Point. Owned by H.M. Shively, Charles W. Johnson, and W.L. Pattison; office Hot Sulphur Springs; claim 300 by 1,500 feet; discovered 1876; located on Baker Mountain, Campbell mining district, 35 miles south of Hot Sulphur Springs; course of vein, north and south, width 3 feet, pay vein 2 to 24 inches, pitch 45 degrees west; nature of ore, galena, copper and carbonates; assay value, 70 ounces silver per ton and 50 per cent lead; 1 shaft 30 feet deep.

Gold Run Placer Diggings. Owned by Edward C. Sumner and George W. Day, of Denver; claim comprises 500 acres of placer ground, located on Gold Run, Willow Creek mining district, 12 miles from Hot Sulphur Springs and 100 miles from Denver.

Grand Lake. Owned by J. Baker, W.N. Brown, and J.H. Markel; office Hot Sulphur Springs; claim 300 by 1,500 feet; discovered 1875; located on Baker Mountain, Campbell mining district, 35 miles from Hot Sulphur Springs; course of vein, east and west, width 4 feet, pay vein 2 to 24 inches, pitch 45 degrees north; nature of ore, gray copper, silver glance, and black sulphurets; average value, $125 per ton; 1 shaft 25 feet deep, 1 tunnel 25 feet in length.

Hahns Peak Gold and Silver Mining Co., of Colorado. Incorporated July 30, 1874; capital stock, $3,000,000, in 3,000 shares, $100 each; office Denver; organized by Daniel C., James. H., and Berty Stover, Charles D. Gurley, Stephen D.N. Bennett, C. Harton, S. Scott, H.S. Moulton, and Emerson Hubbard.

Northwestern Park Gold and Silver Mining Co., of Colorado. Incorporated July 30, 1874; capital stock $200,000, in 2,000 shares, $100 each; office Denver; organized by Daniel C., James H., and Berty Stover, Charles D. Gurley, H.S. Moulton, C. Harton, and S. Scott.

Red Park Mining and Water Co. Incorporated October 1, 1875; capital stock, $12,000, in 120 shares, $100 each; office Denver; organized by Edward H. Rife, Charles B. Sears, and James E. Peterson.

Snake River Gold and Silver Mining Co. Incorporated July 30, 1874; capital stock, $200,000, in 2,000 shares, $100 each; office Denver; organized by Daniel C., James H., and Berty Stover, C. Harton, S. Scott, Charles D. Gurley, and H.S. Moulton.

Crested Mining Co. Incorporated May 29, 1877; capital stock, $10,000, in 100 shares; office Colorado Springs; trustees: R.R. Crawford, Andrew L. Lawton, William L. Wood, Charles Ayer, and Robert Basey.

Gunnison Silver Mining Co., Illinois. Incorporated February 5, 1877; capital stock, $1,000,000, in 10,000 shares, $100 each; office Pekin, Illinois; organized by J.L. Briggs, D.C. Boley, C.N. Priddy, and A.S. Boley.

HINSDALE COUNTY

Thomas B. Corbett reported that Hinsdale County had a population of 5,000 in 1878 and produced bullion valued at $182,000. Mining districts included Burrows Park, Galena, Park, and Uncomprahgre.

Accidental. Owned by John F. Dodds. William N. Ewing, Thomas J. Peter, C.K. Holliday, Jacob Safford, and M.L. Sargent; office Lake City; claim 300 by 1,500 feet; discovered 1878; located on Copper Hill, at the head of Henson Creek, Galena mining district, 15 miles from Lake City; vertical vein, trending northeast and southwest, width 4 feet, pay vein 10 inches; nature of ore, carbonate of copper and sulphurets of silver; average value, $350 per ton; 1 tunnel, 45 feet in length; yield, $4,000.

Alpine Tunnel Co. Incorporated January 10, 1877; capital stock, $100,000, in shares of $100; office Lake City and Alpine Glen; trustees: Rufus C. Vose, James J. Holbrook, T.J. Anderson, John F. Dodds, Henry C. Olney, Daniel Stone, Mark B. Price, Charles H. Kimball, and E.S. Stover.

American. Owned by J. Georke and I.N. Akers; office Lake City; claim 300 by 1,500 feet; discovered 1876; located in Galena mining district, 2 miles from town of Capitol and 10 miles from Lake City; vertical vein, trending north and south, width 7 feet, pay vein 20 inches; nature of ore, gray copper and galena; average assay value, 600 ounces silver per ton; 3 drifts, 20, 25, and 30 feet in length, respectively.

American Eagle. B.J. Smith, proprietor; located at the head of American Basin, Park mining district, 25 miles from Lake City.

Bank of Lake City. L.B. Clay, proprietor; office Ouray; claim 300 by 1,500 feet; discovered 1877; located on Sheep Mountain, Galena mining district, 4 miles from town of Capitol and 15 miles from Lake City; vertical vein, trending northwest and southeast, width 8 feet, pay vein 4 feet; nature of ore, galena; assay value, $36 per ton; 1 drift, 30 feet in length and 40 feet from surface.

Barlow Silver Mining Co. Incorporated February 9, 1876; capital stock, $500,000, in 5,000 shares, $100 each; offices Lake City and Denver, Colorado; organized by Alexander Mesler, Henry Y. Cooper, Frank C. Garbutt, Fiske Farrar, and Mason B. Carpenter.

Belle of the East. J.W. McFerran, Lewis Whipple, and William Peck, proprietors; office Lake City; claim 300 by 1,500 feet; located 8 miles south of Lake City, on the south fork of the Gunnison River; vertical vein, trending northeast and southwest, width 4 feet, pay vein 2 to 15 inches; nature of ore, gray copper and galena; main shaft 100 feet deep, 2 drifts, 110 and 80 feet, respectively.

Belle of the West. Owned by Samuel Wade and Otto Mears; office Lake City; claim 300 by 1,500 feet; located 3 miles south of Lake City, on the Lake Fork of the Gunnison River; vertical vein, trending northeast and southwest, width 5 feet, pay vein 10 inches; nature of ore, gray copper and galena, average value, 85 ounces silver per ton; 4 shafts, 18, 60, 75, and 80 feet in depth, respectively, 4 drifts aggregating 525 feet in length; yield, $50,000. [Otto Mears was best known for the toll roads he built, opening many isolated mountain communities and mining camps. Born in Russia in 1841, he immigrated with his family to America in 1854, settling in California. He served with the 1st Regiment California Volunteers in the Civil War and settled in Colorado's San Luis Valley after the war. As the first treasurer of Saguache County, a very poor area, Mears collected taxes in the form of furs and buckskins and brought them to Denver to sell, enabling the poor county to pay its taxes in cash. He was president of the Rio Grande Southern Railway.]

Belle of the West, No. 2. Owned by Joseph Chambers, J.E. Riley & Co.; office Lake City; claim 300 by 1,500 feet; discovered 1875; located on Hotchkiss Mountain, Lake mining district, 3 miles from Lake City; course of vein, east and west, width 3 feet, pay vein 15 inches, pitch 30 degrees; nature of ore, galena and gray copper, assay value, 80 ounces silver per ton; 1 drift 60 feet, and 1 adit 24 feet in length; accessible by wagon road.

Big Casino. Owned by J.J. Holbrook and W.T. Forrest; office Lake City; claim 300 by 1,500 feet; discovered 1875; located in Galena gulch, Galena mining district, 7 miles from Lake City; course of vein, northeast and southwest, width 3 feet, pitch 20 degrees; nature of ore, gray copper and galena; average value, 65 ounces silver per ton; 1 shaft 50 feet deep, 1 tunnel 90 feet in length.

Black Hornet. B. Snowden, proprietor; office Lake City; claim 300 by 1,500 feet; located in Owl Gulch, Lake mining district, 6 miles from Lake City; course of vein, north and south; width 30 inches, pitch 10 degrees east; nature of ore, horn silver; average value 72 ounces silver per ton; 2 drifts 21 and 35 feet in length, respectively, 1 tunnel 25 feet in length.

Buckeye. Owned by W.N. Marks and G.C. Milner; office Lake City; claim 300 by 1,500 feet; discovered 1876; located on Red Mountain, Lake mining district, 5 miles from Lake City; course of vein north and south, width 9 feet, pay vein, 18 inches; pitch 35 degrees; nature of ore, galena and gray copper; average value, 100 ounces silver per ton; 2 drifts 20 and 35 feet in length, respectively.

Bummer. Owned by Thomas & Mesler; office Capitol City; claim 300 by 1,500 feet; situated on Capitol Mountain, Galena mining district, near town of Capitol, 10 miles from Lake City; course of vein, northeast and southwest, width 6 feet, pay vein 30 inches, pitch 10 degrees southeast; nature of ore, galena; average assay value, $60 per ton; 4 shafts aggregating 70 feet in depth, 1 tunnel 45 feet in length, intersecting vein at a depth of 125 feet.

Cashier. Henry Morey, proprietor; office Ouray; claim 300 by 1,500 feet; discovered 1876; located on Galena Mountain [13,278 feet], near town of Tellurium, Burrows Park mining district, 16 miles from Lake City; course of vein, north and south, width 20 feet, pay vein 4 feet, pitch 30 degrees; nature of ore, galena; average value, 70 ounces silver per ton; 1 shaft 25 feet deep.

Child of Fortune. Owned by William M. Friend. H. Pamperin, and S. Wendell; office Ouray; claim 300 by 1,500 feet; discovered 1876; located on White Cross Mountain, 3 miles from town of Argentum, Burrows Park mining district, 18 miles from Lake City; vertical vein, trending northeast and southwest, width 9 feet, pay vein 10 inches; nature of ore, tellurium and galena, average assay value, $135 per ton; 1 adit, 30 feet in length and 22 feet from surface, at the end of which is sunk a shaft 36 feet deep.

Christone. Owned by Joseph Hamer; office Lake City; claim 300 by 1,500 feet; discovered 1875; located on the Lake Fork of the Gunnison River, Lake mining district, 6 miles from Lake City; vertical vein, trending north and south, width 10 feet, pay vein 24 inches; nature of ore, gray copper and galena; value, 120 ounces silver per ton; development, 1 drift 25 feet in length and 20 feet from surface; yield, $400.

Cleveland. Owned by Charles R. Fitch, Charles O. and J.C. Evarts; office Cleveland, Ohio; claim 300 by 1,500 feet, located on Henson Creek, Lake mining district, 1 mile from Lake City; course of vein, northeast and southwest, width 4 feet, pay vein 1 foot, pitch 25 degrees west; nature of ore, gray copper and galena, average value 60 ounces silver per ton; 1 shaft 30 feet deep.

Comprehensive. Owned by Joseph and Adolph Nathan, John Schreiner, A.R. Bushnell, John G. Clark, Joseph Bock, and C.H. Murray; offices Lake City, Colorado, and Lancaster, Wisconsin; claim 300 by 1,500 feet, located beside Alpine Creek on Park Mountain, Lake mining district, 6 miles from Lake City; course of vein, northwest and southeast, width 4 feet, pay vein 18 inches, pitch east, main shaft 10 feet deep.

Copper Glance. Owned by the Copper Hill Silver Mining Company; claim 300 by 1,500 feet; discovered 1874; located on Copper Hill, Galena mining district, 15 miles from Lake City; nature of ore, gray copper and copper glance; value 150 ounces silver per ton and 35 per cent copper; 2 tunnels, 60 and 150 feet in length, respectively; yield, $2,000.

Copper Hill Silver Mining Co. Incorporated November 25, 1878; capital stock, $200,000, in 2,000 shares, $100 each; office Lake City; organized by W.N. Ewing, John F. Dodds, and John H. Maugham.

Cornucopia. Owned by C. Harding and W. Rummel; office Lake City; claim 300 by 1,500 feet; discovered 1875; located on Cottonwood Creek, Lake mining district, 6 miles from Lake City; vertical vein, trending east and west, width 6 feet, pay vein 18 inches; nature of ore, galena and gray copper; average assay value, 80 ounces silver per ton, 1 shaft 20 feet deep, 1 tunnel 25 feet in length.

Croesus. Owned by Mesler and Co.; office Capitol City; claim 300 by 1,500 feet; located on Capitol Mountain, Galena mining district, near town of Capitol, 10 miles from Lake City; course of vein, east and west, width 7 feet, pay vein 28 inches, pitch 10 degrees south; nature of ore, gray copper, iron and copper pyrites; average value, 60 ounces silver per ton; 1 level, 30 feet in length.

Crooke Mining and Smelting Co. Organized under the laws of the State of New York; John N. Goodwin, president, William H. Smith, secretary, John J. Crooke, agent.

Czar. Owned by Mesler & Co.; office Capitol City; claim 300 by 1,500 feet; located on Yellowstone Mountain, Galena mining district, near town of Capitol, 10 miles from Lake City; course of vein, north and south, width 6 feet, pay vein 18 inches, pitch 10 degrees east; nature of ore, galena, gray copper, and sulphurets; average value, 100 ounces silver per ton; 1 shaft 25 feet deep.

Czarina. Owned by Mesler & Co.; office Capitol City; claim 300 by 1,500 feet; situated on Yellowstone Mountain, Galena mining district, 10 miles from Lake City; course of vein, north and south, width 3 feet, pay vein 14 inches, pitch 10 degrees east; nature of ore, galena, gray copper, and sulphurets; average value, 75 ounces silver per ton; 3 shafts 12, 14, and 20 feet in depth, respectively.

Delta. W. Williams, proprietor; office Lake City; claim 300 by 1,500 feet; discovered 1877; located on Henson Creek, Lake mining district, 5 miles from Lake City; course of vein, northwest and southeast, width 14 feet, pitch 30 degrees; nature of ore, galena and gray copper; average value, 80 ounces silver per ton; 1 shaft 20 feet deep.

Dolly Varden. Peter Houghton, proprietor; office Del Norte; claim 300 by 1,500 feet, patented; discovered 1874; located on Copper Mountain, Galena mining district, 14 miles from Lake City; course of vein, north 30 degrees east and south, 30 degrees west, width 4 feet, pay vein 10 inches; nature of ore, gray copper and copper pyrites; average value, 300 ounces silver per ton; 2 shafts 50 and 30 feet in depth, respectively; yield, $6,000.

Dolphin. Owned by Peter Robertson, E.J. Shaw, and F. Siems; office Lake City; claim 300 by 1,500 feet; located in Lake mining district, 2 miles from Lake City; course of vein, northeast and southwest, width 6 feet, pay vein 18 inches, pitch 20 degrees; nature of ore, galena and gray copper; average value, 150 ounces silver per ton, 2 shafts 25 and 65 feet in depth, respectively.

Duke of Argyle. Owned by Albert Campbell and George Crummey; located in Des Moines Gulch, Park mining district, 25 miles from Lake City; development, 1 tunnel 15 feet in length.

Enterprise Tunnel Co. Incorporated June 12, 1877; capital stock, $500,000, in 10,000 shares, $50 each; office Lake City; organized by J.F. McDonald, William R. Kennedy, A. Danford, J.E. Leonard, J.H. Maugham, James Ecklin, George W. Crummey, Sr., C.B. Evans, H.D. Jones, Rienzi E. Peniston, and William B. Fonda.

Equator. Owned by the Equator Mining Co., of Colorado; incorporated January 26, 1878; capital stock, $1,000,000, in 10,000 shares, $100 each; offices Lake City, Colorado, and Philadelphia, Pennsylvania; organized by John H. and Edwin J. Shaw, James Phillips, Thomas M. Bowen, James H. Casey, John Miles, and George P. Blunt; claim 300 by 1,500 feet, located beside Henson Creek on Equator Mountain, Lake mining district, 6 miles from Lake City; course of vein, northeast and southwest, width 4 feet, pay vein 12 inches, pitch 30 degrees east; nature of ore, gray copper and galena; average value, $75 per ton; 1 shaft 25 feet deep, 1 tunnel 80 feet in length.

Fanny. Owned by the Wisconsin and San Juan Mining Co.; offices Lake City, Colorado, and Whitewater, Wisconsin; claim 300 by 1,500 feet, located on Henson Creek, Galena mining district, 10 miles from Lake City, course of vein, northeast and southwest, width 7 feet, pay vein 10 inches; nature of ore, galena, gray copper, and native silver; average value, 108 ounces silver per ton; 1 drift 25 feet in length.

Fidelia. Owned by Hon. F.W. Pitkin, William Sherman, J. Baum, and D.A. Cowell; office, Lake City; claim 300 by 1,500 feet; located on Red Mountain, Lake mining district, 7 miles from Lake City; course of vein, northeast and southwest, width 5 feet, pay vein 3 feet, pitch 45 degrees; nature of ore, free gold, average assay value, $3,000 per ton; 1 tunnel, 150 feet in length, intersecting vein at a depth of 175 feet.

First National Bank. Owned by George Berry and others; office Lake City; claim 300 by 1,500 feet; located at the head of Henson Creek, foot of Engineer Mountain [12,968 feet], Galena mining district, 18 miles from Lake City; course of vein, northeast and southwest, width 3 feet, pay vein 8 inches, pitch 25 degrees east; nature of ore, gray copper and brittle silver; average value, 100 ounces silver per ton; 1 tunnel 90 feet in length.

Flower of the West. Owned by B.J. Smith; located at the head of American Basin, in Burrows Park mining district, 25 miles from Lake City.

Frank R. Adams. Owned by John F. Dodds, Henry and Frank Adams, and James B. Duval; office Lake City; claim 300 by 1,500 feet; located 1878; located at the head of Henson Creek, Galena mining district, 15 miles from Lake City; course of vein, northeast and southwest, width 18 feet, 2 pay veins, 8 inches and 3 feet, respectively, pitch 8 degrees east; nature of ore, brittle silver; value, 150 to 4,000 ounces silver per ton; 1 shaft 15 feet deep, 1 tunnel 25 feet in length; yield, $500.

George Washington. Owned by F.C. Garbutt & Co.; office Lake City; claim 300 by 1,500 feet; discovered 1877; located in Cataract Gulch, 2 miles from town of Sherman, Burrows Park mining district, 16 miles from Lake City; course of vein, east and west, width 4 feet, pay vein 9 inches, pitch 5 degrees; nature of ore, ruby and brittle silver; average value, 700 ounces silver per ton; 1 drift 75 feet in length and 25 feet from surface; yield, $2,500.

Gladiator. Owned by E.V.B. Hoes, Frank M. Schiedler, J.J. Barrett, and J.M. Hill; office Lake City; claim 300 by 1,500 feet; located on the Lake Fork of the Gunnison River, near the San Cristobal Falls; course of vein, northeast and southwest, width 4 feet, pay vein 30 inches; nature of ore, gray copper and iron pyrites; average value of first-class ore, 315 ounces, and second-class, 80 ounces silver per ton; 1 shaft 40 feet deep, 2 tunnels 130 and 110 feet in length respectively.

Gnome. Owned by James J. and Albert Bernard, Daniel J. and James D. McKay, William Forsyth, and John M. Ross; office Eureka; claim 300 by 1,500 feet; located in American Basin, near town of Tellurium, 16 miles from Lake City; course of vein, northeast and southwest, width 20 feet; nature of ore, galena, gray copper, iron and copper pyrites; average value, 100 ounces silver per ton; 1 tunnel, 40 feet in length.

Gray Copper. Owned by George Crummey, John Williams, William Richards, and Louis Kafka; office Lake City; claim 300 by 1,500 feet; discovered 1876; located on the Lake Fork of the Gunnison River in Burrows Park, Park mining district, 25 miles from Lake City; course of vein, northeast and southwest, width 5 feet, pay vein 15 inches; nature of ore, gray copper and galena; average value, 35 ounces silver per ton; 1 shaft 50 feet deep, 1 tunnel 275 feet in length.

Gunnison Silver Mining Co. Organized under the laws of the State of Illinois, February 5, 1877; capital stock, $1,000,00, in 10,000 shares, $100 each; offices Burrows Park, Colorado, and Pekin, Illinois.

Hidden Treasure. Owned by J.K. Mullen, H. Musgrove, and John S. Hough; office Lake City; claim 300 beet by 1,500 feet; located on Ute Hill, 4 miles from Lake City; course of vein, northeast and southwest, width 4 feet, pay vein 15 inches, pitch 20 degrees northwest; nature of ore, gray copper and galena; 1 tunnel 340 feet in length.

Homestake. Owned by J.J. Barrett, Henry Finley, S. Freeman, and J.M. Hill; office Lake City; claim 300 by 1,500 feet; located on Hotchkiss Mountain, Lake mining district; course of vein, east and west, width 4 feet, pay vein 12 inches, pitch north; nature of ore, gray copper, galena and copper pyrites; average assay value, 75 ounces silver per ton; 1 shaft 35 feet deep, 1 tunnel 27 feet in length.

Hoosier. Located on the Lake Fork of the Gunnison River, Park mining district, 25 miles from Lake City.

Hope. Owned by E.V.B. Hoes, William Phillips, and George Boggs; office Lake City; claim 300 by 1,500 feet; located on Red Mountain, at the head of Alpine Creek, Lake mining district, 8 miles from Lake City; course of vein, northeast and southwest, width 5 feet, pay vein 3 feet, pitch south; nature of ore, tellurium, galena, and gray copper, assay value, 15 ounces gold and 200 ounces silver per ton; 1 tunnel 25 feet, and 1 drift 70 feet in length.

Hope. Robert Kuslick, proprietor; office Burrows Park; claim 300 by 1,500 feet; located on Cooper Creek, in Burrows Park, Park mining district, 25 miles from Lake City; course of vein, northeast and southwest, width 4 feet, pay vein 18 inches, pitch 10 degrees north; nature of ore, gray copper and galena; average value, 40 ounces silver per ton; 1 shaft 20 feet deep.

Hotchkiss. Owned by Fred C. Peck, M.S. Taylor, J.H. Shaw, and George Wilson; office Lake City; claim 300 by 1,500 feet; located on the Lake Fork of the Gunnison River, Lake mining district, 4 miles from Lake City; vertical vein, trending northeast and southwest, pay vein 14 inches; nature of ore, gray copper and tellurium; average value, 400 ounces silver per ton; 2 tunnels 120 and 80 feet in length, respectively; yield, $40,000.

Inez. Owned by the Hard Bros. and Holt; office Lake City; claim 300 by 1,500 feet; discovered 1876; located in Cleveland Gulch, at the head of Burrows Park, 3 miles from Animas Forks and 17 miles from Lake City;

vertical vein, trending northeast and southwest, width 100 feet; nature of ore, native and ruby silver; average value, 80 ounces silver per ton; 1 shaft 25 feet deep, 1 tunnel 25 feet in length.

Inez, No. 2. Owned by Albert Campbell, W.A. Adams, Joseph Keeler, and Col. Terrell; claim 300 by 1,500 feet; located in Cleveland Gulch, at the head of Burrows Park, 25 miles from Lake City; course of vein, northeast and southwest, 1 surface opening 20 feet in length.

Invincible. Henry Morley, proprietor; office Ouray; claim 300 by 1,500 feet; discovered 1877; located on Galena Mountain, near town of Tellurium, Burrows Park mining district, 18 miles from Lake City; vertical vein, trending northwest and southeast, width 30 feet, pay vein 3 feet; nature of ore, galena and gray copper; average value, 80 ounces silver per ton; 1 adit, 12 feet in length.

Iron City. Owned by R. Taylor and M. Curran; office Lake City; claim 300 by 1,500 feet; discovered 1876; located in Galena mining district, 3 miles from town of Capitol and 11 miles from Lake City; course of vein, northwest and southeast, width 4 feet, pay vein 18 inches; nature of ore, galena, iron and copper pyrites; average value, 40 ounces silver per ton; 1 shaft 14 feet deep, 1 adit 18 feet in length and 12 feet from surface.

Joaquin. Owned by John S. Hough, T. Shanley, Clark Peyton, and Alfred Schiffer; office Lake City; claim 300 by 1,500 feet; situated on Henson Creek, on Capitol Mountain, Galena mining district, 12 miles from Lake City; course of vein, northeast and southwest, width 5 feet, pay vein 24 inches; nature of ore, galena; average value, $60 per ton; 1 tunnel 30 feet in length.

Julia. O.A. Mesler, proprietor; office Capitol City; claim 300 by 1,500 feet; located on the north fork of Henson Creek, Galena mining district, near town of Capitol, 15 miles from Lake City; course of vein, north and south, width 7 feet, pay vein 5 feet, pitch 10 degrees east; nature of ore, galena; average value, 56 ounces silver per ton; 3 shafts, aggregating 75 feet in depth.

Kennebec. Owned by John F. Dodds, Mary R. Ewing, and Harry C. Jones; office Lake City; claim 300 by 1,100 feet; located at the head of Henson Creek, Galena mining district, 15 miles from Lake City; course of vein, northeast and southwest, width 18 feet, pay vein 8 inches to 3 feet; value of ore, brittle silver and sulphurets of silver; average value, 150 to

3,000 ounces silver per ton; 1 shaft 15 feet deep, 1 drift 15 feet in length; yield, $100.

La Belle Tunnel. Owned by J.C. Bouton, Harvey Wright, J.C. Hepburn, and J.N. Barrett; located at the foot of White Cross Mountain at the head of Burrows Park, 25 miles from Lake City; course, southerly; length, 150 feet.

Lake City Mining and Smelting Co. Incorporated February 5, 1876; capital stock, $100,000, in 1,000 shares, $100 each; office Lake City; organized by Frank C. Garbutt, John M. Holcombe, Edward P. Wilder, David D. Belden, James W. Abbott, Alba R. Thompson, and Jacob J. Abbott, Jr.

Lee Mining and Smelting Co. Organized under the laws of the State of Illinois, August 9, 1878; capital, paid up stock, $100,000, in 1,000 shares, $100 each; office Capitol City; William H. Turner, president, George S. Lee, manager, Allen M. Culver, secretary.

Little Chief. Owned by Andrew Gill, J.W. Brockett, H. Webber, and M.B. Gerry; office Lake City; claim 300 by 1,500 feet; located on Helen Mountain on Henson Creek; Lake mining district, 2 miles from Lake City; vertical vein, trending northeast and southwest, width 3 feet, pay vein 12 inches; nature of ore, gray copper and sulphurets; average value, 200 ounces silver per ton; 1 shaft 30 feet deep, 1 open cut, 15 feet in length.

Little Giant. Owned by Frank Curtis, George P. Childs, and E.F. Snook; office Lake City; claim 300 by 1,500 feet; discovered 1878; located on Copper Hill, on the north side of Henson Creek, Galena mining district, 14 miles from Lake City; vertical vein, trending northeast and southwest, width 5 feet, pay vein 4 inches; nature of ore, gray copper; development, 1 open cut, with 15-foot face.

Little John. Owned by the Wisconsin and San Juan Mining Company; Hill Witmore, superintendent; offices Lake City, Colorado, and Whitewater, Wisconsin; claim 300 by 1,500 feet; located on Henson Creek, Galena mining district, 10 miles from Lake City; course of vein, northeast and southwest, width 7 feet, pay vein 10 inches; nature of ore, galena, gray copper and native silver; average value 108 ounces silver per ton; 1 drift 35 feet in length.

Little Turk. Owned by Harbottle, Mills & Co.; office Lake City; claim 300 by 1,500 feet; discovered 1876; located in Boulder Basin, Galena mining district, 13 miles from Lake City; course of vein, northeast and southwest, width 4 feet, pay vein 3 to 9 inches; nature of ore, galena and brittle silver; average value, $240 per ton; 1 open cut 20 feet in length.

Maggie Owens. Owned by the Silver Brick Mining Company; John S. Hough, president; office Lake City; claim 300 by 1,500 feet; located half mile south of Mineral City, Uncompahgre mining district, 20 miles from Lake City; course of vein, northeast and southwest, width 25 feet, pitch southeast; nature of ore, galena and gray copper; average value, 30 ounces silver per ton.

May Flower. Owned by J.A. Warner, A.C. and J.B. Cryder; office Lake City; claim 300 by 1.500 feet; located on Hotchkiss Mountain, Lake mining district, 7 miles from Lake City; course of vein, northeast and southwest, width 5 feet, pay vein 15 inches, pitch north; nature of ore, galena, gray copper and iron; average value, 102 ounces silver per ton; 2 tunnels 120 and 180 feet in length, respectively.

Melrose. Owned by Franklin & Co.; office Lake City; claim 300 by 1,500 feet, located 1876; situated 2 miles from town of Capitol, Galena mining district, 8 miles from Lake City, course of vein, east and west, width 8 feet, pitch 30 degrees, pay vein 18 inches; nature of ore, galena and gray copper; average value, 400 ounces silver per ton; 1 adit 30 feet in length.

Moltke. Charles Schaefer, Theodore Dick, and Jacob Gredig, proprietors; office Rose's Cabin; claim 300 by 1,500 feet, located on Sheep Mountain, Galena mining district, 15 miles from Lake City; course of vein, northeast and southwest, width 4 feet, pay vein 20 inches, pitch 10 degrees southeast; nature of ore, galena; average value, 30 ounces silver per ton; 1 tunnel 140 feet in length. [In the 1880 Federal Census for Rose's Cabin, Charles Schaefer was a 34-year old unemployed saloonkeeper from Prussia. Theodore Dick, an unemployed carpenter from Germany, was one of his boarders.]

Myrtle Wreath. Owned by Albert Campbell and W.A. Adams; located in Burrows Park, 25 miles from Lake City.

Monroe. George Zantz, proprietor; office Lake City; claim 300 by 1,500 feet; located southeast of Lake City, in Lake mining district; vertical vein, trending northeast and southwest, width 4 feet, 2 pay veins 18 and 22 inches respectively; nature of ore, iron and copper pyrites, 1 open cut 20 feet in length.

Monster. Owned by Harbottle, Mills & Co.; office Lake City; claim 300 by 1,500 feet; discovered 1875; located in Schaefer Basin, Galena mining district, 13 miles from Lake City; vertical vein trending northeast and southwest, width 30 feet, pay vein 14 inches; nature of ore, galena; development, 2 open cuts, 15 and 20 feet in length respectively.

Mountain Lion. Owned by George A. Kellogg and J.W. Hughes; office Lake City; claim 300 by 1,500 feet; located on Henson Creek, Lake mining district, 1 mile from Lake City; course of vein, northeast and southwest, pitch 25 degrees southwest; 1 shaft 30 feet deep, 1 open cut, 12 feet in length, with 12-foot face.

Napoleon. Owned by Crooke, Posey & Lowe; office Lake City; claim 300 by 1,500 feet; located at the foot of White Cross Mountain, on the Lake Fork of the Gunnison River, Park mining district, 25 miles from Lake City; course of vein, northeast and southwest, width 5 feet, pay vein, 18 inches, pitch 12 degrees north; nature of ore, antimony and silver; average value, 600 ounces silver per ton; 1 shaft 70 feet deep.

Niagara Falls. Owned by E.V.B. Hoes, Otto Mears, J.J. Barrett, J.M. Hill, and James Sparling; office Lake City; claim 300 by 1,500 feet; located in Lake mining district, on the north fork of the Gunnison River, near the falls; course of vein, northeast and southwest, width 4 feet, pay vein, 8 inches, pitch south; nature of ore, galena and gray copper; assay value, 5 ounces gold and 312 ounces silver per ton; 1 tunnel 40 feet in length.

North. Owned by the Wisconsin and San Juan Mining Co.; office Lake City, Colorado, and Whitewater, Wisconsin; claim 300 by 1,500 feet; located on Henson Creek, Galena mining district, 3 miles from Lake City; course of vein, northeast and southwest, width 5 feet, pay vein 3 inches, pitch 10 degrees; nature of ore, galena and gray copper; average value, 65 ounces silver per ton; 1 drift 25 feet in length.

Occidental. Owned by S.P and F.M Truitt and William M. Beers; office Ouray; claim 300 by 1,500 feet; discovered 1875; located on Silver Creek, 1 mile from town of Argentum, Burrows Park mining district, 18 miles from Lake City; course of vein, northeast and southwest, width 30 inches, pay vein 16 inches; nature of ore, galena and gray copper; assay value, 1,100 ounces silver per ton; 1 tunnel 31 feet in length, intersecting vein at a depth of 25 feet from the surface.

Ocean Wave, and Extension. Owned by the Ocean Wave Mining and Smelting Co., of Lake City, Colorado. Incorporated 1877; capital stock, $100,000, in 1,000 shares, $100 each; office Lake City; William R. Bernard, president, James W. Potter, vice president and general manager, A.T. Gunnel, secretary, L.H. Spilker, superintendent and metallurgist; claim 300 by 3,000 feet, patented; located beside Henson Creek, on Red Rover Mountain, Galena mining district, 7 miles from Lake City; course of vein, east and west, width 4 feet, pitch 10 degrees south; nature of ore, galena and gray copper; average value, $200 per ton; development; 4 tunnels, 60, 220, 300, and 420 feet in length, respectively, 3 winzes, aggregating 165 feet; yield to date, $100,000.

Old Hickory Mining Co. Incorporated April 28, 1877; capital stock, $1,000,000, in 10,000 shares, $100 each; offices Lake City, Colorado, and St. Louis, Missouri; trustees: J.D. Angier, N.M. Swansey, Joel N. Angier, L.C. McKenney, John L. Woods, J.L. Sanderson, D.M. Edgerton, T.J. Barnum, and H. Sanderson.

Palmetto. Owned by John S. Hough, Joseph Hense, and Herman Wertzberger; office Lake City; claim 300 by 1,450 feet; located at the head of Henson Creek, at the foot of Engineer Mountain, Galena mining district, 18 miles from Lake City; course of vein, northeast and southwest, width 3 feet, pay vein 8 inches, pitch 25 degrees east; nature of ore, gray copper and brittle silver; value, 100 ounces silver per ton; 1 shaft 53 feet deep, 1 drift 75 feet in length; yield, $500.

Pelican. Owned by M. Rich, Henry Ruggles, and Nye Tuttle; office Lake City; claim 300 by 1,500 feet; located on Sugar Loaf Mountain beside Henson Creek, 2 miles from Lake City; vertical vein, trending northeast and southwest, width 5 feet, pay vein 18 inches; nature of ore, gray copper; average value, 59 ounces silver per ton; 1 shaft 45 feet deep, 1 adit, 10 feet in length; yield, $400.

Philadelphia. Owned by Harbottle, Mills and Gregory; office Lake City; claim 300 by 1,500 feet; located in Schaefer Basin, Galena mining district, 13 miles from Lake City; vertical vein, trending northeast and southwest, width 4 feet, pay vein 6 inches; nature of ore, galena and copper carbonates; average value, 70 ounces silver per ton; 2 open cuts, 20 and 15 feet in length, respectively.

Plutarch. Owned by Nutting, Chambers & Co.; office Lake City; claim 300 by 1,500 feet, patented; discovered 1875; located on Hotchkiss Mountain, Lake mining district, 4 miles from Lake City; vertical vein, trending east and west, width 3 feet, pay vein 18 inches; nature of ore, gray copper and brittle silver; average value, 190 ounces silver per ton; main shaft, 100 feet, and 2 other shafts, aggregating 150 feet in depth, 2 tunnels aggregating 200 feet in length, intersecting the vein at a depth of 60 and 130 feet, respectively, 1 level 130 feet in length and 60 feet from the surface; shaft house, 24 by 16 feet; yield to date, $12,000.

Pocahontas. Owned by Frank Curtis and George A. Kellogg; office Lake City; claim 300 by 1,500 feet, located in Lake mining district, near Lake City; course of vein, northeast and southwest, width 7 feet, pay vein 2 feet, pitch 5 degrees west; nature of ore, galena; assay value, 20 ounces silver per ton; 1 shaft, 36 feet deep.

Pride of America. Owned by A.C. Cryder, Theodore, J.W., and Frank Taylor, and H.F. Smith; office Lake City; claim 300 by 1,500 feet; discovered 1875; located in Galena Gulch, Galena mining district, 7 miles from Lake City; course of vein, northeast and southwest, width 3 feet, pitch 20 degrees; nature of ore, gray copper and galena, value, 65 ounces silver per ton; 1 shaft 50 feet deep, 1 open cut, 15 feet in length.

Providence. Joseph Hamer, proprietor; office Lake City; claim 300 by 1,500 feet; discovered 1874; located on the Lake Fork of the Gunnison River, Lake mining district, 6 miles from Lake City; vertical vein, trending north and south, width 6 feet, pay vein 40 inches; nature of ore, gray copper and galena; assay value, 200 ounces silver per ton; 1 drift, 40 feet in length, and 60 feet from surface.

Putnam Mining Co. Incorporated May 18, 1876; capital stock, $500,000, in 5,000 shares, $100 each; office Denver; organized by Conrad Frick, James J. Rowan, Oliver A. Whittemore, Charles W. Lehman, and August Heckendorf.

Puzzler. Owned by Joseph Chambers and J.E. Riley; office Lake City; claim 300 by 1,500 feet; discovered on Bell of the West Mountain, Lake mining district, 3 miles from Lake City; vertical vein, trending northeast and southwest; width 8 feet; nature of ore, gray copper; assay value, 65 ounces silver per ton; 1 shaft 22 feet deep.

Puzzler. Owned by J. Snowden and A. Rapp; office Lake City; claim 300 by 1,500 feet; located on Red Rover Mountain, on Henson Creek, Galena mining district, 7 miles from Lake City; course of vein, east and west, width 4 feet, pitch 15 degrees south; nature of ore, galena, gray copper, and carbonates; 1 shaft 15 feet deep, 1 open cut 15 feet in length.

Queen of the West. Owned by John F. Dodds, Joseph Lucas, and Uriah Zimmerman; office Lake City; claim 300 by 1,500 feet; discovered 1878; located at the head of Henson Creek, on Poverty Hill, galena mining district, 15 miles from Lake City; course of vein, northeast and southwest, width 3 feet, pay vein 12 inches, pitch 10 degrees west; nature of ore, galena, copper, and brittle silver; average value, 150 ounces silver per ton; 2 shafts 15 and 25 feet in depth, respectively, 1 tunnel, 25 feet in length; yield, $200.

Red Cloud Mining Co. Incorporated March 4, 1876; capital stock, $500,000, in 5,000 shares, $100 each; offices Lake City and Denver, Colorado; organized by Abram S. Rhodes, James H. Lester, Richard E. Higgins, Frank C. Garbutt, James J. Rowan, Mason B. Carpenter, and E.M. Sanford.

Red Rover. Owned by John J. Crooke and J.J. Holbrook; office Lake City; claim 300 by 1,500 feet; situated on Red Rover Mountain, Galena mining district, 7 miles from Lake City; course of vein, northeast and southwest, width 4 feet, pay vein 8 to 20 inches; nature of ore, gray copper and galena; average value, 150 ounces silver per ton.

Rosa. Owned by W.P. McBride, J.P. Michaels, and H.D. McBride; office Lake City; claim 300 by 1,500 feet; located on Cottonwood Hill, Park mining district, 17 miles from Lake City; vertical vein, trending northeast and southwest, width 5 feet, pay vein 6 to 14 inches; nature of ore, silver glance and tellurium; average value, 500 ounces silver per ton; 1 shaft 50 feet deep, 1 tunnel 50 feet in length; yield, $14,000.

Royal. D.D. Durfee and W.N. Grace, proprietors; office Lake City; claim 300 by 1,400 feet; discovered 1876; located on Henson Creek, Lake mining district, 3 miles from Lake City; vertical vein, trending northwest and southeast, width 10 feet; nature of ore, galena, iron, and copper pyrites;

average value, 100 ounces silver per ton; 1 drift 25 feet in length and 22 feet from surface.

Royal Gem. Owned by Albert Campbell and W.A. Adams; office Argentum; claim 300 by 1,500 feet, located in Des Moines Gulch, in Burrows Park, 24 miles from Lake City; vertical vein, trending northeast and southwest; nature of ore, galena; assay value, 100 ounces silver per ton; 1 tunnel 50 feet in length.

Ruby. William M. Friend, proprietor; office Ouray; claim 300 by 1,500 feet; discovered 1876; located on White Cross Mountain, 3 miles from town of Argentum, Burrows Park mining district, 24 miles from Lake City; vertical vein, trending northeast and southwest, width 6 feet; nature of ore, galena; assay value, $130 per ton; 1 shaft 22 feet deep.

Seward County. Owned by Abner Y. Davis, H.C. Beaty, and C.L. Peyton; office Lake City; claim 300 by 1,500 feet; located in Burrows Park, on the Lake Fork of the Gunnison River, Park mining district, 25 miles from Lake City; course of vein, northeast and southwest, width 4 feet, pay vein 12 inches, pitch 10 degrees north; nature of ore, gray copper and galena; average value, 50 ounces silver per ton; 1 shaft, 65 feet deep.

Sherman. Owned by George Zantz, Dennis O'Neil, and P. McGuire; office Lake City; claim 300 by 1,500 feet; located southwest of Lake City, Lake mining district; vertical vein, trending northeast and southwest, width 3 feet, pay vein 15 inches; nature of ore, gray copper; value, $35 per ton; 1 shaft 28 feet deep.

Silver Brick Mining Co. Incorporated March 2, 1876; capital stock, $600,000, shares $10 each; offices Denver and Lake City, Colorado; John S. Hough, president, Samuel T. Thompson, secretary and treasurer.

Silver Coin. Edward Cree, proprietor; office Lake City; claim 300 by 1,400 feet; located 1876; located 2 miles from town of Capitol, Galena mining district, 8 miles from Lake City; course of vein, east and west, width 3 feet, pay vein 12 inches, pitch 20 degrees; nature of ore, galena; average value, 80 ounces silver per ton; 1 shaft 25 feet deep.

Silver Cord. Owned by Mesler & Co.; office Capitol City; claim 300 by 1,500 feet; located in Yellowstone Gulch, 1 mile from town of Capitol, and 10 miles from Lake City; vertical vein, trending northeast and southwest, width 4 feet, pay vein 14 inches; nature of ore, galena and gray copper; average value, 97 ounces silver per ton; 1 shaft 40 feet deep.

Silver Cord Extension. Owned by John H Simmons, A.A. and O.A. Mesler, and A.P. Cook; office Lake City; claim 300 by 1,500 feet; located on Yellowstone Gulch, 1 mile from town of Capitol, and 10 miles from Lake City; vertical vein, trending northeast and southwest, width 4 feet, pay vein 6 to 20 inches; nature of ore, gray copper; average value, 300 ounces silver per ton; 1 shaft 60 feet deep, 1 open cut, 25 feet in length; yield, $1,000.

Silver Ingot. Frank Curtis and George Campbell, proprietors; office Lake City; claim 300 by 1,500 feet; located near the head of Alpine Creek, Lake mining district, 3 miles from Lake City; course of vein, northeast and southwest, width 6 feet; pay vein 20 inches; nature of ore, galena; value, 18 ounces silver per ton, and 73 per cent lead; 1 shaft 65 feet deep.

Sonora. Charles Shaefer; H. Campbell, and Charles [Corydon?] Rose, proprietors; office Rose's Cabin; claim 300 by 1,500 feet; located on Gunnison Mountain, Galena mining district, 15 miles from Lake City; course of vein, northeast and southwest, width 6 feet, pay vein 14 inches, pitch 15 degrees southeast; nature of ore, gray copper; assay value, 60 ounces silver per ton; 1 shaft 25 feet deep.

South. Owned by the Wisconsin and San Juan Mining Co.; offices Lake City, Colorado, and Whitewater, Wisconsin; claim 300 by 1,500 feet; located on Henson Creek, Galena mining district, 3 miles from Lake City; course of vein, northeast and southwest, width 5 feet, pitch 10 degrees; nature of ore, galena and gray copper; average assay value, $85 per ton; 1 drift 100 feet in length.

St. Clair. Benjamin H. Pelton, proprietor; office Silverton; claim 300 by 1,500 feet; discovered 1876; located on Capitol Mountain, Galena mining district, 1 mile from town of Capitol and 9 miles from Lake City; vertical vein, trending northwest and southeast, width 10 feet, pay vein 10 inches; nature of ore, galena and gray copper; average value, 125 ounces silver per ton; 1 shaft 45 feet deep, 1 level 60 feet in length, and 45 feet from surface; yield, $150.

St. Louis and Extension. Owned by Rufus C. Vose, Charles H. Kimball, and John H. Simmons; office Lake City; claim 300 by 3,000 feet, located on Henson Creek, Capitol Mountain, Galena mining district, 14 miles from Lake City; vertical vein, trending northeast and southwest, width 8 feet, pay vein 30 inches; nature of ore, galena; 1 open cut, 25 feet in length, with a 20-foot face.

Sultana. Owned by Albert Campbell and W.A. Adams; office Burrows Park; claim 300 by 1,500 feet, located in Des Moines Gulch, at the head of Burrows Park, 25 miles from Lake City; vertical vein, trending northeast and southwest; nature of ore, galena, iron, and copper; 1 tunnel 15 feet in length.

Tom Benton. Owned by George A. Kellogg and J.W. Hughes; office Lake City; claim 300 by 1,500 feet; located on Henson Creek, Lake mining district, 1 mile from Lake City; vertical vein, trending northeast and southwest, width 5 feet, pay vein 6 inches; nature of ore, galena and gray copper; average value 80 ounces silver per ton; 1 tunnel 50 feet in length, and an open cut with a 12-foot face.

Trowbridge. Owned by Harbottle, Mills & Co.; office Lake City; claim 300 by 1,500 feet, discovered 1876; located in Schaefer's Basin, Galena mining district, 13 miles from Lake City; vertical vein, trending east and west, width 4 feet, pay vein 14 inches; nature of ore, galena and sulphurets of copper; 1 open cut, 15 feet in length.

Twilight, Extension of the Monster. Owned by Albert Campbell and W.A. Adams; claim 300 by 1,500 feet, located in Park mining district, 24 miles from Lake City; course of vein, northeast and southwest, width 20 feet; nature of ore, galena, iron, and copper; 1 surface opening 15 feet in length.

Ute. Crooke & Co., proprietors; office Lake City; claim 300 by 1,500 feet; located on the north side of Henson Creek; Galena mining district, 4 miles from Lake City.

Walker. A.W. Brumfield & Co., proprietors; office Ouray; claim 300 by 1,500 feet; discovered 1876; located on Galena Mountain, near town of Tellurium, 1 mile from Ouray; vertical vein, trending northeast and southwest, average width 4 feet, pay vein 15 inches; nature of ore, galena and gray copper; average value, 250 ounces silver per ton; 1 adit, 20 feet in length.

Wave of the Ocean, and Ocean Wave. Owned by the Ocean Wave Mining and Smelting Co., of Lake City, Colorado; incorporated 1877; capital stock, $100,000, in 1,000 shares, $100 each; office Lake City; William R. Bernard, president, James W. Potter, vice president and general manager; A.T. Gunnel, secretary, L.H. Spilker, superintendent and metallurgist; claim 300 by 3,000 feet, patented; located beside Henson Creek, on Red Rover Mountain, Galena mining district, 7 miles from Lake City; course of vein, east and west, width 4 feet, pay vein 2 inches to 2 feet, pitch 10 degrees south; nature of ore, galena and gray copper; average value, $200 per ton; 4 tunnels, 60, 220, 300, and 420 feet in length, respectively, 3 winzes, aggregating 165 feet in depth; yield to date, $100,000.

Wisconsin and San Juan Mining Co. Incorporated April 1878; capital stock, $30,000, in 300 shares, $100 each; offices Lake City, Colorado, and Whitewater, Wisconsin; Alonzo Elwood, president, James Wintermute, vice president, D.L. Fairchild, secretary, J.D. Lott, treasurer, Hill Witmore, superintendent. This company owns 19 lodes or mineral veins, 300 by 1,500 feet each, and a mill site of 5 acres.

Yankee Blade. Owned by Mesler & Co.; office Capitol City; claim 300 by 1,500 feet; located on Capitol Mountain, Galena mining district, near town of Capitol, 10 miles from Lake City; course of vein, east and west, width 4 feet, pay vein 12 inches, pitch 10 degrees south; nature of ore, gray copper, iron and copper pyrites, average value, 40 ounces silver per ton; 4 shafts, 14 feet in depth each.

Young America. Owned by Henry Adams, A.H. Hurd, and others; office Lake City; claim 300 by 1,500 feet; discovered 1876; located on Yellowstone Hill, Galena mining district, 1 mile from town of Capitol and 10 miles from Lake City; course of vein, east and west. Width 3 feet, pitch 10 degrees north; nature of ore, native silver and sulphurets; average value, 100 ounces silver per ton; 2 shafts, 95 and 65 feet in depth, respectively; 1 drift, 20 feet in length; yield, $4,500.

Hinsdale County Ore Mills

Lake City Mining and Smelting Co. Incorporated February 5, 1876; capital stock, $100,000, in 1,000 shares, $100 each; office Lake City; organized by Frank C. Garbutt, John M. Holcombe, Edward P. Wilder, David D. Belden, James W. and Jacob J. Abbott, and Alba R. Thompson.

Lake City Sampling Works. Titus & McClelland, proprietors; located in Lake City; have a capacity for treating 8 tons of ore daily.

Lee Mining and Smelting Co. Organized under the laws of the State of Illinois, August 9, 1878; capital, paid up stock, $1,000,000, in 10,000 shares, $100 each; George S. Lee, manager, Allen M. Culver, secretary, William H. Turner, president; the works of this company are located near town of Capitol, 8 miles from Lake City.

Lixiviation Works. Owned by W.N. Ewing and associates; located in Lake City; have a capacity for treating 10 tons of ore in 24 hours.

Ocean Wave Mining and Smelting Co., of Lake City, Colorado. Incorporated May 7, 1877, capital stock, $100,000, in 1,000 shares, $100 each; William R. Bernard, president, James W. Potter, vice president and general manager, A.T. Gunnel, secretary, L.H. Spilker, superintendent and metallurgist. The works have a capacity for treating 20 tons of ore in 24 hours; they are run by waterpower and located on the Gunnison River in Lake City. The main building is 48 feet wide by 64 feet in length, and contains all machinery necessary for the successful treatment of ore.

Crooke Concentration Works. Organized under the laws of the State of New York, March 1, 1877; capital stock, $50,000, in 500 shares, $100 each; offices, Lake City, Colorado, and New York City; organized by John J., Robert, and Lewis Crooke, and Oscar E. Schmidt, George V. Tompkins, and William O. Loeschigk. The works are located near Lake City in the Lake Fork of the Gunnison River, and have a capacity for treating 75 tons of ore in 24 hours.

Promontory Point along the Ouray Toll Road.

HUERFANO COUNTY

Thomas B. Corbett reported that Huerfano County had a population of 6,000 in 1878.

Center. George W. Morton, proprietor; office La Veta; claim, 300 by 1,500 feet; located on West Spanish Peak, 12 miles from La Veta, and 65 miles from Pueblo; vertical vein, trending east and west, width 3 feet; development, 1 shaft 20 feet deep.

Boss. Owned by Landis, Mason & Co.; office La Veta; claim 300 by 1,500 feet; located on the east gulch of West Spanish Peak, 12 miles from La Veta, and 65 miles from Pueblo; course of vein, east and west, width 2 feet, pay vein 1 to 9 inches; pitch 5 degrees south; nature of ore, galena and sulphurets; average value, $450 per ton; 3 shafts, 20, 30, and 15 feet in depth, respectively, 1 tunnel 40 feet in length.

La Veta Mining Co. Incorporated May 16, 1878; capital stock, $20,000, in 200 shares, $100 each; office La Veta; F.L. Martin, president, J.M. Francisco, vice president, John F. Moore, treasurer, J.H. McDonald, secretary.

Little Gratz. Owned by Rev. Samuel Lougheed, of La Veta; claim 300 by 1,500 feet; located in the west gulch of West Spanish Peak, 12 miles from La Veta, and 65 miles from Pueblo; course of vein, east and west, width 4 feet, pitch 5 degrees south; nature of ore, galena and sulphurets; assay value, 27 ounces silver and 2 ounces gold per ton; developed by shafts and tunnels.

Setting Sun. Owned by Morton, Patterson, Martin & Co.; office La Veta; claim 300 by 1,500 feet, located on West Spanish Peak, 12 miles from La Veta, and 65 miles from Pueblo; course of vein, northeast and southwest, width 4 feet, pay vein 16 inches, pitch 5 degrees south; nature of ore, galena and sulphurets; assay value, 72 ounces silver and 2 ounces gold per ton; 2 shafts, 12 and 16 feet in depth, respectively, 1 tunnel, 25 feet in length.

Silver Bell. Owned by Louis Crout and Rev. Samuel D. Lougheed; office La Veta; claim, 300 by 1,500 feet, located midway between the east and west gulches of West Spanish Peak, 12 miles from La Veta, and 65 miles from Pueblo; course of vein, nearly east and west, width 2 feet; developed by shafts and tunnels.

Single. Owned by Thomas Miller and George Depp; office La Veta; claim 300 by 1,500 feet, located on the right hand gulch of West Spanish Peak, 12 miles from La Veta, and 65 miles from Pueblo; course of vein, east and west, width 4 feet, pitch; 10 degrees south; nature of ore, galena and gray copper; 1 shaft 20 feet deep, 2 tunnels, 20 and 40 feet in length, respectively.

JEFFERSON COUNTY

Thomas B. Corbett reported that Jefferson County had a population of 7,000 in 1878.

French Smelting Works. Located within the limits of Golden; Jaques Gaillardon, president, M. Sellier, metallurgist and superintendent; these works have a capacity for treating 60 tons of ore daily.

Golden Smelting Co. Incorporated January 5, 1876; capital stock, $40,000, in 400 shares, $100 each; office Golden; organized by William B. Young, William West, and Thomas A. McMorris; the works of this company have a capacity for treating 25 tons of ore daily; are located on the Colorado Central Railroad, within the limits of Golden and consist of a group of buildings covering 2 acres of ground.

Malachite Mining Company of Colorado. Organized under the laws of the State of Maine, March 27, 1877; capital stock, $500,000; offices Golden, Colorado, and Winterport, Maine; Wilmot H. Chapman, agent, Golden, Colorado; the works of this company have a capacity for treating 5 tons of ore daily, and are located on the Colorado Central Railroad, within the limits of Golden.

Malachite Mine. Owned by the Malachite Mining Co; Lyman A. Carr, agent; offices Golden, Colorado, and Boston, Massachusetts; claim comprises 80 acres of land; discovered 1865; located on Bear Creek, 9 miles from Golden; course of vein, east and west, pitch 15 degrees north; average width of crevice, 4 feet; 1 shaft 128 feet deep, 2 tunnels, aggregate length 520 feet.

Trenton Dressing and Smelting Co., Trenton New Jersey. Horace Humphreys, president, William Hancock, vice president, Peter W. Crozer, secretary and treasurer, Daniel Peters, manager, professor Joseph Luce, superintendent; incorporated September 13, 1877; capital stock, $150,000, in 15,000 shares, $10 each; offices, Golden, Colorado, and Trenton, New Jersey; the works of this company are located in town of Golden, on the Colorado Central Railroad, at the entrance to Clear Creek Canyon; they are substantially constructed of brick and stone and have a capacity for treating 20 tons of ore daily.

Valley Smelting Works. Owned by Gregory, Board & Co.; located within the limits of town of Golden; have a capacity for treating 15 tons of ore daily, and contain a 30-horse power engine, 1 blast, and 2 roasting furnaces, with other machinery.

Jefferson County Coal Mines

Colorado Coal Mining Co. Incorporated December 15, 1870; capital stock, $100,000, in 1,000 shares, $100 each; office golden; organized by T.J. Carter, C.C. Welch, and F.E. Everett.

Colorado Co., Limited. Incorporated 1878; capital stock, $500,000; Andrew McKinney, president, William A.H. Loveland, general manager. The property of this company is located on the Colorado Central Railroad, within the limits of Golden, and comprise a strip of land half mile wide by 3 miles in length; the mine has a shaft 300 feet deep, and 2 entries aggregating 1,300 feet in length; the vein is vertical, is 15 feet wide and trends north and south; on the surface is a good shaft house and machinery, capable of raising 100 tons of ore daily.

Hartman Coal Mining Co. Incorporated July 2, 1873; capital stock, $4,000, in 200 shares, $20 each; office Denver; organized by Casper R. Hartman, Daniel H. Pratt, Adam Woeber, John Hausle, William H. Barnes, John A. Conners, and D.C. Crawford.

Murphy Coal Co. Incorporated December 27, 1871; capital stock, $200,000, in 2,000 shares, $100 each; office Denver; John Q. Charles, president, Henry C. Dillon, secretary and treasurer, D.M Murphy, superintendent. The property of this company comprises 480 acres of land, on Ralston Creek, 5 miles from Golden, and 1 mile from Colorado Central Railroad. The mine is well developed, has a shaft 14 feet in width by 18 feet in length and 150 feet deep, on the surface are 2 60-horse power engines, capable of raising 200 tons of coal daily.

Pittsburg [sic]. Owned by John Nichols; located at the mouth of Chimney Gulch, half mile from Golden; the vein is 8 feet in width, trends north and south, and pitches 17 degrees west. The mine is developed by a shaft 200 feet deep, and four entries aggregating 2,000 feet in length; on the surface is an engine house 18 by 32 feet, containing a 16-horse power engine, 1 steam pump and a hoisting apparatus; yield, 80 tons of coal daily.

Thomas. Owned by James M. Thomas and Evan Jones. This property comprises 160 acres of land, located between North Table Mountain and the hogback, 2 miles from Golden; the vein is 8 feet wide and trends northeast and southwest; the mine is developed by 2 entries, aggregating 175 feet in length, and a shaft 8 feet in width by 14 feet in length, and 175 feet deep; on the surface is a shaft house 24 by 32 feet, containing a 35-horse power engine and hoisting apparatus.

White Ash. Owned by Robert D. Hall and Amanzo L. Jones; this property is located near the Colorado Central Railroad, within the limits of Golden; the vein is 12 feet wide, is vertical, and trends north and south; the mine is developed by 2 entries, aggregating 800 feet in length, and a double shaft 7 feet in width by 14 feet in length and 200 feet deep; on the surface is an engine and boiler house, 22 by 33 feet, containing 2 100-horse power engines, hoisting apparatus, sales, etc.

1203. Yak Mills, Leadville, Colo.

The sprawling Yak Mills on the edge of Leadville, Colorado. This photo was taken abut 1908.

The Matchless Gold Mining and Milling Company stock certificate number 96 dated 1896. This certificate was never issued.

LA PLATA COUNTY

Thomas B. Corbett reported that La Plata County had a population of 1,000 in 1878 and a bullion product of $10,000.

Canary. Owned by William Reiley & Co., of Silverton; claim 300 by 1,500 feet; discovered 1878; located on the Dolores River, Pioneer mining district; course of vein, north and south, width 25 feet, pitch 10 degrees, pay vein 3 feet; nature of ore, carbonate of lead; assay value, 40 ounces silver per ton; development, 1 drift 25 feet in length.

Capitol. Owned by W.A. Walker & Co.; office Eureka; claim 300 by 1,500 feet; discovered 1878; located on the Dolores River, Pioneer mining district, near Parrott City; vertical vein, trending east and west, width 3 feet, pay vein 2 feet; nature of ore, galena and carbonate; assay value, 100 ounces silver per ton; 1 shaft 100 feet deep.

Crown Jewel. John Back, proprietor; office Ouray; claim 300 by 1,500 feet; discovered 1878; located on the Dolores River, Pioneer mining district, near Parrott City; vertical vein, trending east and west, width 6 feet; nature of ore, galena and carbonate; average value, 60 ounces silver per ton; 1 adit, 18 feet in length and 14 feet from surface.

Denver and San Juan Mining Co. Incorporated March 7, 1874; capital stock, $50,000, in 20,000 shares, $25 each; office Denver; organized by R.W. Woodbury, E.H. Powers, H.F. Tower, W.E. Broad, A. Huff, J.W. Kendall, and W.T. Flowers.

Gaynor. Owned by E.J. Gaynor and C. Hummel; office Eureka; claim 300 by 1,500 feet; discovered 1874; located on Dolores River, Pioneer mining district; course of vein, east and west, width 40 feet, pay vein 5 feet; nature of ore, carbonate of lead; assay value, 60 ounces silver per ton; main shaft 25 feet deep.

Golden Value. L.B. Clay, proprietor; office Ouray; claim 300 by 1,500 feet; discovered 1878; located on Dolores River, Pioneer mining district, near town of Dolores; vertical vein, trending north and south, width 9 feet; nature of ore, free gold; value 14 ounces gold per ton; 1 adit, 10 feet in length and 10 feet from surface.

Hope. Owned by John H. Schall and A. Campbell; office Silverton; claim 300 by 1,500 feet; discovered 1878; located on Telescope Mountain, Pioneer mining district, 16 miles from town of Ophir, and 14 miles from Ouray; course of vein, northwest and southeast; nature of ore, carbonate of lead; assay value, 500 ounces silver per ton; 1 shaft 12 feet deep, 1 tunnel 25 feet in length.

I.X.L. Thomas Sappington, proprietor; office Lake City; claim 300 by 1,500 feet; discovered 1877; located on the Dolores River, Pioneer mining district, near Parrott City; vertical vein, trending east and west, width 4 feet; nature of ore, galena and carbonate; value, 80 ounces silver per ton; 1 adit 20 feet and 18 feet from surface.

Juniata. U. Jackson, proprietor; office Parrott City; claim 300 by 1,500 feet; discovered 1878; located on the Dolores River, Pioneer mining district, near Parrott City; vertical vein, trending east and west, width 20 feet, pay vein 15 feet; nature of ore; carbonate of lead; average value, 30 ounces silver per ton; 1 adit 20 feet in length and 15 feet from surface.

Major. Charles Humaston, proprietor; office Eureka; claim 300 by 1,500 feet; discovered 1878; located on the Dolores River, Pioneer mining district, 16 miles from the town of Ophir and 14 miles from Ouray; vertical vein, trending east and west; width 10 feet, pay vein 1 foot; nature of ore, carbonate of lead; value, 50 ounces silver per ton; 1 shaft 12 feet deep.

Ocean Wave. Owned by Clabe and Charles Jones and John Glasgow; office Animas Forks; claim 300 by 1,500 feet; discovered 1878; located on the Dolores River, Pioneer mining district, near town of Dolores; vertical vein, trending east and west, width 10 feet, pay vein 6 feet; nature of ore, galena; assay value, $100 per ton; 1 drift 20 feet in length and 25 feet from surface.

Raymond Placer. Owned by Alfred Raymond of Parrott City; claim comprises 120 acres of placer ground; located 12 miles from Parrott City on the Manicos River; has a flume 500 feet in length.

Rio Florida Coal Co. Incorporated November 2, 1877; capital stock, $20,000, in 80 shares, $250 each; office Lake City; trustees: William E.

Taylor, James Hanrahan, A.M. Olds, John McWilliams, W.R. Montieth, E.C. Holmes, F. Fagaly, and J.J. Rourk.

San Juan Silver Mining Co. Incorporated May 17, 1875; capital stock, $50,000, in 500 shares, $100 each; office Titusville, Pennsylvania; organized by Henry Harley, W.H Wallace, J.G. Jackson, Thomas H. Larsen, J.A. Neill, and William H. Abbott.

San Miguel Placer Mining Co. Incorporated November 5, 1875; capital stock, $50,000, in 500 shares, $100 each; office San Juan County, Colorado; organized by William H. Nichols, Edward Jenniss, John H. French, William W. Conner, W.W. Remine, L.M Remine, William M. Leonard, Luke Stanley, John R. Burrows, Thomas Davis, and Andrew M. Bigger.

Silver Belt Mining Co. Incorporated January 27, 1876; capital stock, $300,000, in 3,000 shares, $100 each; office Denver; organized by Charles M. Parker, John H. Stimpson, and Sylvester Cleveland.

Specie Payment. Owned by John Ray and William Clough & Co.; office Lake City; claim 300 by 800 feet; located 1878; located on the Dolores River, near Parrott City; vertical vein, trending north and south, width 5 feet, pay vein, 4 feet; nature of ore, carbonate and galena; average value, 60 ounces silver per ton; 1 adit 18 feet in length and 15 feet from surface.

Streater Gold and Silver Mining Co. Incorporated March 7, 1876; capital stock, $200,000, in 20,000 shares, $10 each; offices Parrott City and Denver, Colorado; organized by Levi Vincent, Edwin E. Williams, T.M. Tripp, J.H. Jones, and Rudolph A. Leimer.

LAKE COUNTY

Thomas B. Corbett reported that Lake County had a population of 15,000 in 1878, and a bullion product of $2,650,000. Mining districts included California, Chalk Creek, and Oro.

Adelaide. Owned by Adelaide Consolidated Silver Mining and Smelting Co.; incorporated May 31, 1878; capital stock, $2,500,000, in 100,000 shares, $25 each; offices Leadville, Colorado, and Washington, D.C.; trustees: John P. Jones, Jay A. Hubbell, Henry D. Cook, John R. Magruder, James L. Hill, Woodbury Blair, and William H. Barnard; claim, 300 by 1,500 feet, discovered 1877; located in Stray Horse Gulch, California mining district, 2 miles from Leadville; course of vein, north and south, width 5 feet, pay vein 2 feet, pitch east; nature of ore, carbonate of lead; average value, 90 ounces silver per ton; developed by 3 shafts aggregating 345 feet in depth; 2 tunnels, 80 and 100 feet respectively, and 1 level 125 feet in length; shaft house, 15 by 25 feet.

Aetna. Owned by the Meyer Mining Co.; August R. Meyer, superintendent, located on Carbonate Hill, half mile from Leadville.

Agassiz. Owned by Aden Alexander and Samuel Morgan; offices Leadville and Denver, Colorado; claim 300 by 1,500 feet; located in Stray Horse Gulch, 1 mile from Leadville; developed by 3 shafts, 20, 45, and 85 feet, respectively; value of ore, $75 to $160 per ton; yield to date, $17,000.

Alama. Owned by Joseph Hudson and others; office Alpine; claim 300 by 1,500 feet; discovered 1877; located in Grizzly Gulch, Chalk Creek mining district, 6 miles from Alpine and 65 miles from Leadville; course of vein, northeast and southwest, width 5 feet, pay vein 8 inches, pitch 55 degrees west; nature of ore, galena; assay value, 145 ounces silver per ton; 1 shaft 42 feet deep.

Alpine Silver Mining Co., of Colorado. Incorporated February 6, 1879; capital stock, $2,000,000, in 200,000 shares, $10 each; office Alpine, Lake County, Colorado; John Splane and Iner Gilbertson, agents.

Alta. Owned by Jones & Shaw; located on Printer Boy Hill, 3 miles from Leadville.

American Eagle. Owned by Herman Hibschle; claim 300 by 1,500 feet; discovered July 3, 1878; located on Fryer Hill, 1 mile from Leadville; developed by 1 shaft, 100 feet.

Amie. Owned by Hardenstein and others, of Leadville; located on Fryer Hill, 1 mile from Leadville; developed by a shaft 125 feet deep.

Amna. Owned by J.R. Riggins, John A.J. Chapman, and J. Splane; office Alpine; claim 300 by 1,500 feet; discovered 1875; located on Franklin Mountain, Chalk Creek mining district, 1 mile from Alpine and 60 miles from Leadville; course of vein, northeast and southwest, width 3 feet, pay vein 15 inches, pitch 45 degrees northwest; nature of ore, galena and iron pyrites; assay value, 75 to 380 ounces silver per ton; 1 shaft 80 feet deep, 1 level 35 feet in length.

Argentine Mining Co. Organized under the laws of the State of Missouri, March 28, 1878; capital stock, $250,000, in 2,500 shares, $100 each; offices Leadville, Colorado, and St. Louis, Missouri; Edwin Harrison, president, John B. Maude, vice president; the mines of this company are located within a radius of 1 mile from the town of Leadville, and comprise the Keystone, Camp Bird, Charlestown, Pine, and Young America; these mines have produced to date, $225,000.

Arkansas River Placer Mining Co. Incorporated October 26, 1876; capital stock, $250,000, in 2,500 shares, $100 each; office Granite; organized by F.A. La Grave, Charles Mater, and Nerii Valle.

Baalbec. Chrysolite Tunnel Co., proprietors; office Alpine; claim 300 by 1,500 feet; discovered 1873; located on Chrysolite Mountain, Chalk Creek mining district, 6 miles from Alpine; course of vein, northeast and southwest, width 8 feet, pitch 55 degrees west; nature of ore, galena; assay value, 500 ounces silver per ton; 1 shaft 15 feet deep.

Belcher. Jones & Shaw, proprietors; office Leadville; claim 300 by 1,500 feet; discovered 1877; located on Long's Hill, California mining district, 5 miles from Leadville; course of vein, north and south, width 4 feet, pay vein 3 feet, pitch 45 degrees east; nature of ore, carbonate of lead; developed by a drift 65 feet in length.

Belgium. Owned by Peter Finerty, Richard Dillon, and others; offices, Leadville and Denver, Colorado; claim 300 by 1,500 feet; discovered June 1878; located in Stray Horse Gulch, 1 mile from

Leadville; developed by 2 shafts 30 and 100 feet deep, respectively; average value of ore, 67 ounces silver per ton.

Belle of Colorado. Owned by Joseph Pearce, William Pearce, and H.A.W. [Horace] Tabor; office Leadville; claim 300 by 1,500 feet; discovered 1877; located on Iron Hill, California mining district, 2 miles from Leadville; course of vein, north and south, width 18 inches; nature of ore, carbonate of lead; average value, 150 ounces silver per ton; main shaft 190 feet, and 1 other shaft 27 feet in depth; 1 level 30 feet in length.

Black Hawk. Owned by the Kansas City Mining and Smelting Co., and A.E. Wright, M.D.; office Alpine; claim 300 by 1,500 feet; located on Pomeroy Mountain, Chalk Creek mining district, 7 miles from town of Alpine, and 66 miles from Leadville; course of vein, northeast and southwest, width 12 feet, pitch 40 degrees southeast; nature of ore, galena and iron pyrites; average value, 35 ounces silver per ton; 1 shaft 30 feet deep, 1 tunnel 160 feet in length and 120 feet from surface.

Bradshaw Placer. Owned by Lucius F. Bradshaw; office Leadville; claim comprises 20 acres of placer ground, located in California Gulch, 1 mile from Oro City and 3 miles from Leadville.

Bullion. Nathan White, proprietor; office Leadville; claim 170 by 1,120 feet; discovered 1878; located in Evans Gulch, California mining district, 2 miles from Leadville; course of vein, north and south, width 7 feet, pay vein 4 feet, pitch 75 degrees east; nature of ore, carbonate of lead; average value 40 ounces silver per ton; 1 shaft 120 feet deep.

Bullseye. Owned by William H. Stevens and Levi Z. Leiter; offices Leadville, Colorado, and Chicago, Illinois; claim 300 by 1,300 feet, patented; discovered 1877; located in Stray Horse Gulch, 2 miles from Leadville.

Camp Bird. Owned by the Argentine Mining Co.; offices Leadville, Colorado, and St. Louis, Missouri; claim 300 by 1,500 feet; discovered 1876; located in Stray Horse Gulch, California mining district, 1 mile from Leadville; course of vein, north and south, width 6 feet, pitch 70 degrees east; nature of ore, carbonate of lead; average value, 85 ounces silver per ton; 1 tunnel 250 feet in length; shaft house 25 by 40 feet.

Carbonate. Nelson Hallock and Albert Cooper, proprietors; office Leadville; claim 300 by 1,500 feet, patented; discovered 1877; located on

Carbonate Hill, California mining district, 1 mile from Leadville; course of vein, north and south, width 4 feet, pay vein 3 feet, pitch 70 degrees east; nature of ore, carbonate of lead; average value, 200 ounces silver per ton; 1 shaft 30 feet deep, 1 drift 500 feet in length, aggregate length of levels 600 feet; yield, $125,000.

Carbonate Gold and Silver Mining Co. Incorporated December 16, 1878; capital stock, $5,000,000, in 500,000 shares, $10 each; offices Leadville and Denver, Colorado; organized by John M. Bradford, J.J. Bush, Charles L. Kusz, Jr., George Woodward, Jr., and P.S. Rice.

Carbonate and Metallic Lead Co., Colorado. Incorporated February 13, 1877; capital stock, $1,000,000, in 100,000 shares, $10 each; office Detroit, Michigan; trustees: Albinus B. Wood, W.H. Stevens, and Albert Cooper.

Carboniferous. Owned by Borden, Tabor & Co.; offices Leadville and Denver, Colorado; located on Fryer Hill, 1 mile from Leadville; developed by a shaft 150 feet deep.

Catalpa. Aden Alexander, proprietor; office Denver; claim 235 by 1,500 feet, patented; discovered 1877; located on Carbonate Hill, California mining district, half mile from Leadville; course of vein, north and south, width 12 feet, pay vein 3 feet; nature of ore, carbonate of lead; average value, 110 ounces silver per ton; main shaft 150 feet, 3 levels aggregating 100 feet in length; shaft house, 16 by 20 feet.

Charlestown. Owned by the Argentine Mining Co.; August R. Meyer, manager; office Leadville; claim 300 by 1,500 feet; located in Stray Horse Gulch, California mining district, 1 mile from Leadville; course of vein, north and south, width 1 foot; nature of ore, carbonate of lead; average value, 85 ounces silver per ton; 2 shafts 40 and 70 feet deep, respectively; aggregate length of drifts, 50 feet.

Chieftain. Owned by W.H. Yankee, J.B. Putnam, and others; office Leadville; claim 300 by 1,500 feet; discovered 1878; located on Little Pittsburgh Hill, California mining district, 1 mile from Leadville; course of vein, north and south, width of pay vein 3 feet, pitch 70 degrees east; nature of ore, carbonate of lead; average value, 100 ounces silver per ton; main shaft 72 feet, and 1 other shaft 40 feet in depth 5 levels aggregating 135 feet in length.

Chrysolite. Owned by Borden, Tabor & Co.; located on Fryer Hill, 1 mile from Leadville; development, 1 shaft 80 feet deep.

Cincinnati. Owned by Sanders, Mater, and Mitchell; office Leadville; claim 300 by 1,500 feet, patented; discovered 1870; located on Weston's Pass, California mining district, 19 miles from Leadville; course of vein, east and west, width 5 feet, pay vein 1 foot to 5 feet; nature of ore, carbonate of lead; assay value, 60 ounces silver per ton; 3 shafts aggregating 144 feet in depth, 1 tunnel 25 feet in length.

Cincinnati Gold Mining Co. Incorporated June 13, 1874; capital stock, $200,000, in 200 shares, $1000 each; office Denver; organized by Emil Loescher, Francis Michael, Fred Cazin, and Ferdinand August Schnell.

Circassian Girl. James G. Gillespie and A.C. Pickens, proprietors; office Leadville; claim 300 by 1,500 feet; discovered 1878; located on Long's Hill, California mining district, 5 miles from Leadville; course of vein, northeast and southwest, width 18 inches, pay vein 6 inches, pitch 30 degrees east; nature of ore, carbonate of lead; average value, 27 ounces silver per ton; 1 shaft 25 feet deep.

Cleora. J.J.B. Dubois and W.J. McDermith, proprietors; office Oro; claim 300 by 1,500 feet; discovered 1877; located on Iron Hill, California mining district, 1 mile from Leadville; course of vein, north and south; nature of ore, carbonate of lead; main shaft 100 feet and 1 other shaft 20 feet deep.

Climax. Owned by Berdell & Wetherell, of Leadville; located on Fryer Hill, 1 mile from Leadville; development, 1 shaft 100 feet deep.

Colorado. Owned by the Topeka Mining and Tunnel Co.; office Alpine; claim 300 by 1,500 feet; located in Baldwin Gulch, Chalk Creek mining district, near town of Alpine; 60 miles from Leadville; vertical vein, trending north and south, width 6 feet, pay vein 4 feet; nature of ore, galena and iron pyrites; average value, 165 ounces silver and 1 ¼ ounces gold per ton; 1 shaft 16 feet deep; operated through Topeka Tunnel.

Consolidated Eureka and Tiger Silver and Gold Mining Co. Incorporated December 5, 1878; capital stock, $1,500,000, in 150,000 shares, $10 each; office Leadville; organized by Charles L. Kusz, Jr., W.W. Coble, L.B. Walters, C.H. Lester, O.F. Cheney, S. Carnahan,

Charles M. Traeger, and Catharine McClellan; Charles L. Kusz, Jr., president, Charles W. Traeger, vice president; W.W Coble, treasurer, O.F. Cheney, superintendent. The property of this company is located in Stray Horse Gulch, California mining district, near town of Leadville.

Continental. Henry Campbell, proprietor; office Alpine; claim 300 by 1,500 feet; discovered 1873; located on Maina Mountain, Chalk Creek mining district, 3 miles from Alpine and 62 miles from Leadville; vertical vein, trending northeast and southwest, width 16 feet, pay vein 30 inches; nature of ore, iron pyrites and galena; assay value, 50 to 385 ounces silver per ton; 1 shaft 45 feet deep, 1 tunnel 25 feet, and 1 level 95 feet in length.

Copper. Owned by the Topeka Mining and Tunnel Co.; office Alpine; claim 300 by 1,500 feet; located in Boulder Gulch, Chalk Creek mining district, near town of Alpine, 60 miles from Leadville; course of vein, north and south, width 3 feet, pay vein 2 feet, pitch 15 degrees west; nature of ore, copper pyrites and galena; assay value, 70 ounces silver and 1 ½ ounces gold per ton; 1 shaft 25 feet deep.

Crescent. Owned by the Meyer Mining Co.; August R. Meyer, manager; office Leadville; claim 300 by 1,500 feet; discovered 1877; located on Carbonate Hill, California mining district, half mile from Leadville; course of vein, north and south, width 12 feet, pay vein 3 feet; nature of ore, carbonate of lead; average value, 175 ounces silver per ton; 4 shafts, 35, 40, 150, and 175 feet in depth, respectively; aggregate length of levels and drifts, 500 feet.

Crown Point. Owned by the Meyer Mining Co.; August R. Meyer, superintendent; located in California Gulch, 1 mile from Leadville.

Crystal Mountain Mining Co. Incorporated March 5, 1877; capital stock, $5,000, in 50 shares, $100 each; office Colorado Springs; trustees: S.I. Smith, S.C. Robinson, E.N. Bartlett, John Shanley, Sr., Charles Ayer, E. Nicholson, R.C. Lyon, J.L. Davis, Joseph Dozier, J.H. Kerr, and William R. Wheeler.

Daisey. Owned by H.C. Chapin, Peter Finerty, and others; discovered 1878; located on Fryer Hill, 1 mile from Leadville; developed by a shaft 140 feet deep.

Darby Lode Mining Co. Incorporated November 9, 1877; capital stock, $21,000, in 2,100 shares, $10 each; office Denver; trustees: John H. Talbott, James C. Langhorne, and Thomas L. Darby.

Decorah. Owned by William A. Fletcher, William Russell, and J.H. Hughes; office Alpine; claim 300 by 1,500 feet; discovered 1873; located on Chrysolite Mountain, Chalk Creek mining district, 6 miles from Alpine and 65 miles from Leadville; vertical vein, trending north and south, width 5 feet, pay vein 18 inches; nature of ore, galena, iron and copper pyrites; assay value, 20 to 50 ounces silver per ton; 1 adit, 20 feet in length, with 20-foot face.

Deer Lodge. Owned by John Kelly and others, of Leadville, located near town of Malta, 2 miles from Leadville.

Dives. Owned by the Little Pittsburg Consolidated Mining Co., located on Fryer Hill, 1 mile from Leadville.

Dolphin. Owned by Peter Finerty, Nelson Hallock, and others; office Leadville; claim 300 by 1,500 feet; discovered April 1879; located on Fryer Hill, 1 mile from Leadville; developed by 2 shafts, 140 and 150 feet deep, respectively.

Dome. Owned by William H. Stevens and Levi Z. Leiter; claim 300 by 1,500 feet, located 2 miles from Leadville.

Double Decker. Owned by James Murray and others; located in Stray Horse Gulch, 2 miles from Leadville; developed by 2 shafts, both 160 feet.

Dunderberg. Dubois, Tooker and Professor Nathaniel P. Hill, proprietors; office Leadville; claim 215 by 1,500 feet; discovered 1878; located in Stray Horse Gulch, California mining district, 2 miles from Leadville; course of vein, north and south; nature of ore, carbonate of lead; assay value, 158 ounces silver per ton; 1 drift, 300 feet in length.

Dyer. H.A.W.[Horace] Tabor and George L. Henderson, proprietors; office Leadville; claim 300 by 1,500 feet; discovered 1872; located on Silver Point Mountain, California mining district, 6 miles from Leadville; course of vein, north and south, width 6 feet, pay vein 18 inches, pitch 30 degrees east; nature of ore, galena and sulphurets; average value, 140 ounces silver per ton; 1 drift 140 feet in length; shaft house, 15 by 30 feet; yield, $15,000. [Judge Elias Dyer discovered the Dyer mine and sold it to

Horace Tabor. Elias Dyer was the son of Rev. John Lewis Dyer, an itinerant Methodist minister, who is often called Colorado's "Snowshoe Itinerant" because he carried the mail over Mosquito Pass to supplement an erratic collection plate. Probate judge for Lake County, Judge Elias Dyer was murdered in his courtroom in 1895, after he issued warrants to arrest violent vigilantes, members of a self-proclaimed "Committee of Safety," that had split the county.]

Eaton. Owned by Borden, Tabor & Co.; located on Fryer Hill, 1 mile from Leadville.

Elizabeth Maher. John Maher and John Sweeney, proprietors; office Alpine; claim 300 by 1,500 feet; discovered 1877; located on Chalk Creek, Chalk Creek mining district, 2 miles from Alpine, and 60 miles from Leadville; vertical vein, trending northeast and southwest, width 16 feet, pay vein 5 feet; nature of ore, galena and pyrites of iron; assay value, 44 to 86 ounces silver per ton; 1 drift 37 feet in length and 35 feet from surface.

Eureka. Daniel Trellinger & Co., proprietors; office Leadville; claim 300 by 1,500 feet; discovered 1877; located on Iron Hill, California mining district, 2 miles from Leadville; course of vein, north and south, width 4 feet, pitch 70 degrees east; nature of ore, carbonate of lead; assay value, 30 to 500 ounces silver per ton; main shaft 200 feet deep, 2 other shafts aggregating 90 feet in depth, 1 level 50 feet in length, and 120 feet from surface.

Eureka Placer. Owned by Aden Alexander and Thomas S. Wells; office Oro; claim comprises 40 acres of placer ground, located in California Gulch, near town of Oro, 3 miles from Leadville; this property is worked by hydraulic pressure and has yielded $50,000.

Evening Star. Owned by William A. Fletcher, William Russell, and J.H. Hughes; office Alpine; claim 300 by 1,500 feet; discovered 1873; located on Chrysolite Mountain, Chalk Creek mining district, 6 miles from Alpine, and 65 miles from Leadville; vertical vein, trending north and south, width 5 feet, pay vein 18 inches; nature of ore, galena; assay value, 25 to 50 ounces silver and 10 to 350 ounces gold per ton; 1 shaft 40 feet deep.

Excelsior. Owned by Meyer Mining Co., H.A.W. [Horace] Tabor, and J.J.B. Dubois; office Leadville; claim 230 by 1,500 feet; discovered 1877; located on Carbonated Hill, California mining district, half mile from Leadville; course of vein, north and south; nature of ore, carbonate of lead; main shaft 450 feet deep.

Faint Hope. Owned by J.F. and J.W. Long and others; office Leadville; claim 300 by 1,500 feet, patented; discovered 1876; located on Long's Hill, California mining district, 5 miles from Leadville; course of vein, northeast and southwest, width 8 feet, pitch 50 degrees east; nature of ore, carbonate of lead; average value, 150 ounces silver per ton; 3 shafts aggregating 142 feet in depth, aggregate length of drifts, 580 feet; total yield, $10,000.

Fairview. Owned by Jacob Freeman and others; claim 300 by 1,500 feet; discovered June 4, 1878; located on the western part of Fryer Hill, 1 mile from Leadville.

Fanny Macon. T.R. Rowe and G.W. Cooke, proprietors; office Alpine; claim 300 by 1,500 feet discovered 1877; located on Chalk Creek, Chalk Creek mining district, 2 miles from Alpine and 61 miles from Leadville; course of vein, north and south, width 12 feet, pay vein 7 feet, pitch 25 degrees west; nature of ore, iron pyrites and galena; assay value, 34 ounces silver per ton; 1 shaft 18 feet deep, 1 adit, 30 feet in length, with 25-foot face.

Flag Staff. Owned by Sydenham Mills and Aden Alexander; office Leadville; claim 300 by 1,500 feet; discovered 1877; located in Stray Horse Gulch, 1 mile from Leadville; course of vein, north and south, width 10 feet, pitch 20 degrees southeast; nature of ore, carbonate of lead; average value, 150 ounces silver per ton; 1 tunnel 185 feet in length.

Flat Fluming Co. Incorporated June 29, 1876; capital stock, $100,000, in 10,000 shares, $10 each; office Langhoff's Ranch, Lake County; organized by Herman and Henry W. Tuerke, and John Langhoff.

Forrest City. Owned by the Small Hope Mining Pool; offices Leadville and Denver, Colorado; claim 300 by 1,500 feet; located in Stray Horse Gulch, 1 mile from Leadville; 1 shaft 60 feet deep.

Frenchman. Owned by Peter Finerty and Richard Dillon; office Leadville; claim 300 by 1,500 feet; discovered October 1878; located on Gallagher Hill, 1 mile from Leadville; 1 shaft 220 feet deep; average value of ore, 67 ounces silver per ton.

Fryer Hill Tunnel Co. Incorporated March 3, 1879; capital stock, $250,000, in 2,500 shares, $100 each; office Leadville; trustees: George W. Davis, Elias Breasley, Henry C. Hewitt, and Edward M. Clark.

G.W. Fish. Chrysolite Tunnel Co., proprietors; office Alpine; claim 300 by 1,500 feet; discovered 1873; located on Chrysolite Mountain, Chalk Creek mining district, 6 miles from Alpine; vertical vein, trending northeast and southwest, width 30 inches, pay vein 10 inches; nature of ore, galena and copper pyrites; 1 shaft 10 feet deep.

Galena. Baxter Stingley and James Benson, proprietors; office Alpine; claim 300 by 1,500 feet; discovered 1876; located on Sugar Loaf Mountain, Chalk Creek mining district, 3 miles from Alpine and 62 miles from Leadville; course of vein, northeast and southwest, width 4 feet, northeast and southwest, width 4 feet, pay vein 1 foot, pitch 45 degrees west; nature of ore, galena; assay value, 11 to 65 ounces silver per ton; 1 shaft 95 feet deep.

Gambetta. Owned by Charles Limberg and Charles Byfield, of Leadville; discovered during the spring of 1878; located on Fryer Hill, 1 mile from Leadville; developed by 2 shafts, 150 and 200 feet respectively.

Golden Age Mining Co., of Colorado Springs. Incorporated November 1, 1878; capital stock, $100,000, in 10,000 shares, $10 each; office Colorado Springs; organized by W.F. Tilton, William Fernsworth, C. and Max Brieglieb, C.W. Sears, F.E. Kimball, and E.F. Johnson.

Gone Abroad. George R. Seney and Co., proprietors; office Leadville; claim 300 by 1,500 feet; discovered 1878; located in Stay Horse Gulch, California mining district, 1 mile from Leadville; course of vein, north and south; nature of ore, carbonate of lead; average value, 150 ounces silver per ton; 1 shaft 40 feet deep, aggregate length of levels 100 feet; yield, $6,000.

Grafton. Owned by William Christian, Gomer Richards, and Lawrence Skelley of Leadville; course of claim, northwest and southeast; discovered December 23, 1878; located on west Fryer Hill, 1 mile from Leadville; development, 1 shaft 70 feet deep.

Grand View. John D. Coon and Judson E. Cole, proprietors; office Leadville; claim 300 by 1.500 feet; discovered 1877; located in Evans Gulch, 3 miles from Leadville; course of vein, north and south; nature of ore, carbonate of lead; assay value, 40 ounces silver per ton; 2 shafts 28 and 58 feet, respectively, 1 drift, 55 feet in length, and 20 feet from surface.

Grand View. Owned by _____ Cohen, of Fairplay; claim 300 by 1,500 feet; located on the west side of Sheridan Peak, on the western slope of Mosquito Range, 11 miles from Fairplay; course of vein, northwest and southeast, width 3 feet, pitch 22 degrees; nature of ore, carbonate and cromate of lead; average value, $40 per ton; 2 tunnels 25 and 60 feet in length, respectively.

Green Mountain. Benjamin Bernard and Edward Breese, proprietors; office Oro; claim 300 by 1,500 feet; discovered 1878; located on Bald Mountain, 4 miles from Leadville; course of vein, north and south; nature of ore, galena and carbonate of lead, average value, 52 ounces silver per ton; 2 shafts 24 and 80 feet, respectively.

Gulnair. Chrysolite Tunnel Co., proprietors; office Alpine; claim 300 by 1,500 feet; discovered 1873; located on Chrysolite Mountain, Chalk Creek mining district, 6 miles from Alpine; course of vein, northeast and southwest; 1 shaft 30 feet deep.

H.D. Owned by William H. Stevens and Levi Z. Leiter; claim 300 by 1,500 feet, located on Carbonate Hill, California mining district, 1 mile from Leadville.

Hannibal. Henry Brittenstein, proprietor; office Alpine; claim 300 by 1,500 feet; discovered 1876; located in Grizzly Gulch, Chalk Creek mining district, 6 miles from Alpine and 60 miles from Leadville; vertical vein, trending east and west, width 12 feet, pay vein 5 feet; nature of ore, galena; average value, 35 ounces silver and $40 gold per ton; 1 drift 25 feet in length, and 15 feet from surface.

Hayes. Owned by Benjamin Riggins, Joseph Hudson, and others; office Alpine; claim 300 by 1,500 feet; discovered 1877; located on Boulder Mountain, Chalk Creek mining district, 3 miles from Alpine and 63 miles from Leadville; course of vein, northwest and southeast, width 4 feet, pay vein 1 foot, pitch 30 degrees west; nature of ore, galena and sulphurets; assay value 30 to 1,350 ounces silver per ton; 2 shafts 20 feet each.

Henrietta. Owned by Joseph Pearce, R.J. Goris, and others; office Leadville; claim 300 by 1,500 feet; discovered 1877; located in Stray Horse Gulch, half mile from Leadville; course of vein, north and south, width 3 feet; nature of ore, carbonate of lead; average value, 250 ounces silver per ton; main shaft 125 feet and 3 other shafts aggregating 110 feet in depth.

Highland Mary. Owned by Joseph Hudson and others; office Alpine; claim 200 by 1,500 feet; discovered 1877; located in Grizzly Gulch, Chalk Creek mining district, 6 miles from Alpine and 65 miles from Leadville; course of vein, northeast and southwest, width 8 feet, pay vein 18 inches, pitch 55 degrees west; nature of ore, galena; 1 shaft 12 feet deep.

Holland. Owned by the Topeka Mining and Tunnel Co.; office Alpine; claim 300 by 1,500 feet; located in Baldwin Gulch, Chalk Creek mining district, 2 miles from town of Alpine, and 60 miles from Leadville; course of vein, north and south, width 4 feet, pitch 15 degrees west; nature of ore, iron pyrites and galena; assay value, 45 ounces silver and 1 ½ ounces gold per ton; 1 shaft 20 feet deep; operated through the Topeka Tunnel.

Home Stake. James Archer, proprietor; office Denver; claim 300 by 3,000 feet; located 1872; located at the head of the Arkansas River, 16 miles from Leadville; course of vein, east and west, width 2 feet, pay vein, 1 foot, pitch 20 degrees; nature of ore, galena and copper pyrites; average value, 150 ounces silver per ton; 2 drifts aggregating 600 feet, and a tunnel 150 feet in length; shaft house, 20 by 40 feet; yield, $10,000.

Hortense. Owned by George Merriam and Estate of E.W. Keyes; office Hortense; claim 300 by 1,500 feet; discovered 1872; located on Mount Princeton, Chalk Creek mining district, 9 miles from Alpine and 50 miles from Leadville; course of vein, northeast and southwest, width 8 feet, pay vein 30 inches, pitch 45 degrees west; nature of ore, galena and sulphurets; average value, 104 ounces silver and 1 ounce gold per ton; main shaft 125 feet and 1 other shaft 50 feet deep, 3 levels aggregating 135 feet in length; yield, $20,000.

Ida. Chrysolite Tunnel Co., proprietors; office Alpine; claim 300 by 1,500 feet; discovered 1877; located on Chrysolite Mountain, Chalk Creek mining district, 6 miles from Alpine and 65 miles from Leadville; course of vein, northeast and southwest, width 8 feet, pay vein 4 feet, pitch 55 degrees west; nature of ore, iron pyrites and galena; assay value, 80 ounces silver per ton.

Ida Nyce. Owned by George Nyce of Leadville; claim 300 by 1,500 feet; located on the south edge of Fryer Hill, 1 mile from Leadville.

Inter-Ocean Mining Co. Incorporated February 11, 1879; capital stock, $10,000, in 100 shares, $100 each; office Leadville, trustees: J.H. Wheeler, W.E. Lee, John E. Fitzgerald, George B. Berger, A.V. Bohm, John A. Clough, and James M. Swem.

Iron. Owned by William H. Stevens and Levi Z. Leiter; offices Leadville, Colorado, and Chicago, Illinois; claim 300 by 1,500 feet, patented; discovered August 1876; located on Carbonate Hill, 1 mile from limits of Leadville; course of vein, north and south, width 2 to 10 feet, pitch 14 degrees; nature of ore, carbonate of lead, chloride of silver, and native silver; average value, $100 per ton; development, 2 incline shafts, 500 and 575 feet in depth, respectively, from which drifts and levels have been run north and south at depths varying from 100 to 300 feet from the surface, with winzes to connect them. On the surface is an office and assay laboratory, a boarding house with accommodations for 100 men, warehouses, ore bins and stables; it is estimated by experts that there is fully $2,000,000 worth of ore in sight in this mine.

Iron Chest. Owned by Darwin, Ray, and others; office Alpine; claim 300 by 1,500 feet, discovered 1877; located on Mary Murphy Mountain, Chalk Creek mining district, 6 miles from Leadville; vertical vein, trending north and south, width 6 feet, pay vein 5 feet; nature of ore, galena; assay value, 100 ounces silver per ton; 1 adit 10 feet in length, with 12-foot face.

J.D. Dana. Owned by J.F. and J.W. Long, and others; office Leadville; claim 300 by 1,500 feet, patented; discovered 1876; located on Long's Hill, California mining district, 5 miles from Leadville; course of vein, northeast and southwest, width 8 feet, pay vein 4 feet, pitch 50 degrees east; nature of ore, carbonate of lead; average value, 290 ounces silver per ton; main shaft 120 feet deep, and 2 other shafts, aggregating 77 feet in depth, 1 tunnel 350 feet in length, aggregate length of levels 285 feet; shaft house 18 by 20 feet; total yield, $17,000.

Jim Wilson. Chrysolite Tunnel Co., proprietors; office Alpine; claim 300 by 1,500 feet; discovered 1877; located on Chrysolite Mountain, Chalk

Creek mining district, 6 miles from Alpine; course of vein, northeast and southwest, width 11 feet; nature of ore, galena.

Kansas City Mining and Smelting Co. Incorporated March 7, 1876; capital stock $1,000,000, in 10,000 shares, $100 each; office town of Alpine; organized by Benjamin L. Riggins, John J. Chapman, and George Holmes. The property of this company is located beside Chalk Creek in town of Alpine, 59 miles from Leadville; it consists of smelting works, several prominent lode claims, and the town site of Alpine.

Keyser. J.F. and J.W. Long, proprietors; office Leadville; claim 300 by 1,500 feet; discovered 1877; located in Iowa Gulch, California mining district, 5 miles from Leadville; vertical vein, trending northeast and southwest, width 30 inches, pay vein 1 foot; nature of ore, galena and sulphurets; assay value, 122 ounces silver per ton; 1 tunnel 25 feet in length.

Keystone. John W. Webb, proprietor; office Leadville; claim 300 by 1,500 feet; discovered 1878; located in Iowa Gulch, California mining district, 6 miles from Leadville; course of vein, northeast and southwest, width 14 inches; nature of ore, galena and sulphurets; assay value, 980 ounces silver per ton; 1 drift 150 feet in length and 150 from surface.

Keystone. Owned by the Argentine Mining Co.; office Leadville; claim 300 by 1,500 feet; discovered 1876; located in Stray Horse Gulch, California mining district, 2 miles from Leadville; course of vein, north and south, width 6 feet, pay vein 5 feet, pitch east; nature of ore, carbonate of lead; average value, 85 ounces silver per ton; 3 shafts, 40, 50, and 70 feet in depth, respectively; aggregate length of drifts 200 feet.

Kit Carson. Owned by Morgan, Jones and others; offices Leadville and Denver, Colorado; claim 300 by 1,500 feet; discovered June 16, 1878; located on the western part of Fryer Hill, 1 mile from Leadville.

Kitt. Owned by Topeka Mining and Tunnel Co.; office Alpine; claim 300 by 1,500 feet; located in Baldwin Gulch, Chalk Creek mining district, near town of Alpine, 60 miles from Leadville; vertical vein, trending north and south, width 5 feet, pay vein 1 foot; nature of ore, galena and iron pyrites; value, 262 ounces silver per ton; 1 shaft 25 feet deep, 1 level 20 feet in length, operated through Topeka Tunnel.

Lady Houghton. Owned by John E. Seeley and William E. Lambert; office Denver; claim 150 by 1,500 feet, patented; located in California Gulch, 1 mile from Leadville; nature of ore, hard carbonate; value, $700 per ton; 1 shaft 70 feet deep, 1 level 35 feet in length; shaft house, 14 by 20 feet; yield to date, $1,400.

Lake County Placer Mining and Ditching Co. Incorporated April 19, 1878; capital stock, $250,000, in 2,500 shares, $100 each; office Granite; trustees: Sydenham Mills, Frank G. and Walter R. Mitchell.

Lake Mining Co. Incorporated August 7, 1871; capital stock, $500 in 30 shares; office Twin Lakes; organized by S.M., C.W., and L.M. Dervy.

La Plata. Owned by Lucius Bradshaw, A.H. Posey, and J. Lightfoot; office Leadville; claim 300 by 1,500 feet; discovered 1873; located in California Gulch, California mining district, 2 miles from Leadville; course of vein, north and south, width 12 feet, pitch east; nature of ore, carbonate of lead; average value, 30 ounces silver per ton; 3 shafts, 16, 25, and 35 feet in depth, respectively; aggregate length of levels, 200 feet; yield, $2,000.

Laurel. Owned by James G. Gillespie, James Donaldson, and F. Hayden; office Leadville; claim 1878; located in Stray Horse Gulch, 2 miles from Leadville; course of vein, north and south, width 6 feet, pitch east; nature of ore, carbonate of lead; average value, 35 ounces silver per ton; 1 shaft 80 feet deep.

Law. Ira Mulock and John Law, proprietors; office Leadville; claim 300 by 1,500 feet; discovered 1877; located on Iron Hill, 2 miles from Leadville; course of vein, north and south, width 5 feet, pitch east; nature of ore, carbonate of lead; assay value, 80 ounces silver per ton; main shaft 150 feet deep, 2 other shafts aggregating 230 feet in depth; shaft house 25 by 25 feet.

Leadville Mining Co. Incorporated February 11, 1879; capital stock, $2,000,000, in 200,000 shares, $10 each; offices Leadville, Colorado, and New York City; A.W. Gill, president, J.S. Lockwood, secretary, Theodore F. Van Wagenen, agent.

Leadville Mining and Manufacturing Co. Incorporated July 12, 1878; capital stock, $1,000,000, in 10,000 shares, $100 each; office

Leadville; organized by Nelson Hallock, J.D. and D.T. Griffith, Thomas Starr, A.P. Hereford, George W. Trimble, and Charles Cavender.

Leetonia. Owned by J.W. Gildersleeve, Stephen Joyce, John C. Stallcup, and D.C. Lyles; offices Leadville and Denver, Colorado; discovered June 8, 1878; located on west Fryer Hill, 1 mile from Leadville; claim 300 by 1,500 feet; course of vein, northwest and southeast; value of ore, 44 ounces silver per ton; developed by a shaft 70 feet deep.

Legal Tender. Owned by C.C. Kellogg, John C. Stallcup, and D.C. Lyles; office Denver; claim 300 by 1,500; located on Carbonate Mountain, in Oro mining district, 2 miles from Leadville; developed by a shaft 75 feet deep.

Lilly. Henry Brittenstein and Henry Orear, proprietors; office Alpine; claim 300 by 1,500 feet; discovered 1878; located in Grizzly Gulch Chalk Creek mining district, 6 miles from Alpine and 65 miles from Leadville; course of vein, northeast and southwest, width 5 feet, pay vein 18 inches, pitch 22 degrees north; nature of ore, galena, copper pyrites and native silver; assay value, 458 ounces silver per ton; 1 drift 75 feet in length and 35 feet from surface.

Lime. Owned by Stevens and Leiter; offices Leadville, Colorado, and Chicago, Illinois; claim 300 by 1,500 feet, patented; located on Carbonate Hill, 1 mile from limits of Leadville.

Little Chief. Owned by John V. Farwell & Co.; office Chicago, Illinois; claim 300 by 1,500 feet; located on Fryer Hill, 1 mile from Leadville.

Little Ella. Owned by George R. Fisher and others, of Leadville; claim 300 by 1,500 feet; discovered May 27, 1878; located on Fryer Hill, 1 mile from Leadville.

Little Eva. Owned by Borden, Tabor, and Co.; claim 300 by 1,500 feet; discovered May 27, 1878; located on Fryer Hill, 1 mile from Leadville.

Little Mattie. Joseph E. McClure, proprietor; office Alpine; claim 300 by 1,500 feet; located on Mama Mountain, Chalk Creek mining district, 2 miles from Alpine and 61 miles from Leadville; vertical vein, trending northeast and southwest; width, 12 feet, pay vein 18 inches;

nature of ore, galena, iron and copper pyrites; assay value, 50 to 75 ounces silver per ton; 1 drift 35 feet in length and 30 feet from surface.

Little Pittsburg. Owned by the Little Pittsburg Consolidated Mining Co.; J.C. Wilson, manager; this property is located on Fryer Hill, 1 mile from Leadville, and comprises the Little Pittsburg, Winnemuc, Dives, and New Discovery mines; the property is well developed and has produced to date, $500,000.

Log Cabin. Owned by the Topeka Mining and Tunnel Co.; office Alpine; claim 300 by 1,500 feet; located in Baldwin Gulch, Chalk Creek mining district, 2 miles from Alpine and 60 miles from Leadville; course of vein, north and south, width 3 feet, pay vein 2 feet, pitch 15 degrees west; nature of ore, galena and iron pyrites; 1 shaft 10 feet deep.

Loker. Owned by the Meyer Mining Co.; August R. Meyer, superintendent; located on the Arkansas River, 3 miles from Leadville.

Lower Printer Boy. Benjamin Bernard and Co., proprietors; office Leadville; claim 50 by 950 feet; discovered 1868; located on Printer Boy Hill, California mining district, 3 miles from Leadville; course of vein, north and south, width 2 feet; nature of ore, free gold; average value, 8 ounces gold per cord; main shaft 260 feet deep, 4 other shafts aggregating 250 in depth, aggregate length of levels 250 feet.

Maggie Anderson. Chrysolite Tunnel Co., proprietors; office Alpine; claim 300 by 1,500 feet, discovered 1873; located on Chrysolite Mountain, Chalk Creek mining district, 6 miles from Alpine and 65 miles from Leadville; vertical vein, trending northeast and southwest, width 30 inches, pay vein 10 inches; nature of ore, galena and copper pyrites; assay value, 65 ounces silver per ton; 1 shaft 10 feet deep.

Maid of the Mist. Chrysolite Tunnel Co., proprietors; office Alpine; claim 300 by 1,500 feet; discovered 1873; located on Chrysolite Mountain, Chalk Creek mining district, 6 miles from Alpine; course of vein, northeast and southwest, width 12 feet, pitch 55 degrees west; nature of ore, galena; assay value, 262 ounces silver per ton; 1 adit 10 feet in length.

Malta Smelting and Mining Co. Incorporated February 1, 1876; capital stock, $125,000, in 250 shares, $500 each; office Malta; organized by Henry G. Meyer, Rudolph Rheinholdt, Louis Krohn, B. Bettman, and J. Buss.

Mary Murphy. Owned by A.E. Wright, M.D., and the Kansas City Mining and Smelting Co.; office Alpine; claim 300 by 1,500 feet; located on Pomeroy Mountain, Chalk Creek mining district, 66 miles from Leadville; course of vein, northeast and southwest, width 5 feet, pay vein, 4 feet, pitch 40 degrees southeast; nature of ore, galena and carbonate of lead; average value, 75 ounces silver and 2 ounces gold per ton; 2 shafts 40 feet deep each, 2 levels 25 feet in length, respectively.

Meyer Mining Co. Organized under the laws of the State of Missouri, November 12, 1877; capital stock, $220,000, in 2,200 shares, $100 each; office Leadville; G.W. Chadburne, president, John B. Maude, vice president, James R. Loker, secretary and treasurer, August R. Meyer, resident director. The mines owned by this company are located within a radius of 1 mile from the town of Leadville, and comprise the Crescent, Aetna, Rough and Ready, Pinnacle, Crown Point, and Loker; these mines have produced to date, $150,000.

Mike. Owned by Thomas Starr, Michael and John Haynes; office Leadville; claim 300 by 3,000 feet; discovered 1867; located in Stray Horse Gulch, 2 miles from Leadville; course of vein, north and south, width 26 feet, pitch east; nature of ore, carbonate of lead; average value, 300 ounces silver per ton; 1 shaft 65 feet, 1 tunnel 500 feet in length, and 150 feet from surface.

Minnetoka. Joseph Hudson and G.W. Fisher, proprietors; office Alpine; claim 300 by 1,500 feet; discovered 1873; located on Chrysolite Mountain, Chalk Creek mining district, 6 miles from Alpine; vertical vein, trending northeast and southwest, width 60 feet, pay vein 3 feet; nature of ore, galena, iron and copper pyrites; assay value, 3 to 305 ounces silver and 2 ounces gold per ton; 1 shaft 105 feet deep.

Mint of the Mount. Owned by the Topeka Mining and Tunnel Co.; office Alpine; claim 300 by 1,500 feet; discovered September 1875; located in Baldwin Gulch, Chalk Creek mining district, near town of Alpine, 60 miles from Leadville; vertical vein, trending north and south, width 5 feet, pay vein 4 feet; nature of ore, copper pyrites and galena; average value, 40 ounces silver per ton and 24 per cent copper; 2 shafts 15 feet each; operated through Topeka Tunnel.

Monte Christo. Owned by Samuel Morgan and Aden Alexander; claim 200 by 1,500 feet; located on Carbonate Hill, 1 mile from Leadville; developments, 2 shafts, 30 and 45 feet, respectively.

Morgan Placer. Owned by Louis Dugal, F.M. Case, and Witter Estate; claim comprises 76 acres of patented ground; located on the Arkansas River, in town of Granite, 20 miles from Leadville.

Morning Star. Owned by John L. Routt [a three-time governor of Colorado], of Denver; claim 202 by 1,500 feet; discovered 1877; located on Carbonate Hill, California mining district, half mile from Leadville; course of vein, north and south, width 4 feet, pay vein 3 feet, pitch east; nature of ore, carbonate of lead; development, 1 shaft 100 feet deep and 1 drift 200 feet in length; shaft house 20 by 30 feet.

Nevada. Owned by T.D. Anthony, F.H. Foss, and others; office Leadville; claim 300 by 1,500 feet; discovered 1878; located in Evans Gulch, California mining district, 2 miles from Leadville; course of vein, north and south, width 3 feet, pitch east; nature of ore, carbonate of lead; average value, 100 ounces silver per ton; 1 tunnel 30 feet and 1 level 20 feet in length.

New Discovery. Owned by Chaffee, Tabor and Co.; claim 300 by 1,500 feet; discovered 1878; located in Stray Horse Gulch, California mining district, half mile from Leadville; course of vein, north and south, width 18 feet, pitch east; nature of ore, carbonate of lead; average value, $50 per ton; 1 shaft 60 feet deep, aggregate length of drifts 75 feet; yield, $30,000.

Nisi Prius. Owned by George E. and Stephen Pease; office Leadville; claim 300 by 1,500 feet; discovered 1878; located in Iowa Gulch, California mining district, 3 miles from Leadville; course of vein, north and south, width 10 feet, pitch east; nature of ore, carbonate of lead; assay value, 300 ounces silver per ton; main shaft 60 feet and 1 other shaft 40 feet deep, 3 drifts aggregating 700 feet in length, 3 levels 19, 30, and 35 feet from surface, respectively, aggregating 250 feet in length.

Northern Light. John Maher and John Sweeney, proprietors; office Alpine; claim 300 by 1,500 feet; discovered 1876; located in Chalk Creek mining district, 2 miles from Alpine and 61 miles from Leadville; vertical vein, trending northeast and southwest, width 8 feet, pay vein 30 inches;

nature of ore, galena; assay value, 100 ounces silver and 1 ½ ounces gold per ton; 1 adit 10 feet in length with 10-foot face.

Norway Boy. Owned by Fletcher, Hughes & Russell; office Alpine; claim 300 by 1,500 feet; discovered 1873; located on Chrysolite Mountain, Chalk Creek mining district, 6 miles from Alpine; vertical vein, trending north and south, width 3 feet, pay vein 15 inches; nature of ore, galena and iron pyrites; assay value, 200 ounces silver and 4 ounces gold per ton; 1 adit 20 feet in length with a 15-foot face.

O.K. Owned by Peter Finerty, William Smith, P.O. Day, and Charles Donnelly; offices Leadville and Denver, Colorado; claim 300 by 1,500 feet; discovered May 1878; located on Fryer Hill, half mile from Leadville; development, 2 shafts, 80 and 130 feet, respectively.

Oro. Owned by Daniel Trellinger, Peter Loye, and John McDonald; office Leadville; claim 300 by 1,500 feet; discovered 1877; located on Iron Hill, California mining district, 2 miles from Leadville; course of vein, north and south, pitch east; nature of ore, carbonate of lead; 1 shaft 100 feet deep.

Oro-La Plata. Located on Carbonate Hill, California mining district, 1 mile from limits of Leadville.

Oro Mining Ditch and Fluming Co. Incorporated May 26, 1874; capital stock, $500,000, in 100,000 shares, $5 each; offices Oro City, Colorado, and Chicago, Illinois; organized by William H. Stevens, Sullivan D. Breece, J. Marshall Paul, and A.B. Wood. The property owned by this company comprises 225 acres of placer ground located in Pawnee, Georgia, Oregon, and California Gulches, near town of Leadville.

Pandora. Owned by Borden, Tabor & Co.; claim 200 by 800 feet; discovered July 1878; located on Fryer Hill, 1 mile from Leadville; developed by 2 shafts, 80 and 114 feet, respectively.

Pat Murphy. Owned by A.E. Wright, M.D., and the Kansas City Mining and Smelting Co.; office Alpine; claim 300 by 1,500 feet; located on Pomeroy Mountain, Chalk Creek mining district, 7 miles southwest of town of Alpine and 66 miles from Leadville; course of vein, northeast and southwest, width 5 feet, pay vein 4 feet, pitch 40 degrees southwest; nature of ore, galena and carbonate of lead; assay value, $50 to $150 per ton; 1 level 30 feet in length and 25 feet from surface.

Philadelphia and Boston Gold and Silver Mining Co. Incorporated February 20, 1871; capital stock, $200,000, in 2,000 shares, $100 each; office Philadelphia; organized by John Rodman Paul, Henry N. Paul, and James Marshall Paul.

Pilot. Benjamin Bernard & Co., proprietors; office Oro; claim 50 by 300; discovered 1868; located on Printer Boy Hill, California mining district, 3 miles from Leadville; course of vein, north and south; nature of ore, free gold; average value, $65 per ton; 1 shaft 50 feet deep, 3 drifts, 100 and 250 feet from surface, respectively, aggregating 1,350 feet in length; yield, $35,000.

Pine. Owned by the Argentine Mining Co.; office Leadville; claim 300 by 1,500 feet; discovered 1876; located in Stray Horse Gulch, 2 miles from Leadville; course of vein, north and south, width 6 feet, pay vein 5 feet, pitch east; nature of ore, carbonate of lead; average value, 85 ounces silver per ton; 3 shafts, 20, 50, and 100 feet in depth, respectively; aggregate length of drifts 200 feet; shaft house, 15 by 20 feet.

Pinnacle. Owned by the Meyer Mining Co.; August R. Meyer, manager; office Leadville; claim 300 by 1,500 feet; discovered 1876; located on Stevens Hill, 2 miles from Leadville; course of vein, north and south, width 8 feet, pitch east; nature of ore, carbonate of lead; average value, 15 ounces silver per ton; 1 shaft 150 feet deep, aggregate length of drifts 125 feet; shaft house, 18 by 25 feet.

Pittsburgh. Owned by John McCombe and others; office Leadville; claim 300 by 1,500 feet; discovered 1878; located in Stray Horse Gulch, California mining district, 1 mile from Leadville; course of vein, north and south, width 5 feet, pay vein 42 inches, pitch east; nature of ore, carbonate of lead; average value, 124 ounces silver per ton; 1 shaft 135 feet deep.

Porphyry. Owned by J.F. and J.W. Dana and others; office Leadville; claim 300 by 1,500 feet; discovered 1878; located on Long's Hill, California mining district, 5 miles from Leadville; course of vein, northeast and southwest, width 7 feet, pay vein 6 inches, pitch east; nature of ore, oxide of iron; assay value, 20 ounces silver per ton; 1 shaft 160 feet deep.

Printer Boy. Owned by Estate of J. Marshall Paul; located in Oro mining district.

Quincy. Henry Brittenstein, proprietor; office Alpine; claim 300 by 1,500 feet; discovered 1876; located on Chrysolite Mountain, Chalk Creek mining district, 6 miles from Alpine; vertical vein, trending north and south, width 12 feet, pay vein 4 feet; nature of ore, galena; assay value, 75 to 340 ounces silver per ton; 1 shaft 40 feet deep.

Ranchero. Owned by the Small Hope Mining Pool; offices Denver and Leadville, Colorado; claim 220 by 1,500 feet; located in Stray Horse Gulch, 1 mile from Leadville; value of ore, 46 to 700 ounces silver per ton; 1 shaft 50 feet deep, 2 levels, 30 and 130 feet in length, respectively.

Ready Cash. Owned by Houx and Tucker; office Leadville; claim 300 by 1,500 feet; discovered 1878; located in Iowa Gulch, California mining district, 6 miles from Leadville, vertical vein, trending northeast and southwest, width 3 feet, pay vein 10 inches; nature of ore, galena and sulphurets; average value, $567 per ton; 1 shaft 15 feet deep, tunnel 85 feet, 1 level 60 feet in length.

Result. Owned by the Small Hope Mining Pool; offices Leadville and Denver, Colorado; claim 300 by 1,500 feet; located in Stray Horse Gulch, 1 mile from Leadville.

Riggins. Owned by the Kansas City Mining and Smelting Co.; office Alpine; claim 300 by 1,500 feet; located in Baldwin Gulch, Chalk Creek mining district, 4 miles from the town of Alpine and 63 miles from Leadville; course of vein, 15 degrees west of north and south of east, width 5 feet, pay vein 3 to 4 feet, pitch 45 degrees west; nature of ore, galena, sulphurets and carbonate of lead; value, 40 to 60 ounces silver per ton; main shaft 140 feet deep, 1 drift 30 feet in length.

Robert E. Lee. Henry Brittenstein, proprietor; office Alpine; claim 300 by 1,500 feet; discovered 1876; located in Grizzly Gulch, Chalk Creek mining district, 6 miles from Alpine and 65 miles from Leadville; vertical vein, trending northeast and southwest, width 10 feet, pay vein 3 feet; nature of ore, galena; assay value, 170 to 200 ounces silver per ton; 1 shaft 18 feet deep.

Robert Emmet. Owned by the Small Hope Mining Pool; offices Denver and Leadville, Colorado; claim 300 by 1,500 feet; discovered 1878; located in Stray Horse Gulch, 1 mile from Leadville; nature of ore, carbonate of lead; value $75 to $100 per ton; 1 shaft 50 feet deep, 1 drift 50 feet in length.

Rock. Owned by William H. Stevens and Levi Z. Leiter; claim 300 by 1,500 feet, patented; located on Carbonate Hill, 1 mile from limits of Leadville.

Rough and Ready. Owned by the Meyer Mining Co.; August R. Meyer, superintendent; located on Carbonate Hill, half mile from Leadville.

Routt. Owned by Peter Finerty and Mason B. Carpenter; offices Leadville and Denver, Colorado; claim 300 by 1,500 feet; discovered October 1878; located on Gallagher Hill, 1 mile from Leadville; developed by 2 shafts, 80 and 100 feet, respectively.

Royal George. Owned by Joel W. Shackelford, Henry Burns, George Wilder, and William Clark; offices Leadville and Denver, Colorado; located in Big Evans Gulch, 2 miles from Leadville; developed by a shaft 65 feet deep.

Savage. Owned by James M. Lynch; office Leadville; claim 300 by 1,500 feet, patented; discovered 1877; located in Iowa Gulch, California mining district, 5 miles from Leadville; course of vein, northeast and southwest, width 7 feet, pay vein 4 feet, pitch 45 degrees north; nature of ore, carbonate of lead and sulphurets of silver; value, 200 ounces silver per ton; development, 1 drift 50 feet in length and 45 feet in depth; yield $450.

Senator. James M. Lynch, proprietor; office Leadville; claim 300 by 1,300 feet; discovered 1878; located on Green Mountain, California mining district, 4 miles from Leadville; course of vein, north and south, width 4 feet, pay vein 32 inches, pitch 15 degrees; nature of ore, galena and carbonate; assay value, 80 ounces silver per ton; 1 shaft 18 feet deep, 1 level 25 feet in length and 18 feet from surface.

Seventy-Six. Owned by William H. Stevens and Levi Z. Leiter; located in California Gulch, 1 mile from Leadville.

Shamrock. Owned by Thomas S. Wells, Albert Cooper, and Nelson Hallock; office Leadville; claim 300 by 1,500 feet, patented; discovered 1878; located on Carbonate Hill, California mining district, half mile from Leadville; course of vein, north and south, width 8 feet, pay vein 18 inches, pitch east; nature of ore, carbonate of lead; average value, 150 ounces silver per ton; 1 shaft 35 feet deep, 1 level 25 feet in length; yield, $3,200.

Sherman. Robert N. Scott, proprietor; office Cleora; claim 300 by 1,500 feet; located on South Arkansas River, 70 miles from Leadville; vertical vein, trending northeast and southwest, width 4 feet, pay vein 2 feet; nature of ore, gray copper and oxide of copper; 2 shafts, 25 feet each.

Shoo Fly. Owned by William Russell, William A. Fletcher, and J.H. Hughes; office Alpine; claim 300 by 1,500 feet; discovered 1873; located on Chyrsolite Mountain, Chalk Creek mining district, 6 miles from Alpine and 66 miles from Leadville; vertical vein, trending north and south, width 3 feet, pay vein 5 inches; nature of ore, galena; assay value, 250 to 350 ounces silver per ton; 1 adit 20 feet in length with 15-foot face.

Silver Wave. Andrew W. Gill & Co., proprietors; office Leadville; claim 270 by 1,500 feet; discovered 1877; located on Iron Hill, California mining district, 1 ½ miles from Leadville; course of vein, north and south, width 4 feet, pay vein 2 feet, pitch east; nature of ore, carbonate of lead; average value, 150 ounces silver per ton; 2 shafts, 43 and 75 feet, respectively; 1 drift 100 feet in length; yield, $10,000.

Small Hope Mining Pool. W.H Williams, M.D., M.H. Slater, J.F. Sanders, W.S. Williams, and others; offices Leadville and Denver, Colorado; the mines owned by this company are the Ranchero, Robert Emmett, Forrest City, and Result; located one mile from Leadville, in Stray Horse Gulch.

Smoky Hill. James G. Gillespie and James Donaldson, proprietors; office Leadville; claim 300 by 1,500 feet; discovered 1878; located on Long's Hill, California mining district, 5 miles from Leadville; course of vein, northeast and southwest, width 2 feet, pay vein 5 inches, pitch 30 degrees east; nature of ore, carbonate of lead; average value, 250 ounces silver per ton; 1 shaft 45 feet deep.

Smuggler. Judson H. Dudley and George T. Clark, proprietors; office Leadville; claim 300 by 1,500 feet; discovered 1877; located on Iron Hill, 2 miles from Leadville; course of vein, north and south, width 4 feet, pay vein 3 feet, pitch east; nature of ore, carbonate of lead; average value, 140 ounces silver per ton; main shaft 100 feet and 1 other shaft 43 feet in depth, 1 level 175 feet in length; shaft house 15 by 30 feet.

Smuggler. Max Brookes and Robert Ray, proprietors; office Alpine; claim 300 by 1,500 feet; located in Poplar Gulch, Chalk Creek mining district, 3 miles from Alpine and 57 miles from Leadville, vertical vein, trending northeast and southwest, width 12 feet, pay vein 8 inches; nature of ore, galena; assay value, 60 to 300 ounces silver and 2 ½ ounces gold per ton; 1 drift 40 feet in length and 50 feet from surface.

Starr Placer. Owned by Thos. Starr and George S. Henderson; office Leadville; claim 100 by 1,000 feet, patented; located in town of Leadville.

Stevens and Leiter Placer. Owned by William H. Stevens and Levi Z. Leiter; located in town of Leadville.

Stone. Owned by William H. Stevens and Levi Z. Leiter; claim 300 by 1,500 feet; located on Carbonate Hill, 1 mile from the limits of Leadville.

Stray Horse. Located in the southern portion of Fryer Hill, 1 mile from Leadville.

St. Louis Smelting and Refining Co. Organized under the laws of the State of Missouri, May 10, 1871; capital stock, $250,000, in 2,500 shares, $100 each; Edwin Harrison, president.

Sunday. Lucius Bradshaw and John Bergman, proprietors; office Leadville; claim 300 by 1,500 feet; discovered 1878; located on Bald Mountain, 3 miles from Leadville; vertical vein, trending north and south, width 4 feet; nature of ore, galena and oxide of iron; assay value, 50 ounces silver and 4 ounces gold per ton; main shaft 65 feet deep, and 2 other shafts aggregating 20 feet in depth.

Talisman. Owned by Lucius Bradshaw, Joseph Richards, and others; office Leadville; claim 300 by 1,500 feet; discovered 1877; located on Bald Mountain, 3 miles from Leadville; course of vein, north and south, width 3 feet, pay vein 30 inches, pitch 45 degrees east; nature of ore, galena, gray copper, and ruby silver; average value, 520 ounces silver per ton; 1 shaft 16 feet deep, 1 drift 100 feet in length and 46 feet from surface; yield, $1,000.

Terrible. Owned by the Adelaide Consolidated Silver Mining and Smelting Co.; offices Leadville, Colorado, and Washington, D.C.; claim 300 by 1,500 feet, discovered 1877; located in Stray Horse Gulch, 2 miles from Leadville; course of vein, north and south, width 5 feet, pitch east; nature of ore, carbonate of lead; average value, 19 ounces silver per ton; 2 shafts, 50 and 75 feet in length, respectively; 1 level 50 feet in length.

Thatcher Mining Co. Incorporated March 12, 1879; capital stock, $1,000,000, in 100,000 shares, $10 each; offices Leadville, Colorado, and Kansas City, Missouri; trustees: A.P. Fonda, Stephen D. Thatcher, George E. Stevens, James Gibson, James H. Ralston, E.J. Gump, and W.H. Hurlbut.

Tilden. Owned by Charles A. Shanks, John S. Shanks, James Shumate, and George W. Knox; office Alpine; claim 300 by 1,500 feet; discovered 1876; located on Boulder Mountain, Chalk Creek mining district, 3 miles from Alpine and 61 miles from Leadville; course of vein, north and south, width 9 feet; nature of ore, galena and native silver; average value of first-class ore 2,310 ounces and second class 290 ounces silver per ton; 1 shaft 35 feet deep, 1 drift 65 feet in length.

Topeka Mining and Tunnel Co. Incorporated June 26, 1877; capital stock, $20,000, in 40 shares, $500 each; office Alpine; trustees: Henry F. Brittenstein, George W. Knox, and C.W. Angle.

Trio. Owned by Daniel Trellinger, Peter Love and John McDonald; office Leadville; claim 300 by 1,500 feet; discovered 1877; located on Iron Hill, 2 miles from Leadville; course of vein, north and south, pitch east; nature of ore, carbonate of lead, 1 shaft 30 feet deep.

Twin Lake Metallurgic Company. Incorporated August 31, 1878; capital stock, $12,000, in 24 shares, $500 each; office, near Twin Lakes; trustees: T.C. Wetmore, J.C. Carrera, and P. Suerreri.

Ute. Chrysolite Tunnel Co., proprietors; office Alpine; claim 300 by 1,500 feet; discovered 1873; located on Chrysolite Mountain, Chalk Creek mining district, 6 miles from Alpine; course of vein, northeast and southwest, width 5 feet, pay vein 23 inches, pitch 55 degrees west; nature of ore, galena; assay value, 444 ounces silver per ton; 1 shaft 15 feet deep.

Virginia. Henry Brittenstein, proprietor; office Alpine; claim, 300 by 1,500 feet; locate 1875; located in Grizzly Gulch, Chalk Creek mining district, 6 miles from Alpine and 65 miles from Leadville; vertical vein, trending north and south, width 4 feet, pay vein 18 inches; nature of ore, galena; assay value, $675 to $28,820 per ton; 2 drifts, 20 and 45 feet in length, respectively.

Vulture. Owned by Borden, Tabor & Co.; claim 300 by 1,500 feet; located in Stray Horse Gulch, 1 mile from Leadville.

Washington Gulch Mining, Fluming and Ditching Co. Incorporated March 16, 1870; capital stock, $2,000, in 200 shares, $10 each; organized by Lewis Hayden, William Kraft, and J.H. Johnson.

Wells Placer. Owned by Thomas S. Wells and William Mayer; office Oro City; claim comprises 190 acres of placer ground; located in California Gulch, in town of Oro City, 3 miles from Leadville; this property is worked by hydraulic pressure and has a flume 700 feet in length; yield to date, $600,000.

Winnemuc. Owned by the Little Pittsburg Consolidated Mining Co.; J.C. Wilson, manager; located on Fryer Hill, 1 mile from Leadville.

Wolf Tone. Owned by Samuel Morgan and Aden Alexander; claim 300 by 1,500 feet; discovered 1876; located on Carbonate Hill, 1 mile from Leadville; value of ore, $250 per ton; development, 5 shafts ranging in depth from 30 to 80 feet.

Wyman. Owned by Lucius Bradshaw and A. Niles; office Leadville; claim 300 by 1,000 feet; discovered 1878; located on Printer Boy Hill; 3 miles from Leadville; course of vein, east and west, width 2 feet, pitch east; nature of ore, galena and sulphurets; assay value, 90 ounces silver per ton; 3 shafts 26 feet each, 2 levels, 30 and 50 feet in length, respectively.

Yankee Boy. Owned by the Yankee Boy Mining Co.; incorporated November 6, 1875; capital stock, $10,000, in 100 shares, $100 each; office Colorado Springs; organized by J.D. McCloskey, E.N. Bartlett, A.J. Belcher, A.A. McGooney, and Arthur Peck; claim 300 by 1,500 feet; discovered 1875; located in Chalk Creek mining district, 2 miles from Alpine and 61 miles from Leadville; horizontal vein, trending northeast and southwest, width 8 feet, pay vein 30 inches; nature of ore, galena, iron and copper pyrites; assay value, 40 to 75 ounces silver per ton; 1 shaft 70 feet deep.

Yellow Jacket. Owned by Charles A. Shanks, John S. Shanks, and John Gwinn; office Alpine; claim 300 by 1,500 feet; discovered 1876; located on Summit Mountain, Chalk Creek mining district, 3 miles from Alpine and 62 miles from Leadville; vertical vein, trending northeast and southwest, width 6 feet, pay vein 15 inches; nature of ore, galena; assay value, 30 ounces silver and 9 ounces gold per ton; 1 drift 90 feet in length and 70 feet from surface.

Young America. Owned by the Argentine Mining Co.; offices Leadville, Colorado, and St. Louis, Missouri; claim 300 by 1,500 feet; discovered 1876; located in Stray Horse Gulch, California mining district, 1 mile from Leadville; course of vein, north and south, width 12 inches, pitch east; nature of ore, carbonate of lead; value, 35 to 75 ounces silver per ton; 1 tunnel, 546 in length.

Lake County Ore Mills

Adelaide Consolidated Silver Mining and Smelting Co. Incorporated May 31, 1878; capital stock, $2,500,000, in 100,000 shares, $25 each; offices Washington, D.C., and Leadville, Colorado; trustees: John P. Jones, Jay A. Hubbell, Henry D. Cooke, John R. Magruder, James L. Hill, Woodbury Blair, and William H. Barnard. The works are located in Stray Horse Gulch, 2 miles from Leadville, and have a capacity for treating 30 tons of ore in 24 hours.

Berdell and Witherell's Sampling and Smelting Works. Located in California Gulch, near limits of Leadville. These works have a capacity for treating 50 tons of ore in 24 hours. The main building is a frame 50 feet wide by 70 feet in length, containing an 80-horse power engine, Blake crusher, Cornish rollers, steam pump, etc.

Grant's Smelting Works. James and James B. Grant, proprietors; located in town of Leadville; have a capacity for treating 50 tons of ore in 24 hours. The main building is a frame 72 feet wide by 100 feet in length, and contains a 25-horse power engine, Blake crusher, Cornish rollers, blower, etc.

Harrison Reduction Works. Branch of St. Louis Smelting and Refining Co.; Edwin Harrison, president, A.V. Weise, superintendent. These works were erected 1877 and are located within the corporate limits of Leadville; the building is a wooden structure 50 feet in width by 132 feet in length and contains 1 calcining and 2 blast furnaces, 2 Baker blowers, Blake crusher, 50-horse power engine, and all other necessary machinery; have a capacity for treating 50 tons of ore in 24 hours.

Kansas City Mining and Smelting Co. Incorporated March 7, 1876; capital stock, $100,000, in 1,000 shares, $100 each; office town of Alpine; organized by Benjamin L. Riggins, John J. Chapman, and George Holmes. The works were commenced in 1874 by Riggins and Chapman and completed the following year; they have a capacity for treating 10 tons of ore in 24 hours, are run by water power, and located on Chalk Creek, near town of Alpine, 59 miles from Leadville. The main building is a frame 70 feet wide by 130 feet in length.

Lake County Sampling Co. Incorporated May 4, 1878; capital stock, $6,000, in 60 shares, $100 each; office Leadville; organized by Thomas Starr, James M. and W.F. Patrick.

Leadville Sampling Works. Eddy & James, proprietors; located in Leadville; have a capacity for treating 80 tons of ore in 24 hours; the main building is a frame 40 by 70 feet long.

Malta Smelting and Mining Co. Incorporated February 1, 1876; capital stock, $125,000, in 250 shares, $500 each; office Malta; organized by Henry G. Meyer, Rudolph Rheinholdt, Louis Krohn, B. Bettman, and J. Buss. The works of this company are located in California Gulch, in town of Malta, 5 miles from Leadville. They have a capacity for treating 10 tons of ore in 24 hours.

Meyer's Sampling Works. August R. Meyer & Co., proprietors; these works consist of a group of buildings which cover 1 acre of ground and contain a 50-horse power engine, Blake crusher, Cornish rollers, etc. They are located in the City of Leadville and have a capacity for treating 80 tons of ore in 24 hours.

Storm-battered wreckage of the once fabulous Mary Murphy Mine.

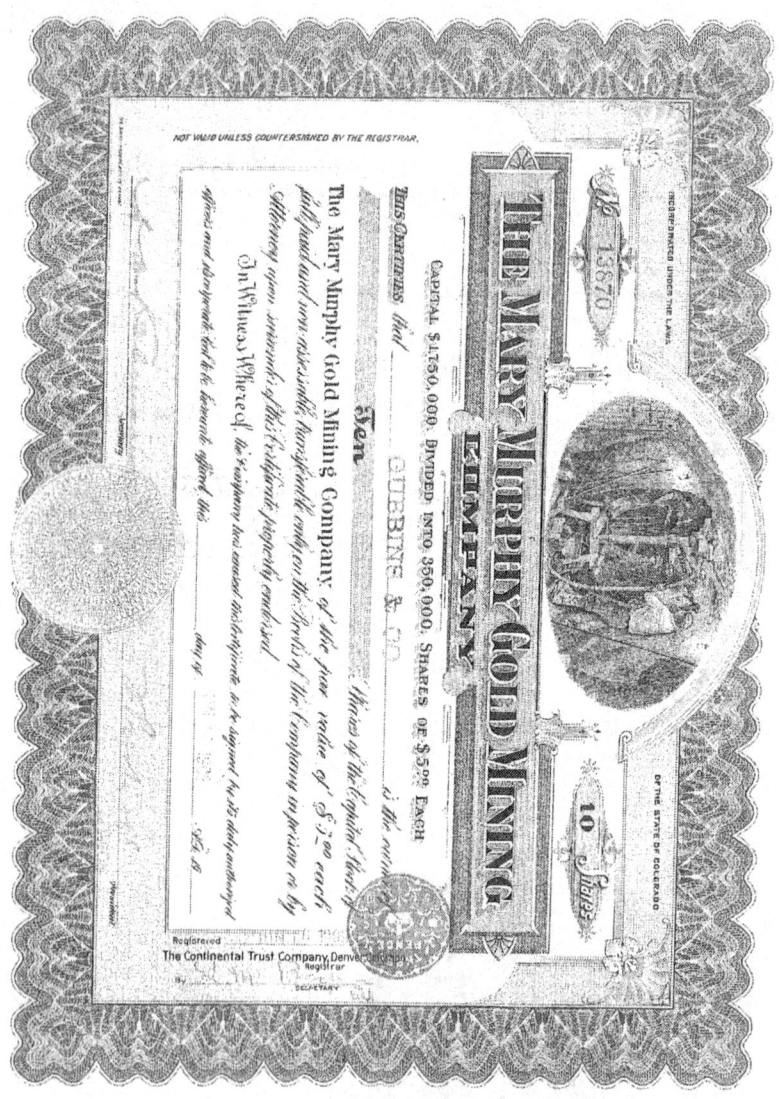

Stock certificate issued in 1909 for 10 shares of the Mary Murphy gold Mining Company.

OURAY COUNTY

Thomas B. Corbett reported that Ouray County had a population of 1,000 in 1878 and a bullion product of $124,125 for that year. Ouray county mining districts included Iron Springs, Mount Sneffels, Uncompahgre, and Upper San Miguel.

A. Owned by Ward, Snyder & Neumeyer; office Ophir; claim 300 by 1,500 feet; discovered 1877; located on Silver Mountain, Iron Springs mining district, near town of Ophir, 12 miles from Ouray; course of vein, northeast and southwest, width 3 feet, pay vein 18 inches; nature of ore, galena and free gold; average value $200 per ton; 1 adit 20 feet in length.

Alexis. D. McKenzie, proprietor; office Ouray; claim 300 by 1,500 feet; discovered 1874; located on Engineer Mountain, Uncompahgre mining district, 7 miles from town of Ouray; course of vein, east and west, width 7 feet, pitch 10 degrees; nature of ore, iron and copper pyrites and galena; average value 40 ounces silver per ton; 1 drift 40 feet in length and 30 feet from surface.

Amazon. E. Thayer and J. Van Doren, proprietors; office Ouray; claim 300 by 1,500 feet; discovered 1878; located on Bear Creek, Uncompahgre mining district, 4 miles from Ouray; course of vein, north and south, width 7 feet, pay vein 15 inches, pitch 10 degrees; nature of ore, galena and gray copper; assay value, 60 ounces silver per ton; 1 adit 25 feet in length with 12-foot face.

American Flag. Owned by A.F. Kreuss, J. Godin, and C. May; office Ouray; claim 300 by 1,500 feet; discovered 1878; located on Old Stony Mountain, Mount Sneffels mining district, 9 miles from Ouray; vertical vein, trending east and west, width 4 feet, pay vein 22 inches; nature of ore, galena and ruby silver; assay value, 125 ounces silver per ton; 1 shaft 20 feet deep.

Angelina. E. McCarty, proprietor; office Silverton; claim 300 by 1,500 feet; discovered 1876; located on Mount Sneffels [this is one of Colorado's "fourteeners" at 14,150 feet], 9 miles from Ouray; course of vein, northwest and southeast, width 4 feet, pitch 20 degrees; nature of ore, gray copper; value, 100 ounces silver per ton; 2 adits, 20 and 25 feet in length, respectively.

Apache. W. Parrott and C.S. Barker, proprietors; office Ouray; claim 300 by 1,500 feet; discovered 1877; located on Red Mountain, Uncompahgre mining district, 3 miles from Ouray; vertical vein, trending north and south, width 28 inches; nature of ore, galena, copper and iron pyrites; assay value, 80 ounces silver per ton; 1 drift 20 feet in length and 15 feet from surface.

Atlantic. T. Resor & Co., proprietors; office Ouray; claim 300 by 1,500 feet; discovered 1878; located on Gold Run, Upper San Miguel mining district, 9 miles from Ouray; vertical vein, trending north and south, width 3 feet; nature of ore, gray copper, iron and copper pyrites; average value, 120 ounces silver per ton; 1 adit 12 feet in length.

B. Owned by A.W. Neumeyer and J.B. Ward; office Ophir; claim 300 by 1,500 feet; discovered 1878; located on Silver Mountain, Iron Springs mining district, 1 mile from town of Ophir and 12 miles from Ouray; course of vein, northeast and southwest, width 5 feet, pay vein 1 foot, pitch 30 degrees; nature of ore, sulphurets and carbonate of silver; 1 adit 14 feet in length and 10 feet from surface.

Banner. Owned by F. Herbst, N.H. Love, and others; office Ouray; claim 300 by 1,500 feet; discovered 1876; located on Mount Hendricks, Mount Sneffels mining district, 9 miles from Ouray; course of vein, northwest and southeast, width 6 feet, pay vein 18 inches; nature of ore, galena and gray copper; assay value, 200 ounces silver per ton; 1 drift 20 feet in length and 18 feet from surface.

Battle Flag. D. McKenzie, proprietor; office Ouray; claim 300 by 1,500 feet; discovered 1877; located on Red Mountain, Uncompahgre mining district, 6 miles from Ouray; course of vein, northeast and southwest, width 4 feet, pay vein 20 inches, pitch 25 degrees; nature of ore, gray copper and iron pyrites; average value, 70 ounces silver per ton; 1 adit 15 feet in length.

Bay State. George Peters and C. Brock, proprietors; office Ouray; claim 300 by 1,200 feet; located 1877; located on Buckeye Mountain, Mount Sneffels mining district, 8 miles from Ouray; course of vein, east and west, width 4 feet, pay vein 30 inches, pitch 40 degrees; nature of ore, gray copper and brittle silver; assay value, 200 ounces silver per ton; 1 drift 15 feet in length and 35 feet from surface.

Beaver. Owned by E.C. Abbott & Co., office Ouray; claim 300 by 1,500 feet; located 1876; located on Bear Creek, Uncomprahgre mining district, 4 miles from Ouray; course of vein, northwest and southeast, width 5 feet, pitch 20 degrees; nature of ore, gray copper; assay value, 300 ounces silver per ton; 2 shafts, 15 and 20 feet in depth, respectively.

Begole Mineral Farm. Owned by Norfolk and Ouray Mining and Smelting Co.; M.S. Corbett, superintendent; claim 1,050 by 1,500 feet, patented; discovered 1875; located on Canyon Creek, 2 miles from Ouray; course of vein, northeast and southwest, width 300 feet; nature of ore, galena and gray copper; value, 160 ounces silver per ton; development, 2 shafts 20 feet each, 11 adits 12 feet each; yield, $3,000.

Belle of the West. J.L. Haines and M.C. McCormick, proprietors; office Silverton; claim 300 by 1,500 feet; discovered 1876; located on Yellow Mountain, Iron Springs mining district, 1 mile from town of Ophir and 12 miles from Ouray; course of vein, northeast and southwest, width 8 feet, pay vein 3 feet, pitch 20 degrees; nature of ore, galena and gray copper; assay value, 150 ounces silver per ton; 1 adit 25 feet in length with 25-foot face.

Belvidere. Owned by John McElroy, J.M. McCormick, and others; claim comprises 100 acres of placer ground; located on the San Miguel River, 2 miles from Placerville and 25 miles from Ouray.

Betsey. Owned by F. Fodde & Co.; office Ophir; claim 300 by 1,500 feet; discovered 1878; located on Yellow Mountain, Iron Springs mining district, 1 mile from town of Ophir, and 13 miles from Ouray; vertical vein, trending northeast and southwest, width 4 feet, pay vein 15 inches; nature of ore, gray copper and galena; average value, 94 ounces silver per ton; 1 drift 30 feet in length; intersected by the Pike County Tunnel at a depth of 185 feet; yield, $750.

Big Bear. M. Phelps and J. Britton, proprietors; office Silverton; claim 300 by 1,500 feet; discovered 1874; located on Gold Run, Upper San Miguel mining district, 9 miles from Ouray; vertical vein, trending northeast and southwest, width 6 feet, pay vein 2 feet; nature of ore, galena and gray copper; 1 adit 16 feet in length with 12-foot face.

Big Casino. W.A. Walker and Co., proprietors; office Upper San Miguel City; claim 300 by 1,500 feet; discovered 1878; located on Gold Run, Upper San Miguel mining district, 9 miles from Ouray; course of vein, northeast and southwest, width 15 feet, pitch 10 degrees; nature of ore, gray copper and ruby silver; assay value, 1,600 ounces silver per ton; 1 drift 30 feet in length and 20 feet from surface.

Bismark. William Frazer, proprietor; office Ouray; claim 300 by 1,500 feet; discovered 1877; located in Bear Creek Basin, Upper San Miguel mining district, 14 miles from Ouray; vertical vein, trending north and south, width 6 feet, pay vein 18 inches; nature of ore, gray copper and galena; average value, 160 ounces silver per ton; 1 drift 28 feet in length and 35 feet from surface; yield, $300.

Black Diamond. Owned by W.H. Lilley & Co.; office Ouray; claim 300 by 1,500 feet; discovered 1876; located in Sneffels Basin, Mount Sneffels mining district, 9 miles from Ouray; vertical vein, trending northwest and southeast, width 12 feet, pay vein 4 feet; nature of ore, gray copper and ruby silver; average value, 70 ounces silver per ton; 1 shaft 40 feet deep, 1 drift 60 feet in length.

Black Prince. C. Andrews & Co., proprietors; office Ouray; claim 300 by 1,500 feet; discovered 1878; located on Uncompahgre Creek, Uncompahgre mining district, 4 miles from Ouray; course of vein, east and west, width 8 feet, pay vein 30 inches, pitch 30 degrees; nature of ore, galena and gray copper; assay value, 100 ounces silver per ton; 1 shaft 13 feet deep.

Black and Brown Bear. R.E. Scott & Co., proprietors; office Ophir; claim 300 by 3,000 feet; discovered 1878; located on Silver Mountain, Iron Springs mining district, 5 miles from Ophir and 16 miles from Ouray; vertical vein, trending north and south, width 8 feet; nature of ore, gray copper and galena; assay value, 80 ounces silver per ton; 1 adit 15 feet in length.

Black Swan. Owned by L.J. Rustin & Co., of Ouray; claim 300 by 1,500 feet; discovered 1876; located on Uncompahgre Creek, 5 miles from Ouray; course of vein, northeast and southwest, width 10 feet; nature of ore, galena; assay value, 45 ounces silver per ton; 1 shaft 12 feet and 1 tunnel 20 feet in length.

Blazer. Owned by A.W. Neumeyer and J.B. Ward; office Ophir; claim 300 by 1,500 feet; discovered 1878; located on Silver Mountain, Iron Springs mining district, 1 mile from town of Ophir and 12 miles from Ouray; course of vein, northeast and southwest, width 4 feet, pay vein 8 inches, pitch 30 degrees; nature of ore, gray copper and galena; 1 adit 15 feet in length.

Blue Grass. Owned by L.W. Tevis, M.D. Crow, and others; office Ouray; claim 300 by 1,350 feet; discovered 1877; located in Virginia Basin, Mount Sneffels mining district, 9 miles from Ouray; vertical vein, trending northeast and southwest, width 4 feet, pay vein 14 inches; nature of ore, galena and gray copper; assay value, 250 ounces silver per ton; 2 shafts, 14 and 40 feet, respectively.

Blue Jacket. R.W. Moffat and W. Chandler, proprietors; office Ouray; claim 300 by 1,500 feet; discovered 1877; located on Cline Mountain, Uncompahgre mining district, 3 miles from Ouray; course of vein, northwest and southeast, width 3 feet, pay vein 14 inches, pitch 20 degrees; nature of ore, galena and gray copper; assay value, 68 ounces silver per ton; 1 drift 20 feet in length and 60 feet from surface.

Blue Lick. M. Phelps and J. Britton, proprietors; office Silverton; claim 300 by 1,500 feet; discovered 1877; located on Bear Creek, Uncompahgre mining district, 4 miles from Ouray; vertical vein, trending northeast and southwest, width 10 feet, pay vein 30 inches; nature of ore, galena and gray copper; assay value, 82 ounces silver per ton; 1 adit 15 feet in length with a 12-foot face.

Bowling Green. Owned by L.B. Clay and Norfolk and Ouray Mining Co.; office Ouray; claim 300 by 1,500 feet; discovered 1877; located in Defiance Gulch, Mount Sneffels mining district, 9 miles from Ouray; course of vein, north and south, width 7 feet, pay vein 6 feet; nature of ore, gray copper, carbonate and galena; average value, 180 ounces silver per ton; 2 drifts, 30 and 60 feet in length, respectively.

Buckeye Girl. Owned by Francis G. and J.H. King, Leonard C. Kilham, and D.D. Mallory; office Denver; claim 300 by 1,500 feet; discovered September 11 and recorded October 30, 1875; located on the east side of Buckeye Mountain, 1 mile west of Imogene Gulch, in Mount Sneffels mining district, 8 miles from Ouray; vertical vein, trending northwest and southeast, width 5 to 12 feet, pay vein 6 to 20 inches; nature of ore, galena and gray copper; assay value, 20 to 500 ounces silver per ton; developed by a tunnel 30 feet and a surface opening 15 feet in length. This is the oldest of 6 other claims, located on this lode. It is easy of access, has a good waterpower, and the surrounding country is covered with timber. The mountain rises above the mine at an angle of about 40 degrees, affording a fine opportunity to gain depth by drifting and tunneling.

Byron. Charles Thurmond, proprietor; office Lake City; claim 300 by 1,500 feet; discovered 1877; located on Engineer Mountain, Uncompahgre mining district, 8 miles from Ouray; vertical vein, trending northwest and southeast, width 4 feet, pay vein 10 inches; nature of ore, galena; assay value, 260 ounces silver per ton; 2 drifts 40 feet each.

Calaveras. Mullen & Gilchrist, proprietors; office Silverton; claim 300 by 1,500 feet; located 1877; located 4 miles from Upper San Miguel City and 15 miles from Ouray; vertical vein, trending east and west, width 4 feet, pay vein 10 inches; nature of ore, gray copper; average value, 90 ounces silver per ton; 1 adit 22 feet in length and 12 feet from surface.

Cash. Owned by E.C. Abbott; office Ouray; claim 300 by 1,500 feet; discovered 1877; located on Marshall Creek, Upper San Miguel mining district, 10 miles from Ouray; vertical vein, trending east and west, width 23 feet; nature of ore, gray copper and galena; assay value, 2,100 to 2,240 ounces silver per ton; 1 tunnel 30 feet in length, intersecting vein at a depth of 35 feet, 1 level 40 feet in length.

Cash. Owned by R.E. Scott & Co.; office Ophir; claim 300 by 1,500 feet; discovered 1878; located on Silver Mountain, Upper San Miguel mining district, 5 miles from Miguel City and 14 miles from Ouray; vertical vein, trending north and south, width 5 feet, pay vein 14 inches; nature of ore, galena; assay value, 75 ounces silver per ton; 1 adit 14 feet in length.

Celt. Owned by W. Weston & Co., of Ouray; claim 300 by 1,500 feet; discovered 1876; located on Bear Creek, 4 miles from Ouray; course of vein, north and south, width 10 feet; nature of ore, galena and gray copper; assay value, 60 to 185 ounces silver per ton; development, 1 adit 30 feet in length and 20 feet from surface.

Censor. Owned by J. Rinker & Co., of Ouray; claim 300 by 1,500 feet; discovered 1876; located on Sneffels Creek, 8 miles from Ouray; course of vein, east and west, width 6 feet, pay vein 2 feet; nature of ore, galena and gray copper; assay value, 200 ounces silver per ton; development, 1 adit 40 feet in length and 25 feet from surface.

Centuple. Owned by S.M. Ransom, W. Turner, and C.A. Ward; office Ouray; claim 300 by 1,500 feet; located 1875; located on Bear Mountain, Uncompahgre mining district, 4 miles from Ouray; vertical vein, trending northeast and southwest, width 20 inches, pay vein 14 inches; nature of ore, galena; 1 shaft 20 feet deep.

Charles Morris. Charles Morris, proprietor; office Ouray; claim 300 by 1,500 feet; discovered 1876; located on Uncompahgre Creek, Uncompahgre mining district; vertical vein, trending northwest and southeast, width 6 feet, pay vein 18 inches; nature of ore, galena; average value, 25 ounces silver per ton; 1 drift 20 feet in length and 150 feet from surface.

Charlotte Hill. Owned by S.M Ransom, T. Turner, and C.A. Ward; office Ouray; claim 300 by 1,500 feet; discovered 1875; located on Bear Mountain, Uncompahgre mining district, 4 miles from Ouray; course of vein, northeast and southwest, width 2 feet, pay vein 18 inches; nature of ore, galena and gray copper; assay value, 36 ounces silver per ton; 2 adits, 15 feet each.

Chief Deposit and Caribou. Owned by Francis G. and J.H. King, Leonard C. Kilham, and D.D. Mallory; office Denver; claim 300 by 2,900 feet; the former mine was discovered September 15 and recorded October 30, 1875; the latter was discovered October 2, and recorded October 30, 1875; these claims join each other, are located on the spur of Buckeye Mountain between Imogene Gulch and Silver Creek, 8 miles from Ouray, and in the center of the largest lode in the district, having every advantage for working and development. The vein is vertical, trends northwest and southeast, width 8 to 16 feet, pay vein 3 to 8 feet; nature of ore, gray copper, galena and sulphurets of iron and copper; assay value 50 to 1,500 ounces silver per ton; development 2 tunnels 20 feet each, and 7 open cuts aggregating 130 feet in length, ranging from 9 to 20 feet in depth. This is a well-defined true fissure vein and lies in a trachyte formation. It is located for a distance of over 2 miles, showing good ore wherever opened. The surface ground of these mines and the adjoining country is covered with timber, suitable for mining and other purposes, and within 1,000 feet of the Caribou is a good waterpower.

Cincinnati. Owned by Samuel Wing, E.L. Hess, and others; office Ouray; claim 300 by 1,500 feet; discovered 1877; located in Defiance Gulch, Mount Sneffels mining district, 9 miles from Ouray; course of vein, northeast and southwest, width 8 feet, pitch 10 degrees; nature of ore, gray copper and ruby silver; assay value, 340 ounces silver per ton; 1 shaft 15 feet deep.

Circassian. Owned by Francis G. and J.H. King, Leonard C. Kilham, and D.D. Mallory; office Denver; claim 300 by 1,500 feet; discovered may 30, and recorded August 28, 1876; located on Old Stony Mountain, 1 mile up Mount Sneffels Creek, 8 miles from Ouray; vertical vein, trending east and west, width 12 feet, pay vein 4 feet; nature of ore, galena, gray copper, brittle silver, iron and copper pyrites, and silver glance; assay value, 60 to 10,000 ounces silver per ton; development, a tunnel 24 feet in length, which intersects the vein at a depth of 82 feet and an open cut 15 feet in length, with a 14-foot face. This property is easy of access, convenient to a fine waterpower; and the country surrounding it is covered with timber suitable for all purposes.

Cliff. John Billings & Co., proprietors; office Ouray; claim 300 by 1,500 feet; discovered 1875; located on Uncompahgre Creek, Uncompahgre mining district, 7 miles from Ouray; vertical vein, trending northwest and southeast, width 3 feet, pay vein 4 inches; nature of ore, galena, gray copper and iron pyrites; assay value, 240 ounces silver per ton; 1 drift 28 feet in length and 40 feet from surface.

Cliff. Owned by L. Coplen & Co.; office Lake City; claim 300 by 1,500 feet; discovered 1877; located on Engineer Mountain, Uncompahgre mining district, 6 miles from Ouray; vertical vein, trending north and south, width 20 inches, pay vein 9 inches; nature of ore, copper pyrites and galena; assay value, 140 to 700 ounces silver per ton; 1 drift 40 feet in length and 25 feet from surface.

Climax. Albert Colvin and J.B. Craige, proprietors; office Ouray; claim 300 by 1,500 feet; discovered 1876; located on Mount Abram, Uncompahgre mining district, 7 miles from Ouray; course of vein, north and south, width 10 feet, pitch 10 degrees; nature of ore, gray copper and brittle silver, average value, 100 ounces silver per ton, 1 adit 25 feet in length, with 20-foot face.

Climax. Owned by A. Danforth, J. Russell, and J. Coplen; office Ouray; claim 300 by 1,500 feet; located on Cline Mountain, Uncompahgre mining district, 5 miles from Ouray; course of vein, north and south, width 20 feet, pay vein 6 feet, pitch 10 degrees east; nature of ore, gray copper and galena; average value, 50 ounces silver per ton; main shaft, 20 feet and 1 other shaft 15 feet deep.

Clipper. Owned by M.J. Alkire and John Meek; office Ophir; claim 300 by 1,500 feet; discovered 1878; located on Silver Mountain, Iron Springs mining district, 1 mile from town of Ophir and 12 miles from Ouray; course of vein, northeast and southwest, width 5 feet, pay vein 15 inches, pitch 30 degrees; nature of ore, free gold; 1 shaft 20 feet deep, 1 adit 25 feet, and 1 tunnel 25 feet in length.

Cloud Drift. Charles Hopkins and E.N. Groupe, proprietors; office Ouray; claim 300 by 1,500 feet; discovered 1876; located on Canyon Creek, Uncompahgre mining district, 6 miles from Ouray; vertical vein, trending northwest and southeast, width 7 feet, pay vein 15 inches; nature of ore, gray copper and galena; assay value, 400 ounces silver per ton; 2 adits, 12 and 15 feet in length, respectively.

Comet. Owned by T. Terrill and M. Rush; office Ophir; claim 300 by 1,500 feet; discovered 1878; located on Silver Mountain, Upper San Miguel mining district, 4 miles from Miguel City and 15 miles from Ouray; vertical vein, trending north and south, width 5 feet, pay vein 20 inches; nature of ore, gray copper and sulphurets; assay value, 200 ounces silver per ton; 1 shaft 15 feet deep.

Commander. Owned by Henry Thomas and others; claim 300 by 1,500 feet; discovered 1877; located on Eagle Mountain, 12 miles from Ouray; course of vein, north and south, width 20 inches, pitch 15 degrees; nature of ore, brittle silver and gray copper; assay value, 2,300 ounces silver per ton; development, 1 shaft 25 feet deep.

Comstock. T. Shaw & Co., proprietors; office Ouray; claim 300 by 1,500 feet; discovered 1876; located on cline Mountain, Uncompahgre mining district, 3 miles from Ouray; vertical vein, trending north and south, width 4 feet, pay vein 18 inches; nature of ore, gray copper and galena; average value, 45 ounces silver per ton; 1 adit 15 feet in length, with 10-foot face.

Continental. James Tolliver and G. Brown, proprietors; office Ouray; claim 300 by 1,500 feet; discovered 1876; located on Cline Mountain, Uncompahgre mining district, 6 miles from Ouray; course of vein, northeast and southwest, width 5 feet, pay vein 15 inches, pitch 15 degrees; nature of ore, gray copper and galena; assay value, 125 ounces silver per ton; 2 shafts, 10 and 16 feet, respectively.

Crescent. Charles Morris, proprietor; office Ouray; claim 300 by 750 feet; discovered 1876; located on Bear Creek, Uncompahgre mining district, 2 miles from Ouray; course of vein, north and south, width 5 feet, pay vein 23 inches, pitch 15 degrees; nature of ore, copper pyrites and galena; average value, 40 ounces silver per ton; 1 drift 22 feet in length and 18 feet from surface.

Creve Cour. Owned by F. Fodde, W. Bruner, F.B. Walton, and H. Clark; office Ophir; claim 300 by 1,500 feet; discovered 1878; located on Yellow Mountain, Iron Springs mining district, 1 mile from town of Ophir and 12 miles from Ouray; vertical vein, trending northeast and southwest, width 8 feet, pay vein 28 inches; nature of ore, gray copper and galena; assay value, 200 ounces silver per ton; main shaft 25 feet deep, 1 adit 10 feet in length.

Cumberland. F. Herbst & Co., proprietors; office Ouray; claim 300 by 1,500 feet; discovered 1876; located on Mount Hendricks, Mount Sneffels mining district, 9 miles from Ouray; vertical vein, trending northwest and southeast, width 7 feet, pay vein 18 inches; nature of ore, gray copper, ruby silver, and galena; average value, 400 ounces silver per ton; 1 shaft 20 feet deep.

Denver. Owned by M.W. Dresser and G.H. Smith; office Ouray; claim 300 by 1,500 feet; discovered 1876; located on Abraham Mountain, Uncompahgre mining district, 3 miles from Ouray; vertical vein, trending northeast and southwest, width 6 feet, pay vein 1 foot; nature of ore, gray copper and galena; average value, 100 ounces silver per ton; 1 drift 130 feet in length and 100 feet from surface.

Dictator. Owned by Thomas Wallace and George Autrie, of Ouray; claim 300 by 1,500 feet; discovered 1876; located on Red Mountain, 3 miles from Ouray; course of vein, north and south, width 6 feet, pay vein 15 inches; nature of ore, gray copper and galena; assay value, 175 ounces silver per ton; development, 2 drifts, 20 and 28 feet in length, respectively.

Dora. M. Hendricks and G.H. Martin, proprietors; office Ophir; claim 300 by 1,500 feet; discovered 1878; located on the Lake Fork of the San Miguel River, Iron Springs mining district, 2 miles from town of Ophir and

12 miles from Ouray; vertical vein, trending east and west, width 40 feet, pay vein 6 inches; nature of ore, galena and ruby silver; assay value, 200 ounces silver per ton; 1 adit 15 feet in length.

Dora. Owned by L.B. Clay and Co.; office Ouray; claim 300 by 1,500 feet; discovered 1878; located in Defiance Gulch, Mount Sneffels mining district, 9 miles from Ouray; course of vein, each and west, width 6 feet, pay vein, 8 inches, pitch 15 degrees; nature of ore, gray copper and galena; assay value, $500 per ton; 1 drift, 30 feet in length and 20 feet from surface.

Dutch Boy. R.B. Town & Co., proprietors; office Ouray; claim 300 by 1,500 feet; discovered 1875; located on Mount Sneffels, Mount Sneffels mining district, 9 miles from Ouray; vertical vein, tending northwest and southeast, width 5 feet, pay vein 18 inches; nature of ore, ruby and brittle silver; assay value, 500 to 1,500 ounces silver per ton; 1 tunnel 70 feet in length, intersecting vein at a depth of 45 feet, 1 level 45 feet in length and 40 feet from surface.

E Pluribus Unum. Owned by Charles Hill, J.W. Mills & Co.; office Howardsville; claim 300 by 1,500 feet; discovered 1877; located on Silver Mountain, Iron Springs mining district, 1 mile from town of Ophir and 12 miles from Ouray; course of vein, northwest and southeast, width 3 feet, pay vein 18 inches, pitch 15 degrees; nature of ore, galena and sulphurets; average value, $50 per ton; 1 shaft 12 feet deep.

Eclipse. M.P. Nutting and E. Brolaskie, proprietors; office Ouray; claim 300 by 1,500 feet; discovered 1876; located on Engineer Mountain, Uncompahgre mining district, 7 miles fro Ouray; vertical vein, trending northwest and southeast, width 30 inches, pay vein 1 foot; nature of ore, galena and gray copper; assay value 500 ounces silver per ton; 1 tunnel 20 feet in length.

El Dorado. Owned by James K. Reid & Bros.; office Miguel City; claim 300 by 1,500 feet; discovered 1877; located on Marshall Creek, Upper San Miguel mining district, 14 miles from Ouray; course of vein, north and south, width 4 feet, pay vein 30 inches, pitch 20 degrees; nature of ore, galena and gray copper; assay value, 600 ounces silver per ton; 1 adit 20 feet in length with 15-foot face.

Eldorado. William Shultz and J.W. Rinker, proprietors; office Ouray; claim 300 by 1,500 feet; discovered 1877; located on Old Stony Mountain,

Uncompahgre mining district, 9 miles from Ouray; course of vein, northwest and southeast; width 7 feet, pay vein 30 inches; nature of ore, ruby and brittle silver; average value, 640 ounces silver per ton; 1 drift 75 feet in length, and 40 feet from surface; yield, $3,000.

Emma. O.W. Jameson and H.E. Volkman, proprietors; office Ouray; claim 300 by 1,500 feet; discovered 1876; located on Mount Emma, Mount Sneffels mining district, 9 miles from Ouray; vertical vein, trending east and west, width 3 feet pay vein 2 feet; nature of ore, galena and gray copper; 1 adit, 20 feet in length, with 25-foot face.

Etta. Owned by Joseph C. Tevis and Bros., and M. Dolphin; office Ouray; claim 300 by 1,500 feet; discovered 175; located on Bear Mountain, Uncompahgre mining district, 12 miles from Ouray; course of vein, northeast and southwest, width 4 feet, pay vein 2 feet, pitch 20 degrees; nature of ore, galena and gray copper; average value, 160 ounces silver per ton; 1 tunnel 12 feet in length intersecting vein at a depth of 25 feet, 1 level 15 feet in length and 25 feet from surface.

Eureka. Owned by M. McGuigan and G. Holt; office Miguel City; claim 300 by 1,500 feet; discovered 1878; located on the Upper San Miguel River, 5 miles from the Upper San Miguel River, 5 miles from Miguel City and 13 miles from Ouray; course of vein, north and south; width 8 feet, pay vein 30 inches; nature of ore, gray copper and galena; assay value, 100 ounces silver per ton; 1 shaft 14 feet deep.

Eureka. Charles Kirk, proprietor; office Silverton; claim 300 by 1,500 feet; discovered 1878; located on Yellow Mountain, Iron Springs mining district, 2 miles from town of Ophir and 14 miles from town of Ouray; vertical vein, trending north and south, width 30 inches, pay vein 6 inches; nature of ore, gray copper; assay value, 300 ounces silver per ton; 1 adit 10 feet in length, with 10-foot face.

Exedos [sic]. Owned by G.W. Hight, C.C. Gaines, and O. Wright; office Ophir; claim 300 by 1,500 feet; discovered 1877; located on Silver Mountain, Iron Springs mining district, near town of Ophir, 12 miles from Ouray; vertical vein, trending east and west, width 22 inches, pay vein 14 inches; nature of ore, free gold; assay value, $2,500 per ton; 1 drift 30 feet in length and 25 feet from surface.

Eyrie. M. Williams and C.C. Jorden, proprietors; office Ouray; claim 300 by 1,500 feet; discovered 1876; located on Red Mountain,

Uncompahgre mining district, 5 miles from Ouray; vertical vein, trending east and west, width 10 feet; nature of ore, galena and copper pyrites; assay value, 80 ounces silver per ton; 2 shafts, 10 and 18 feet in depth, respectively.

Fidelity. John Ray & Co., proprietors; office Ouray; claim 300 by 1,500 feet; located 1877; located on Mount Sneffels, Mount Sneffels mining district, 9 miles from Ouray; course of vein, north, width 30 inches, pay vein 6 inches, pitch 25 degrees; nature of ore, gray copper and galena, assay value, 400 ounces silver per ton; 1 adit 25 feet in length with 20-foot face.

Fire Fly. Owned by William Frazer & Co.; office Ouray; claim 300 by 1,500 feet; discovered 1877; located on Bear Creek, Upper Miguel mining district, 14 miles from Ouray; vertical vein, trending northwest and southeast, width 3 feet, pay vein 10 inches; nature of ore, gray copper and brittle silver; assay value, 1,000 ounces silver per ton; 1 tunnel 30 feet in length, intersecting vein at a depth of 20 feet.

Flag. Owned by J.L. Arnold, J.M Mulholland & Co.; office Ophir; claim 300 by 1,500 feet; discovered 1878; located on Yellow Mountain, Iron Springs mining district, 1 mile from town of Ophir and 12 miles from Ouray; vertical vein, trending northwest and southeast, width 2 feet, pay vein 6 inches; nature of ore, gray copper and galena; average value, 80 ounces silver per ton; 1 drift 43 feet in length and 50 feet from surface.

Flagg. Owned by J.S. Resor and W. Mitchell; office Silverton; claim 300 by 1,500 feet; discovered 1878; located on Turkey Creek, Upper San Miguel mining district, 2 miles from Miguel City and 14 miles from Ouray; vertical vein, trending northeast and southwest, width 7 feet, pay vein 15 inches; nature of ore, gray copper and galena; 1 shaft 18 feet deep.

Flagstaff. Owned by W.R. Watson, W. Storey, and James H. Lester; office Ouray; claim 300 by 1,500 feet; discovered 1877; located on Bear Creek Range, Uncompahgre mining district, 4 miles from Ouray; vertical vein, trending north and south, width 16 feet, pay vein 18 inches; nature of ore, gray copper and sulphurets; average value, 90 ounces silver per ton; 1 tunnel 30 feet in length, interesting the vein at a depth of 25 feet, 1 level 15 feet in length.

Flagstaff. Owned by N. Peterson and E. McGinnis; office Ouray; claim 300 by 1,500 feet; discovered 1878; located on Stony Mountain,

Mount Sneffels mining district, 9 miles from Ouray; vertical vein, trending north and south, width 8 feet, pay vein 20 inches; nature of ore, ruby and brittle silver; 1 drift 70 feet in length and 40 feet from surface.

Flagstaff. Owned by F. Herbst and William A. Clark; office Ouray; claim 300 by 1,500 feet; discovered 1876; located on Mount Hendricks, Mount Sneffels mining district, 9 miles from Ouray; vertical vein, trending northwest and southeast, width 7 feet, pay vein, 18 inches; nature of ore, galena and gray copper; average value, 250 ounces silver per ton; 1 drift 25 feet in length and 18 feet from surface.

Free Gold. Owned by William B. Fonda and F. Gardner & Co.; office Ophir; claim 300 by 1,500 feet; discovered 1878; located on Silver Mountain, Upper San Miguel mining district, 3 miles from Miguel City and 12 miles from Ouray; course of vein, north and south, width 3 feet, pay vein 5 inches; nature of ore, free gold; assay value 22 ounces gold per ton; 1 shaft 12 feet deep, 1 drift 100 feet in length and 200 feet from surface.

Fulton. Owned by L. Sears and H.F. Shanks; office Ophir; claim 300 by 1,500 feet; discovered 1877; located on Silver Mountain, Iron Springs mining district, 1 mile from town of Ophir and 12 miles from Ouray; vertical vein, trending northeast and southwest, width 10 feet, pay vein 1 foot; nature of ore, gray copper and galena; assay value, 75 ounces silver per ton; 1 adit 12 feet in length with 12-foot face.

Geneva. Owned by the Staatsburg Silver Mining Co.; office Silverton; claim 300 by 1,500 feet; discovered 1878; located on Silver Mountain, Iron Springs mining district, 1 mile from town of Ophir and 12 miles from Ouray; vertical vein, trending northeast and southwest, width 4 feet, pay vein 8 inches; nature of ore, galena; average value, 100 ounces silver per ton; 1 adit 20 feet in length, 1 level 50 feet and 20 feet from surface; yield, $600.

Gold King. Owned by John and David A. Munn, and Joseph E. Lacome; office Silverton; claim 300 by 1,500 feet, patented; discovered 1878; located on Silver Mountain, Iron Springs mining district, near town of Ophir and 12 miles from Ouray; vertical vein, trending north and south, width 6 feet, pay vein 6 inches; nature of ore, free gold; average value, $2,000 per ton; 1 shaft 25 feet deep, 1 tunnel 200 feet in length, intersecting vein at a depth of 200 feet; yield, $5,000.

Gold Queen. Owned by B.B. Haddox, J.B. Laffoon, and Munn Bros.; office Ophir; claim 300 by 1,500 feet; discovered 1878; located on Silver Mountain, Upper San Miguel mining district, 2 miles from town of Ophir and 12 miles from Ouray; course of vein, north and south, width 6 feet, pay vein 6 inches; nature of ore, free gold; value, $1,200 per ton; 1 shaft 40 feet deep, 1 tunnel 40 feet in length, intersecting vein at a depth of 35 feet, 1 level 150 feet in length and 35 feet from surface; yield, $5,000.

Golden Anchor Mining and Tunneling Co. Incorporated April 15, 1878; capital stock, $250,000, in 2,500 shares, $100 each; office Silverton; trustees: Dempsey Reese, Emile Charest, G.D. Gove, F.M. Mindenhall, and H.W. Morrison.

Gramont. Owned by A. Peck, C. Peck, and J.N. Rolfe; office Ouray; claim 300 by 1,500 feet; discovered 1877; located on Bear Creek, Uncompahgre mining district, 5 miles from Ouray; vertical vein, trending north and south, width 4 feet, pay vein 8 inches; nature of ore, gray copper and galena; assay value, 200 ounces silver per ton; 1 shaft 20 feet deep.

Grand Era. Owned by W.R. Watson, W. Storey, and James H. Lester; office Ouray; claim 300 by 1,500 feet; discovered 1877; located on Bear Creek Range, Uncompahgre mining district, 4 miles from Ouray; course of vein, northwest and southeast, width 15 feet, pay vein 2 feet, pitch 25 degrees, nature of ore, galena and iron pyrites; 1 drift 15 feet in length and 15 feet from surface.

Grand View. Owned by Munn Bros.; office Silverton; claim 300 by 1,500 feet; discovered 1875; located on Silver Mountain, Iron Springs mining district, 35 miles from Ouray; vertical vein, trending northeast and southwest, width 8 feet; nature of ore, sulphurets and galena, average value, 200 ounces silver per ton; 2 shafts, 15 and 25 feet in depth, respectively; yield, $2,000.

Greenback. Owned by Philip Clark, of Silverton; claim 300 by 1,500 feet; discovered 1878; located on Yellow Mountain, Iron Springs mining district, 2 miles from town of Ophir and 14 miles from Ouray; vertical vein trending east and west, width 10 feet, pay vein 20 inches; nature of ore, carbonate and sulphurets; assay value, 80 ounces silver per ton; 1 shaft 15 feet deep.

Greenback. Owned by M.J. Alkire, J.H. Sears, and J. Meek; office Ophir; claim 300 by 1,500 feet; discovered 1878; located on Silver

Mountain, Iron Springs mining district, 1 mile from town of Ophir and 12 miles from Ouray; vertical vein, trending northeast and southwest, width 6 feet, pay vein 20 inches; nature of ore, free gold; 1 drift 80 feet in length and 75 feet from surface.

Grub Stake. W. Morgan and E. Van Pelt, proprietors; office Ouray; claim 300 by 1,500 feet; discovered 1876; located on Bear Creek Range, Uncompahgre mining district, 4 miles from Ouray; vertical vein, trending east and west, width 7 feet, pay vein 15 inches; nature of ore, galena and iron pyrites; assay value, 100 ounces silver per ton; 2 drifts aggregating 42 feet in length, 10 and 24 feet from surface, respectively.

Grub Stake. Owned by M.W. Dresser and G.H. Smith; office Ouray; claim 300 by 1,500 feet; discovered 1877; located on the Uncompahgre River, 1 mile from Ouray; vertical vein, trending north and south, width 5 feet, pay vein 20 inches; nature of ore, copper pyrites and galena, assay value, 100 ounces silver per ton; 1 adit 30 feet and 1 tunnel 50 feet in length.

Hawkeye. A.K. Prescott & Co., proprietors; office Del Norte; claim 300 by 1,500 feet; discovered 1877; located on Mount Sneffels, Mount Sneffels mining district, 8 miles from Ouray; course of vein, northeast and southwest, width 3 feet, pay vein 18 inches, pitch 20 degrees; nature of ore, ruby silver and gray copper; 1 adit 20 feet in length.

Hercules. James O'Boyle and G.F. Cork, proprietors; office Ouray; claim 300 by 1,500 feet; discovered 1877; located on Buckeye Mountain, Mount Sneffels mining district, 8 miles from Ouray; vertical vein, trending north and south, width 30 inches; nature of ore, gray copper and ruby silver; assay value, 800 ounces silver per ton; 1 drift 20 feet in length.

Hidden Treasure. Owned by George A. Brantley and Robert W. Bell; office Ouray; claim 300 by 1,500 feet; discovered 1878; located in Imogene Basin, Mount Sneffels mining district, 7 miles from Ouray; vertical vein, trending northeast and southwest, width 5 feet, pay vein 4 feet; nature of ore, gray copper and galena; average value, 172 ounces silver per ton; 1 drift 125 feet in length and 100 feet from surface; yield, $3,000.

Highland Boy. Owned by R.M. Mercer, H. Blake, and others; office Ouray; claim 300 by 1,500 feet; discovered 1876; located on Red Mountain, Uncompahgre mining district, 3 miles from Ouray; course of vein, northeast and southwest, width 7 feet, pay vein 2 feet, pitch 25 degrees; nature of ore, galena and gray copper; assay value, 200 ounces silver per ton; 1 shaft 12 feet deep, 1 adit 15 feet in length.

Highland Chief. Owned by D.P. Quinn and A.W. Richardson; office Denver; claim 300 by 1,500 feet; located near the mouth of Silver Creek, Mount Sneffels mining district, 7 miles from Ouray; vertical vein, trending northwest and southeast, width 3 to 4 feet, pay vein 6 to 18 inches; nature of ore, gray copper and galena; assay value, 40 to 800 ounces silver per ton; developed by a tunnel 50 feet in length.

Highland Lassie. Owned by Francis G. and J.H. King, Leonard C. Kilham, and D.D. Mallory; office Denver; claim 300 by 1,500 feet; located in Mount Sneffels mining district, 7 miles southwest from Ouray; vertical vein, trending west 30 degrees north, width 3 feet, pay vein 2 feet; nature of ore, gray copper, galena, iron and copper pyrites; average value, 20 to 1,000 ounces silver per ton; developed by a tunnel, 15 feet and an open cut, 20 feet in length. This lode is over 1 mile in length, and is one of the most promising in the district. The surface is covered with timber suitable for all purposes and beside it is a good waterpower sufficient to run an air compressor or other machinery.

Hindoo. Owned by W.B. Fonda, David A. Munn, and E.L. Crim; office Ophir; claim 300 by 230 feet; discovered 1877; located on Silver Mountain, Iron Springs mining district, 2 miles from town of Ophir and 13 miles from Ouray; vertical vein, trending northeast and southwest, width 5 feet, pay vein 1 foot; nature of ore, galena and sulphurets; average value, 194 ounces silver per ton; 1 shaft 15 feet deep.

Home Stake. Owned by J. Resor, R.C. King, and others; office Ouray; claim 300 by 1,500 feet; discovered 1878; located on Gold Run, Upper San Miguel mining district, 9 miles from Ouray; course of vein, north and south, width 8 feet, pay vein 2 feet, pitch 15 degrees; nature of ore, gray copper and ruby silver; assay value, 250 ounces silver per ton; 1 adit 12 feet in length.

Home Stake. Charles Johnson, proprietor; office Ouray; claim 300 by 1,500 feet; discovered 1876; located on Uncompahgre Creek, Uncompahgre mining district, 4 miles from Ouray; course of vein, northwest and southeast, width 8 feet, pay vein 3 feet, pitch 15 degrees; nature of ore, galena and sulphurets; assay value, 90 ounces silver per ton; 1 adit 15 feet in length with 12-foot face.

Hoosier Boy. George W. Spencer and Adam Beidle, proprietors; claim 300 by 1,500 feet; discovered 1878; located in Imogene Basin, Mount Sneffels mining district, 7 miles from Ouray; course of vein, northwest and southeast, width 3 feet, pay vein 1 foot, pitch 10 degrees; nature of ore, gray copper and galena; assay value, 800 ounces silver per ton; 1 shaft 10 feet deep.

Hoosier Girl. George W. Spencer, proprietor; office Ouray; claim 300 by 1,500 feet; discovered 1875; located in Imogene Basin, Mount Sneffels mining district, 7 miles from Ouray; course of vein, northwest and southeast, width 3 feet, pay vein 1 foot, pitch 10 degrees; nature of ore, gray copper and galena; assay value, 263 ounces silver per ton; 1 shaft 12 feet deep, 1 drift 100 feet in length; yield, $500.

Hornet. Joseph C. Tevis and Bros., proprietors; office Ouray; claim 300 by 1,500 feet; discovered 1875; located in Imogene Basin, Mount Sneffels mining district, 8 miles from Ouray; vertical vein, trending northwest and southeast, width 18 inches, pay vein 15 inches; nature of ore, galena, gray copper, and brittle silver; average value, 285 ounces silver per ton; 2 adits, 12 and 15 feet in length, respectively.

Hudson. Owned by M. Crane, J.C. Bartlett, and others; office Ouray; claim 300 by 1,500 feet; discovered 1875; located on Cline Mountain, Uncompahgre mining district, 6 miles from Ouray; vertical vein, trending northeast and southwest, width 4 feet, pay vein 3 feet; nature of ore, galena and gray copper; assay value, 25 to 200 ounces silver per ton; 1 shaft 20 feet deep, 1 adit 15 feet in length.

Ida. J.L. Haines and H.O. Wing, proprietors; office Silverton; claim 300 by 1,500 feet; discovered 1878; located on Silver Mountain, Iron Springs mining district, 1 mile from town of Ophir and 12 miles from Ouray; course of vein, northeast and southwest, width 3 feet, pay vein 1 foot, pitch 30 degrees; nature of ore, gray copper and iron pyrites; assay value, 200 ounces silver per ton; 2 shafts, 10 and 25 feet in depth, respectively.

Idaho. Owned by L. Sears and M. McCormick; office Ophir; claim 300 by 1,500 feet; discovered 1876; located on Yellow Mountain, Iron Springs mining district, near town of Ophir, 12 miles from Ouray; vertical vein, trending east and west, width 5 feet; nature of ore, gray copper and galena; 2 adits, 12 and 15 feet from surface, respectively, aggregating 40 feet in length.

Imogene. Owned by Francis G. and J.H. King, Leonard C. Kilham, and D.D. Mallory; office Denver; claim 300 by 1,500 feet; discovered September 13, and recorded September 30, 1878; located on the southeast side of Buckeye Mountain, 1,800 feet west of Imogene Gulch, in Mount Sneffels mining district, 9 miles from Ouray; vertical vein, trending northwest and southwest, width 4 feet, pay vein 1 to 4 inches; nature of ore, galena, gray copper, iron and copper pyrites; assay value, 56 to 1,378 ounces silver per ton; developed by a tunnel 103 feet in length, which intersects the vein at a depth of 60 feet. About 250 yards distant form the mine is a comfortable cabin, which, with the development of the property, cost $2,200. The mine is surrounded by timber and has a good waterpower. It was worked during the winters of 1875 and 1876, the altitude is 10,500 feet above the level of the sea, and the thermometer ranged from 8 degrees below to 30 degrees above zero.

Inca. M.P. Nutting & Co., proprietors; office Ouray; claim 300 by 1,500 feet; discovered 1875; located on Engineer Mountain, Uncompahgre mining district, 7 miles from Ouray; course of vein, northwest and southeast, width 8 feet, pitch 10 degrees; nature of ore, galena and gray copper; assay value, 260 ounces silver per ton; 2 drifts, 20 and 22 feet in length, respectively.

Inter-Ocean. Ernest Galt and Charles B. Frye, proprietors; office Ouray; claim 300 by 1,500 feet; discovered 1876; located on Poughkeepsie Mountain, Uncompahgre mining district, 5 miles from Ouray; course of vein, north and south, width 7 feet, pay vein 15 inches, pitch 10 degrees; nature of ore, galena and copper pyrites; assay value, 150 ounces silver per ton; 1 drift 28 feet in length.

Inverness. John Billings, proprietor; office Ouray; claim 300 by 1,500 feet; discovered 1876; located on Mount Abram, Uncompahgre mining district, 6 miles from Ouray; course of vein, northeast and southwest, width 10 feet, pitch 30 degrees; nature of ore, gray copper and brittle silver; assay value, 1,500 ounces silver per ton, 1 shaft 21 feet deep.

Iowa. J.W. Patterson and R. Kearns, proprietors; office, Ouray; claim 300 by 1,500 feet; discovered 1876; located on Bear Creek, Uncompahgre mining district, 4 miles from Ouray; course of vein, northwest and southeast, width 8 feet, pay vein 2 feet, pitch 25 degrees; nature of ore, galena and gray copper; assay value, 90 ounces silver per ton; 1 shaft 7 feet deep, 1 adit 12 feet in length with a 10-foot face.

Iron Sides. D. Durfee and J.R. Rounds, proprietors; office Ophir; claim 300 by 1,500 feet; discovered 1877; located on Silver Mountain, Iron Springs mining district, 2 miles from town of Ophir and 12 miles from Ouray; course of vein, east and west, width 20 feet, pitch 15 degrees; nature of ore, gray copper and galena; assay value, 300 ounces silver per ton; 1 shaft 4 feet deep.

Irving. J.B. Andrews and E.B. Cope, proprietors; office Ouray; claim 300 by 500 feet; discovered 1877; located on Treasure Hill, Mount Sneffels mining district, 9 miles from Ouray; vertical vein, trending northwest and southeast, width 6 feet; nature of ore, gray copper; assay value, 160 ounces silver per ton; 1 adit 30 feet in length.

Josephine. Owned by W. Ingersoll and J.C. Rankin; office Ouray; claim 300 by 1,500 feet; discovered 1877; located on Uncompahgre Creek, Uncompahgre mining district, 5 miles from Ouray; course of vein, north and south, width 24 feet, pay vein 3 feet, pitch 25 degrees; nature of ore, galena, zinc, copper and iron pyrites; assay value, 160 ounces silver per ton; 1 tunnel 20 feet in length, which intersects the vein 25 feet from the surface.

Junction. Owned by M.N. Funk & Co.; office Ouray; claim 300 by 1,500 feet; discovered 1875; located on Hardin Mountain, Uncompahgre mining district, 3 miles from Ouray; vertical vein, trending northwest and southeast, width 8 feet, pay vein 5 feet; nature of ore, gray copper and galena; assay value, 40 to 800 ounces silver per ton; 1 adit 20 feet in length with a 25-foot face.

Kentucky Belle. Andrew Cumtons, proprietor; office Ouray; claim 300 by 1,500 feet; discovered 1876; located on Uncompahgre Creek, Uncompahgre mining district, 5 miles from Ouray; course of vein, northeast and southwest, width 20 feet, pitch 15 degrees; nature of ore, galena; assay value, 80 ounces silver per ton; 1 shaft 14 feet deep, 1 adit 20 feet in length, with a 16-foot face.

Kentucky Giant. Owned by G. Purcells, J.A. White, and P. Cusick; office Ouray; claim 300 by 1,350 feet; discovered 1875; located on Abraham Mountain, Uncompahgre mining district, 6 miles from town of Ouray; course of vein, northeast and southwest, width 12 feet, pay vein 4 feet, pitch 10 degrees; nature of ore, brittle silver and iron pyrites; average value, 65 ounces silver per ton; 1 shaft 70 feet deep, 2 adits, 12 and 18 feet in length, respectively.

Keystone Placer. Located on the north fork of the San Miguel River, 3 miles from town of San Miguel and 23 miles from Silverton; comprises 60 acres of ground, worked by hydraulic pressure; has a flume 2,600 feet in length.

King of the Mountain. Owned by J.F. Robison and G.C. Horner; office Silverton; claim 300 by 1,500 feet; discovered 1877; located on Silver Mountain, Iron Springs mining district, near town of Ophir, 12 miles from Ouray; course of vein, north and south, width 2 feet, pay vein 1 foot, pitch 20 degrees southeast; nature of ore, free gold; average value, $40 per ton; 1 drift 25 feet in length and 18 feet from surface.

Lamar. Owned by L.B. Clay & Co.; office Ouray; claim 300 by 1,500 feet; discovered 1878; located in Defiance Gulch, Mount Sneffels mining district, 9 miles from Ouray; vertical vein, trending north and south, width 20 feet, pay vein 2 feet; nature of ore, gray copper and galena; 1 drift 30 feet in length and 30 feet from surface.

Last Chance. E. Morton, proprietor; office Ouray; claim 300 by 1,500 feet; discovered 1877; located on Bear Creek, Uncompahgre mining district, 4 miles from Ouray; vertical vein, trending north and south, width 5 feet, pay vein 15 inches; nature of ore, galena and gray copper; average value, 80 ounces silver per ton; 1 shaft 20 feet deep.

Lawrence. Owned by J.L. Arnold, J.M. Mulholland & Co.; office Ophir; claim 300 by 1,500 feet; discovered 1878; located on Silver Mountain, Iron Springs mining district, 1 mile from town of Ophir and 12 miles from Ouray; course of vein, northeast and southwest, width 7 inches, pitch 20 degrees; nature of ore, gray copper and galena; value 18 ounces silver and 2 ounces gold per ton; 1 tunnel 20 feet in length.

Lincoln Boy. Owned by O.W. Strouse and S.S. Austin; office Ouray; claim 300 by 1,500 feet; discovered 1875; located on Abraham Mountain, Uncompahgre mining district, 6 miles from Ouray; course of vein, northeast and southwest, width 6 feet, pay vein 2 feet, pitch 20 degrees; nature of ore, iron pyrites, gray copper, and antimony; assay value, 100 ounces silver per ton; 1 shaft 22 feet deep, 1 adit 12 feet in length and 12 feet from surface; shaft house, 14 by 14 feet.

Little Nellie. Owned by J.F. Robison and G.C. Horner; office Silverton; claim 300 by 1,500 feet; discovered 1877; located on Silver

Mountain, Iron Springs mining district, 1 mile from town of Ophir and 12 miles from Ouray; vertical vein, trending north and south, width 6 inches, pay vein 6 inches; nature of ore, gray copper and brittle silver; average value, 15 ounces silver per ton; 1 drift 24 feet in length and 16 feet from surface.

Little Robbie. Owned by William M. Friend & Co.; office Ouray; claim 300 by 1,500 feet; discovered 1877; located on Ruby Mountain, Mount Sneffels mining district, 7 miles from Ouray; course of vein, northeast and southwest, width 12 feet; pitch 40 degrees; nature of ore, gray copper and galena; assay value, 260 ounces silver per ton; 2 adits aggregating 40 feet in length, 1 drift 30 feet in length and 30 feet from surface.

Lone Star. Owned by Terrill & Co.; office Ophir; claim 300 by 1,500 feet; discovered 1878; located on Turkey Creek, San Miguel mining district, 3 miles from Miguel City and 14 miles from Ouray; course of vein, northwest and southeast; width 3 feet, pay vein 10 inches; pitch 20 degrees; nature of ore, copper pyrites and gray copper; assay value, 150 ounces silver per ton; 1 adit 10 feet in length with 10-foot face.

Lookout. R.W. Moffat, proprietor; office Ouray; claim 300 by 1,500 feet; discovered 1878; located on Canyon Creek, Uncompahgre mining district; near Ouray course of vein, northwest and southeast, width 40 feet, pitch 15 degrees; nature of ore, galena, brittle silver, and gray copper; assay value, 500 ounces silver per ton; 1 adit 20 feet in length with a 15-foot face.

Lookout. George Thomas, proprietor; office Ouray; claim 300 by 1,500 feet; discovered 1877; located on Engineer Mountain, Uncompahgre mining district, 7 miles from Ouray; vertical vein, trending northeast and southwest, width 15 feet, pay vein 3 feet; nature of ore, galena and iron pyrites; assay value, 100 ounces silver per ton; 1 adit 20 feet in length with a 20-foot face.

Lookout. A. Clare, proprietor; office Ouray; claim 300 by 1,500 feet; discovered 1878; located on Bear Creek; Uncompahgre mining district, 4 miles from Ouray; vertical vein, trending east and west, width 7 feet, pay vein 2 feet; nature of ore, gray copper and galena; assay value, 135 ounces silver per ton; 1 adit 15 feet in length and 12 feet from surface.

Louisiana. C. Coogan and G.O. Roberts, proprietors; office Ouray; claim 300 by 1,500 feet; discovered 1876; located on Treasure Hill, Mount Sneffels mining district, 9 miles from Ouray; course of vein, northwest and southeast, width 50 feet, pay vein 2 feet; pitch 30 degrees; nature of ore, galena, ruby and brittle silver; assay value, 100 to 400 ounces silver per ton; 3 shafts aggregating 130 feet in depth.

Mack. Owned by M. McCormick & Co.; office Ophir; claim 300 by 1,500 feet; discovered 1876; located on Yellow Mountain, Iron Springs mining district, 3 miles from town of Ophir and 14 miles from Ouray; course of vein, east and west, width 15 feet, pay vein 3 feet; nature of ore, iron pyrites and gray copper; assay value, 260 ounces silver per ton; 1 drift 45 feet in length and 30 feet from surface.

Magnet. Owned by William Frazer & Co.; office Ouray; claim 300 by 1,500 feet; discovered 1877; located in Bear Creek Basin, Upper San Miguel mining district, 14 miles from Ouray; course of vein, north and south, width 10 feet, pitch 10 degrees; nature of ore, gray copper; assay value, 260 ounces silver per ton; 1 tunnel 80 feet in length.

Magnolia. Owned by Clark, Sears & Co.; office Ophir; claim 300 by 1,500 feet; discovered 1878; located on Yellow Mountain, Iron Springs mining district, 1 mile from town of Ophir and 12 miles from Ouray; vertical vein, trending east and west, width 15 feet, pay vein 3 feet; nature of ore, gray copper and galena; assay value, 200 ounces silver per ton; 2 drifts, 27 and 40 feet in length, respectively.

Malta. C. Germansen and George Wood, proprietors; office Ouray; claim 300 by 1,500 feet; discovered 1875; located on Red Mountain, Uncompahgre mining district, 5 miles from Ouray; course of vein, northeast and southwest, width 12 feet, pitch 15 degrees; nature of ore, gray copper and brittle silver; average value, 200 ounces silver per ton; 1 tunnel 25 feet and 1 adit 15 feet in length.

Mary. Owned by Harry Diehl and G. Mish; office Ophir; claim 300 by 1,500 feet; discovered 1878; located on Silver Mountain, Iron Springs mining district, near town of Ophir and 12 miles from Ouray; course of vein, northeast and southwest, width 2 feet, pay vein 1 foot; nature of ore, free gold; assay value, 100 ounces gold per ton; 1 adit 25 feet in length with a 20-foot face.

Mary Belle. James W. Platt & Co., proprietors; office Eureka; claim 300 by 1,500 feet; discovered 1877; located on Stony Mountain, Mount Sneffels mining district, 9 miles from Ouray; vertical vein, trending east and west, width 4 feet, pay vein 1 foot; nature of ore, gray copper; 1 adit 18 feet in length and 20 feet from surface.

Mary Jane. J.C. Anderson & Bro.; proprietors; office Ouray; claim 300 by 1,500 feet; discovered 1875; located on Uncompahgre Creek, Uncompahgre mining district, 7 miles from Ouray; course of vein, north and south, width 12 feet, pay vein 3 feet, pitch 16 degrees; nature of ore, galena and gray copper; assay value, 80 ounces silver per ton; 1 shaft 15 feet deep, 1 drift 35 feet in length.

Mary Jane. Owned by J.D. Case and William Mercer; office Upper San Miguel City; claim 300 by 1,500 feet; discovered 1878; located on Gold Run, Upper San Miguel mining district, 9 miles from Ouray; course of vein, north and south, width 2 feet, pay vein 6 inches, pitch 15 degrees; nature of ore, gray copper; average value, 50 ounces silver per ton; 1 shaft 12 feet deep.

Melissa. J.P. Barrows and Co., proprietors; office Ouray; claim 300 by 1,500 feet; discovered 1878; located on Canyon Creek, Uncompahgre mining district, near town of Ouray; course of vein, northeast and southwest, width 20 feet, pitch 15 degrees; nature of ore, galena and gray copper; 1 adit 15 feet in length with an 11-foot face.

Mineral Farm. Owned by Norfolk and Ouray Mining Co.; M.S. Corbett, superintendent; office Ouray; claim 300 by 1,500 feet; discovered 1877; located on Canyon Creek, Uncompahgre mining district, 3 miles from Ouray; course of vein, north and south, width 3 feet, pay vein 16 inches; nature of ore, iron pyrites, gray copper, and galena; assay value, 300 ounces silver per ton; 1 drift 30 feet and 20 feet from surface.

Minnie. Owned by Theodore & Clark, of Ouray; claim 300 by 1,500 feet; discovered 1876; located in Uncompahgre Canyon, 6 miles from Ouray; course of vein, northeast and southwest, width 12 feet, pay vein 1 foot; nature of ore, gray copper; assay value, $438 per ton; development, 1 drift 80 feet deep.

Mishawaka. Owned by S.J. Dickey, W.H Rogers, and A. Hill of Colorado Springs; claim 300 by 1,500 feet; located on Engineer Mountain, Uncompahgre mining district, 7 miles from town of Ouray; course of vein, northeast and southwest, width 8 feet, pitch 45 degrees northeast, pay vein 14 inches; nature of ore, gray copper and copper pyrites; average value, 100 ounces silver per ton; 1 drift 100 feet in length.

Mocking Bird. Owned by M.W. Johnson; office Ophir; claim 300 by 1,500 feet; discovered 1878; located on Yellow Mountain, 1 mile from town of Ophir and 12 miles from Ouray; course of vein, northeast and southwest, width 4 feet, pay vein 18 inches; nature of ore, gray copper and galena; assay value 160 ounces silver per ton; 1 shaft 15 feet deep.

Mogul. James West & Son, proprietors; office Ouray; claim 300 by 1,500 feet; discovered 1878; located on Turkey Creek, 12 miles from Ouray; course of vein, north and south, width 4 feet, pitch 10 degrees; nature of ore, gray copper and brittle silver; assay value, 410 ounces silver per ton; 2 shafts, 19 and 24 feet in depth, respectively.

Mohave. William Rayner and N.J. Lee, proprietors; office Silverton; claim 300 by 1,500 feet; discovered 1878; located on Silver Mountain, 12 miles from Ouray; vertical vein, trending northeast and southwest, width 8 feet, pay vein 18 inches; nature of ore, bismuth and native silver, assay value, 100 to 2,400 ounces silver per ton; 1 shaft 16 feet deep.

Mohawk. Owned by E. McCarty & Co.; office Silverton; claim 300 by 1,500 feet; discovered 1876; located on Treasure Hill, Mount Sneffels mining district, 9 miles from Ouray; course of vein, northwest and southeast, width 4 feet, pay vein 3 feet, pitch 20 degrees; nature of ore, gray copper; assay value, 126 ounces silver per ton; 1 shaft 10 feet deep, 1 adit 30 feet in length.

Monument. J.P. Barrows & Co., proprietors; office Ouray; claim 300 by 1,500 feet; discovered 1877; located in Imogene Basin, Mount Sneffels mining district, 7 miles from Ouray; course of vein, northwest and southeast, width 9 feet, pay vein 4 feet, pitch 30 degrees; nature of ore, galena and gray copper; assay value, 50 to 200 ounces silver per ton; 1 drift 20 feet in length and 20 feet from surface.

Monument. R. Cusick, proprietor; office Ouray; claim 300 by 1,500 feet; discovered 1877; located on Old Stony Mountain, Mount Sneffels mining district, 8 miles from Ouray; vertical vein, trending east and west, width 4 feet, pay vein 10 inches; nature of ore, gray copper and ruby silver; assay value, 150 ounces silver per ton; 2 adits, 12 feet each.

Monitor. Owned by Harry Diehl & Co.; office Ophir; claim 300 by 1,500 feet; discovered 1878; located on Silver Mountain, Iron Springs mining district, near town of Ophir and 12 miles from Ouray; course of vein, northeast and southwest, width 2 feet, pay vein 10 inches; nature of ore, galena and gray copper; assay value, 100 ounces silver per ton; 1 drift 25 feet in length and 15 feet from surface.

Montezuma. Owned by M. Rich, Harry Beattie, and Abner Y. Davis; office Ophir; claim 300 by 1,500 feet; discovered 1878; located at the Howard Fork of the San Miguel River, near town of Ophir and 15 miles from Silverton; vertical vein, trending east and west, width 5 feet, pay vein 8 inches; nature of ore, gray copper and galena; average value, 135 ounces silver per ton; 1 shaft 26 feet deep; yield, $800.

Monticello. Owned by G.W. Hight and O. Wright; office Ophir; claim 300 by 1,500 feet; discovered 1878; located on Silver Mountain, Iron Springs mining district, near town of Ophir and 12 miles from Ouray; vertical vein, trending northeast and southwest, width 30 inches, pay vein 3 inches; nature of ore, free gold; assay value, $1,800 per ton; 1 shaft 10 feet deep, 2 adits, 10 and 40 feet in length, respectively.

Mountain Ram. Charles Josephs & Co., proprietors; office Ouray; claim 300 by 1,500 feet; discovered 1877; located on Engineer Mountain, Uncompahgre mining district, 7 miles from town of Ouray; vertical vein, trending east and west, width 10 feet, pay vein 5 feet; nature of ore, galena and gray copper; assay value, 80 ounces silver per ton; 1 adit 30 feet in length with a 22-foot face.

Myrtle. Owned by N.J. Downer, M.D. Cooper, P. Brown, and G.R. Porter; office Ouray; claim 300 by 1,500 feet; discovered 1877; located on Ruby Mountain, Mount Sneffels mining district, 7 miles from Ouray; course of vein, northeast and southwest, width 6 feet, pay vein 18 inches, pitch 15 degrees; nature of ore, sulphurets; assay value, 1,100 ounces silver per ton; 1 drift 60 feet in length and 20 feet from surface.

Nashville. Owned by Herbst, Love & Co.; office Ouray; claim 300 by 1,500 feet; discovered 1876; located on Mount Hendricks, Mount Sneffels mining district, 9 miles from Ouray; vertical vein, trending northwest and southeast, width 7 feet, pay vein 18 inches; nature of ore, galena and ruby silver; assay value, 300 ounces silver per ton; 1 drift 35 feet in length and 25 feet from surface.

Neptune. N.J. Downer and S.H. Day, proprietors; office Ouray; claim 300 by 1,500 feet; discovered 1878; located on Potosi Mountain, Mount Sneffels mining district, 7 miles from Ouray; vertical vein, trending northwest and southeast, width 4 feet, pay vein 22 inches; nature of ore, galena and gray copper; assay value, 250 ounces silver per ton; 1 adit 30 feet in length, 1 shaft 30 feet deep.

New Castle. Owned by E. Henry and N.C. Taylor, of Ouray; claim 300 by 1,500 feet; discovered 1876; located on Poughkeepsie Creek, 3 miles from Ouray; course of vein, north and south, width 6 feet, pay vein 20 inches; nature of ore, gray copper and galena; assay value, 90 ounces silver per ton; development, 1 drift 40 feet in length.

New Castle. Thomas Carleton, proprietor; office Upper San Miguel City; claim 300 by 1,500 feet; discovered 1878; located on Marshall Mountain, Upper San Miguel mining district, 9 miles from Ouray; vertical vein, trending east and west; width 3 feet, pay vein 5 inches; nature of ore, gray copper and ruby silver; assay value, 1,000 ounces silver per ton; 1 drift 30 feet in length.

Nimbus. Owned by George Hughes, L. Clark, and others; office Upper San Miguel City; claim 300 by 1,500 feet; discovered 1878; located in upper San Miguel mining district, 16 miles from Ouray; course of vein, northeast and southwest, width 4 feet, pay vein 20 inches, pitch 10 degrees; nature of ore, gray copper and ruby silver; assay value, 120 to 850 ounces silver per ton; 1 drift 20 feet in length and 18 feet from surface.

Nipsey. L.B. Clay, proprietor; office Ouray; claim 300 by 1,500 feet; discovered 1878; located in Defiance Gulch, Mount Sneffels mining district, 9 miles from Ouray; course of vein, east and west, width 15 feet, pitch 15 degrees; nature of ore, gray copper and galena; 1 drift 20 feet in length and 20 feet from surface.

No Name. C. Ogle and E.M. Smith, proprietors; office Upper San Miguel City; claim 280 by 1,500 feet; discovered 1877; located on Gold Run, Upper San Miguel mining district, 9 miles from Ouray; course of vein, east and west, width 24 feet, pitch 20 degrees; nature of ore, brittle silver and gray copper; assay value, 200 to 700 ounces silver per ton; 1 shaft 14 feet deep.

None Such. C. Morris and G. Van Pelt, proprietors; office Ouray; claim 300 by 1,500 feet; discovered 1876; located on Canyon Creek;

Uncompahgre mining district, 1 mile from Ouray; course of vein, east and west, width 12 feet, pay vein 10 inches; nature of ore, gray copper and galena; assay value, 150 ounces silver per ton; 2 adits 30 feet each.

None Such. Owned by J.C. Resor and W. Calely; office Silverton; claim 300 by 1,500 feet; discovered 1878; located on Bear Creek, Upper San Miguel mining district, 3 miles from Miguel City and 14 miles from Ouray; course of vein, northeast and southwest, width 4 feet, pay vein 20 inches; nature of ore, brittle silver and gray copper; assay value, 50 to 200 ounces silver per ton; 1 adit 12 feet in length with a 12-foot face.

Norfolk and Ouray Mining and Smelting Co. Incorporated March 28, 1878; capital stock, $500,000, in 5,000 shares, $100 each; offices Ouray, Colorado, and Norfolk, Virginia; John L. Roper, president, George E. Maltby, treasurer, R.D. Parrott, secretary, M.S. Corbett, superintendent, W.H. Strout, metallurgist, Howard N. Johnson, Samuel McWilliams, Fred Greenwood, Theodore F. Rogers, and L.C. Kilby, trustees.

Norma. Owned by W. Weston & Co., of Ouray; claim 300 by 1,500 feet; discovered 1877; located in Imogene Basin, Mount Sneffels mining district, 9 miles from Ouray; course of vein, north and south, width 12 feet, pay vein 3 feet, pitch 10 degrees; nature of ore, gray copper; assay value, 160 ounces silver per ton; development, 1 tunnel 100 feet in length.

North Star. Owned by Charles J. and A. Owens; office Ouray; claim 300 by 1,500 feet; discovered 1876; located on Red Mountain, Uncompahgre mining district, 4 miles from Ouray; course of vein, northwest and southeast, width 8 feet, pay vein 6 inches, pitch 25 degrees; nature of ore, galena and iron pyrites; assay value, 100 ounces silver per ton; 1 drift 40 feet in length and 25 feet from surface.

Northern Light. O.W. Jameson and H.E. Volkman, proprietors; office Ouray; claim 300 by 1,500 feet; discovered 1876; located on Mount Emma, Mount Sneffels mining district, 9 miles from Ouray; vertical vein, trending northwest and southeast, width 4 feet, pay vein 15 inches; nature of ore, gray copper and brittle silver; assay value, 400 ounces silver per ton; 1 shaft 12 feet deep, 1 adit 10 feet in length.

Occident Mining Co. Incorporated February 2, 1879; capital stock, $25,000, in 2,500 shares, $10 each; office Denver; organized by Mark Curran, John Curran, William S. Marshall, F.J. Elbert, Frederick L. Hahn, and Edward P. Greenough.

Ophir. John Thomas & Co., proprietors; office Ouray; claim 300 by 1,500 feet; discovered 1877; located on Engineer Mountain, Uncompahgre mining district, 6 miles from Ouray; vertical vein, trending northeast and southwest, width 22 inches, pay vein 4 inches; nature of ore, gray copper and galena; assay value, 400 ounces silver per ton; 1 shaft 18 feet deep.

Ophir. Owned by J.F. Robison and J.E. Horner; claim 300 by 1,500 feet; discovered 1877; located on Silver Mountain, Iron Springs mining district, near town of Ophir, 12 miles from Ouray; vertical vein, trending north and south, width 10 inches, pay vein 6 inches; nature of ore, free gold; value $20 gold and $10 silver per ton; 1 drift 20 feet in length and 20 feet from surface.

Ophir, Placer. Owned by C. Googen & Co., of Ouray; claim comprises 20 acres of placer ground; located 5 miles from Miguel City, on the Lower Miguel River.

Osceola. Owned by W.S. Fink and Mrs. Martha Roberts; office Silverton; claim 300 by 1,500 feet; discovered 1877; located on Silver Mountain, Iron Springs mining district, near town of Ophir, 12 miles from Ouray; course of vein, east and west, width 4 feet, pay vein 10 inches, pitch 12 degrees; nature of ore, free gold; value $12,500 per ton; 1 shaft 50 feet deep; yield $3,000.

Ouray. E. Thayer and Philip E. Morehouse, proprietors; office Ouray; claim 300 by 1,500 feet; discovered 1877; located on Buckeye Mountain, Mount Sneffels mining district, 8 miles from Ouray; course of vein, north and south, width 5 feet, pay vein 2 feet, pitch 10 degrees; nature of ore, gray copper and galena; average value, 200 ounces silver per ton; 2 adits 20 feet each.

Ouray Discovery and Mining Co. Incorporated February 7, 1877; capital stock, $50,000, in 500 shares, $100 each; office Ouray; trustees: L.J. Worden, E.S. Eldridge, A.D. Seal, W.G. Melville, and James E. Watson.

Ouray Mining Co. Incorporated September 24, 1877; capital stock, $30,000, in 300 shares, $100 each; offices Ouray, Colorado, and Oil City, Pennsylvania; trustees: Wesley Chambers, Jacob Ohlwiler, J.T. Jones, and E.C. Bradley.

Pacific. Owned by S.P. and F.M. Truitt, and M.L. Wimber; office Ouray; claim 300 by 1,500 feet; discovered 1877; located in Defiance Gulch, Mount Sneffels mining district, 9 miles from Ouray; course of vein, east and west, width 30 inches, pay vein 14 inches; nature of ore, gray copper; assay value, 440 ounces silver per ton; 1 drift 20 feet in length and 12 feet from surface.

Pacific Tunnel. Owned by W.S Fink and Mrs. Martha Roberts; office Silverton; claim 1,500 feet by 3,000 feet; established 1878; located on Silver Mountain, Iron Springs mining district, near town of Ophir, 12 miles from Ouray; course of tunnel, north and south, length 40 feet, designed to cut the Powhattan and Oscoela veins.

Pategranda. Owned by S.M. Ransom, W. Turner, and C.A. Ward; office Ouray; claim 300 by 1,500 feet; discovered 1875; located on Bear Mountain, Uncompahgre mining district, 4 miles from Ouray; vertical vein, trending northeast and southwest, width 10 feet; nature of ore, galena and gray copper; average value, 30 ounces silver per ton; 1 shaft 20 feet deep.

Paymaster. J.P. Barrows, proprietor; office Ouray; claim 300 by 1,500 feet; discovered 1878; located on Canyon Creek, Uncompahgre mining district, half mile from Ouray; course of vein, northwest and southeast, width 15 feet, pay vein 10 inches, pitch 25 degrees; nature of ore, iron and copper pyrites and galena; 1 adit 12 feet in length with an 11-foot face.

Pike County. Owned by F. Fodde & Co.; office Ophir; claim 300 by 1,500 feet; discovered 1878; located on Yellow Mountain, Iron Springs mining district, 1 mile from town of Ophir and 12 miles from Ouray; course of vein, north and south, width 4 feet, pay vein 18 inches; nature of ore, free gold; assay value, $494 per ton; 1 drift 250 feet in length and 250 feet from surface.

Pocahontas. Owned by Francis G. and J.H. King, Leonard C. Kilham,, and D.D. Mallory; office Denver; claim 300 by 1,500 feet, patented; discovered September 13, and recorded October 30, 1875; located near the junction of Imogene and Pocahontas Gulches, Mount Sneffels mining district, 8 miles from Ouray; vertical vein, trending northwest and southeast, width 5 feet, pay vein 18 inches; nature of ore, galena and gray copper; assay value, 100 to 500 ounces silver per ton; developed by a tunnel 14 feet and a drift 15 feet in length. The country in the vicinity of this property is covered with timber, suitable for milling and other purposes. Judging from the amount of ore to be seen in the workings and on the surface of this lode, there can be no doubt but what it will eventually become the sterling mine of the district.

Pocahontas. Owned by G.W. Hight and John Eddy; office Ophir; claim 300 by 1,500 feet; discovered 1878; located on Yellow Mountain, Iron Springs mining district, near town of Ophir, 12 miles from Ouray; vertical vein, trending east and west, width 22 inches, pay vein 6 inches; nature of ore, galena; assay value, 100 ounces silver per ton; 1 shaft 12 feet deep.

Polar Star. Charles Kirk, proprietor; office Silverton; claim 300 by 1,500 feet; discovered 1877; located on Yellow Mountain, Iron Springs mining district, 2 miles from town of Ophir and 13 miles from Ouray; vertical vein, trending north and south, width 8 feet, pay vein 30 inches; nature of ore, galena and gray copper; assay value, 100 ounces silver per ton; 1 adit 30 feet in length with a 20-foot face.

Potomac. Owned by M. Funk and Bro.; office Ouray; claim 300 by 1,500 feet; discovered 1875; located on Hardin Mountain, Uncompahgre mining district, 2 miles from Ouray; vertical vein, trending northwest and southeast, width 5 feet, pay vein 31 inches; nature of ore, iron pyrites and gray copper; average value, 85 ounces silver and 2 ounces gold per ton; 3 adits, 12, 18, and 20 feet in length, respectively.

Potosi, Extension of Seven-Thirty. Owned by Francis G. and J.H. King, Leonard C. Kilham, and D.D. Mallory; office Denver; claim 300 by 1,500 feet, patented; discovered September 13, and recorded October 30, 1875; located on the south side of Potosi Mountain, Mount Sneffels mining district, 7 miles from Ouray; course of vein, northeast and southwest, width 6 feet, pay vein 12 to 30 inches, pitch southeast; nature of ore, gray copper; assay value, 225 ounces silver per ton; developed by a tunnel 25 feet and an open cut 20 feet in length; the surface is well timbered, and convenient to the property is a good cabin and water power suitable for driving all kinds of heavy mining machinery.

Powhattan. Owned by W.S. Fink and Mrs. Martha Roberts; office Silverton; claim 300 by 1,500 feet; discovered 1878; located on Silver Mountain, Iron Springs mining district, near town of Ophir, 12 miles from Ouray; vertical vein, trending northeast and southwest, width 5 feet, pay vein 20 inches; nature of ore, free gold; value, $60 to $5,000 per ton; 2

drifts, 20 and 75 feet from surface, respectively, aggregating 125 feet in length; yield, $3,500.

Prince Consort. J. Thornton, proprietor; office Ouray; claim 300 by 1,500 feet; discovered 1876; located on Red Mountain, Uncompahgre mining district, 5 miles from Ouray; vertical vein, trending east and west, width 6 feet, pay vein 18 inches; nature of ore, gray copper and iron pyrites; 1 adit 15 feet in length with a 12-foot face.

Provedencia Groupe. Owned by J.D. Coplen, J. Russell, J. Scitz, and others; office Ouray; claims comprise 6 lodes, or mineral veins, 300 by 1,500 feet each; located on Republican Mountain [12,386 feet], Uncompahgre mining district, 6 miles from town of Ouray; main shaft 40 feet deep, 1 drift 100 feet in length.

Raleigh. Owned by L.W. Tevis, H.F. Blythe, and others; office Ouray; claim 300 by 1,500 feet; discovered 1876; located in Virginia Basin, Mount Sneffels mining district, 9 miles from Ouray; vertical vein, trending northeast and southwest, width 5 feet, pay vein 8 inches; nature of ore, galena and gray copper; 2 adits, 14 feet each.

Red Cloud. L. Quinn & Co., proprietors; office Ouray; claim 300 by 1,500 feet; discovered 1876; located on Canyon Creek, Uncompahgre mining district, 4 miles from Ouray; course of vein, east and west, width 5 feet, pitch 10 degrees; nature of ore, gray copper and iron pyrites; assay value, 140 to 1,000 ounces silver per ton; 1 drift 25 feet in length.

Revenue Mining and Milling Co. Incorporated February 28, 1878; capital stock, $500,000, in 50,000 shares, $10 each; office Denver, Colorado; trustees: Harry Hunt, Joseph C. Wilson, Howard B. and Lorenzo A. Jeffries, and E.B. Kephart.

Ringdove. Owned by H.W. Stingley and E.R. Jameson, of Ouray; claim 300 by 1,500 feet; discovered 1876; located on Canyon Creek, Uncompahgre mining district, 3 miles from Ouray; course of vein, east and west, width 3 feet, pitch 30 degrees; nature of ore, gray copper and galena, assay value, 600 ounces silver per ton; development, 1 tunnel 50 feet in length.

River Treasure, and Stony. Owned by William M. Friend, J.W. Vance, and G. Kimball; office Ouray; claim 300 by 3,000 feet; discovered 1878; located on Ruby Mountain, Mount Sneffels mining district, 7 miles from Ouray; vertical vein, trending northeast and southwest, width 3 feet, pay vein 8 inches; nature of ore, gray copper and galena; 2 adits, 10 and 15 feet in length, respectively.

Roanoke. Owned by H. Le Clair and W.S. Stinson; office Ouray; claim 300 by 1,500 feet; discovered 1875; located on Engineer Mountain, Uncompahgre mining district, 6 miles from Ouray; course of vein, east and west, width 15 feet, pay vein 3 feet, pitch 25 degrees; nature of ore, gray copper and galena; assay value, 125 ounces silver per ton; 1 tunnel 25 feet and 1 level 10 feet in length.

Rob Roy. Owned by John C. Munn and W.B. Fonda; office Ophir; claim 300 by 1,500 feet; discovered 1878; located on Silver Mountain, Upper San Miguel mining district, 3 miles from Miguel City and 12 miles from Ouray; vertical vein, trending northeast and southwest, width 4 feet, pay vein 18 inches; nature of ore, galena and sulphurets; average value, 150 ounces silver per ton; 1 shaft 15 feet deep.

Rocky Mountain Lion. Charles Josephs & Co., proprietors; office Ouray; claim 300 by 1,500 feet; discovered 1878; located on Uncompahgre Creek, Uncompahgre mining district, 5 miles from Ouray; vertical vein, trending north and south, width 30 feet; nature of ore, galena and gray copper; assay value, 250 ounces silver per ton; 1 shaft 15 feet deep.

S.B. Kenrick. Owned by J.A. Beattie and A.A. Klingensmith; office Ophir; claim 300 by 1,500 feet; discovered 1877; located on Horn Mountain, Iron Springs mining district, 4 miles from town of Ophir and 12 miles from Ouray; vertical vein, trending northeast and southwest, width 6 feet, pay vein 18 inches; nature of ore, gray copper and sulphurets; average value, 106 ounces silver per ton, 1 tunnel 25 feet in length, intersecting vein at a depth of 25 feet; yield, $400.

Safe Deposit. T. Resor & Co., proprietors; office Ouray; claim 300 by 1,500 feet; discovered 1878; located on Gold Run, Upper San Miguel mining district, 9 miles from Ouray; vertical vein, trending northwest and southeast, width 9 feet, pay vein 30 inches; nature of ore, galena and gray copper; assay value, 90 ounces silver per ton; 1 adit 15 feet in length.

San Juan. Owned by Joseph Cuenin, C.J. Rouech, and Howard Home; office Del Norte; claim 300 by 1,500 feet; discovered 1877; located on Yellow Mountain, Iron Springs mining district, 4 miles from Miguel City and 12 miles from Ouray; vertical vein, trending northeast and southwest, width 4 feet, pay vein 18 inches; nature of ore, gray copper; average value, $245 per ton; 1 drift 160 feet in length.

San Juan Hydraulic Mining Co. Incorporated June 5, 1877; capital stock, $1,000,000, in 100,000 shares, $10 each; office Denver; trustees: John A. Luce, Dennis B. Harris, Nicholas S. Snyder, William F. Shanks, Robert Connely, John M. Lord, Edward Ohio, R.T. Brown, and Becker Folsom.

San Juan and St. Louis Mining and Smelting Co. Incorporated October 23, 1877; capital stock, $500,000, in 5,000 shares, $100 each; office Ouray; Robert T. Long, president, Claude P. Hardin, treasurer.

San Miguel Hydraulic Mining Co. Incorporated June 7, 1877; capital stock, $12,000, in 1,200 shares, $10 each; office Denver; trustees: Dennis B. Harris, Robert Connely, and William F. Shanks.

Santa Clara. W.H. Lilley & Co., proprietors; office Ouray; claim 300 by 1,500 feet; discovered 1876; located in Sneffels Basin, Mount Sneffels mining district, 9 miles from Ouray; vertical vein, trending northeast and southwest, width 4 feet, pay vein 20 inches; nature of ore, gray copper and ruby silver; assay value 261 ounces silver per ton; 1 drift 55 feet in length and 35 feet from surface.

Security. N.J. Downer and M.D. Cooper, proprietors; office Ouray; claim 300 by 1,500 feet; discovered 1877; located on Ruby Mountain, Mountain Sneffels mining district, 7 miles from Ouray; course of vein, northwest and southeast, width 4 feet, pay vein 20 inches, pitch 15 degrees; nature of ore, sulphurets, brittle and ruby silver; average value, 900 ounces silver per ton; 1 drift 40 feet in length and 18 feet in depth.

Senator. W. Reid and M.J. Buskirk, proprietors; office Ouray; claim 300 by 1,500 feet; discovered 1876; located on Wilson Mountain, Mount Sneffels mining district, 10 miles from Ouray; course of vein, east and west, width 3 feet, pay vein, 1 foot, pitch 30 degrees; nature of ore, gray copper and galena; average value 80 ounces silver per ton; 2 drifts, 20 feet each, being 15 and 20 feet from surface, respectively.

Seven-Thirty. Owned by Francis G. and J. H. King, Leonard C. Kilham, and D.D. Mallory; office Denver; claim 300 by 1,200 feet; discovered September 13, and recorded October 30, 1875; located at the base and on the north side of Ruby Mountain, Mount Sneffels mining district, 7 miles from Ouray; vertical vein, trending north and south, width 5 feet, pay vein 8 to 12 inches; nature of ore, gray copper and ruby silver; average value, 200 ounces silver per ton; developed by a tunnel, 40 feet in length, and 2 open cuts, 13 and 14 feet in length, respectively. At a convenient distance from the mine is a good waterpower, cabin, and blacksmith shop. The mountain rises at an angle of about 40 degrees from the vertical, affording a fine opportunity to gain depth by tunneling. The surface is covered with timber suitable for mining and other purposes.

Seven-Thirty. Owned by B.B. Shutts, W.R. Hart, and W.B. Fonda; office Ophir; claim 300 by 1,500 feet; discovered 1878; located on Silver Mountain, Iron Springs mining district, 1 mile from town of Ophir and 12 miles from Ouray; course of vein, northeast and southwest, width 6 feet, pay vein 20 inches; nature of ore, free gold and chloride of silver; average value, $240 per ton; main shaft 25 feet deep, 1 drift 25 feet in length and 30 feet from surface.

Silver Belt. L.B. Clay, proprietor; office Ouray; claim 300 by 1,500 feet; located 1878; located on Virginus Mountain, Mount Sneffels mining district, 6 miles from Ouray; course of vein, northeast and southwest, width 9 feet, pay vein 3 feet, pitch 15 degrees; nature of ore, brittle silver and gray copper, 1 drift, 15 feet in length and 12 feet from surface.

Silver Crown. H. Dunton and Co., proprietors; office Ouray; claim 300 by 1,500 feet; discovered 1877; located on Mineral Point Mountain, Uncompahgre mining district, 6 miles from Ouray; vertical vein, trending northeast and southwest, width 6 feet, pay vein 6 inches; nature of ore, galena and gray copper; 1 adit, 12 feet in length with a 15-foot face.

Silver Gray and San Juan Mining Co. Incorporated January 30, 1878; capital stock, $100,000, in 10,000 shares, $10 each; office Denver; organized by William C. Roberts, Samuel E. Collyer, and John P. Oglesby; E.R. Dougherty, president, Thomas R. Burch, vice president, A.C. Fisk, secretary, Andrew Templeton, treasurer, Samuel E. Collyer, superintendent. The property of this company is located in the Uncompahgre mining district, on the south side of Yellow Stone Gulch and on Red Mountain in Gray Copper Gulch, 3 miles from Dow & Waters' reduction works, and 12 miles north of Silverton. It comprises the Silver Gray and Broad Gauge lodes, which were discovered by Samuel E. Collyer, the former being located July 15, and the latter August 2, 1876. Each of these lodes is 300 feet in width by 1,500 feet in length, and run parallel to each other. The Silver Gray is developed by a drift on the vein

150 feet in length, showing a crevice 3 feet wide; the nature of ore is gray copper, iron and copper pyrites, and the average value 128 ounces silver per ton. The Broad Gauge is developed by 1 open cut 50 feet in length; course of vein, north and south, width 50 feet; nature of ore, gray copper, iron and copper pyrites, average value, 50 ounces silver per ton; there is an abundance of timber, suitable for all purposes, and a good supply of water in the immediate vicinity of this property.

Silver King. Owned by W. H. Rogers, of Colorado Springs; claim 300 by 1,500 feet; located on Engineer Mountain, Uncompahgre mining district, 7 miles from Ouray; course of vein, northeast and southwest, width 7 feet, pitch 45 degrees northeast, pay vein 1 foot; nature of ore, gray copper and galena; average value, 200 ounces silver per ton; 1 drift 100 feet in length.

Silver King. W. and G. Mitchell, proprietors; office Ouray; claim 300 by 1,500 feet; discovered 1877; located on Uncompahgre Creek, Uncompahgre mining district; vertical vein, trending north and south, width 20 feet, pay vein 3 feet; nature of ore, galena and iron pyrites, assay value, 90 ounces silver per ton; 1 drift, 15 feet in length and 40 feet from surface.

Silver King. W.S. Corey, proprietor; office Ouray; claim 300 by 1,500 feet; discovered 1878; located in Virginius Basin, Mount Sneffels mining district, 9 miles from Ouray; course of vein, northwest and southeast, width 7 feet, pay vein, 30 inches; nature of ore, gray copper, ruby and brittle silver; assay value, 800 ounces silver per ton; 1 adit 20 feet in length with an 11-foot face.

Silver Leaf. Owned by J.D. Waring & Co.; office Silverton; claim 300 by 1,500 feet; discovered 1876; located on Ruby Mountain, Mount Sneffels mining district, 10 miles from Ouray; vertical vein, trending northwest and southeast, width 5 feet, pay vein 18 inches; nature of ore, gray copper and galena; assay value, 20 to 500 ounces silver per ton; 1 drift 30 feet in length and 25 feet from surface; 1 adit, 12 feet in length, with a 12-foot face.

Silver Link. Owned by J. Russell, W. Page, J. Coplen, and S. Wade; office Ouray; claim 300 by 1,500 feet; located on Cline Mountain, Uncompahgre mining district, 5 miles from Ouray; course of vein, north and south, width 24 feet, pitch 10 degrees east, pay vein 12 feet; nature of ore, gray copper and galena; average value, 75 ounces silver per ton; main shaft 15 feet deep.

Silver Mountain Tunnel. Owned by J. Orr, John R. Curry, F. Morey, and F.H. Whitney; office Silverton; claim 1,500 by 3,000 feet; established 1878; located on Silver Mountain, Iron Springs mining district, near town of Ophir, 12 miles from Ouray; course of tunnel, north and south, length 100 feet.

Silver Plume. Owned by A. Peck & Bro. and others; office Ouray; claim 300 by 1,500 feet; discovered 1876; located on Cline Mountain, Uncompahgre mining district, 4 miles from Ouray; course of vein, north and south, width 4 feet, pay vein 6 inches, pitch 15 degrees; nature of ore, galena and gray copper; assay value, 90 ounces silver per ton; 1 adit 30 feet in length, with a 25-foot face.

Silver Wave. W. Caldwell & Co., proprietors; office Ouray; claim 300 by 1,500 feet; discovered 1877; located on Uncompahgre Creek, Uncompahgre mining district, 5 miles from Ouray; course of vein, north and south, width 12 feet, pitch 15 degrees; nature of ore, galena and gray copper; assay value, 150 ounces silver per ton; 1 adit 15 feet in length.

Snow Bird. C. Coogan and G.O. Roberts, proprietors; office Ouray; claim 300 by 1,500 feet; discovered 1878; located on Treasure Hill, Mount Sneffels mining district, 9 miles from Ouray; course of vein, northwest and southeast, width 4 feet, pay vein 20 inches, pitch 30 degrees; nature of ore, ruby and brittle silver and galena; assay value, 200 ounces silver per ton; 1 drift 16 feet in length and 16 feet from surface.

Snow Flake. C. Coogan and G.O. Roberts, proprietors; office Ouray; claim 300 by 1,500 feet; discovered 1876; located on Treasure Hill, Mount Sneffels mining district, 9 miles from Ouray; course of vein, northwest and southwest, width 6 feet, pay vein 2 feet, pitch 30 degrees; nature of ore, ruby and brittle silver and galena; assay value, 200 ounces silver per ton; 1 adit 30 feet in length with an 18-foot face.'

Sonora. C. Coogan and G.O. Roberts, proprietors; office Ouray; claim 300 by 1,500 feet; located on Mount Sneffels, Mount Sneffels mining district, 9 miles from Ouray; vertical vein, trending northwest and southeast; width 8 feet, pay vein 30 inches; nature of ore, galena and gray copper; assay value, 50 to 300 ounces silver per ton; 1 shaft 40 feet deep.

Spar. Owned by L. Sears and H.F. Shanks; office Ophir; claim 300 by 1,500 feet; discovered 1877; located on Silver Mountain, Iron Springs mining district, 1 mile from town of Ophir and 12 miles from Ouray; vertical vein, trending northeast and southwest, width 10 feet, pay vein 1 foot; nature of ore, gray copper and galena; assay value, 80 ounces silver per ton; 1 drift 20 feet in length.

Specie Payment. Charles Johnson, proprietor; office Ouray; claim 300 by 1,500 feet; discovered 1877; located on Canyon Creek; Mount Snefffels mining district, 7 miles from Ouray; vertical vein, trending north and south, width 3 feet, pay vein 10 inches; nature of ore, galena and gray copper; assay value, 100 ounces silver per ton; 1 drift 20 feet in length and 15 feet from surface.

Spotted Tail. W.S. Corey, proprietor; office Ouray; claim 300 by 1,500 feet; discovered 1878; located on Uncompahgre Creek, Uncompahgre mining district, 9 miles from Ouray; vertical vein, trending northeast and southwest, width 2 feet, pay vein 1 foot; nature of ore, galena and gray copper; assay value, 200 ounces silver per ton; 2 adits, each 14 feet in length.

Staatsburg. Owned by the Staatsburg Silver Mining Co.; incorporated 1878; capital stock, $600,000, in 6,000 shares, $100 each; office Silverton; J.H. Brink, president, H. Staats, vice president, A.A. Fain, secretary, J.M. Stuart, treasurer; claim 300 by 1,500 feet; discovered 1877; located on Silver Mountain, Iron Springs mining district, 1 mile from town of Ouray; vertical vein, trending east and west, width 15 feet, pay vein 10 inches; nature of ore, sulphurets and galena; average value, 150 ounces silver per ton; 1 tunnel 140 feet in length, which intersects the vein at a depth of 80 feet, 1 level 50 feet in length and 80 feet from surface; yield, $1,000.

Staatsburg Tunnel. Owned by the Staatsburg Silver Mining Co.; office Silverton; claim 1,500 by 3,000 feet; established 1877; located on Silver Mountain, Iron Springs mining district, 1 mile from town of Ophir, and 12 miles from Ouray; course of tunnel, north and south, length 200 feet; will intersect 5 lodes within a distance of 600 feet from the mouth.

Star of Hope. Owned by W. Caldwell, S.S. Evens, and others; office Ouray; claim 300 by 1,500 feet; discovered 1876; located on Red Mountain, Uncompahgre mining district, 7 miles from Ouray; vertical vein, trending northeast and southwest, width 4 feet, pay vein 20 inches;

nature of ore, galena and gray copper; assay value, 90 ounces silver per ton; 1 shaft 8 feet deep, 1 drift 12 feet in length.

Starlight. J. Cooper and S. Smith, proprietors; office Ouray; claim 300 by 1,500 feet; discovered 1876; located on Bear Creek, Uncompahgre mining district, 3 miles from Ouray; vertical vein, trending northwest and southeast, width 5 feet, pay vein 30 inches; nature of ore, galena; assay value, 80 ounces silver per ton; 1 drift 30 feet in length and 18 feet from surface.

St. Louis. Joseph Beck and C.C. Crawford, proprietors; office Ouray; claim 300 by 1,500 feet; discovered 1876; located on Canyon Creek, Uncompahgre mining district, 6 miles from Ouray; vertical vein, trending northeast and southwest, width 6 feet, pay vein 10 inches; nature of ore, gray copper and galena; assay value, 90 ounces silver per ton; 1 adit 15 feet in length with a 15-foot face.

Sulphurets. Owned by W.C. Hess, Joseph E. Lacome, and others; office Silverton; claim 300 by 1,500 feet; discovered 1877; located on Silver Mountain, Iron Springs mining district, 1 mile from town of Ophir and 12 miles from Ouray; vertical vein, trending northeast and southwest, width 8 feet, pay vein 20 inches; nature of ore, sulphurets; average value, 175 ounces silver per ton; 1 drift 150 feet in length and 125 feet from surface; yield, $3,000.

Summit. Owned by George Green & Co., Alexander Fleming, Timothy Cart, and Thomas Stanton; office Silverton; claim 300 by 1,500 feet; discovered 1877; located on Silver Mountain, Iron Springs mining district, 1 mile from town of Ophir and 12 miles from Ouray; vertical vein, trending northeast and southwest, width 4 feet, pay vein 4 inches; nature of ore, galena and gray copper; average value, 200 to 250 ounces silver per ton; yield, $5,000.

Sunny Side. Owned by Thomas E. Breckenridge and Sons; office Ophir; claim 300 by 1,500 feet; discovered 1877; located on Silver Mountain, Upper San Miguel mining district, 3 miles from town of San Miguel and 12 miles from Ouray; vertical vein, trending east and west, width 5 feet, pay vein 4 inches; nature of ore, gray copper and sulphurets; average value, 200 ounces silver per ton; 1 drift 25 feet in length and 25 feet from surface.

Surprise. Benjamin B. Haddox, L. B. Laffoon, and David A. Munn; office Ophir; claim 300 by 1,500 feet; discovered 1878; located on Silver Mountain, Upper San Miguel mining district, 2 miles from town of Ophir and 12 miles from Ouray; course of vein, north and south, width 2 feet, pay vein 6 inches; nature of ore, free gold; 1 shaft 12 feet deep.

Tecumseh. Owned by W.S. Fink and Mrs. Martha Roberts; office Silverton; claim 300 by 1,500 feet; discovered 1877; located on Silver Mountain, Iron Springs mining district, near town of Ophir, 12 miles from Ouray; course of vein, east and west, width 14 inches, pay vein 1 foot, pitch 12 degrees; nature of ore, free gold; average value $60 per ton; 1 shaft 20 feet deep.

Terrible. James K. Reed & Co., proprietors; office Miguel City; claim 300 by 1,500 feet; discovered 1878; located on Bear Creek, Upper San Miguel mining district, 3 miles from Miguel City and 14 miles from Ouray; vertical vein, trending east and west, width 10 feet, pay vein 3 feet; nature of ore, sulphurets and carbonate; average value, 75 ounces silver per ton; 1 adit 10 feet in length with a 10-foot face.

Terrible. Owned by M. McCormick and P. Peterson; office Ophir; claim 300 by 1,500 feet; discovered 1877; located on Yellow Mountain, Iron Springs mining district, 3 miles from town of Ophir and 13 miles from Ouray; course of vein, east and west, width 5 feet, pay vein 3 feet; nature of ore, gray copper and galena; value, 220 ounces silver and 20 ounces gold per ton; 1 adit 12 feet, and 1 drift 22 feet in length.

Terrible. W. McIntire and F. Coombes, proprietors; office Ophir; claim 300 by 1,500 feet; discovered 1878; located on Yellow Mountain, Iron Springs mining district, 2 miles from town of Ophir and 12 miles from Ouray; vertical vein, trending north and south, width 10 feet, pay vein 20 inches; nature of roe, galena and gray copper; assay value, 90 ounces silver per ton; 1 adit 15 feet in length.

Texas. Owned by James J. and Albert Bernard, Daniel J. and James C. McKay, William Forsyth, and John M. Ross; office Howardsville; claim 300 by 1,500 feet; discovered 1874; located in Poughkeepsie Gulch, Uncompahgre mining district, 7 miles from Ouray; course of vein northeast and southwest; width 6 feet, pay vein 3 feet; nature of ore, galena and gray copper; assay value, 557 ounces silver per ton; and tunnel 25 feet in length.

Tilden. M.C. Clingham & Co., proprietors; office Silverton; claim 300 by 1,500 feet; discovered 1878; located on Turkey Creek, Upper San Miguel mining district, 12 miles from Ouray; vertical vein, trending east and west, width 20 feet, pay vein 4 feet; nature of ore, gray copper and galena; assay value, 300 ounces silver per ton; ore shaft 12 feet deep.

Tornado and Bank of San Juan. Owned by O.W. Jameson and H.E. Volkman; office Ouray; claim 300 by 3,000 feet; discovered 1876; located on Mount Emma, Mount Sneffels mining district, 9 miles from Ouray; vertical vein, trending northwest and southeast, width 4 feet, pay vein 18 inches; nature of ore, gray copper and ruby silver; assay value, 600 ounces silver per ton; 2 drifts, 20 and 30 feet in length, respectively.

Traverse. Owned by J.J. Crane and J. Benton; office Ouray; claim 300 by 1,500 feet; discovered 1877; located on the Uncompahgre River, in town of Ouray; vertical vein, trending northeast and southwest; width 5 feet; nature of ore, gray copper and galena; 1 adit 20 feet in length and 15 feet from surface.

Treasure Hill. C. Coogan and G.O. Roberts, proprietors; office Ouray; claim 300 by 1,500 feet; discovered 1876; located on Treasure Hill, Mount Sneffels mining district, 9 miles from Ouray; course of vein, northwest and southeast, width 8 feet, pitch 30 degrees; nature of ore, galena and gray copper; assay value, 30 to 150 ounces silver per ton; 1 shaft 10 feet deep; 1 adit 30 feet in length, with a 16-foot face.

Treasure Vault. Owned by M. Williams & Bro.; office Animas Forks; claim 300 by 1,500 feet; discovered 1878; located on Bear Mountain, Upper San Miguel mining district, 4 miles from Miguel City and 13 miles from Ouray; course of vein, northeast and southwest, width 20 feet, pay vein 30 inches, pitch 10 degrees; nature of ore, gray copper and galena; average value, 75 ounces silver per ton; 1 adit 20 feet in length.

Triumph, Placer. Owned by R. Bennet and L. Stacey, of Ouray; claim comprises 80 acres of placer ground, located on the Miguel River, 2 miles from Ouray.

U.S. Depository. John C. Waring, proprietor; office Ouray; claim 300 by 1,500 feet; discovered 1875; located on Buckeye Mountain, Mount Sneffels mining district, 9 miles from Ouray; course of vein, southeast and northwest, width 6 feet, pay vein 28 inches, pitch 20 degrees; nature of ore, gray copper and galena; assay value, 50 to 3,000 ounces silver per ton; 1

drift 200 feet in length and 160 feet from surface, 1 adit 20 feet in length; tunnel house 16 by 18 feet.

Umpecaw. Owned by S.P and F.M. Truitt and W. Beers; office Ouray; claim 300 by 1,500 feet; discovered 1877; located in Imogene Basin, Uncompahgre mining district, 6 miles from Ouray; course of vein, northeast and southwest, width 8 feet, pay vein 16 inches; nature of ore, brittle silver; average value, 100 ounces silver per ton; 1 drift 20 feet in length and 20 feet from surface.

Union. Owned by M.N. Funk & Co.; office Ouray; claim 300 by 1,500 feet; discovered 1875; located on Harden Mountain, Uncompahgre mining district, 3 miles from Ouray; course of vein, northeast and southwest, width 11 feet, pay vein 8 inches; nature of ore, brittle silver and gray copper; average value, $150 per ton; 1 shaft 100 feet deep, 1 tunnel 20 feet in length, intersecting vein at a depth of 20 feet, 1 drift 50 feet in length and 50 feet from surface, 1 level 100 feet in length; yield, $3,500.

Ute. Owned by W. Mitchell, J. Cole, and R. Ronk; office Ouray; claim 300 by 1,500 feet; discovered 1876; located on Uncompahgre Creek, 4 miles from Ouray; vertical vein, trending east and west, width 30 inches, pay vein 8 inches; nature of ore, galena; assay value, 75 ounces silver per ton; 1 shaft 8 feet deep, 1 adit 20 feet in length.

Valley View. Owned by David A. and John Munn, Benjamin B. Haddox, and Joseph E. Lacome; office Silverton; claim 300 by 1,500 feet, patented; discovered 1877; located on Silver Mountain, Iron Springs mining district, near town of Ophir, 12 miles from Ouray; course of vein, northeast and southwest, width 20 feet, pay vein 3 feet, pitch 4 degrees; nature of ore, carbonate of lead; average value, 250 ounces silver per ton; main shaft 50 feet deep; to be intersected by the Gold King tunnel at a depth of 600 feet; yield, $6,000.

Victor. R.M. Mercer & Co., proprietors; office Ouray; claim 300 by 1,500 feet; discovered 1877; located on Red Mountain, Uncompahgre mining district, 3 miles from Ouray; course of vein, northeast and southwest, width 12 feet, pay vein 30 inches, pitch 25 degrees; nature of ore, gray copper and galena; assay value, 120 ounces silver per ton; 1 drift 20 feet in length and 15 feet form surface.

Virginius. Owned by C.C. Alvord, of Denver; claim 300 by 1,500 feet; discovered 1876; located in Virginius Basin, Mount Sneffels mining district, 9 miles from Ouray and 30 miles from Lake City; vertical vein, trending northeast and southwest, width 4 feet, pay vein 20 inches; nature of ore, gray copper and galena; average value, 300 ounces silver per ton; 2 tunnels aggregating 500 feet in length, 70 and 140 feet from surface, respectively.

Washington. J. Frye & Co., proprietors; office Ouray; claim 300 by 1,500 feet; discovered 1877; located on Canyon Creek, Uncompahgre mining district, 6 miles from Ouray; vertical vein, trending east and west, width 30 inches, pay vein 8 inches; nature of ore, gray copper; assay value, 150 ounces silver per ton; 1 adit 15 feet in length, with a 10-foot face.

Water Fall. Owned by Harry Diehl & Co.; office Ophir; claim 300 by 1,500 feet; discovered 1876; located on Yellow Mountain, Iron Springs mining district, near town of Ophir, 12 miles from Ouray; course of vein, northeast and southwest, width 4 feet, pay vein 1 foot; nature of ore, gray copper and galena; assay value, 240 ounces silver per ton; 1 drift 20 feet in length.

Western. Owned by W.R. Watson & Co., office Ouray; claim 300 by 850 feet; discovered 1877; located on Bear Creek Range, Uncompahgre mining district, 4 miles from Ouray; vertical vein, trending north and south, width 15 feet, pay vein 15 inches; nature of ore, brittle silver, gray copper and iron pyrites; 2 adits, each 12 feet in length.

Western Star. Owned by M. McCormick and P. Peterson; office Ophir; claim 300 by 1,500 feet; discovered 1878; located on Yellow Mountain, Iron Springs mining district, 3 miles from town of Ophir and 12 miles from Ouray; course of vein, northwest and southeast, width 6 feet, pay vein 5 feet; nature of ore, gray copper and galena; 1 adit 15 feet in length and 12 feet from surface.

Windham Silver Mining and Smelting Co. Incorporated April 2, 1878; capital stock, $50,000, in 500 shares, $100 each; offices Ouray, Colorado, and Windham, Connecticut; organized by Frank C. Garbutt, Sarah C., Richard W., and P. Henry Woodward, and Thomas Ramsdell.

Yankee Boy. Charist, Morrison, & Co., proprietors; offices Ouray and Silverton, Colorado; claim 300 by 1,500 feet; discovered 1875; located on Mount Sneffels, Mount Sneffels mining district, 10 miles from Ouray; vertical vein, trending southeast and northwest, width 5 feet, pay vein 20 inches; nature of ore, ruby and brittle silver; average value, 400 ounces silver per ton; 1 tunnel 60 feet in length, intersecting vein at a depth of 75 feet, 1 drift, 200 feet in length, and 150 feet from surface, 1 level, 75 feet in length, and 180 feet from surface; yield, $15,000.

Ouray County Ore Mills

Norfolk and Ouray Mining and Smelting Co. Incorporated March 28, 1878; capital stock, $500,000, in 5,000 shares, $100 each; John L. Roper, president, George E. Maltby, treasurer, R.D. Parrott, secretary, M.S. Corbett, superintendent; W.H. Strout, metallurgist. The works of this company are located 2 miles north of Ouray, and have a capacity for treating 40 tons of ore daily.

San Juan and St. Louis Mining and Smelting Co. R.F Long, manager. The works of this company are located 1 mile from town of Ouray and have a capacity for treating 20 tons of ore daily.

Windham Silver Mining and Smelting Co. Incorporated April 2, 1878; capital stock, $50,000, in 500 shares, $100 each; offices Ouray, Colorado, and Windham, Connecticut; organized by Frank C. Garbutt, Sarah C., Richard W., and P. Henry Woodward, and Thomas Ramsdell. The works of this company have a capacity for treating 12 tons of ore in 24 hours; they are run by water and situated on the Uncompahgre River, 3 miles from Ouray.

PARK COUNTY

Thomas B. Corbett reported that Park County had a population of 6,000 in 1878 and a bullion product of $479,000 for that year. Mining districts included: Buckskin, Montgomery, Hall Valley, Horse Shoe, and Sacramento.

Alma Pool Association. Incorporated 1874; capital stock, $60,000, in 2,400 shares, $25 each; office Alma; John T. Brownlow, president, James V. Dexter, secretary and treasurer; Assyria Hall, George W. Brunk, and James F. Flanagan, directors.

Alpine. Owned by the Alma Pool Association; office Alma; claim 50 by 1,500 feet; located on the southeast spur of Mount Lincoln [at 14,268 feet, this is one of Colorado's "fourteeners"], Consolidated Montgomery mining district, 6 miles from Alma and 12 miles from Fairplay; nature of ore, galena and sulphurets; average value, $50 per ton; 1 shaft 25 feet deep.

Argente. Owned by C.S. Eyster and Thomas Willey; office Denver; claim 200 by 1,500 feet; discovered 1874; located in Sacramento Gulch, 7 miles from Fairplay; horizontal vein, width 12 to 18 inches; nature of ore, carbonate of lead; value, 100 ounces silver per ton.

At Last. Owned by Robert A. Kirker and Elisha Bass; office Fairplay; claim 300 by 1,500 feet; located on divide between Horse Shoe Creek and Platte River, 10 miles from Fairplay; course of vein, northeast and southwest, width 20 feet, pitch 12 degrees southeast; nature of ore, iron and carbonate of lead; 1 shaft 15 feet deep, 1 tunnel 15 feet in length.

Atlantic. William Bemrose and L.L. Higgins, proprietors; office Alma; claim 300 by 1,500 feet; discovered 1876; located on North Star Mountain, Consolidated Montgomery mining district, 6 miles from Alma and 12 miles from Fairplay; course of vein, north and south, width 8 feet, pay vein 4 feet, pitch 15 degrees; nature of ore, oxide of iron; average value, $45 gold per ton; 1 shaft 210 feet deep, 1 level 150 feet in length; yield, $30,000; shaft house 30 by 50 feet.

Baltic. John R. Lindgren, proprietor; office Chicago; claim 300 by 1,500 feet, patented; located on the southern slope of Mount Lincoln, Consolidated Montgomery mining district, 6 miles from Alma and 12 miles from Fairplay; course of vein, northeast and southwest; nature of ore, galena and sulphurets; 1 tunnel 75 feet in length.

Barrett, Hall & Co.'s Placer. Owned by George W. Barrett, Assyria Hall, and W.L. Wilson. This property comprises 380 acres of ground, located on Tarryall creek, 2 miles above Hamilton, and 17 miles from Fairplay. [Hamilton, now a ghost town, was founded in 1859 by prospectors who had been turned away from the town of Tarryall. In 1877, about 100 people lived in Hamilton.] It has been worked since 1859, and the yield for two years, 1876 and 1877, was $17,000.

Barrett Placer. George W. Barrett, proprietor; office Hamilton; claim comprises 75 acres of ground, located on Tarryall Creek, 2 miles above Hamilton, and 17 miles from Fairplay; yield, $15,000.

Bass. Owned by the Horse Shoe Mining Co.; Robert A. Kirker, superintendent; office Fairplay; claim 300 by 1,500 feet; located on the south fork of Four Mile Creek, Horse Shoe mining district, 10 miles from Fairplay; course of vein, northeast and southwest, width 12 feet; nature of ore, iron and carbonate of lead; 1 shaft 45 feet deep.

Belle Fountain. Moses Hall, proprietor; office Alma; claim 300 by 1,500 feet; discovered 1875; located in Mosquito Gulch, 3 miles from Alma and 6 miles from Fairplay; course of vein, northwest and southeast, width 22 inches, pay vein 8 inches, pitch 45 degrees east; nature of ore, sulphurets and zinc; average value, 75 ounces silver per ton; 1 shaft 10 feet deep, 1 drift 65 feet in length and 45 feet from surface.

Blue Jacket. Owned by the Horse Shoe Mining Co.; Robert A. Kirker, superintendent; office Fairplay; claim 300 by 1,500 feet; located on the flats between the north and south fork of Four Mile Creek, 10 miles from Fairplay; course of vein, north 45 degrees east, width 1 foot, pitch 20 degrees; nature of ore, carbonate of lead; main shaft 30 feet deep.

Brownlow. Owned by the Horse Shoe Mining Co.; Robert A. Kirker, superintendent; office Fairplay; claim 300 by 1,500 feet; located in Horse Shoe mining district, 10 miles from Fairplay; course of deposit, north 45 degrees east, width 5 feet, pitch 70 degrees; nature of ore, carbonate of lead and copper; average value, $45 per ton; main shaft 40 feet deep, 1 tunnel 70 feet in length; yield, $12,000.

Brownlow Extension. Owned by the Horse Shoe Mining Co.; Robert A. Kirker, superintendent; office Fairplay; claim 300 by 1,500 feet; located on the eastern slope of Horse Shoe Mountain, 10 miles from Fairplay; course of vein, north 45 degrees east, width 2 feet, pitch 70 degrees; nature of ore, carbonate of lead; average value, $25 per ton; main shaft 20 feet deep, 1 open cut 30 feet in length; yield, $10,000.

Bruchinell Placer. Owned by W.K. Burchinell and A.B. Crook, of Fairplay; claim comprises 600 acres of placer ground, located in Sacramento Gulch, Sacramento mining district, 5 miles from Fairplay.

Buckeye. Owned by Americus Loudy Pogue, Henry Williams, and D.H. Dougan, M.D.; office Alma; claim 300 by 1,500 feet; located on North Star Mountain, Consolidated Montgomery mining district, 6 miles from Alma, and 12 miles from Fairplay; course of vein, northeast and southwest, width 4 feet, pitch 10 degrees west; nature of ore, free gold; average value, $100 per ton; 1 tunnel 50 feet in length. [Americus Loudy Pogue was an investor who lived with his family in Richmond, Indiana, and frequently traveled to Colorado to manage his mining investments. His older brother, Christopher Columbus Pogue, fought in the Civil War and spent the final eight months of the war in Andersonville prison; like many of the soldiers imprisoned there, he never fully recovered and died a few years later. According to family stories, their father, Loudy J. Pogue, was a history buff who often left his family for days at a time, retreating to a small cabin in the woods to read his history books. Americus Pogue died in Chicago in 1908, on his way home from a trip to Colorado.]

Bullion. Owned by William J. Curtice and Henry Crow; offices Dudley and Denver, Colorado; claim 400 by 1,500 feet; located on the north spur of Mount Lincoln, Consolidated Montgomery mining district, 6 miles from Alma and 12 miles from Fairplay; nature of ore, galena and sulphurets; average value, 200 ounces silver per ton; 3 tunnels, aggregating 150 feet in length; these are run through large bodies of low grade ore, from which eventually a handsome fortune will be realized. The buildings comprise an ore house, blacksmith shop, and dwelling house; yield to date, $5,000,000.

Carbonate Mining Co., of Colorado Springs. Incorporated October 29, 1877; capital stock, $15,000, in 3,000 shares, $50 each; office Colorado Springs; trustees: O.C. Knox, T.A. Benbow, F.E. Kimball, Conrad Brieglieb, Arthur Peck, J.H. Mathers, and William Fensworth.

Champaigne. Benjamin F. Shoemaker and Co., proprietors; office Alma; claim 300 by 1,500 feet; discovered 1876; located in Mosquito Gulch, 7 miles from Alma and 10 miles from Fairplay; course of vein, northeast and southwest, width 6 feet, pay vein 3 feet, pitch 15 degrees; nature of ore, galena sulphurets, and carbonates; average value, 150 ounces silver per ton; 2 drifts, 150 and 200 feet in length, respectively; yield, $2,000.

Chespeake. Thomas B. Beauchamp, proprietor; office Leadville; claim 50 by 1,500 feet; located on the north spur of Mount Lincoln, 7 miles from Alma and 13 miles from Fairplay; nature of ore, galena, sulphurets, and carbonate of lead; average value, 130 ounces silver per ton; 1 tunnel, 25 feet in length and 25 feet from surface; yield, $300.

Chicago. John S. Borders, proprietor; office Alma; claim 200 by 1,500 feet; located on the north spur of Mount Lincoln, Consolidated Montgomery mining district, 7 miles from Alma and 13 miles from Fairplay; nature of ore, galena and gray copper; average value, 120 ounces silver per ton; 2 tunnels, 15 and 25 feet in length, respectively; yield, $4,000.

Chicago. Owned by Americus Loudy Pogue, Henry Williams, and D.H. Dougan, M.D.; office Alma; claim 300 by 1,500 feet, patented; located on North Star Mountain, Consolidated Montgomery mining district, 6 miles from Alma and 12 miles from Fairplay; course of vein, northeast and southwest, width 4 feet, pitch 10 degrees west; nature of ore, iron and copper pyrites; average value, $75 per ton; main shaft 250 feet deep, 1 tunnel 175 feet in length; boarding house 20 by 40 feet; yield to date, $100,000.

Chicago and New York Mining and Smelting Co. Incorporated February 27, 1874; capital stock, $500,000, in 5,000 shares, $100 each; offices, town of Holland, Colorado, and Chicago, Illinois; organized by Charles, Dwight G., and Park Holland, Jeremiah Laming, Richard S. Thompson, and Elijah W. Bontecon. The works of this company are located in the town of Holland, 3 miles from Alma and 4 miles from Fairplay.

Chicago Pool Gold and Silver Mining Co. Incorporated October 2, 1874; capital stock, $25,000, in 1,000 shares, $25 each; office Chicago; organized by Elisha E. Hundley, William H. Stevens, Henry Paul, John D.

Best, and Albert A. Cooper; James H. Rees, president, L.H. Pierce, secretary and treasurer.

Chicago Mica, Silver and Gold Mining Corporation. Incorporated May 4, 1878; capital stock, $500,000, in 50,000 shares, $10 each; office Pueblo; organized by John Fidler, W.W. Hartsell, and Joseph Bardine.

Columbia, Extension of the Chicago. Owned by Americus Loudy Pogue, D.H. Dougan, M.D., and Henry Williams; office Alma; claim 300 by 1,500 feet; located on North Star Mountain, Consolidated Montgomery mining district, 6 miles from Alma and 12 miles from Fairplay; course of vein, northeast and southwest, width 4 feet; pitch 10 degrees west; average value of ore, $75 per ton; 1 shaft, 250 feet deep, 4 tunnels, aggregating 600 feet in length, shaft house 40 by 60 feet.

Comet. Hall Valley Mining Co., proprietors; office Hall Valley; claim 300 by 1,500 feet, patented; discovered 1875 located on Bullion Mountain, Hall Valley mining district, 25 miles from Fairplay; course of vein, northeast and southwest, width 2 feet; nature of ore, galena and gray copper; 1 shaft 50 feet deep, 1 tunnel 200 feet in length.

Commercial. Elisha Bass and M.H. Mahaney, proprietors; office Fairplay; claim 300 by 1,500 feet; discovered 1878; located in Buckskin Gulch, Mosquito mining district, 5 miles from Alma and 10 miles from Fairplay; vertical vein, trending north and south, width 4 feet; nature of ore, galena; assay value, 20 ounces silver and $15 gold per ton; 1 shaft 21 feet deep.

Como. Owned by Horse Shoe Mining Co. and Lyman Fay; office Alma; claim 300 by 1,500 feet; discovered 1876; located on Buckskin Mountain, 3 miles from Alma and 9 miles from Fairplay; course of vein, northwest and southeast, width 1 foot; nature of ore, sulphurets and galena; average value, 158 ounces silver per ton; 3 tunnels, 26, 60, and 105 feet in length, respectively.

Compromise. Assyria Hall and George W. Brunk, proprietors; office Alma; located on Mount Bross, Montgomery mining district, 4 miles from Alma and 10 miles from Fairplay.

Condor. John R. Lindgren, proprietor; office Chicago; claim 300 by 1,500 feet, patented; located on the southern slope of Mount Lincoln, 6 miles from Alma and 12 miles from Fairplay; course of vein, northwest and southeast; nature of ore, galena and sulphurets; 1 shaft 15 feet deep, 1 tunnel 80 feet in length, 2 drifts 15 feet each.

Confidence. Owned by James Redman, William F. Redman, and others; office Alma; claim 300 by 1,500 feet; discovered 1877; located in Mosquito Gulch; 6 miles from Alma and 9 miles from Fairplay; vertical vein, trending north and south; width, 15 feet, pay vein 4 feet; nature of ore, galena and iron pyrites; 1 adit 15 feet in length.

Cora Silver Mining Co. Incorporated November 5, 1872; capital stock, $500,000, in 5,000 shares, $100 each; office Golden; organized by H.H. Given, Charles Fleury, G.D. Root, Stephen Z. Hoyle, and George Westman.

D.H. Hill. Owned by W.H. Weir; office Alma; claim 300 by 1,500 feet; located on the north slope of Mount Lincoln, 4 miles from town of Alma and 10 miles from Fairplay.

Danville. Owned by John S. Borders and Thomas Beauchamp; office Alma; claim 300 by 1,490 feet; located on the north spur of Mount Lincoln, Consolidated Montgomery mining district, 7 miles from Alma and 13 miles from Fairplay; nature of ore, gray copper and galena; average value, 200 ounces silver per ton, 5 tunnels, aggregating 200 feet in length; yield, $12,000.

Democrat. Owned by Benjamin Hoil and Dr. E.F. Cameron; office Dudley; claim 300 by 1,500 feet; located at the eastern base of Mount Lincoln, Consolidated Montgomery mining district, 5 miles from Alma and 11 miles from Fairplay; vertical vein, trending northeast and southwest, width 18 inches; nature of ore, oxide of iron; average value, $200 per ton; 1 shaft 25 feet deep, 1 tunnel 175 feet in length; yield, $10,000.

Dolly Varden. Owned by the Hall & Brunk Silver Mining Co.; George W. Brunk, superintendent; office Alma; claim 300 by 1,500 feet, patented; located on Mount Bross, 4 miles from Alma and 10 miles from Fairplay; course of vein, southwest and northeast; width 4 feet, pay vein 3 feet, pitch 15 degrees west; nature of ore, sulphurets of silver; average value for 6 months, $276.40 per ton; main shaft 120 feet deep, 6 tunnels, aggregating 1,600 feet in length, 12 levels, 80 to 100 feet from surface, aggregating 1,700 feet in length; this property is yielding largely.

Drum Major. Owned by the Hall & Brunk Silver Mining Co.; George W. Brunk, superintendent; located on Mount Bross, Consolidated Montgomery mining district, 4 miles from Alma and 10 miles from Fairplay.

Eagle. Owned by the Alma Pool Association; office Alma; claim 300 by 1,500 feet, patented; located on the southeastern spur of Mount Lincoln, 6 miles from Alma and 12 miles from Fairplay; nature of ore, galena and sulphurets; average value, $200 per ton; main shaft 100 feet deep, 5 levels aggregating 150 feet in length.

Eclipse. Henry R. Horace Wolcott and James V. Dexter, proprietors; office Alma; claim 150 by 1,400 feet; discovered 1872; located in Mosquito Gulch, 6 miles from Alma and 9 miles from Fairplay; course of vein, northeast and southwest; width 18 inches, pitch 12 degrees southeast; nature of ore, sulphurets; average value, 200 ounces silver per ton; 1 shaft 200 feet deep, 4 levels aggregating 245 feet in length.

Eclipse. Owned by Cady, Mullen & Frazer; office Dudley; claim 300 by 1,500 feet; located on the north fork of Four Mile Creek, 10 miles from Fairplay; course of vein, northeast and southwest, width 8 feet, pitch 75 degrees; nature of ore, carbonate of lead; 1 shaft 12 feet deep.

Ella. Owned by Dexter & Shields; office Alma; claim 300 by 1,500 feet; located on Mount Bross, Consolidated Montgomery mining district, 4 miles from Alma and 10 miles from Fairplay.

Eureka Gold and Silver Mining Co., Colorado. Incorporated November 24, 1874; capital stock, $5,000, in 200 shares, $25 each; office Chicago; organized by Albert Cooper, L.H. Pierce, W.H. Stevens, A.B. Wood, and William H. Loomis.

Fairplay Gold Mining Co. Incorporated June 30, 1874; capital stock, $1,000,000, in 10,000 shares, $100 each; offices Fairplay and Denver, Colorado; organized by John W. Smith, Frederick A. Clark, and Henry Crow. This property extends along the bed of the Platte River for a distance of 6 miles. It is located at the town of Fairplay, and has a flume 1 mile in length. It is operated by E.L. Thayer, and yielded in four months of 1877, $30,000.

Ford. Owned by the Ford Consolidated Gold and Silver Mining Co.; incorporated June 2, 1877; capital stock, $1,000,000, in 100,000 shares,

$10 each; offices Fairplay, Colorado, and New York City; trustees: James E. Broome, James A. Betts, James Plant, Charles K. Bill, George L. Hooper, Oliver H. Pierson, and James McNassar; claim 300 by 1,500 feet, located on the east spur of Mount Lincoln, 6 miles from Alma and 12 miles from Fairplay; nature of ore, galena and sulphurets; average value, 200 ounces silver per ton; 2 tunnels, 100 and 150 feet in length, respectively, 1 drift 40 feet in length.

Forlorn Hope. Owned by Dr. E.F. Cameron and M.A. Mahany; office Dudley; claim 300 by 1,500 feet; located at the eastern base of Mount Lincoln, 5 miles from Alma and 11 miles from Fairplay; nature of ore, sulphurets of silver; average value, 150 ounces silver per ton; 1 open cut 12 feet in length with a 15-foot face; yield, $130.

Fortunatus. Owned by Dr. E.F. Cameron, M.A. Mahany, and E. Bass; office Fairplay; claim 300 by 1,500 feet; discovered 1877; located in Montgomery mining district, 6 miles from Alma and 12 miles from Fairplay; width of vein, 18 inches; nature of ore, oxide of iron; average value, 3 ounces gold per ton; 1 shaft 18 feet deep.

Fritz. John S. Borders, proprietor; office Alma, claim 210 by 2,100 feet; located on the north spur of Mount Lincoln, Montgomery mining district, 7 miles from Alma and 13 miles from Fairplay; nature of ore, galena and gray copper; value 250 ounces silver per ton; 2 tunnels, 45 and 65 feet in length, respectively, cabin 20 by 24 feet; yield, $1,200.

Gertrude. Owned by Felix McLaughlin, Mrs. Shallcross, and Harris & Bass; office Alma; claim 50 by 1,500 feet; located on the summit of Mount Lincoln, 8 miles from Alma and 12 miles from Fairplay; nature of ore, galena and sulphurets; average value, 150 ounces silver per ton; 1 tunnel 100 feet in length, from which a shaft is sunk to a depth of 60 feet.

Gilbert. William F. Kendrick, proprietor; office Alma; claim 300 by 1,500 feet; discovered 1878; located on Mount Bross, 3 miles from Alma and 9 miles from Fairplay; course of vein, northeast and southwest, width from 4 to 40 inches, pitch 70 degrees; nature of ore, sulphurets and galena; assay value, 60 to 350 ounces silver per ton; 1 drift 20 feet in length and 15 feet from surface; yield, $1,000.

Gold Placer Mining Co. Organized under the laws of the State of Connecticut, April 16, 1877; capital stock, $5,000,000; offices Fairplay,

Colorado, and Hartford, Connecticut; H.G. Angle, president, C. Burnham, secretary, Sydenham Mills, agent.

Gould and Curry. Theodore King, of Georgetown, proprietor; claim 150 by 1,500 feet; discovered 1873; located on Bullion Mountain, 20 miles from Georgetown, in Hall Valley mining district; vertical vein, running north and south, width 12 feet, pay vein 2 feet; nature of ore, galena and iron pyrites; average value, $75 per ton; 3 shafts, aggregate depth 65 feet, 2 levels, aggregate length 190 feet.

Gray Horse. Owned by the Horse Shoe Mining Co.; Robert A. Kirker, superintendent; office Fairplay; claim 300 by 1,500 feet; located near the south fork of Four Mile Creek, 10 miles from Fairplay; course of vein, east and west, width 12 feet, pitch 75 degrees; nature of ore, carbonate of lead and iron; 1 shaft 20 feet deep, 1 tunnel, 25 feet in length.

Greeley Mining, Ditch and Fluming Co. Incorporated July 20, 1870; capital stock, $24,000, in 2,400 shares, $10 each; office Hamilton; organized by John Foot, A.M. Greenleaf, and George W. Leas.

Gregory. Owned by William F. Kendrick; office Alma; claim 300 by 1,500 feet; discovered 1878; located on Mount Bross, 3 miles from Alma and 9 miles from Fairplay; course of vein, northeast and southwest, width 6 to 36 inches; nature of ore, galena and sulphurets; value, 60 to 300 ounces silver per ton; 1 drift 60 feet in length and 15 feet from surface.

Guinea Pig. Owned by Henry R. Wolcott, James V. Dexter, and George Sidell; office Alma and Fairplay; claim 300 by 1,500 feet, patented; discovered 1873; located on the western slope of Mount Bross, 6 miles from Alma and 12 miles from Fairplay; course of vein, northeast and southwest, pitch 15 degrees; nature of ore, galena and sulphurets, value 80 ounces silver per ton; 1 drift 100 feet in length.

Hall and Brunk Silver Mining Co. Incorporated May 14, 1877; capital stock, $1,000,000, in 40,000 shares, $25 each; offices Alma, Colorado, and New York City; E. Boudinot Colt, president, Assyria Hall, vice president, John T. Brownlow, secretary, George W. Brunk, superintendent. The property of this company is located on Mount Bross, Consolidated Montgomery mining district, 4 miles from town of Alma and 10 miles from Fairplay. It comprises 80 acres of the most prominent lode claims of this mountain, together with 160 acres of placer ground, which is located on the Platte River, 1 mile from town of Dudley. The abovementioned property is patented and in a well developed and paying condition. The principal mines owned by this company are the Dolly Varden, German, Polaris, Iron Dyke, Friday, and Undercliff.

Hard to Beat. L. Ford and T. Pusks, proprietors; office Fairplay; claim 300 by 1,500 feet; discovered 1874; located in Mosquito Gulch, 7 miles from Alma and 10 miles from Fairplay; course of vein, northeast and southwest, width 16 feet, pay vein 5 feet, pitch 30 degrees; nature of ore, carbonate of lead; average value, 3 ounces gold and 30 ounces silver per ton; 2 shafts, 14 and 30 feet in depth, respectively.

Hathaway, Placer. Charles G. Hathaway and George W. Brunk, proprietors; office Fairplay; claim comprises 320 acres of placer ground, located in the Consolidated Montgomery mining district, 6 miles from Alma and 12 miles from Fairplay.

Haynes. Owned by the Alma Pool Association; office Alma; claim 50 by 1,500 feet, patented; located on the southeast spur of Mount Lincoln, 6 miles from Alma and 12 miles from Fairplay; 3 shafts, 15, 25, and 50 feet in depth, respectively.

Hiawatha. Assyria Hall and George W. Brunk, proprietors; office Alma; claim 2,000 by 3,000 feet; located on Mount Bross, 4 miles from Alma and 10 miles from Fairplay; course of vein, northeast and southwest, average width 2 feet, pitch 15 degrees; nature of ore, black sulphurets; average value, $200 per ton; development, 8 tunnels, aggregating 300 feet in length.

Hidden Treasure. W.K. Burchinell and A.B. Crook, proprietors; office Fairplay; claim 300 by 1,500 feet; discovered 1876; located in Mosquito Gulch, Mosquito mining district, 6 miles from Alma and 9 miles from Fairplay; course of vein, northwest and southeast, width 7 feet, pay vein 2 feet; nature of ore, sulphurets and galena; 2 shafts, 35 and 65 feet in depth, respectively.

Hoil. Owned by Benjamin Hoil and E.F. Cameron; office Dudley; claim 300 by 1,500 feet; located at the eastern base of Mount Lincoln, 5 miles from Alma and 11 miles from Fairplay; nature of ore, galena and sulphurets; average value, 175 ounces silver per ton; 4 tunnels, 35, 60, 70, and 90 feet in length, respectively, and 4 levels aggregating 350 feet in length; yield, $7,000.

Horse Shoe Mining Co. Incorporated 1874; A.W. Gill, president, Robert A. Kirker, superintendent; office Fairplay. The property of this company comprises 13 lodes or mineral veins, 300 by 1,500 feet each, and 1,200 acres of patented ground on which, of late, discoveries of carbonates have been made; located in Horse Shoe mining district, 10 miles from Fairplay.

Joe Chaffee. John H. Smith, proprietor; office Fairplay; claim 300 by 1,500 feet; discovered 1875; located in Mosquito Gulch, 5 miles from Alma and 8 miles from Fairplay; course of vein, northeast and southwest, width 12 feet, pay vein, 2 feet; nature of ore, gray copper and galena; assay value, 500 ounces silver per ton; 2 shafts, 20 and 30 feet, respectively, 1 tunnel 80 feet in length.

Joe Thatcher. Assyria Hall and George W. Brunk, proprietors; office Alma; located on Mount Bross, Montgomery mining district, 5 miles from Alma and 10 miles from Fairplay.

Juniata. Assyria Hall and George W. Brunk, proprietors; office Alma; located on Mount Bross, Montgomery mining district, 4 miles from Alma and 10 miles from Fairplay.

Kansas, and Extension. Jackson & Shelhamer, proprietors; office Alma; claim 300 by 3,000 feet; discovered 1876; located on Buckskin Mountain, 3 miles from Alma and 9 miles from Fairplay; course of vein, northeast and southwest, width 6 feet, pitch 15 degrees; nature of ore, sulphurets and galena; average value, 180 ounces silver per ton; 1 shaft 12 feet deep, 3 drifts, 40, 75, and 200 feet in length, respectively; yield, $15,000.

Ketsby. Owned by Benjamin Hoil and Dr. E.F. Cameron; office Dudley; claim 300 by 1,500 feet; located at the eastern base of Mount Lincoln, 5 miles from Alma and 11 miles from Fairplay; vertical vein, trending north and south, width 14 inches; nature of ore, iron and copper pyrites, average value, $52 per ton; 1 shaft 25 feet deep, 1 tunnel 30 feet in length; yield, $500.

Keystone. Owned by D. Sheahan, J.G. Marshall, and R.B. Whilston; office Dudley; claim 300 by 1,500 feet, patented; located in Mosquito Gulch, 6 miles from Alma and 9 miles from Fairplay; course of vein, northeast and southwest, width 8 inches, pitch 45 degrees southeast; nature of ore, galena and sulphurets; average value, 200 ounces silver per ton; 1

shaft 100 feet deep, 1 level 100 feet, and 1 drift 45 feet in length; yield, $5,000.

K.P. T.J. Hughes and S.P. Lunt, proprietors; office Alma; claim 300 by 1,500 feet; located in Mosquito Gulch, 4 miles from Alma and 7 miles from Fairplay; course of vein, north and south, width 2 feet, pitch 40 degrees; nature of ore, free gold; 2 drifts, 60 and 130 feet in length, respectively.

Last Chance. Owned by Samuel McMillan, Benjamin Graham, Charles Defebaugh, and James Brennan; office Fairplay; claim 300 by 1,500 feet; located at the head of the north fork of Four Mile and Horse Shoe creeks, 10 miles from Fairplay; course of vein, northeast and southwest, width 3 feet, pitch 70 degrees; nature of ore, carbonate of lead; average value, $40 per ton; 1 shaft 20 feet deep, 1 tunnel 75 feet in length.

Leftwick. Hall Valley Mining Co., proprietors; office Hall Valley; claim 300 by 1,500 feet, patented; discovered 1872; located on Bullion Mountain, Hall Valley mining district, 25 miles from Fairplay; course of vein, northeast and southwest, width 3 feet; nature of ore, galena and gray copper.

Lime. Owned by Assyria Hall and George W. Brunk; office Alma; claim 50 by 1,500 feet; located on the north spur of Mount Lincoln, 7 miles from Alma and 13 miles from Fairplay; nature of ore, galena and sulphurets; value, $250 per ton; 1 surface opening 70 feet in length; yield, $7,000.

Lincoln Silver Mining Co., of Colorado. Incorporated May 6, 1874; capital stock, $100,000, in 5,000 shares, $20 each; offices Dudley and Denver, Colorado; organized by Judson H. Dudley, Frederick ("Fred") A. Clark, Jas. B. Cass, Elihu C. Wilson.

Little Nell. Owned by Thomas and Alfred Willey; office Fairplay; claim 300 by 1,500 feet; discovered 1877; located on Horse Shoe Mountain, Mosquito mining district, 7 miles from Fairplay; course of vein, northeast and southwest, width 6 feet, pay vein 6 inches; nature of ore, galena and gray copper; average value, 61 ounces silver per ton; 1 shaft 25 feet deep.

London. L. Ford and T. Pusks, proprietors; office Fairplay; claim 300 by 1,500 feet; discovered 1874; located in Mosquito Gulch, 7 miles from

Alma and 10 miles from Fairplay; course of vein, northeast and southwest, width 16 feet, pay vein 5 feet, pitch 30 degrees; nature of ore, carbonate of lead, average value, 3 ounces gold and 30 ounces silver per ton; 2 drifts, 103 and 285 feet in length, respectively.

Lone Star. Ira and Joshua Mulock, proprietors; office Alma; claim 300 by 1,500 feet; discovered 1876; located in Mosquito Gulch, 7 miles from Alma and 10 miles from Fairplay; vertical vein, trending north and south, width 6 feet, pay vein 4 feet; nature of ore, sulphurets; average value, 200 ounces silver per ton; 1 shaft, 230 feet deep, 5 levels, aggregating 290 feet in length; yield, $12,000; shaft house 30 by 30 feet.

Madagascar. Hall Valley Mining Co., proprietors; office Hall Valley; claim 300 by 1,500 feet; discovered 1872; located on Bullion Mountain, Hall Valley mining district, 25 miles from Fairplay; course of vein, northeast and southwest; 1 drift 150 feet in length.

Magnolia. Lyman Fay, proprietor; office Alma; claim 300 by 1,500 feet; discovered 1876; located on Silver Lake Hill, Consolidated Montgomery mining district, 7 miles from Alma and 13 miles from Fairplay; course of vein, north and south, width 4 feet, pay vein 2 feet, pitch 10 degrees; nature of ore, free gold; average value, $200 gold per cord; 1 shaft 30 feet deep, 1 drift 100 feet in length, 1 level 50 feet in length and 20 feet from surface; yield, $10,000.

Mannion. Owned by William Mannion and Co.; office Alma; claim 300 by 1,500 feet; located on the northeast spur of Mount Bross, 4 miles from town of Alma and 10 miles from town of Fairplay; horizontal deposit, trending southeast and northwest; nature of ore, sulphurets of silver and galena; value, 75 to 250 ounces silver per ton; developed by a shaft 15 feet deep and drift 55 feet in length.

Marie De Los Reyes. Owned by the Horse Shoe Mining Co.; Robert A. Kirker, superintendent; office Fairplay; claim 250 by 1,500 feet; located on the summit of Horse Shoe Mountain, 10 miles from Fairplay; course of vein, northeasterly; width 5 feet, pitch 70 degrees; nature of ore, carbonate of lead; average value, $45 per ton; main shaft 40 feet; yield, $4,000.

McDonald. Miles McDonald and James Moynahan, proprietors; office Alma; claim 300 by 1,500 feet; discovered 1877; located on Iron Hill, California mining district, 1 mile from Leadville; course of vein, north and south, width 22 inches, pitch 70 degrees east; nature of ore, carbonate of lead; average value, 82 ounces silver per ton; main shaft 132 feet deep, and 2 other shafts, 50 and 76 feet in depth, respectively.

Miller. Owned by Assyria Hall and George W. Brunk; office Alma; claim 50 by 1,500 feet; located on the north spur of Mount Lincoln, 7 miles from Alma and 13 miles from Fairplay; nature of ore, galena and sulphurets; value, $200 per ton; 1 tunnel 45 feet in length; yield, $7,000.

Millionaire Mining and Tunneling Co. Incorporated January 19, 1876; capital stock, $250,000, in 25,000 shares, $10 each; office Denver; William F. McClelland, president, Jacob L. Peabody, vice president, John Kiefer, treasurer, Frank M. Davis, secretary, James S. Dillon, assistant secretary. The property of this company is located on Grand View Mountain, Hall Valley mining district, 25 miles from Fairplay. It comprises the Millionaire, Lavinia, Central, and Lucky lodes; developed by a tunnel 300 feet in length.

Mills & Co.'s Placer. Owned by Sydenham Mills, Henry W. Hodges, John and Crawford Reed, Alfred Ward, and Clark S. Topping; located in town of Alma, at the confluence of the Platte and Buckskin rivers. This property has been successfully worked since 1870, comprises 600 acres of patented ground, and has 3 miles of ditch, carrying 1,000 inches of water. It is worked by hydraulic pressure and has yielded to date, $80,000.

Milwaukee. Assyria Hall and George W. Brunk, proprietors; office Alma; located on Mount Bross, Montgomery mining district, 4 miles from Alma and 10 miles from Fairplay.

Minerva. Owned by Thomas Willey and C.S. Eyster; office Denver; claim 200 by 1,500 feet; discovered 1874; located in Sacramento Gulch, 7 miles from Fairplay; nature of ore, carbonate of lead; value, 100 ounces silver per ton; 1 shaft 20 feet deep.

Missouri. Owned by F.M. Rainey and John Ifinger; office Fairplay; claim 300 by 1,500 feet; located on the flats, near south fork of Four Mile Creek, 10 miles from Fairplay; course of vein, northeast and southwest, pitch 70 degrees; nature of ore, carbonate of lead; 1 shaft 15 feet deep.

Moose. Owned by the Moose Mining Co.; incorporated April 15, 1872; capital stock, $2,000,000, in 200,000 shares, $10 each; offices Dudley, Colorado, and 62 Broadway, New York; organized by Daniel Plummer, Richard B. Ware, Joseph H. Myers, Andrew W. Gill, and Judson

H. Dudley; Daniel B. Allen, president, Charles M. Stead, vice president, Harry Allen, secretary, Jacob Houghton, superintendent. The property of this company is well developed and located on the north slope of Mount Bross, 4 miles from town of Dudley and 10 miles from Fairplay. It comprises 25 lodes or mineral veins covering 160 acres of the surface. The vein is a contact deposit; the ore is galena, sulphurets, and zinc blende, ranging in value from 40 to 700 ounces silver per ton. The buildings consist of a blacksmith shop, ore houses, a boarding house with accommodations for 80 men, the office, and the works. The works have a capacity for treating 10 tons of ore per day, they cover 1 acre of ground and are located beside the Platte River, in the town of Dudley, 1 mile from Alma, and 7 miles from Fairplay.

Moss Vale. Owned by John S. Borders and Thomas B. Beauchamp; office Alma; claim 210 by 1,200 feet; located on the north spur of Mount Lincoln, Consolidated Montgomery mining district, 7 miles from Alma and 13 miles from Fairplay; nature of ore, galena, sulphurets, and carbonate of lead; average value, 150 ounces silver per ton; 2 tunnels, 45 and 100 feet in length, respectively; yield in 1877, $20,000.

Mother. L. Ford and T. Pusks, proprietors; office Fairplay; claim 300 by 1,500 feet, patented; discovered 1873; located in Mosquito Gulch, 7 miles from Alma and 10 miles from Fairplay; course of vein, northeast and southwest, width 16 feet, pay vein 5 feet, pitch 30 degrees; nature of ore, carbonate of lead; average value, 30 ounces silver and 3 ounces gold per ton; 2 drifts, 20 and 130 feet in length, respectively.

Mount Bross Tunneling Co. Organized under the laws of the State of New York, May 14, 1877; capital stock, $2,000,000, in 200,000 shares, $10 each; offices Alma, Colorado, and New York City; George W. Brunk, president, Henry S. Vanderbilt, secretary, George W. Brunk, manager; Assyria Hall, John T. Brownlow, and Aaron H. Cragin, directors; claim 300 by 3,000 feet; located on Mount Bross, 4 miles from town of Alma and 10 miles from Fairplay; length of tunnel, 300 feet.

Mount Lincoln Improvement Co. Incorporated August 10, 1875; capital stock, $50,000, in 5,000 shares, $10 each; office town of Alma; organized by William A. Hawkins, William E. Musgrove, Henry R. Wolcott, and James V. Dexter.

Mountain Boy. Owned by William Sigafus, M.W., W.K., and C.S. Smith.

Mountain Lion. Owned by S.I. and H.F. Smith; office Alma; claim 300 by 1,500 feet; located on Mosquito Mountain, 6 miles from Alma and 13 miles from Fairplay; course of vein, northeast and southwest, width 35 feet, pitch 10 degrees southwest; nature of ore, oxide of iron and iron pyrites; average value, $30 per ton; 1 shaft 65 feet deep, 1 tunnel 35 feet in length.

Mudsill. Owned by C.S. Eyster and Thomas Willey; office Denver; claim 300 by 1,500 feet; located in Sacramento Gulch, Montgomery mining district, 7 miles from Fairplay; width of vein, 2 to 18 inches; nature of ore, carbonate of lead; 1 shaft 10 feet deep.

Musk Ox. Owned by the Musk Ox Gold and Silver Mining Co., of Colorado; incorporated January 21, 1878; capital stock, $500,000, in 50,000 shares, $10 each; office Denver, Colorado, and New York City; trustees: Charles K. Bill, Alexander R. Chisholm, Oliver H. Pierson, Jonas O. Molander, John J. Van Demoer, Edward P. Fowler, William L. Woodruff, James McNassar, and Theodore D. Jervey; claim 300 by 1,500 feet; located on the eastern spur of Mount Lincoln, 6 miles from Alma and 12 miles from Fairplay; nature of ore, galena and sulphurets; average value, 200 ounces silver per ton; 3 tunnels, 30, 60, and 120 feet in length, respectively.

New Jersey. Owned by W. P. Cook and Windham S. Sayer and James Moynahan; office Alma; claim 300 by 1,500 feet; located in Mosquito Gulch, Mosquito mining district, 2 miles from Alma and 8 miles from Fairplay; course of vein northeast and southwest; nature of ore, iron and copper pyrites, value, $60 per ton; main shaft 50 feet deep, 3 drifts, 85, 100, and 150 feet in length, respectively; yield, $10,000.

Nova Zembla. George W. Brunk and Charles G. Hathaway, proprietors; office Fairplay; claim 300 by 1,500 feet, patented; discovered 1875; located in Montgomery mining district, 6 miles from Alma and 12 miles from Fairplay; course of vein, northwest and southeast, width 6 feet, pay vein 3 feet; nature of ore, free gold; average value, $100 gold per ton; main shaft 185 feet deep, and 2 other shafts aggregating 145 feet in depth, 1 tunnel 70 feet in length.

Occidental. Owned by Nelson Hallock and Samuel Davidson; office Alma; claim 50 by 1,500 feet; located on the eastern slope of Mount Lincoln, Montgomery mining district, 7 miles from Alma and 13 miles from Fairplay; nature of ore, galena and sulphurets; average value, $100 per ton; 2 tunnels, aggregating 150 and 300 feet in length, respectively; shaft house 14 by 16 feet.

Ohio. Owned by the Horse Shoe Mining Co., Robert A. Kirker, superintendent; office Fairplay; claim 300 by 1,500 feet, located on the flats between the forks of Four Mile Creek, ten miles from Fairplay; course of vein, 52 degrees, 45 minutes east, width 2 feet, pay vein 1 foot, pitch 70 degrees southeast; nature of ore, carbonate of lead; average value, $45 per ton; main shaft 60 feet deep; yield, $2,000.

Orphan Boy Mining and Tunnel Co. Incorporated December 4, 1874; capital stock, $250,000, in 2,500 shares, $100 each; offices, town of Mosquito and Denver, Colorado; organized by Henry Crow, John W. Smith, Charles M. Taylor, Samuel Leach, and William U. Johnston.

Pans. L. Ford and T. Pusks, proprietors; office Fairplay; claim 300 by 1,500 feet, patented; discovered 1874; located in Mosquito Gulch, 7 miles from Alma and 10 miles from Fairplay; course of vein, northeast and southwest, width 16 feet, pay vein 5 feet; pitch 30 degrees; nature of ore, carbonate of lead; average value, 3 ounces gold and 30 ounces silver per ton; 4 drifts aggregating 193 feet in length.

Park Pool Co. Incorporated December 17, 1875; capital stock, $100,000, in 1,000 shares, $100 each; office Black Hawk; organized by Prof. Nathaniel P. Hill, J.A. Thatcher, H.R. Wolcott, William H. Stevens, and H. Bieger.

Peabody, Placer. Lelon Peabody, proprietor; located on Tarryall Creek, at town of Hamilton, 15 miles from Fairplay.

Peerless. Owned by James H.B. McFerran of Colorado Springs; claim 300 by 3,000 feet; located on Horse Shoe Mountain, Horse Shoe mining district, 5 miles from town of Eureka, and 10 miles from Fairplay; course of vein, northeast and southwest, width 5 feet, pitch 50 degrees; nature of ore, galena and carbonates; average value, 30 ounces silver and 1 ¾ ounces gold per ton; main shaft 40 feet deep, aggregate length of tunnels 180 feet; yield, $10,000.

Phillips. N.J. Bond, proprietor; office Council Bluffs, Iowa; claim 300 by 1,500 feet; located in Buckskin Gulch, Buckskin mining district, 1½ miles from Alma and 7 miles from Fairplay; vertical vein, trending north and south, width 50 feet; nature of ore, free gold; average value,

$100 gold per ton; 1 shaft 200 feet deep, 3 drifts aggregating 62 feet in length; yield, $500,000.

Platte Valley Placer Gold Mining Co. Incorporated March 6, 1878; capital stock, $3,000,000, in 300,000 shares, $10 each; offices Denver, Colorado, and New York City; trustees: Henry Crow, Walter S. Cheesman, John W. Smith, Edward E. Hartwell, Samuel Leach, T.W.B. Hughes, and H.G. Angle.

Pleasant Valley Placer Mining Co. Incorporated October 29, 1872; capital stock, $8,000; organized by Stephen H. Pease, John Mechling, Thomas J. Freeman, George E. Pease, Albert B. Crook, Aden Alexander, Sylvestus A. Safford, and Daniel H. Wilson.

Present Help. Owned by Louis Dugal and the estates of Breece and Witter; office Denver; claim 300 by 1,500 feet; discovered 1870; located near the summit of Mount Lincoln, Montgomery mining district, 4 miles from Alma and 10 miles from Fairplay; nature of ore, gray copper; average value $100 per ton; main shaft 50 feet deep, 2 levels, 60 and 180 feet in length, respectively; yield to date, $2,300.

Prince Albert Tunnel. Owned by William Sigafus, M.W., W.K., and C.S. Smith; office Alma; claim 300 by 6,000 feet; located in Mosquito mining district, 6 miles from Alma and 12 miles from Fairplay; nature of ore, gray copper and galena; average value, 100 ounces silver per ton.

Queen of the Hills. Nelson Hallock & Co., proprietors; office Alma; claim 300 by 1,500 feet; discovered 1876; located in Mosquito Gulch, 7 miles from Alma and 10 miles from Fairplay; course of vein, north and south, width 2 feet, pay vein 15 inches, pitch 60 degrees; nature of ore, galena and carbonate of lead; average value, $160 silver per ton; 1 shaft 65 feet in depth, 2 levels aggregating 160 feet in length; yield, $15,000.

Queen of the West. Owned by James G. Brooks, Frank Kaley, and W.N. Dameron; office Alma; claim 300 by 1,500 feet; located on the eastern slope of Mount Lincoln, Montgomery mining district, 6 miles from Alma and 12 miles from Fairplay; vertical vein, trending northeast and southwest, width 6 feet; nature of ore, galena and iron pyrites; average value, 15 ounces silver per ton; 1 tunnel 100 feet in length.

Rendezvous. John S. Borders, proprietor; office Alma; claim 150 by 1,500 feet; located on the north spur of Mount Lincoln, Montgomery mining district, 7 miles from Alma and 13 miles from Fairplay; nature of ore, sulphurets, galena and carbonate; value, 95 to 400 ounces silver per ton; 1 tunnel 100 feet in length and 20 feet from surface, 2 levels, 40 and 60 feet in length, respectively; this mine has yielded in 12 months, $23,000.

Rip Van Winkle. Lyman Fay, proprietor; office Alma; claim 300 by 1,500 feet; discovered 1877; located in Montgomery mining district, 6 miles from Alma and 12 miles from Fairplay; course of vein, north and south, width 3 feet, pay vein 2 feet, pitch 10 degrees; nature of ore, free gold; average value, $200 per cord; 2 shafts, 25 and 30 feet, respectively, 1 adit 56 feet in length.

Rob Roy. Owned by the Musk Ox Gold and Silver Mining Co.; offices Denver, Colorado, and New York City; claim 300 by 1,500 feet; located on the eastern spur of Mount Lincoln, Montgomery mining district, 6 miles from Alma and 12 miles from Fairplay; nature of ore, galena and sulphurets; average value, 100 ounces silver per ton; 1 shaft 30 feet deep.

Russia. Owned by the Russia Silver Mining Co.; organized under the laws of the State of New York, December 5, 1876; capital stock, $600,000, in 60,000 shares, $10 each; offices Alma, Colorado, and Chicago, Illinois; Americus L. Pogue, president, John H. Webber, vice president, E. Graham Haight, secretary and treasurer, D.H. Dougan, M.D., superintendent; claim 400 by 1,500 feet, patented; located on Mount Lincoln, Montgomery mining district, 5 miles from Alma and 11 miles from Fairplay; nature of ore, sulphurets and galena; average value, 140 ounces silver per ton; development, 2 tunnels, 40 and 500 feet in length, respectively, 9 levels aggregating 300 feet in length; yield, $100,000.

St. Marys. Moses Hall, proprietor; claim 300 by 1,500 feet; discovered 1874; located in Mosquito Gulch, 5 miles from Alma and 8 miles from Fairplay; course of vein, northeast and southwest, width 14 inches, pitch 40 degrees; nature of ore, sulphurets and galena; average value, 45 ounces silver per ton; 1 drift 35 feet in length and 35 feet from surface.

Saltiel Mica and Porcelain Co. Incorporated June 17, 1878; capital stock, $1,000,000, in 100,000 shares, $10 each; office Denver; trustees: John Elsner, Emanuel H. Saltiel, David I. Ezekiel, A.M. Ghost, and A.C. Fisk.

Schiller. Otto Mangold and C.C. McCarthy, proprietors; office Alma; claim 300 by 1,500 feet; discovered 1877; located in Mosquito Gulch, 6 miles from Alma and 9 miles from Fairplay; vertical vein, trending east and west, width 8 feet, pay vein 30 inches; nature of ore, gray copper and galena, average value, 90 ounces silver per ton; 1 shaft 75 feet deep, 1 drift 20 feet in length, 1 level 58 feet in length and 45 feet from surface; yield, $3,500.

Security. Owned by the Chicago Pool, Gold and Silver Mining Co.; incorporated October 2, 1875; capital stock, $25,000, in 1,000 shares, $25 each; office Chicago, Illinois; organized by Elisha E. Hundley, William H. Stevens, Henry Paul, John D. Best, and Albert A. Cooper, James H. Rees, president, L.H. Pierce, secretary and treasurer; claim 300 by 1,500 feet, patented; located on Mount Bross, Montgomery mining district, 3 miles from Alma and 9 miles from Fairplay; nature of ore, iron pyrites; average value, $80 per ton; 1 tunnel 700 feet in length and 180 feet from surface.

Sedgwick. Owned by the Horse Shoe Mining Co.; Robert A. Kirker, superintendent; office Fairplay; claim 300 by 1,500 feet; located on the north fork of Four Mile Creek, 10 miles from Fairplay; course of vein, north 52 degrees and 30 minutes east, width 5 feet, pitch 68 degrees; nature of ore, carbonate of lead and copper; average value, $45 per ton; main shaft 70 feet; yield, $10,000.

Shamrock Property. Owned by John T. Brownlow, William H. James, and P.H. Barron; office Alma; claim comprises 35 acres of mineral land; located on the northwest slope of Mount Bross, Consolidated Montgomery mining district, 6 miles from Alma and 12 miles from Fairplay; nature of ore, sulphurets and oxide of iron; average value, 50 ounces silver per ton; 2 tunnels, 30 and 50 feet in length, respectively.

Shamrock. Owned by Hanrahan and McAndrews; office Alma; claim 300 by 1,500 feet; located on Mount Bross Consolidated Montgomery mining district, 4 miles from Alma and 10 miles from Fairplay.

Silver Chief. Owned by Estate of Henry Knight; claim 300 by 1,500 feet; discovered 1873; located on Bullion Mountain, Park mining district, near town of Montezuma; course of vein, northeast and southwest, width 7 feet, pay vein 1 foot; development, 1 shaft 51 feet and 1 tunnel 45 feet.

Silver Exchange. Owned by S.I. and H.F. Smith; office Alma; claim 300 by 1,500 feet; located on Mosquito Mountain, 6 miles from Alma and

12 miles from Fairplay; course of vein, northeast and southwest, width 4 feet, pay vein 6 inches, pitch 10 degrees southeast; nature of ore, gray copper and galena; value, 300 to 1,500 ounces silver per ton; 1 shaft 15 feet deep, 1 tunnel 70 feet in length; yield, $300.

Silver Gem. Owned by James Moynahan, William H. Stevens, and Elisha E. Hundley; office Alma; claim 300 by 1,500 feet, patented; located on the western slope of Mount Bross, 6 miles from Alma and 12 miles from Fairplay; course of vein, northeast and southwest; nature of ore, galena and sulphurets; value, 80 ounces silver per ton; 1 shaft 30 feet deep, 1 tunnel 60 feet, and 1 drift 59 feet in length; ore house, blacksmith, and dwelling house; yield, $7,000.

Silver Queen. Owned by George H. Rummell, S.G. Hayden, and R.H. Wright; office Alma; claim 300 by 1,500 feet; discovered 1878; located on Mosquito Gulch, 6 miles from Alma and 12 miles from Fairplay; course of vein, northeast and southwest, width 4 feet, pay vein 1 foot, pitch 45 degrees east; nature of ore, sulphurets and galena; average value, 77 ounces silver per ton; 1 adit 10 feet in length with a 12-foot face.

Silver Saddle. Owned by Williams and Trevan; office Alma; claim 300 by 1,500 feet; located on Mount Bross, 4 miles from Alma and 10 miles from Fairplay; value of ore, $75 per ton; 1 shaft 100 feet deep.

Spar. Owned by the Horse Shoe Mining Co.; Robert A. Kirker, superintendent; office Fairplay; claim 300 by 1,500 feet; located on the north fork of Four Mile Creek, 10 miles from Fairplay; course of vein, northeast and southwest, width 6 feet, pitch 70 degrees; nature of ore, galena spar, iron, and carbonate of lead; 1 shaft 20 feet deep, 1 surface opening 15 feet in length; yield $500.

Spotted Hornet. Owned by the Horse Shoe Mining Co.; Robert A. Kirker, superintendent; office Fairplay; claim 300 by 1,500 feet; located on the flats between the north and south forks of Four Mile Creek, 10 miles from Fairplay; course of vein, north 45 degrees east, width 1 foot, pitch 70 degrees; nature of ore, carbonate of lead; average value, $25 per ton; 2 shafts, 20 feet in depth each.

Tarryall Creek Mining Co. Incorporated August 7, 1874; capital stock, $100,000, in 1,000 shares, $100 each; office Fairplay; organized by James Luttrell, Charles L. Hall, William Christian, Thomas T. Welley, James R. Olliver, Hugh Murdock, L.W. Lewis, and Assyria Hall.

Tarryall Diggings. Owned by Assyria Hall and others; claim comprises 380 acres of placer ground, situated near town of Hamilton, 80 miles from Denver; this property has been worked since 1859; the deposit is from 7 to 40 feet in depth and the improvements consist of cabins, ditches, and flumes.

Ten-Forty Silver Co. Incorporated February 7, 1870; capital stock, $80,000, in 800 shares, $100 each; office Denver; organized by Charles M. Mullen, William J. Palmer, Horace A. Gray, Walter S. Cheesman, David H. Moffat Jr., and John Evans.

Treasury. Owned by the Horse Shoe Mining Co.; Robert A. Kirker, superintendent; office Fairplay; claim 300 by 1,500 feet; located in Horse Shoe mining district, 10 miles from Fairplay; course of deposit, 52 degrees north and 30 degrees east, width 3 feet, pitch 70 degrees; nature of ore, carbonate of lead; average value, $20 per ton; main shaft 40 feet deep; yield, $1,000.

True Blue. Owned by the Alma Pool Association; office Alma; claim 50 by 1,500 feet; located on the southeast spur of Mount Lincoln, 6 miles from Alma and 12 miles from Fairplay; nature of ore, galena and sulphurets; average value, $100 per ton; 1 shaft 25 feet deep, 3 tunnels, 40, 50, and 100 feet in length, respectively.

U.P. T.J. Hughes and S.P. Lunt, proprietors; office Alma; claim 300 by 1,500 feet; located in Mosquito Gulch, 4 miles from Alma and 7 miles from Fairplay; vertical vein, trending north and south, width 18 inches; nature of ore, galena and iron pyrites, 1 drift 126 feet in length.

Whale. Hall Valley Mining Co., proprietors; office Hall Valley; claim 300 by 1,500 feet; discovered 1872; located on Bullion Mountain, Hall Valley mining district, 25 miles from Fairplay; course of vein, northeast and southwest, width 3 feet, pitch 45 degrees; nature of ore, galena and gray copper; average value, 115 ounces silver per ton; 4 drifts aggregating 1,960 feet in length.

Wilson. Owned by Assyria Hall and George W. Brunk; office Alma; claim 50 by 3,000 feet, patented; located on the north spur of Mount Lincoln, 7 miles from Alma and 13 miles from Fairplay; nature of ore, galena and sulphurets, value 80 to 400 ounces silver per ton; 4 tunnels, 10, 40, 50, and 60 feet in length, respectively; yield in 1878, $5,000.

Ypsilanti. Hall Valley Mining Co., proprietors; office Hall Valley; claim 300 by 1,500 feet; located on Bullion Mountain, Hall Valley mining district, 25 miles from Fairplay; course of vein, northeast and southwest, width 3 feet; nature of ore, galena and gray copper; average value, 85 ounces silver per ton; 1 shaft 40 feet deep, 1 tunnel 72 feet in length, 1 drift 120 feet in length.

Park County Ore Mills

Alma Branch, Boston and Colorado Smelting Co. Erected in 1874; Henry Williams, superintendent, located in town of Alma, 100 miles from Denver; have a capacity for treating 25 tons of ore daily.

Chicago and New York Mining and Smelting Co. Incorporated February 27, 1874; capital stock, $500,000, in 5,000 shares, $100 each; offices, town of Holland, Colorado, and Chicago, Illinois; organized by Charles, Dwight G., and Park Holland, Jeremiah Laming, Richard S. Thompson, and Elijah W. Bontecon; the works of this company are located in town of Holland, 3 miles from Alma and 4 miles from Fairplay.

Hall Valley Smelting Works. Owned by the Hall Valley Mining Co.; J.G. Jebb, agent; located in Hall Valley; have a capacity for treating 10 tons of ore in 24 hours.

Mount Lincoln Sampling Works. Located in town of Alma, 6 miles from Fairplay.

Moose Mining Co.'s Works. Have a capacity for treating 10 tons of ore daily; they cover 1 acre of ground and are located beside the Platte River, in town of Dudley, 1 mile from Alma and 7 miles from Fairplay.

Sampling Works of A.R. Meyer Co. Philip T. Amm, manager; located in town of Alma, 6 miles from Fairplay; have capacity for treating 20 tons of ore in 10 hours.

South Park Smelting and Reduction Works. Owned by James H.B. McFerran, J.S. Wolfe, and C.H. White; located in town of Eureka; have a capacity for treating 10 tons of ore in 24 hours; the main building is 60 by 100 feet and contains an engine, crushers, roller, etc.

PUEBLO COUNTY

Pueblo Reduction Works. Owned by the Pueblo Lixiviation Co., located in Pueblo; have a capacity for treating 5 tons of ore daily.

Pueblo Smelting Works. Owned by J.C. Mather and A.W. Geist; located at the crossing of the Atchison, Topeka, & Santa Fe and the Rio Grande railroads, 1 mile from Pueblo; have a capacity for the treatment of 40 tons of ore daily.

Rocky Mountain Mica Co. Incorporated September 15, 1877; capital stock, $1,000,000, divided into 100,000 shares, $10 each; offices Pueblo, Colorado, and New York City; trustees: George M. Chilcott, James Carlisle, James McDonald, Joseph Bardine, Robert H. Speer, Emanuel H. Saltiel, and Helen M. Ghost.

RIO GRANDE COUNTY

Thomas B. Corbett reported that Rio Grande County had a population of 5,000 in 1878, and the bullion product for that year was $78,000.

American Flag. Owned by John R. Burrows and Lewis C. Smith; offices Summit, Del Norte, and Denver, Colorado; claim 300 by 1,500 feet; discovered 1876; located on the north slope of South Mountain, near town of Summit, 27 miles from Del Norte; course of vein, northeast and southwest, width 5 feet, pitch 15 degrees south; nature of ore, iron pyrites; average value, $20 per ton; development, 1 surface opening, 20 feet in length; yield $600.

Baker. Owned by the Baker Mining Co., William H. Baker, John R. Burrows, and John H. Young; offices summit and Del Norte; claim 300 by 1,500 feet; discovered 1876; located on the north slope of South Mountain, near town of Summit, 27 miles from Del Norte; course of vein, northwest and southeast, width 9 feet, pay vein 5 feet, pitch 15 degrees south; nature of ore, iron pyrites and free gold; average value, $32 per ton; 2 surface openings, 20 and 75 feet in length, respectively; yield, $5,000.

Columbia. Owned by Pear J. Peterson and W.H. Van Gieson; office Summit and Del Norte; claim 300 by 1,500 feet; discovered 1874; located on the north slope of South Mountain, near town of Summit; 27 miles from Del Norte; vertical vein, trending east and west, width 12 feet; nature of ore, free gold quartz; average value, $35 per ton; 1 shaft 25 feet deep.

Del Norte. Owned by the Little Annie Gold Mining Co.; Pear J. Peterson, superintendent, Charles E. Robins treasurer; office Summit; claim 300 by 1,500 feet; discovered 1873; located on the north slope of South Mountain, Summit mining district, near town of Summit; nature of ore, free gold; average value, $44 per ton.

Eighth Wonder. Owned by the Golden Star Gold and Silver Mining Co.; offices Chicago, Illinois, and New York City; claim 300 by 1,500 feet; discovered June 1874; located on South Mountain, near town of Summit.

Golden Queen. Owned by Johnston Livingston, John J. Crooke, Adams & Posey, Arthur Burton, Peter Beeker, Joseph S. Reef, John A. McDonald, Lucius A. Winchester, J.S. Partridge, Lewis Crooke, William Beck, James L. Hill, Henry B. Clark, F.C. Day, L.C. Smith, R.C. Sheppard, and L.C. Baker; offices Summit, Colorado; claim 300 by 1,500 feet; discovered 1873; located on South Mountain, near town of Summit, 27 miles from Del Norte.

Golden Star. Owned by the Golden Star Gold and Silver Mining Co.; offices Chicago, Illinois, and New York City; claim 300 by 1,500 feet; discovered June 1874; located on South Mountain, Summit mining district, 27 miles from Del Norte; nature of ore, free gold, assay value, $20 per ton.

Highland Mary. Owned by the Highland Mary Mining Co.; incorporated under the laws of the State of Iowa, May 1878; capital stock, $2,500,000, in 125,000 shares, $20 each; offices Summit and Del Norte, Colorado, Creston, Iowa, and Chicago, Illinois; R.P. Smith, president, John Gibson, secretary, J.H. Patt, treasurer; claim 300 by 1,500 feet; discovered 1874; located on the northeast side of South Mountain, near town of Summit 27 miles from Del Norte; nature of ore, white iron; average value, $100 per ton; 1 shaft 30 feet deep, 1 tunnel 40 feet in length, 3 adits 25 feet each.

John. Owned by John R. Burrows and Louis C. Smith; office town of Summit, and Del Norte; claim 300 by 1,500 feet; discovered 1876; located on the north slope of South Mountain, near town of Summit, 27 miles from Del Norte; course of vein, north and south; nature of ore, iron pyrites; development, 1 surface opening, 15 feet in length; yield, $200.

Keystone. Owned by Golden Star Gold and Silver Mining Co.; office Chicago, Illinois, and New York City; claim 300 by 1,500 feet; discovered July 1874; located on South Mountain, near town of Summit.

Little Annie. Owned by Ferdinand Henry Brandt, Johnston Livingston, Pear J. Peterson, John J. Crooke, Winchester and Partridge, Le Grand Dodge, Henry S. Hoyt, Jr., Lewis Crooke, Eliza S. Winchester, Prof. C.E. Robins, and Frank W. Winchester; Pear J. Peterson, superintendent, Prof C.E. Robins, treasurer; office town of Summit, Colorado. The property of this company is located on the north slope of South Mountain, and on Wightman's Fork of the Alamosa, in the town of Summit, 27 miles from Del Norte and 57 miles from the present terminus of the Denver and Rio Grande railway. It comprises the Little Annie, Del Norte, and

Margaretta gold miles, and 40 acres of placer ground, a 10-stamp mill, an ore house, office and assay laboratory, store, 2 dwelling houses, retort and charcoal houses, blacksmith shop, mine and mill dumps, tramway, cars and 2,500 feet of steel cable, dams, flumes, sluices, etc. The Little Annie was discovered September 13, 1873, and located by Ferdinand and Henry Brandt and Pear J. Peterson. The claim is 50 to 300 feet in width, by 1,500 feet in length; 600 feet of the superficial length is below and 900 feet above the timberline. The trend of the vein is north 15 degrees, 15 minutes east and south 15 degrees, 15 minutes west. The ore is oxidized iron and free gold, averaging from $50 to $75 per ton; development, a shaft 90 feet deep and 5 tunnels, 30, 45, 50, 54, and 150 feet in length, respectively. The main mill building is a 2-story wooden structure, 32 by 42 feet, with an ell 24 by 28 feet. It was erected in August 1875, at a cost of $38,578. The mill has a capacity for treating 6 tons of ore per day and contains 10 stamps, weighting 530 pounds each, a 25-horse power engine, Blake crusher, 4 Dolly tubs, an agitator, and all other necessary machinery. The tramway was built in the spring of 1876 for the purpose of conveying ore from the mine to the mill, a distance of 2,135 feet. It has a capacity for delivering 80 tons of ore per day, is furnished with a steel cable 2,500 feet in length and 2 cars, with a carrying capacity of 4,800 pounds. The operation is automatic, the loaded car pulling up the empty one, being governed by a brake stationed at the mouth of the mine.

Little Ida. Owned by the San Juan consolidated Gold and Silver Mining Co; offices Del Norte and Denver, Colorado; claim 300 by 1,500 feet, patented; discovered 1874; located at town of Summit, 27 miles from Del Norte, immediately west and parallel to the "Little Annie," vertical vein, trending north and south, width 25 feet; nature of ore, oxidized iron; average value, $20 per ton; 2 shafts, 25 and 70 feet deep, respectively; 2 drifts from bottom of main shaft 60 feet each.

Little Jessie, Southern Extension of the Little Annie. Owned by Thomas M. Bowen and F.F. Riggs; office Del Norte; claim 300 by 1,500 feet; discovered 1874; located on South Mountain, near town of Summit, 27 miles from Del Norte; vertical vein, trending north and south, width 25 feet; development, 2 shafts, 30 feet each, and surface opening 30 feet in length.

Margaretta. Owned by the Little Annie Gold Mining Co.; Pear J. Peterson, superintendent, Charles E. Robins, treasurer; office Summit; claim 300 by 1,500 feet; discovered 1873; located on the north slope of

South Mountain, near town of Summit, 27 miles from Del Norte; nature of ore, free gold; average assay value, $25 per ton.

Moltke. Owned by the Highland Mary Mining Co.; offices Del Norte, Colorado, Creston, Iowa, and Chicago, Illinois; claim 300 by 1,500 feet; discovered 1874; located on the north side of South Mountain, near town of Summit, 27 miles from Del Norte; vertical vein, trending northeast and southwest, width 7 feet; nature of ore, iron pyrites and galena; average value, $60 per ton; development, 1 shaft 40 feet deep, surface opening 40 feet in length.

Odin. Thomas M. Bowen, proprietor; office Del Norte; claim 300 by 1,500 feet; discovered 1874; located on the north slope of South Mountain, Summit mining district, near town of Summit, 27 miles from Del Norte; vertical vein, trending southeast and northwest, width 25 feet, pay vein 18 inches to 4 feet; development, 2 shafts, 22 and 80 feet in depth, respectively, 1 tunnel 120 feet in length, and 2 drifts from main shaft aggregating 50 feet in length.

Poor Man. Owned by Archie C. Fisk, Charles H. Toll, Otto Ruossou, and others; office Denver; claim 300 by 1,500 feet; located on South Mountain, Summit mining district, 30 miles from Del Norte and within 1,000 feet of the justly celebrated Little Annie mine; the development consists of an open cut, 75 feet long, 20 feet in width and 27 feet deep, showing a pay vein of 4 feet in width, the ore averaging $50 per ton. In doing the above amount of work, there was taken from this mine 800 tons of ore that is at present on the dump, with it is thought will average $50 per ton.

San Juan Consolidated Gold Mining Co. Incorporated August 18, 1875; capital stock, $3,000,000, in 30,000 shares, $100 each; offices Denver and Del Norte, Colorado; organized by Charles W. Tankersley, Charles F. Gillett, Thomas M. Bowen, and George M. Binckley. The property of this company is located on South Mountain, and on Wightman's Fork of the Alamosa, at and near the town of Summit, 27 miles from the present terminus of the Denver and Rio Grande Railway. It comprises 15,000 linear feet of lode claims, and a 30-stamp mill.

Silver Brick. Owned by Charles A. Stringham, Thomas Donovan, and Russell A. Hibbard; offices Summit, Colorado, and Creston, Iowa; claim 300 by 1,500 feet; located on the west side of South Mountain, near town of Summit, 27 miles from Del Norte; course of vein, northwest and southeast, width 6 feet, pay vein 18 inches, pitch 10 degrees south; nature of ore, iron pyrites and galena; average value, $50 per ton; main shaft 25 feet deep.

Summit. Owned by R.F Adams, Lewis Crooke, and Le Grand Dodge; office Del Norte; claim 300 by 1,500 feet; discovered 1873; located on the northeastern slope of South Mountain, near town of Summit, assay value of ore, $10 to $200 per ton.

Rio Grande County Ore Mills

Bowen's Mill. Thomas M. Bowen, proprietor; office Del Norte; located on Wightman's Fork of the Alamosa, at town of Summit, 27 miles from Del Norte; has a capacity for treating 36 tons of ore daily; contains 24 stamps, a 40-horse power engine, and all other necessary machinery.

Golden Queen Gold and Silver Mining Co.'s Mill. Located on Wightman's Fork of the Alamosa, at town of Summit, 27 miles from Del Norte; has a capacity for treating 10 tons of ore daily; contains 10 stamps, a 20-horse power engine, etc.

Golden Star Gold and Silver Mining Co.'s Mill. Located in town of Summit, Summit mining district, 27 miles from Del Norte; contains 10 stamps and a 40-horse power engine, has a capacity for treating 20 tons of ore in 24 hours.

Little Annie Mill. Owned by the Little Annie Gold Mining Co.; Pear J. Peterson, superintendent, Charles E. Robins, treasurer. The mill building is a 2-story wooden structure, 32 by 42 feet, with an ell 24 by 28 feet. It has cost $38,578, and was erected in August 1875. It has a capacity for treating 6 tons of ore daily and contains 10 stamps weighing 530 pounds each, a 25-horse power engine, Blake crusher, 4 dolly tubs, an agitator, and all other necessary equipment.

San Juan Consolidated Gold and Silver Mining Co.'s Mill. Located on Wightman's Fork of the Alamosa, in town of Summit, 27 miles from Del Norte; contains 30 stamps, a 40-horse power engine, and has a capacity for treating 40 tons of ore in 24 hours.

San Juan Smelting Works. Incorporated December 5, 1873; capital stock, $50,000, in shares of $100 each; offices Del Norte and Denver,

Colorado; organized by Oscar L. Mathews, Ellis Smith, John J. Epley, Horace H. Powers, James and Samuel McFarland, Orange B. Stoddard, Leonard B. Reid, Cornelius Donovan, and William H. Weinrich.

Summit Mill. Owned by R.F. Adams, Lewis Crooke, and Le Grand Dodge; this mill was erected in 1874 on Wightman's Fork of the Alamosa, near town of summit, 27 miles from Del Norte. It is run by steam and waterpower, and has a capacity for treating 20 tons of ore per day.

SAN JUAN COUNTY

In 1878, San Juan County had a population of about 5,000 and a bullion product of $201,050, according to Thomas Corbett. Mining districts included: Animas, Eureka, and Uncompahgre.

Addie May. Owned by Samuel Watson, J.G. Morris, and B.F. McKinzie; office, Animas Forks; claim 300 by 1,500 feet; discovered 1876; located in Picayune Gulch, Eureka mining district, 2 miles from town of Animas Forks and 11 miles from Silverton; course of vein, northeast and southwest, width 6 feet, pitch 35 degrees; nature of ore, galena and gray copper; assay value, $171 silver per ton; 1 adit 18 feet in length with a 23-foot face.

Adelia. Edward W. and Charles H. McIntire, proprietors; office Ouray; claim 300 by 1,500 feet; discovered 1875; located on Mineral Point Mountain, eighteen miles from Silverton; vertical vein, trending northwest and southeast, width 30 feet, pay vein 3 feet; nature of ore, galena; assay value, 200 ounces silver per ton; 1 shaft 12 feet deep.

Adelphi. Edward Cree, proprietor; office Lake City; claim 300 by 1,500 feet; discovered 1877; located on Engineer Mountain, 18 miles from Silverton; vertical vein, trending north and south, width 7 feet, pay vein 3 feet; nature of ore, galena and gray copper; assay value, 200 ounces silver per ton; 1 drift 28 feet in length and 20 feet from surface.

Admiral. Charles H. McIntire and George N. Propper, proprietors; office Ouray; claim 300 by 1,500 feet; discovered 1878; located on Mineral Point Mountain, 1 mile from Mineral City and 18 miles from Silverton; vertical vein, trending northeast and southwest, width of pay vein, 16 inches; nature of ore, galena; 1 shaft 18 feet deep.

Alabama. Charles Wickman and J. Swanson, proprietors; office Silverton; claim 300 by 1,500 feet; discovered 1875; located on Poughkeepsie Gulch, Uncompahgre mining district, 14 miles from Silverton; vertical vein, trending northeast and southwest, width 30 feet, pay vein 2 feet; nature of ore, galena and gray copper; average value, 200 ounces silver per ton; 2 adits, 10 and 15 feet in length, respectively; yield, $120.

Alaska. Owned by Peterson and Johnson, of Silverton; claim 300 by 1,500 feet; discovered 1876; located in Poughkeepsie Basin, 12 miles from Silverton; course of vein, north and south, width 8 feet, pay vein 3 feet; nature of ore, gray copper; value, 200 ounces silver per ton; development, 1 shaft 40 feet deep, and 1 adit 15 feet in length; yield, $8,000.

Alice. Owned by Frederick Blaisdale, John Neiswanger, M. Rich, and John Spadling; office Eureka; claim 300 by 1,500 feet; discovered 1873; located on the Animas River, near town of Eureka, 9 miles from Silverton; vertical vein, trending north and south, width 5 feet, pay vein 28 inches; nature of ore, galena and gray copper; average value 45 ounces silver per ton; 1 drift 50 feet, and 1 tunnel 35 feet in length.

Alice. Owned by E. McCarty and Mark Biedell; office Silverton; claim 50 by 1,500 feet, discovered 1873; located 2 miles from town of Eureka, on south fork of Eureka Gulch; vertical vein, trending east and west, width 3 feet, pay vein 1 foot; nature of ore, galena; assay value, $200 per ton; 1 drift 30 feet in length and 25 feet from surface.

Alleghany. J. Williamson and E. Howcutt, proprietors; office Lake City; claim 300 by 1,500 feet; discovered 1877; located on Woods Mountain, 2 miles from Mineral City and 17 miles from Silverton; vertical vein, trending northeast and southwest, width 7 feet, pay vein 30 inches; nature of ore, galena; assay value, 75 ounces silver per ton; 1 drift 25 feet in length.

Alma. Henry LeClair, proprietor; office Ouray; claim 300 by 1,500 feet; discovered 1876; located on Poughkeepsie Mountain, 7 miles from Ouray; vertical vein, trending northwest and southeast, width 20 feet, pay vein 15 inches; nature of ore, gray copper and brittle silver; assay value, 100 ounces to 1,200 ounces silver per ton; 1 shaft 19 feet deep.

Almaden. Owned by J.C. Peterson and M. Guinn, of Eureka; claim 300 by 1,500 feet; discovered 1876; located on Brown Mountain, 7 miles from Eureka; course of vein, east and west, width 73 feet, pay vein 28 inches; nature of ore, galena and iron pyrites; development, 1 shaft 20 feet deep, and 1 adit 12 feet in length.

Alta. Owned by John Goodin & Co., of Ouray; claim 300 by 1,500 feet; located on Cement Creek, 6 miles from Eureka; curse of vein, east and west, width 50 feet; development, 1 shaft 60 feet.

Altoff. J.F. Robison, proprietor; office Silverton; claim 300 by 1,500 feet; discovered 1878; located on Sultan Mountain [13,370 feet], Animas mining district, 3 miles from Silverton; course of vein, northeast and southwest, width 1 foot, pitch 15 degrees south; nature of ore, galena and gray copper; assay value 15 to 852 ounces silver per ton; 1 adit 25 feet in length, with a 16-foot face.

American. C.J. Bilyien & Co., proprietors; office Eureka; claim 300 by 1,500 feet; discovered 1874; located in Niagara Gulch, Eureka mining district, 1 mile from town of Eureka and 7 miles from Silverton; vertical vein, trending east and west, width 40 feet, pay vein 9 feet; nature of ore, galena and gray copper; assay value of ore, 237 ounces silver per ton; 1 drift 35 feet in length.

American Boy. Thomas Clege, proprietor; office Ouray; claim 300 by 1,500 feet; discovered 1876; located in Poughkeepsie Gulch, 14 miles from Silverton; course of vein, northwest and southeast, width 10 feet, pay vein 18 inches, pitch 25 degrees; nature of ore, galena and gray copper; assay value, 560 ounces silver per ton; 2 adits, 10 and 15 feet in length, respectively.

Animas Forks Mining Co. Incorporated May 3, 1877; capital stock, $250,000; organized by Angus Smith, William Young, William Beck, E.B. Greenleaf, and David Vance.

Anna. Owned by J.O. Cessna, M. McGraw, and J.T. Bell; office Silverton; claim 300 by 1,500 feet; discovered 1875; located on Mineral Mountain, Animas mining district, 3 miles from Silverton; course of vein, northwest and southeast, width 3 feet; nature of ore, gray copper and brittle silver; 1 drift 40 feet in length and 35 feet from surface.

Annie Laurie. Benjamin R. Eaton, proprietor; office Silverton; claim 300 by 1,500 feet; discovered 1876; located on Galena Mountain, Animas mining district, 4 miles from Silverton; vertical vein, trending northwest and southeast, width 2 feet; nature of ore, galena and iron pyrites; assay value, 100 ounces silver per ton; 1 shaft 30 feet deep.

Arctic. James M. Marshall, proprietor; office Animas Forks; claim 300 by 1,500 feet; discovered 1874; located on Hourding Peak, Eureka mining district, 3 miles from Animas Forks and 7 miles from Silverton; course of vein, northeast and southwest, width 40 feet, pay vein 3 feet; nature of ore, gray copper; assay value, 600 ounces silver per ton; 1 adit 40 feet in length and 20 feet from surface.

Arctic. James W. Platt and Co., proprietors; office Eureka; claim 300 by 1,500 feet; discovered 1874; located on Hurricane Peak, Eureka mining district, 6 miles from town of Eureka and 14 miles from Silverton; vertical vein, trending northeast and southwest, width 4 feet, pay vein 1 foot; nature of ore, gray copper; assay value, 100 to 800 ounces silver per ton; 1 drift 25 feet in length and 20 feet from surface.

Arctic. Owned by M. Green, S.G. Jeffries, and others; office Animas Forks; claim 300 by 1,500 feet; discovered 1876; located on the Animas River, Eureka mining district, 2 miles from town of Animas Forks and 14 miles from Silverton; course of vein, northwest and southeast, width 8 feet, pay vein 30 inches, pitch 18 degrees; nature of ore, galena and iron pyrites; assay value, 65 ounces silver per ton; 1 adit 30 feet in length and 20 feet in depth.

Aspen. Owned by George Green and Co., William Mulholland, Thomas Blair, and D. Reese; office Silverton; claim 300 by 1,500 feet; located 1874; located on Hazelton Mountain, Animas mining district, 2 miles from Silverton; course of vein, northeast and southwest, width 4 feet, pay vein 6 to 10 inches, pitch 15 degrees south; nature of ore, gray copper and galena; average value, 140 ounces silver per ton and 60 per cent lead; 2 shafts, 150 and 125 feet in depth, respectively, 1 tunnel 300 feet in length, intersecting vein at a depth of 250 feet; 2 levels 150 feet in length each; yield, $100,000.

Badger State. Owned by James Gillett, E. Ray, and Dr. Ankney; office Silverton; claim 300 by 1,500 feet; discovered 1876; located in Poughkeepsie Gulch, Uncompahgre mining district, 13 miles from Silverton; course of vein, northeast and southwest, width 12 feet, pay vein 18 inches, pitch 15 degrees east; nature of ore, gray copper and iron pyrites; assay value, 70 ounces silver per ton; 1 adit 12 feet, and 1 drift 40 feet in length.

Bald Eagle. N. Garrett and W.J. Brewster, proprietors; office Mineral City; claim 300 by 1,500 feet; located 1876; located on Houghton

Mountain, Uncompahgre mining district, 18 miles from Silverton; vertical vein, trending northwest and southeast, width 60 feet; nature of ore, galena; assay value, 45 ounces silver per ton; 1 shaft 20 feet deep, 1 adit 12 feet in length.

Bald Eagle. William Girardin, proprietor; office Silverton; claim 300 by 1,500 feet; discovered 1878; located on Kendall Mountain [13,068 feet] Animas mining district, 2 miles from Silverton; course of vein, northwest and southeast, width 40 feet, pay vein 20 inches, pitch 30 degrees; nature of ore, gray copper and galena; 1 shaft 10 feet deep, 1 drift 20 feet in length and 40 feet from surface.

Baldwin. Owned by the J. Baldwin Mining and Reduction Co.; incorporated July 21, 1878; capital stock, $500,000, in 50,000 shares, $10 each; offices Animas Forks, Colorado, and Sharon, Pennsylvania; Peter Kimberly, president, M. Rich, vice president, W.B Dunham, superintendent; claim 300 by 1,500 feet; located at the head of the south fork of the Animas River, Eureka mining district, 3 miles from Animas Forks and 16 miles from Silverton; course of vein, northeast and southwest; width 240 feet, pitch 15 degrees south; nature of ore, gray copper and galena; main shaft 25 feet deep.

Baltic. George Wood, proprietor; office Ouray; claim 300 by 1,500 feet; discovered 1877; located on Mount Abram, 14 miles from Silverton; vertical vein, trending northeast and southwest, width 3 feet, pay vein 18 inches; nature of ore, gray copper and brittle silver; assay value, 100 ounces silver per ton; 1 drift 25 feet in length.

Baltic. G. Ferguson and A.T. Terrill, proprietors; office Ouray; claim 300 by 1,500 feet; discovered 1876; located on Mineral Point Mountain, 2 miles from Mineral City and 9 miles from Ouray; vertical vein, trending northwest and southeast, width 40 feet; nature of ore, galena; assay value, 70 ounces silver per ton; 3 shafts aggregating 34 feet in depth.

Baltimore. Owned by George N. Propper and A.W. Burrows; office Del Norte; claim 250 by 1,500 feet; discovered 1874; located in Uncompahgre mining district, near town of Mineral City and 18 miles from Silverton; vertical vein, trending northeast and southwest, width 12 feet, pay vein 26 inches; nature of ore, galena; assay value, 60 to 200 ounces silver per ton; 1 shaft 15 feet deep.

Bavarian, Muskegon, and Dunderberg. Owned by M.J. Condon & Co.; office Eureka; claim 300 by 4,500 feet; discovered 1875; located in Eureka Gulch, 2 miles from Eureka and 10 miles from Silverton; course of vein, northeast and southwest, width 200 feet; nature of ore, galena and copper pyrites; assay value, 400 ounces silver per ton; 2 adits, 20 and 30 feet in length, respectively, 1 drift 12 feet in length.

Belcher. Owned by John D. Murphy, George Purcell, and J.A. White; office Ouray; claim 300 by 1,500 feet; discovered 1876; located on Oberto Mountain, 12 miles from Silverton; vertical vein, trending northeast and southwest, width 100 feet, pay vein 7 feet; nature of ore, galena and gray copper; average value, 194 ounces silver per ton; 2 drifts aggregating 100 feet in length, 12 and 37 feet from surface, respectively.

Belcher. Owned by George W. Thompson and L.C. Dunn; office Denver; claim 300 by 1,500 feet; discovered 1875; located in Picayune Gulch, 3 miles from Animas Forks and 11 miles from Silverton; course of vein, northeast and southwest, width 100 feet, pitch 30 degrees, nature of ore, gray copper and carbonates; assay value, 100 ounces silver per ton; 1 shaft 25 feet deep.

Bell Wether. Olaf Sandstone & Co., proprietors; office Silverton; claim 300 by 1,500 feet; discovered 1877; located on Hurricane Peak, Uncompahgre mining district, 12 miles from Silverton; vertical vein; trending northeast and southwest, width 8 feet, pay vein 18 inches; nature of ore, gray copper; assay value, 44 ounces silver per ton; 1 drift 20 feet in length and 20 feet from surface.

Belle Isle. William Girardin, proprietor; office Silverton; claim 300 by 1,500 feet; discovered 1878; located on Sultan Mountain, Animas mining district, 4 miles from Silverton; course of vein, northwest and southeast, width 6 feet, pay vein 14 inches; nature of ore, gray copper and galena; assay value, 321 ounces silver per ton; 1 drift 100 feet in length and 50 feet from surface.

Belle of the West. George Green & Co., proprietors; office Silverton; claim 300 by 1,500 feet.

Big Giant. Charles W. McIntire, proprietor; office Ouray; claim 300 by 1,500 feet; discovered 1874; located on Mineral Point Mountain, Uncompahgre mining district, 1 mile from Mineral City and 18 miles from Silverton; vertical vein, trending northeast and southwest, width 100 feet;

nature of ore, gray copper; assay value, 500 ounces silver per ton; 1 tunnel 30 feet in length.

Black Crook. Owned by W.R. Mitchell, Winfield Scott Stratton, and W. B. Sherman; office Colorado Springs; claim 300 by 1,500 feet; located on Green Mountain, Animas mining district, 4 miles from Howardsville and 8 miles from Silverton; vertical vein, trending northwest and southeast, width 5 feet, pay vein 14 inches; nature of ore, gray copper and galena, value, 160 ounces silver per ton; 3 adits aggregating 100 feet in length. [Winfield Scott Stratton was a carpenter who came to Colorado in 1872. He prospected in Silver Cliff, Red Cliff, Aspen, and Tin Cup, without any real success. He took an assaying and mineral course at Colorado College. On the Fourth of July 1891, he staked the claims in the Cripple Creek area, the Independence and the Martha Washington, which would make him a Bonanza King. He bought the house that he had built years earlier and used his vast wealth to help others. He later partnered in the discovery of the Portland Mine. At his death in 1902, he left most of his fortune to establish the Myron Stratton Home.]

Black Prince. William Jackson & Co., proprietors; office Animas Forks; claim 300 by 1,500 feet; discovered 1876; located on California Mountain, Eureka mining district, 2 miles from town of Animas Forks and 17 miles from Silverton; vertical vein, trending east and west, width 7 feet; nature of ore, galena and gray copper; assay value, 300 ounces silver per ton; 1 drift 25 feet in length.

Blacksmith. George N. Propper and A.W. Burrows, proprietors; office Del Norte; claim 282 by 1,500 feet; discovered 1874; located on Seigle Mountain, Uncompahgre mining district, Uncompahgre mining district, 1 mile from Mineral City and 18 miles from Silverton; vertical vein, trending northeast and southwest, width 30 feet, pay vein 3 feet; nature of ore, galena; assay value 100 ounces silver per ton; 1 drift 34 feet in length.

Blue Bird. Charles Clase, proprietor; office Eureka; claim 300 by 1,500 feet; discovered 1878; located in Poughkeepsie Gulch, Uncompahgre mining district, 9 miles from Ouray; course of vein, northeast and southwest, width 7 feet, pay vein 3 feet, pitch 15 degrees; nature of ore, iron pyrites and gray copper; assay value, 200 ounces silver per ton; 1 adit 10 feet in length and 10 feet from surface.

Blue Jay. H. Leroy, J.M Crissy, and R. Grant, proprietors; office Animas Forks; claim 300 by 1,500 feet; discovered 1877; located on Hurricane Peak, Eureka mining district, 4 miles from Animas Forks and 10 miles from Silverton; vertical vein, trending east and west, width 4 feet, pay vein 2 feet; nature of ore, galena; assay value, 80 ounces silver per ton; 2 adits 10 feet in length each.

Blue Jay. T. Shaw and Co., proprietors; office Ouray; claim 300 by 1,500 feet; discovered 1877; located on Lake Mountain, Uncompahgre mining district, 1 mile from Mineral City and 18 miles from Silverton; vertical vein, trending north and south, width 30 feet; nature of ore, galena; assay value, 60 ounces silver per ton; 1 shaft 18 feet deep.

Bonanza. Antonio Aberto [also called Anthony Oberto?] and Mark Biedell, proprietors; office Silverton; claim 372 by 1,500 feet, patented; discovered 1877; located in Poughkeepsie Gulch, Uncompahgre mining district, 12 miles from Silverton; course of vein, northeast and southwest, width 372 feet, pay vein 22 feet; nature of ore, gray copper and brittle silver; average value, 200 ounces silver per ton; 4 shafts, aggregating 58 feet in depth; 1 tunnel 75 feet in length; total yield, $2,500.

Bonanza Silver Mining Co. Incorporated December 4, 1878; capital stock, $5,000,000, in 200,000 shares, $5 each; offices Animas Forks, Colorado, and Chicago, Illinois; organized by Andrew Forbes, John B. Robinson, and John R. Curry. The property of this company is located on California, Houghton, and Hurricane mountains, at the head of the south fork of the Animas River, in Eureka mining district, and comprise the Philadelphia, Washington, Columbus Extension, and other lodes.

Bonanza Tunnel. Owned by Bonanza Tunnel Co., of New York; claim 1,500 by 3,000 feet; located on Houghton Mountain, near Animas Forks; course, east and west, length 700 feet.

Bonnie Cord. J.W. Spencer & Co., proprietors; office Ouray; claim 300 by 1,500 feet; discovered 1874; located on King Solomon Mountain, Animas mining district, 3 miles from Silverton; vertical vein, trending northwest and southeast, width 3 feet, pay vein 18 inches; nature of ore, gray copper; average value, 125 ounces silver per ton; 1 shaft 14 feet deep, 1 drift 30 feet in length, 3 adits 15 feet in length each.

Boomerang. J.R. McKinnie and Co., proprietors; office Eureka; claim 300 by 1,500 feet; discovered 1875; located in Niagara Gulch,

Eureka mining district, 1 mile from town of Eureka and 9 miles from Silverton; course of vein, northwest and southeast, width 8 feet, pay vein 1 foot, pitch 12 degrees; nature of ore, gray copper and copper pyrites; average value, 200 ounces silver per ton; 1 shaft 14 feet deep, 1 adit 13 feet in length.

Boone. James F. Rawling & Co., proprietors; office Denver; claim 300 by 1,500 feet; discovered 1875; located on the Animas River, Eureka mining district, 2 miles from town of Animas Forks and 14 miles from Silverton; course of vein, east and west, width 7 feet, pay vein 18 inches, pitch 20 degrees north; nature of ore, galena; assay value, 82 ounces silver per ton; 1 drift 30 feet in length and 25 feet from surface.

Boss. John Lattimore & Co., proprietors; office Lake City; claim 300 by 1,500 feet, discovered 1876; located in Burns Gulch, Eureka mining district, 2 miles from town of Eureka and 10 miles from Silverton; course of vein, northeast and southwest, width 20 feet, pay vein, 4 feet; nature of ore, galena and iron pyrites; assay value, 30 ounces silver per ton; 2 shaft aggregating 25 feet in length.

Boss Boy. Owned by C.W. Burris, Charles Bales, and I.S. Calder; office Del Norte; claim 300 by 1,500 feet; discovered 1875; located in Boulder Gulch, Animas mining district, 3 miles from Silverton; course of vein, northeast and southwest, width 30 feet, pitch 15 degrees; nature of ore, pyrites of iron; value, $200 per ton.

Bounty. Owned by James J. and Albert Bernard, Daniel J. and James D. McKay, William Forsyth, and John M. Ross; office Eureka; claim 300 by 1,500 feet; discovered 1874; located in Eureka Gulch, 6 miles from Silverton; course of vein, northeast and southwest, average width 16 feet, pay vein 6 feet; nature of ore, gray copper, iron and copper pyrites; average value, $105 per ton; 1 shaft 50 feet deep.

Brazilian. M. Green & Co., proprietors; office Eureka; claim 300 by 1,500 feet; discovered 1876; located in Eureka Gulch, 5 miles from town of Eureka and 10 miles from Silverton; vertical vein, trending east and west, width 20 inches; nature of ore, galena and gray copper; assay value, 80 ounces silver per ton; 1 adit 15 feet in length.

British Consuls Silver Mining Co. Incorporated April 27, 1878; capital stock, $100,000, in 1,000 shares, $100 each; office Silverton; trustees: John M. Stuart, John Elliott, and John H.P. Voorhies.

Broad Gauge and Extension. George Green & Co., proprietors; office Silverton; claim 300 by 3,000 feet.

Buckeye State. F. Herbst and Co., proprietors; office Ouray; claim 300 by 1,500 feet; discovered 1875; located in Ohio Gulch, Animas mining district, 3 miles from Silverton; course of vein, northeast and southwest, width 15 feet, pay vein 2 feet, pitch 30 degrees; nature of ore, galena and gray copper; average value, 75 ounces silver per ton; 1 shaft 30 feet deep, 1 tunnel 175 feet in length, which intersects the vein at a depth of 160 feet.

Buell. Colorado Mining and Land Co., proprietors; office Buffalo, N.Y.; claim 300 by 1,500 feet; discovered 1876; located on Middle Creek, Uncompahgre mining district, 1 mile from Mineral City and 17 miles from Silverton; vertical vein running northeast and southwest, width 10 feet; nature of ore, gray copper; assay value, 100 ounces silver per ton, 1 tunnel 18 feet and 1 adit 48 feet in length.

Burrows. Charles H. McIntire and A.W. Burrows, proprietors; office Ouray; claim 300 by 1,500 feet; discovered 1874; located on Mineral Point Mountain, Uncompahgre mining district, half mile from Mineral City, and 18 miles from Silverton; vertical vein, trending northeast and southwest, width 70 feet, pay vein 7 feet; nature of ore, galena; average value, 30 ounces silver per ton; 1 shaft 40 feet deep.

Burrows, No. 2. Charles H. McIntire and A.W. Burrows, proprietors; office Ouray; claim 120 by 1,500 feet; discovered 1874; located on Mineral Point Mountain, Uncompahgre mining district, 18 miles from Silverton; vertical vein, trending northeast and southwest, width 50 feet, pay vein 28 inches; nature of ore, galena and gray copper; assay value, 200 ounces silver per ton; 1 adit 40 feet in length.

Byron Gold and Silver Mining Co. Incorporated February 18, 1878; capital stock, $1,000,000, in 20,000 shares, $50 each; office Eureka; trustees: Joseph B. and Joseph R. Fay, Robert S. Cross, Royal C. Bradshaw, and John Martin.

C. and C. John Clark and M.J. Cross, proprietors; office Eureka; claim 300 by 1,500 feet; discovered 1878; located on Monument Mountain, Eureka mining district, 2 miles from town of Eureka and 9 miles from Silverton; course of vein, northeast and southwest, width 4 feet, pay vein 16 inches, pitch 15 degrees; nature of ore, galena; average value, 30 ounces silver per ton; 1 adit 15 feet in length.

California. Owned by J. Gillette, E. Ray, and W.W. Ross; office Silverton; claim 300 by 1,500 feet; discovered 1874; located in Poughkeepsie Basin, Uncompahgre mining district, 13 miles from Silverton; course of vein, east and west, width 25 feet, pay vein 6 feet, pitch 15 degrees south; nature of ore, gray copper and galena; average value, 50 ounces silver per ton; 1 adit and 2 drifts, 10, 15, and 20 feet in length, respectively.

California Belle. James M. Marshall & Co., proprietors; office Animas Forks; claim 300 by 1,500 feet; discovered 1875; located on California Mountain, 3 miles from Animas Forks and 18 miles from Silverton; vertical vein, trending east and west, width 3 feet, pay vein 10 inches; nature of ore, gray copper; assay value, 300 ounces silver per ton; 1 adit 15 feet in length and 15 feet from surface, 1 shaft 10 feet deep.

Calypso. Owned by George Cooke, M.V. Brookes, and others; office Eureka; claim 300 by 1,500 feet; discovered 1876; located on Ross Basin, Eureka mining district, 10 miles from Silverton; course of vein, northeast and southwest, width 6 feet, pay vein 4 feet, pitch 18 degrees; nature of ore, galena and iron pyrites; average value, 30 ounces silver per ton; 2 shafts 10 feet each.

Candia. Owned by John Stuart, of Silverton; claim 300 by 1,500 feet; discovered 1877; located in Poughkeepsie Basin, 12 miles from Silverton; course of vein, northeast and southwest, width 7 feet, pay vein 18 inches; nature of ore, gray copper; assay value, 250 ounces silver per ton; development, 1 drift 30 feet in length.

Caribou. Moyle, Robbins & Co., proprietors; office Silverton; claim 300 by 1,500 feet; discovered 1877; located on Sultan Mountain, Animas mining district; 3 miles from Silverton; course of vein, northeast and southwest, width 4 feet, pay vein, 14 inches; nature of ore, galena and gray copper; average value, 100 ounces silver per ton; 2 adits, 14 and 25 feet in length, respectively.

Carlisle. J. Peterson & Co., proprietors; office Ouray; claim 300 by 1,500 feet; discovered 1876; located on Mount Abram, Uncompahgre mining district, 14 miles from Silverton; vertical vein, trending east and west, width 3 feet; nature of ore, gray copper, iron and copper pyrites; assay value, 300 ounces silver per ton; 1 drift, 20 feet in length.

Cashier. John Hammill & Bro., proprietors; office Ouray; claim 300 by 1,500 feet; discovered 1876; located on Lake Mountain, 1 mile from Mineral City, and 17 miles from Silverton; vertical vein, trending north and south, width 50 feet, pay vein 4 feet; nature of ore, gray copper and galena; assay value, 80 ounces silver per ton; 2 adits 10 feet each.

Cashier. Owned by William and A.E. Norris; office Eureka; claim 300 by 1,500 feet; discovered 1878; located in Eureka Gulch, 2 miles from town of Eureka and 9 miles from Silverton; course of vein, north and south, width 12 feet, pitch 20 degrees; nature of ore, gray copper and galena; 1 adit 12 feet, and 1 tunnel 16 feet in length, the latter intersecting vein at a depth of 40 feet.

Caucausus. J.B. Ross & Co., proprietors; office Eureka; claim 300 by 1,500 feet; discovered 1873; located at the south base of Hurricane Peak, Eureka mining district, 10 miles from Silverton; course of vein, east and west, width 60 feet, pay vein 20 feet, pitch 30 degrees; nature of ore, galena and gray copper; assay value, 340 ounces silver per ton; 2 adits, 15 and 20 feet from surface, respectively, aggregating 50 feet in length.

Centennial Tunnel. Moyle, Robbins & Co., proprietors; office Silverton; claim 1,500 by 3,000 feet; discovered 1876; located on Sultan Mountain, Animas mining district, near limits of Silverton; course of tunnel, north and south, length 200 feet; tunnel house 26 by 36 feet.

Charles IX. Henry Carley, proprietor; office Eureka; claim 300 by 1,500 feet; discovered 1878; located on Eureka Mountain, Eureka mining district, 2 miles from town of Eureka and 10 miles from Silverton; course of vein, northwest and southeast, width 5 feet, pay vein 20 inches, pitch 30 degrees; nature of ore, gray copper and galena; assay value, 600 ounces silver per ton; 1 adit 10 feet in length.

Charles Morris. Charles H. McIntire, proprietor; office Ouray; claim 300 by 1,500 feet; discovered 1874; located in Uncompahgre Canyon, 3 miles from Mineral City and 18 miles from Silverton; course of vein, northeast and southwest, width 10 feet, pay vein 6 feet; nature of ore, galena and gray copper; assay value, 75 ounces silver per ton; 1 adit 10 feet, and 1 drift 15 feet in length.

Charlton. James W. Platt, proprietor; office Eureka; claim 300 by 1,500 feet; discovered 1874; located on Parson's Gulch, Eureka mining district, 2 miles from town of Eureka and 10 miles from Silverton; course of vein, northwest and southeast, width 4 feet, pay vein 15 inches; nature of ore, iron and copper pyrites; 1 adit, 20 feet in length with an 18-foot face.

Chautauqua and San Juan Mining Co. Incorporated August 30, 1878; capital stock, $200,000, 2,000 shares, $100 each; office Animas Forks, Eureka mining district; trustees: R.B. Landon, Joseph Westley, Gains C. Howard, Henry A. Rowley, Richard C. Bristol, Squire L. Brown, and Edward P. Howard.

Chenango. Charles Raymond, proprietor; office Ouray; claim 300 by 1,450 feet; discovered 1876; located on Poughkeepsie Creek, Uncompahgre mining district, 12 miles from Silverton; course of vein, north and south, width 20 feet, pay vein 22 inches, pitch 10 degrees; nature of ore, gray copper and brittle silver; assay value, 340 ounces silver per ton; 2 adits, 10 and 25 feet in length, respectively.

Chicago. Owned by C.N. Cooper and Clabe and Charles Jones; office Animas Forks; claim 300 by 1,500 feet; discovered 1877; located on Mineral Point Mountain, near Mineral City, 18 miles from Silverton; vertical vein, trending northeast and southwest, width 6 feet, pay vein 30 inches; nature of ore, brittle silver, gray copper, and galena; average value, 125 ounces silver per ton; main shaft 25 feet deep; yield $200.

Chief. Albert M. Knight and Charles Slocum, proprietors; office Silverton; claim 300 by 1,500 feet; discovered 1875; located on Sultan Mountain, Animas mining district, 1 mile from Silverton; vertical vein, trending northwest and southeast, width 6 feet, pay vein 20 inches; nature of ore, gray copper and galena; assay value, 200 ounces silver per ton; 1 shaft 27 feet deep, 1 tunnel 50 feet in length.

Clara. George and Mrs. Clara McIntire, proprietors; office Ouray; claim 300 by 1,500 feet; discovered 1874; located on Mineral Point Mountain, half mile from Mineral City and 18 miles from Silverton; vertical vein, trending northeast and southwest, width 8 feet, pay vein 4 feet; nature of ore, galena; 1 adit 25 feet in length.

Cleopatra. Albert Marzetti, proprietor; office Howardsville; claim 300 by 1,500 feet; discovered 1877; located on Tower Mountain, 2 miles from town of Eureka and 9 miles from Silverton; course of vein, east and west, width 5 feet, pitch 25 degrees; nature of ore, galena, copper pyrites, and brittle silver; assay value, 240 ounces silver per ton; 2 tunnels, 19 and 32 feet, and 1 drift 26 feet in length.

Cleveland and Dayton. R.C. Luesley & Co., proprietors; office Silverton; claim 300 by 3,000 feet; discovered 1876; located on Sultan Mountain, Animas mining district, 3½ miles from Silverton; course of vein, northeast and southwest, width 3 feet, pay vein 10 inches; nature of ore, gray copper; average value, 600 ounces silver per ton; 1 drift 50 feet in length and 50 feet from surface; yield, $1,200.

Cleveland and Divide. Owned by J.M. King, E.S. Finch, and others; office Silverton; claim 300 by 3,000 feet; discovered 1878; located on Pioneer Mountain, Animas mining district, 8 miles from Silverton; course of vein, east and west, width 15 feet, pay vein 3 feet, pitch 35 degrees; nature of ore, galena and gray copper; average value, 50 ounces silver per ton; 2 adits 10 feet each.

Cocktail. R.C. Luesley, proprietor; office Silverton; claim 300 by 1,500 feet; discovered 1874; located on Sultan Mountain, Animas mining district, 1 mile from Silverton; course of vein, northeast and southwest, width 3 feet, pay vein 6 inches; nature of ore, gray copper; assay value, 500 ounces silver per ton; 1 tunnel 90 feet in length, intersecting the vein at a depth of 100 feet, 1 level 140 feet in length.

Colorado Mining and Land Co. Organized under the laws of the State of New York, March 25, 1875; capital stock, $1,000,000, in 40,000 shares, $25 each; offices, Buffalo, New York, and Mineral City, Colorado; incorporators, J.S. Buell, Robert D. Hamlen, Orville Parker, and John O'Brien.

Columbia. Owned by Ogle, Scott, and others; office Eureka; claim 300 by 1,500 feet; discovered 1876; located on Crown Mountain, Eureka mining district, 2 miles from Silverton; vertical vein, trending north and south, width 2 feet, pay vein 10 inches; nature of ore, galena; assay value, 200 ounces silver per ton; 1 drift 40 feet in length.

Columbia. Owned by Peterson & Johnson, of Silverton; claim 300 by 1,500 feet; discovered 1876; located on Hurricane Peak, Eureka mining district, 5 miles from Eureka; course of vein, northwest and southeast, width 12 feet, pay vein 4 feet, pitch 20 degrees; nature of ore, gray copper; value, 160 ounces silver per ton; development, 1 tunnel 140 feet in length and 125 feet from surface.

Comet. J. Williamson, proprietor; office Ouray; claim 300 by 1,500 feet; discovered 1877; located on the Animas River, Eureka mining district, 2 miles from town of Animas Forks and 14 miles from Silverton; vertical vein, trending north and south, width 5 feet, pay vein 10 inches; nature of ore, gray copper and galena; assay value, 175 ounces silver per ton; 1 drift 20 feet in length.

Commodore. M. Green and S. Traverse, proprietors; office Ouray; claim 300 by 1,500 feet; discovered 1877; located on Mineral Point Mountain, 1 mile from Mineral City and 18 miles from Silverton; course of vein, north and south, width 9 feet; nature of ore, gray copper and galena; assay value, 70 ounces silver per ton; 1 drift 23 feet in length.

Como. W.U. Jackson, proprietor; office Mineral City; claim 300 by 1,500 feet; discovered 1877; located on Seigle Mountain, 2 miles from Mineral City and 18 miles from Silverton; vertical vein, trending north and south, width 8 feet; nature of ore, galena; assay value, 40 ounces silver per ton; 1 adit 15 feet in length.

Conder. D. McConnell, proprietor; office Animas Forks; claim 300 by 1,500 feet; discovered 1877; located in Burns Gulch, Eureka mining district, 2 miles from town of Eureka and 10 miles from Silverton; vertical vein, trending north and south, width 12 feet, pay vein 2 feet; nature of ore, galena and gray copper; assay value, 120 ounces silver per ton; 1 shaft 20 feet deep.

Consolidated Corn Exchange. Owned by R.C. Luesley and E.W. Hodges; office Silverton; claim 300 by 6,000 feet; discovered 1876; located on Lookout Mountain, Animas mining district, 7 miles from Silverton; course of vein, east and west, width 100 feet, pay vein 2 feet; nature of ore, carbonate of lead, average value, 69 ounces silver per ton; 1 shaft 18 feet deep, 1 tunnel 140 feet in length, intersecting vein at a depth of 130 feet.

Consolidated Virginia. J.N. Southgate, proprietor; office Howardsville; claim 300 by 1,500 feet, discovered 1876; located on Boulder Mountain, Animas mining district, 2 miles from Silverton; vertical vein, running northeast and southwest, width 15 feet; nature of ore, galena, gray copper and iron pyrites; assay value, 80 to 600 ounces silver per ton; 3 adits, 12, 14, and 15 feet in length, respectively.

Consolidated Virginia. Owned by J.M. Donnel, J. Watson, and W. Waldo; office Animas Forks; claim 300 by 1,500 feet; discovered 1877; located on California Mountain, Eureka mining district, 2 miles from Animas Forks and 18 miles from Silverton; course of vein, northeast and southwest, width 100 feet, pay vein 3 feet, pitch 30 degrees; nature of ore, galena, gray copper, and copper pyrites; average value, 110 ounces silver per ton; 1 adit 20 feet in length; yield, $500.

Covenant. Owned by Frank Herrod and John M. Stuart; office Mineral City; claim 300 by 1,500 feet; discovered 1877; located on Abraham Mountain, 12 miles from Silverton; vertical vein, trending northeast and southwest, width 35 feet, pay vein 4 feet; nature of ore, galena and copper pyrites; assay value, 60 ounces silver per ton; 1 drift 50 feet in length and 50 feet from surface.

Crescent. L. Quinn & Co., proprietors; office Silverton; claim 300 by 1,500 feet; discovered 1876; located on Ruby Creek, Animas mining district, 7 miles from Silverton; vertical vein, trending north and south, width 8 feet, pay vein 3 feet; nature of ore, gray copper and carbonate of lead; assay value, 160 ounces silver per ton; 2 adits 15 feet each.

Crispin. Owned by J.P. Johnson, R.J. Carley, and J.W. Wallace; office Eureka; claim 250 by 1,500 feet; discovered 1873; located on Eureka Mountain, near town of Eureka, 8 miles from Silverton; course of vein, northwest and southeast, width 8 feet, pay vein 5 feet, pitch 30 degrees; nature of ore, gray copper and galena; value of first-class ore, 80 to 150 ounces, and second-class 40 ounces silver per ton; 1 drift 55 feet in length and 75 feet from surface.

Crown Jewel. E.M. King and J.J. Hance, proprietors; office Silverton; claim 300 by 1,500 feet; discovered 1877; located on Red Mountain, Eureka mining district, 4 miles from Eureka and 12 miles from Silverton; vertical vein, trending northeast and southwest, width 6 feet; nature of ore, gray copper; 1 adit 40 feet in length.

Crown Point. C.S. Roe and F. Frick, proprietors; claim 300 by 1,500 feet; discovered 1877; located on Mineral Creek, Animas mining district, 3 miles from Silverton; vertical vein, trending east and west, width 4 feet, pay vein 3 feet; nature of ore, galena; assay value, 30 to 170 ounces silver per ton; 1 drift, 20 feet in length with a 14-foot face.

Crown Point. Owned by F. Marlow, G. Miller, and S. Regan; office Silverton; claim 300 by 1,500 feet; discovered 1876; located on Boulder Mountain, Animas mining district, 3 miles from Silverton; vertical vein, trending east and west, width 7 feet, pay vein 4 feet; nature of ore, galena and copper pyrites; assay value, 60 ounces silver per ton; 1 drift 30 feet in length.

Crown Point. Owned by J.M. Donnel, W. Waldo, and J. Watson; office Animas Forks; claim 300 by 1,500 feet; discovered 1877; located on California Mountain, Eureka mining district, 2 miles from Animas Forks and 18 miles from Silverton; course of vein, northeast and southwest, width 100 feet, pitch 30 degrees; nature of ore, galena; 1 adit 20 feet in length and 12 feet from surface.

Crystal. George N. Propper and A.W. Burrows, proprietors, office Del Norte; claim 300 by 1,500 feet; discovered 1878; located on Mineral Point Mountain, Uncompahgre mining district, near Mineral City, 18 miles from Silverton; vertical vein, trending northeast and southwest, width 25 feet, pay vein 30 inches; nature of ore, galena; 2 adits, 10 and 20 feet in length, respectively.

Custer. George A. Castle and John M. Custer, proprietors; office Silverton; claim 300 by 1,500 feet; discovered 1877; located on California Mountain, Eureka mining district, 2 miles from town of Animas Forks and 18 miles from Silverton; course of vein, northwest and southeast, width 150 feet; nature of ore, galena and gray copper; assay value, 150 ounces silver per ton; 1 drift, 30 feet in length and 20 feet from surface.

Dakota. Owned by Charles H. McIntire, A.W. Burrows, and others; office Del Norte; claim 150 by 1,500 feet; discovered 1874; located on Mineral Point Mountain, 18 miles from Silverton; vertical vein, trending northeast and southwest, width 50 feet, pay vein 3 feet; nature of ore, galena; assay value, 90 ounces silver per ton; 1 adit 30 feet in length.

Dardenelle. Owned by A.O. Terry, B. Pope, and J.W. Tiernan; office St. Louis, Missouri; claim 300 by 1,350 feet; discovered 1874; located on Watson Mountain, near town of Eureka, 8 miles from Silverton; vertical vein, trending northeast and southwest, width 6 feet, pay vein 2 feet; nature of ore, gray copper and sulphurets; average value, 162 ounces silver per ton; 1 shaft 10 feet deep, 1 drift 22 feet; yield, $200.

Dean Swift. Owned by George Purcell, J.A. White, and P. Cusick; office Ouray; claim 300 by 1,500 feet; discovered 1877; located on Oberto Mountain, Uncompahgre mining district, 12 miles from Silverton; vertical vein, trending east and west, width 8 feet, pay vein 2 feet; nature of ore, galena and gray copper; assay value, 65 ounces silver per ton; 2 adits, aggregating 50 feet in length.

Deganfels. O.F. Krauss & Co., proprietors; office Ouray; claim 300 by 1,500 feet; discovered 1876; located on Green Mountain, Animas mining district, 8 miles from Silverton; course of vein, northwest and southeast, width 3 feet, pay vein 32 inches; nature of ore, galena and gray copper; assay value, 90 ounces silver per ton; 1 tunnel 130 feet in length.

Democrat. Henry Promise & Co., proprietors; office, Animas Forks; claim 300 by 1,500 feet; discovered 1875; located on Treasure Mountain, Eureka mining district, 1 ½ miles from town of Animas Forks, and 17 miles from Silverton; vertical vein, trending northeast and southwest, width 15 feet, pay vein 3 feet; nature of ore, galena and gray copper; assay value, 900 ounces silver per ton; 1 tunnel 95 feet in length, intersecting vein 80 feet from surface.

Democrat. Owned by C. Rugg, J.W. Peirce, and others; office Eureka; claim 300 by 1,500 feet; discovered 1876; located on Silver Mountain, Eureka mining district, 3 miles from Silverton; course of vein, north and south, width 20 feet, pitch 15 degrees; nature of ore, galena and iron and copper pyrites; assay value, 80 ounces silver per ton; 1 shaft 8 feet deep, 2 adits, 10 and 12 feet in length, respectively.

Democrat. Owned by John H.P. Voorhies and others; office Silverton; claim 300 by 1,500 feet; discovered 1876; located on Ruby Hill, Animas mining district, 4 miles from Silverton; vertical vein, trending northeast and southwest, width 4 feet, pay vein 18 inches; nature of ore, galena and gray copper; average value, 200 ounces silver per ton; 1 shaft 50 feet deep, 1 tunnel 70 feet and 1 level 140 feet in length; shaft house 18 by 24 feet.

Deposit. Owned by Charles H. McIntire, A.W. Burrows, and others; office Ouray; claim 150 by 1,500 feet; discovered 1874; located on Mineral Point Mountain, Uncompahgre mining district, 18 miles from Silverton; vertical vein, trending northeast and southwest, width 50 feet, pay vein 3 feet; nature of ore, gray copper and iron pyrites; assay value, 125 ounces silver per ton; 1 adit 20 feet in length.

Detroit. Henry Promise, proprietor; office Animas Forks; claim 300 by 1,500 feet; discovered 1876; located on Eureka Mountain, 10 miles from Silverton; course of vein, northwest and southeast, width 12 feet, pay vein, 2 feet, pitch 45 degrees; nature of ore, galena and copper pyrites; 1 shaft 10 feet deep; 1 adit 14 feet in length.

Diamond L. James J. and Albert Bernard, Daniel J. and James D. McKay, William Forsyth, and John M. Ross; proprietors; office Howardsville; claim 300 by 1,500 feet; discovered 1873; located in Arastra Basin, 3 miles from Silverton; course of vein, northeast and southwest, width 3 feet; nature of ore, galena and gray copper; average assay value, $700 per ton; 1 tunnel 50 feet in length.

Diamond Tunnel. Owned by the Silver Producing Mining Co.; incorporated October 12, 1877; capital stock, $150,000, in 1,500 shares, $100 each; office Silverton; Edmund Higginbotham, president, Thomas Campau, vice president and treasurer, John G. Heid, secretary; claim 1,500 by 3,000 feet; discovered 1876; located on Sultan Mountain, Animas mining district, 1 mile from Silverton; course of tunnel, east and west, length 350 feet.

Digger Injun. J. Roos & Co., proprietors; office Eureka; claim 300 by 1,500 feet; discovered 1874; located in Poughkeepsie Gulch, Uncompahgre mining district, 12 miles from Silverton; vertical vein, trending east and west, width 3 feet, pay vein 1 foot; nature of ore, galena; assay value, 50 to 100 ounces silver per ton; 1 drift 10 feet in length and 10 feet from surface.

Dolly. Ellen M. McIntire, proprietor; office Ouray; claim 300 by 1,500 feet; discovered 1874; located on Mineral Point Mountain, half mile from Mineral City and 18 miles from Silverton; course of vein, east and west, width 4 feet, pay vein 2 feet; nature of ore, galena and iron pyrites; 1 adit 25 feet in length.

Dom Pedro. R.P. Bukey and Co., proprietors; office Animas Forks; claim 300 by 1,500 feet; discovered 1876; located on California Mountain, 2 miles from Animas Forks and 18 miles from Silverton; course of vein, northeast and southwest, width 5 feet, pay vein 16 inches; nature of ore, galena; 1 shaft 12 feet deep.

Eagle. Henry Carley & Co., proprietors; office Animas Forks; claim 300 by 1,500 feet; discovered 1875; located on Treasure Mountain, Eureka mining district, 1 mile from Eureka and 11 miles from Silverton; course of vein, northeast and southwest, width 35 feet, pay vein 30 inches; nature of ore, gray copper and galena; assay value, 60 ounces silver per ton; 2 tunnels, 18 and 25 feet in length, respectively.

Eclipse. Owned by Galbraith, Rockefellow & Co.; office Silverton; claim 300 by 1,500 feet; discovered 1878; located on King Solomon Mountain, Animas mining district, 6 miles from Silverton; course of vein, northeast and southwest, width 4 feet, pay vein 18 inches, pitch 45 degrees; nature of ore, gray copper; average value, 300 ounces silver per ton; 2 shafts, 20 and 30 feet in depth, respectively; yield, $2,000.

Eclipse, No. 2. Owned by M. Boulware, D. Keyes, C.J. Gibbons, and C. Hart; office Lake City; claim 300 by 1,500 feet; discovered 1878; located on North Lookout Mountain, Animas mining district, 8 miles from Silverton; vertical vein, trending northeast and southwest, width 25 feet, pay vein 2 feet; nature of ore, carbonate of lead; average value, 200 ounces silver per ton; 1 shaft 35 feet deep.

Edna C. M. Pray & Co., proprietors; office Eureka; claim 300 by 1,500 feet; discovered 1876; located 2 miles from town of Eureka and 10 miles from Silverton; course of vein, northeast and southwest, width 12 feet, pay vein 2 feet, pitch 10 degrees; nature of ore, galena, iron and copper pyrites; assay value, 75 ounces silver per on; 1 drift 25 feet in length.

Eighth Wonder. O. Larson & Co., proprietors; claim 300 by 1,500 feet; discovered 1876; located on Seigle Mountain, 1 mile from Mineral City and 18 miles from Silverton; vertical vein, trending northeast and southwest, width 6 feet, pay vein 22 inches; nature of ore, gray copper and sulphurets; assay value, $300 ounces silver per ton; 1 tunnel 32 feet in length, 2 adits aggregating 27 feet in length.

Elephant. Colorado Mining and Land Co., proprietors; office Buffalo New York; claim 300 by 1,350 feet; discovered 1874; located on Lake Mountain, Uncompahgre mining district, 17 miles from Silverton; vertical vein, trending northeast and southwest, width 50 feet; nature of ore, gray copper; assay value, 46 ounces silver per ton; 1 tunnel 32 feet, and 1 adit 50 feet in length.

Elephant. P. Crout and R. Hanson, proprietors; office Mineral City; claim 300 by 1,500 feet; discovered 1875; located on Wood Mountain,

Eureka mining district, 4 miles from town of Eureka and 12 miles from Silverton; course of vein, northeast and southwest, width 8 feet; pay vein 22 inches; nature of ore, galena; assay value, 65 ounces silver per ton; 1 shaft 10 feet deep, 1 drift 23 feet in length and 25 feet from surface.

Ellen. Ellen M. McIntire, proprietor; office Ouray; claim 300 by 1,500 feet; discovered 1874; located on Mineral Point Mountain, half mile from Mineral City and 18 miles from Silverton; course of vein, north and south, width 10 feet, pay vein 5 feet; nature of ore, galena and gray copper; assay value, $100 silver per ton; 1 adit 30 feet in length.

Emerald. D. O'Connell, proprietor; office Animas Forks; claim 300 by 1,500 feet; discovered 1878; located on Silver Mountain, 3 miles from town of Eureka, and 11 miles from Silverton; course of vein, northeast and southwest, width 2 feet, pay vein 6 inches; pitch 30 degrees; nature of ore, galena; assay value, 35 ounces silver per ton; 1 adit 14 feet in length.

Emma N. L. Woodbury & Co., proprietors; office Animas Forks; claim 300 by 1,500 feet; discovered 1877; located on Wood Mountain, Eureka mining district, 1 mile from Animas Forks and 17 miles from Silverton; course of vein, northwest and southeast, width 6 feet; nature of ore, galena; 1 adit 10 feet deep and 10 feet from surface.

Empire. J. Georke & Co., proprietors; office Eureka; claim 300 by 1,500 feet; discovered 1876; located on Crown Mountain, 3 miles from town of Eureka and 11 miles from Silverton; course of vein, north and south, width 12 feet, pitch 20 degrees; nature of ore, galena and gray copper; average value 40 ounces silver per ton; 1 adit 10 feet and 1 drift 30 feet in length.

Empire. Knight, Slocum & Co., proprietors; office Silverton; claim 300 by 1,500 feet; discovered 1874; located on Sultan Mountain, Animas mining district, near limits of Silverton; course of vein, northwest and southeast, width 6 feet, pay vein 30 inches, pitch 5 degrees; nature of ore, galena and gray copper; value, 150 ounces silver per ton; 2 shafts, 65 and 70 feet in length, respectively, 1 tunnel 219 feet in length, intersecting vein at a depth of 125 feet, 2 levels aggregating 60 feet in length, 55 and 65 feet from surface, respectively; shaft house 16 by 20 feet.

Empire. T.V. Brock & Co., proprietors; office Eureka; claim 300 by 1,500 feet; discovered 1878; located on Jones Mountain, Eureka mining district, 3 miles from town of Eureka and 11 miles from Silverton; vertical vein, trending north and south, width 30 inches, pay vein 15 inches; nature of ore, galena; assay value, 140 ounces silver per ton; 1 drift 20 feet in length and 14 feet from surface.

Empress. Thomas Dunton, proprietor; office Eureka; claim 300 by 1,500 feet; discovered 1877; located in Picayune Gulch, Eureka mining district, 3 miles from town of Eureka and 11 miles from Silverton; course of vein, north and south, width 15 feet, pay vein 4 feet, pitch 30 degrees; nature of ore, galena; assay value, 85 ounces silver per ton; 1 shaft 14 feet deep.

Equator. A. Trellinger and C.W. Ross, proprietors; office Ouray; claim 300 by 1,500 feet; discovered 1875; located on Mineral Point Mountain, 18 miles from Silverton; vertical vein, trending northeast and southwest, width 75 feet; nature of ore, galena, iron and copper pyrites; assay value, 60 ounces silver per ton; 3 shafts, 10, 12, and 15 feet in depth, respectively.

Eudora. F. Franklin and I. Gross, proprietors; office Lake City; claim 300 by 1,500 feet; discovered 1877; located on California Mountain, 2 miles from town of Animas Forks and 18 miles from Silverton; vertical vein, trending northeast and southwest, width 20 feet, pay vein 5 feet; nature of ore, galena and iron pyrites; assay value, 80 ounces silver per ton; 1 shaft 16 feet deep.

Eureka and Grand. J.R. McKinnie & Co., proprietors; office Eureka; claim 300 by 1,500 feet; discovered 1875; located on Eureka Mountain, Eureka mining district, 1 mile from town of Eureka and 9 miles from Silverton; course of vein, north and south, width 50 feet, pay vein 16 feet, pitch 30 degrees; nature of ore, galena; assay value, 21 to 144 ounces silver per ton, 3 adits, 10, 15, and 25 feet, respectively, and 1 tunnel 12 feet in length.

Eureka Tunnel. Owned by the Eureka Gold and Silver Mining co.; organized under the laws of Indiana in the year 1878; capital stock, $30,000, in 1,000 shares, $30 each; offices Eureka, Colorado, and Columbus, Ohio; J.H. Wasson, president, J.F. Miller, vice president, T.R. Wing, treasurer, C.H. Cole, secretary, E.S. Armstrong, superintendent; claim 1,500 by 3,000 feet; discovered 1874; located beside the Animas River in town of Eureka 8 miles from Silverton; course of tunnel, northwest and southeast; length 75 feet.

Eureka Tunnel. Owned by James J. and Albert Bernard, Daniel J. and James D. McKay, William Forsythe, and John M. Ross; office Eureka; claim 1,500 by 3,000 feet; established 1874; located in the bed of the Animas River near town of Eureka; length of tunnel 75 feet.

Evening Star. Owned by S. Allen and R. Skinner; office Animas Forks; claim 300 by 1,500 feet; discovered 1876; located on California Mountain, Eureka mining district, 2 miles from Animas Forks and 18 miles from Silverton; vertical vein, trending north and south, width 10 feet, pay vein 3 feet; nature of ore, galena and gray copper; assay value, 43 to 110 ounces silver per ton; 1 shaft 20 feet deep.

Excelsior. W. McIntire, proprietor; office Silverton; claim 300 by 1,500 feet; discovered 1874; located in Ohio Gulch, Animas mining district, 6 miles from Silverton; course of vein, northeast and southwest, width 2 feet, pay vein 20 inches, pitch 40 degrees; nature of ore, galena and gray copper; assay value, 50 ounces silver per ton; 4 adits, 8, 10, 15, and 16 feet in length, respectively.

Excelsior. W.P. Hartman & Co., proprietors; office Silverton; claim 300 by 1,500 feet; discovered 1876; located in Cunningham Gulch, Animas mining district, 5 miles from Silverton; vertical vein, trending east and west, width 30 inches; pay vein 8 inches; nature of ore, gray copper and galena; assay value, 300 ounces silver per ton; 1 drift 25 feet in length.

Exchequer. G. Ferguson & Co., proprietors; office Ouray; claim 300 by 1,500 feet; discovered 1877; located on Mineral Point Mountain, 2 miles from Mineral City and 9 miles from Ouray; vertical vein, trending northeast and southwest, width 5 feet, pay vein 6 inches; nature of ore, gray copper and brittle silver; assay value, 180 ounces silver per ton; 2 adits, 10 and 15 feet in length, respectively.

Fortune. O. Larson and S. Larson, proprietors; office Del Norte; claim 300 by 1,500 feet; discovered 1875; located on Lake Mountain, near Mineral City, 18 miles from Silverton; course of vein, northeast and southwest, width 30 inches, pitch 25 degrees; nature of ore, galena and gray copper; assay value, $75 silver per ton; 1 adit 20 feet in length.

Frank Barber. Owned by John Meyer, James Dooley, and George Rohwer; office Ouray; claim 300 by 1,500 feet; discovered 1877; located on Poughkeepsie Mountain, 9 miles from Ouray; course of vein, northeast and southwest, width 15 feet, pay vein 20 inches; nature of ore, sulphurets and gray copper; assay value, 50 to 1,500 ounces silver per ton; 2 drifts, 11 and 20 feet in length, respectively.

Frederica. A.K. Prescott & Co., proprietors; office Del Norte; claim 300 by 1,500 feet; discovered 1873; located in Burns Gulch, 3 miles from town of Eureka and 9 miles from Silverton; vertical vein, trending east and west, width 7 feet; nature of ore, galena and gray copper; average value, $112 per ton; 2 adits, 15 and 25 feet in length, respectively.

Free Coinage. Owned by T. Byron and M. Hogan; office Animas Forks; claim 300 by 1,500 feet; discovered 1878; located on Treasure Mountain, Eureka mining district, 1 mile from town of Eureka and 10 miles from Silverton; course of vein, northeast and southwest, width 50 feet; nature of ore, galena and sulphurets; 1 shaft 11 feet deep.

Gem. James Thornton and W. Bell, proprietors; office Eureka; claim 300 by 1,500 feet; discovered 1877; located in Minnie Gulch, Eureka mining district, 2 miles from town of Eureka and 7 miles from Silverton; course of vein, northeast and southwest, width 8 feet, pay vein 3 feet, pitch 40 degrees; nature of ore, galena and iron pyrites; assay value, 100 ounces silver per ton, 1 shaft 8 feet deep; 1 adit 15 feet in length.

Georgia. W.L. Guinn and J. Long, proprietors; office Eureka; claim 300 by 1,500 feet; discovered 1875; located in Burns Gulch, Eureka mining district, 2 miles from Eureka, and 10 miles from Silverton; course of vein, northwest and southeast, width 10 feet, pay vein 3 feet, pitch 10 degrees; nature of ore, galena and gray copper; assay value, 15 to 300 ounces silver per ton; 1 adit 30 feet in length.

Georgia Clarke. P. McEnany & Co., proprietors; office Animas Forks; claim 300 by 1,500 feet; discovered 1874; located on the Animas River, Eureka mining district, 1 mile from town of Animas Forks and 16 miles from town of Silverton; course of vein, northeast and southwest, width 12 feet, pay vein 22 inches, pitch 25 degrees; nature of ore, galena; assay value, 75 ounces silver per ton; 1 drift 40 feet in length, and 30 feet from surface.

German. Owned by G. Griffin, H.B. Jenkins, and others; office Animas Forks; claim 300 by 1,500 feet; discovered 1875; located on California Mountain, Eureka mining district, 2 miles from town of Animas Forks and 16 miles from Silverton; course of vein, north and south, width 60 feet, pitch 15 degrees; nature of ore, galena and iron pyrites; assay value, 80 ounces silver per ton; 1 shaft 12 feet deep, 1 tunnel 30 feet in length.

Gypsy. P. Crout, proprietor; office Mineral City; claim 300 by 1,500 feet; discovered 1877; located in Poughkeepsie Gulch, Uncompahgre mining district, 12 miles from Silverton; course of vein, northeast and southwest, width 23 feet, pay vein 2 feet, pitch 30 degrees; nature of ore, sulphurets and brittle silver; average value, 600 ounces silver per ton; 1 shaft 50 feet deep.

Gypsy Maid. S.H. Day and D. Brown, proprietors; office Ouray; claim 300 by 1,500 feet; discovered 1877; located in Poughkeepsie Gulch, 12 miles from Silverton; vertical vein, trending northeast and southwest, width 20 feet, pay vein 22 inches; nature of ore, galena and sulphurets; assay value, 800 ounces silver per ton; 1 shaft 12 feet deep, 1 tunnel 25 feet in length.

Golden Eagle. J.D. Resor & Co., proprietors; office Silverton; claim 300 by 1,500 feet; discovered 1876; located in Poughkeepsie Gulch, 12 miles from Silverton; vertical vein, trending northwest and southeast, width 10 feet, pay vein 30 inches; nature of ore, gray copper and galena; assay value, 125 ounces silver per ton; 1 drift 18 feet in length.

Good Enough. D. McGillivray & Co., proprietors; office Eureka; claim 300 by 1,500 feet; discovered 1877; located in Eureka mining district, 1 mile from town of Eureka and 9 miles from Silverton; vertical vein, trending east and west, width 2 feet, pay vein 15 inches; nature of ore, galena and gray copper; assay value, 65 ounces silver per ton; 1 drift 30 feet in length and 25 feet from surface.

Good Enough. Owned by Thomas A. Breckenridge & Co.; office Silverton; claim 300 by 1,500 feet; discovered 1877; located on Lookout Mountain, Animas mining district, 6 miles from Silverton; vertical vein, trending east and west, pay vein 14 inches; nature of ore, carbonate of lead; assay value, 80 ounces silver per ton; 1 shaft 35 feet deep.

Gould and Curry. John W. Patterson, proprietor; office Howardsville; claim 300 by 1,500 feet; discovered 1877; located in Boulder Gulch, Animas mining district, 2 miles from Silverton; vertical vein, trending northwest and southeast, width 10 feet, pay vein, 18 inches; nature of ore, galena and gray copper; assay value, 140 ounces silver per ton; 1 drift 15 feet in length.

Governor Hayes. Owned by George W. Osborn, George N. Propper, and W.W. Straight, office Pueblo; claim 230 by 1,500 feet; discovered 1877; located on Lake Mountain, 1 mile from Mineral City and 18 miles from Silverton; vertical vein, running northeast and southwest, width 12 feet, pay vein 22 inches; nature of ore, gray copper; assay value, 100 to 700 ounces silver per ton; 1 adit 30 feet in length.

Grand Central. Owned by John C. Dunn, of Denver; claim 300 by 1,500 feet; discovered 1875; located on Eureka Mountain, 3 miles from town of Eureka and 11 miles from Silverton; vertical vein, trending northeast and southwest, width 60 feet, pay vein 40 feet; nature of ore, gray copper and galena; assay value, 40 to 340 ounces silver per ton; 2 drifts, 20 and 50 feet in length, respectively.

Grand Duke. E. Morton and R. N. Lake, proprietors; office Eureka; claim 300 by 1,500 feet; discovered 1876; located on Silver Mountain, Eureka mining district, 3 miles from town of Eureka and 11 miles from Silverton; vertical vein, trending northwest and southeast, width 14 feet; nature of ore, galena and iron pyrites; assay value, 65 ounces silver per ton; 1 adit 30 feet in length.

Grand Duke. Owned by Charles Humaston, George Howard, and M.M. Engleman; office, Eureka; claim 300 by 1,500 feet; discovered 1875; located on Hurricane Peak, Eureka mining district, 3 miles from town of Eureka and 12 miles from Silverton; course of vein, northwest and southeast, width 20 feet, pay vein 2 feet, pitch 10 degrees; nature of ore, gray copper; assay value, 200 ounces silver per ton; main shaft 15 feet deep, 1 tunnel 25 feet in length.

Grand Republic. Owned by M. Hogan and T. Byron; office Animas Forks; claim 300 by 1,500 feet; discovered 1878; located on Treasure Mountain, Eureka mining district, 1 mile from town of Eureka and 10 miles from Silverton; course of vein, east and west, width 50 feet, pitch 20 degrees; nature of ore, galena and copper pyrites; 1 shaft 12 feet deep.

Grand Trunk Mining Co. Incorporated November 29, 1876; capital stock, $500,000, in 5,000 shares, $100 each; office Denver; organized by Archie C. Fisk, William Smedley and Edward W. Pierce; Archie C. Fisk, president, William Smedley, treasurer. The property of this company is located on Galena Mountain, Animas mining district, 4 miles from Silverton; claim 300 y 3,000 feet; discovered 1873; course of vein, northeast and southwest, width 30 feet, pay vein 6 feet; nature of ore, galena and gray copper; average value, $125 per ton; development, 6 shafts, aggregating 100 feet in depth and 1 tunnel 55 feet in length. The surface of this property and the adjoining country is covered with timber suitable for mining and other purposes.

Granger. S. Peterson, proprietor; office Eureka; claim 300 by 1,500 feet; discovered 1874; located on Eureka Mountain, 2 miles from town of Eureka and 10 miles from Silverton; course of vein, northeast and southwest, width 30 feet, pitch 15 degrees, nature of ore, galena and copper pyrites; 2 shafts, 11 and 12 feet deep, respectively.

Gray Copper Falls. George Green & Co., proprietors; office Silverton; claim 300 by 1,500 feet; located at the head of Cement Creek, Eureka mining district, 15 miles from Silverton; course of vein, northeast and southwest, width 6 feet, pay vein 4 feet; nature of ore, gray copper and galena; operated by a tunnel, 300 feet in length.

Gray Eagle. J. Orr and F. Morey, proprietors; office Silverton; claim 200 by 1,010 feet; discovered 1874; located on Hazelton Mountain, Animas mining district, 3 miles from town of Silverton; vertical vein, trending northeast and southwest, width 2 feet, pay vein 1 foot; nature of ore, galena and gray copper; average value, 100 ounces silver per ton; 1 shaft 60 feet deep, 2 drifts, 50 and 150 feet in length, respectively; yield, $1,000.

Great American. George N. Propper and A.W. Burrows, proprietors; office Del Norte; claim 300 by 1,500 feet; discovered 1874; located on Lake Mountain, near Mineral City, 18 miles from Silverton; vertical vein, trending northeast and southwest, width 30 feet, pay vein 5 feet; nature of ore, galena; assay value, 80 ounces silver per ton; 1 shaft 16 feet deep, 1 adit 50 feet in length.

Great Eastern. E. McCarty, proprietor; office Silverton; claim 300 by 1,500 feet; discovered 1876; located on Silver Mountain, 2 miles from town of Eureka and 8 miles from Silverton; course of vein, north and south, width 15 feet, pay vein 3 feet, pitch 20 degrees; nature of ore, gray copper and copper pyrites; assay value, 200 ounces silver per ton; 1 drift 60 feet in length.

Gritiza. Gust. Johnson, proprietor; office Silverton; claim 300 by 1,500 feet; discovered 1877; located in Ross Basin, Eureka mining district, 10 miles from Silverton; course of vein, northwest and southeast, width 4 feet, pay vein 18 inches, pitch 20 degrees; nature of ore, gray copper; value, 106 ounces silver per ton; 1 shaft 15 feet deep.

Grub Stake. James T. Rawlings & Co., proprietors; office Denver; claim 300 by 1,500 feet; discovered 1874; located on the Animas River, Eureka mining district, 2 miles from town of Animas Forks and 14 miles from Silverton; course of vein, northwest and southeast, width 5 feet, pay vein 20 inches; nature of ore, gray copper and iron pyrites; assay value, 416 ounces silver per ton; 1 drift 70 feet in length and 80 feet from surface; yield, $225.

Hafer. Charles H. McIntire, proprietor; office Ouray; claim 300 by 1,500 feet; discovered 1874; located on Mineral Point Mountain, 1 mile from Mineral City and 18 miles form Silverton; vertical vein, trending northeast and southwest, width 30 feet, pay vein 3 feet; nature of ore, gray copper; 1 adit 30 feet in length and 10 feet from surface.

Hard to Beat. J. Frye & Co., proprietors; office Silverton; claim 300 by 1,500 feet; discovered 1877; located on Mineral Creek, Animas mining district, 5 miles from Silverton; vertical vein, trending east and west, width 3 feet, pay vein 20 inches; nature of ore, gray copper and carbonate of lead; assay value, 90 ounces silver per ton; 1 adit 24 feet in length.

Hard to Beat. W. Williams and J. Richards, proprietors; office Howardsville; claim 300 by 1,500 feet; discovered 1876; located in Cunningham Gulch, Animas mining district, 7 miles from Silverton; course of vein, east and west; width 8 feet, pay vein 2 feet, pitch 20 degrees; nature of ore, gray copper and galena; assay value, 150 ounces silver per ton; 1 drift 20 feet in length.

Hattie. J. Williamson & Co., proprietors; office Silverton; claim 300 by 1,500 feet; discovered 1877; located on Kendall Mountain, Animas mining district, 2 miles from Silverton; course of vein, north and south, width 4 feet, pitch 15 degrees; nature of ore, gray copper and galena; assay value, 90 ounces silver per ton; 1 adit 18 feet in length.

Hawkeye. Owned by the Silver Producing Mining Co.; office Silverton; claim 300 by 1,500 feet; discovered 1873; located on Sultan Mountain, Animas mining district, 1 mile from Silverton; vertical vein, trending northeast and southwest; width 4 feet, pay vein 20 inches; nature of ore, gray copper and galena; value, 200 ounces silver per ton; 1 shaft 60 feet deep; shaft house 16 by 20 feet.

Hawkeye State. William Morgan and J.L. Price, proprietors; office Howardsville; claim 300 by 1,500 feet; discovered 1878; located in Cunningham Gulch, Animas mining district, 6 miles from Silverton; course of vein, northwest and southeast, width 3 feet, pay vein 20 inches, pitch 30 degrees; nature of ore, galena and gray copper; assay value, 60 ounces silver per ton; 3 adits, 12, 14, and 24 feet in length, respectively.

Hazelton Mountain Tunnel and Mining Co. Incorporated August 15, 1877; capital stock, $16,000, in 16 shares, of $1,000 each; office Silverton; trustees: Wesley and G.A. Tinker, J.S. Hall, Gustavus Johnson, Julius E. Bates, A.B. Hart, J.H. Brink, and C.J. Marrh.

Hendricks. J. McKinzie and R. Struby, proprietors; office Animas Forks; claim 300 by 1,500 feet; discovered 1877; located on Treasure Mountain, Eureka mining district, 2 miles from town of Animas Forks and 18 miles from Silverton; course of vein, northwest and southeast, width 12 feet, pay vein 30 inches, pitch 20 degrees; nature of ore, galena and copper pyrites; assay value, 75 ounces silver per ton; 1 shaft 10 feet deep, 1 adit 15 feet in length.

Hercules. Knight, Slocum & Co., proprietors; office Silverton; claim 300 by 1,500 feet; discovered 1874; located on Sultan Mountain, Animas mining district, near limits of Silverton; course of vein, northwest and southeast, width 7 feet, pay vein 5 feet, pitch 15 degrees; nature of ore, gray copper, galena and copper pyrites; average value, $100 per ton; 1 shaft 26 feet deep, 1 tunnel 235 feet in length, intersecting vein at a depth of 100 feet; shaft house 16 by 22 feet.

Hidden Treasure. Charles Galbraith & Co., proprietors; office Silverton; claim 300 by 1,500 feet; discovered 1878; located on King Solomon Mountain; Animas mining district, 6 miles from Silverton; course of vein, northeast an southwest, width 3 feet, pay vein 2 feet; nature of ore, gray copper; assay value, 500 to 2,000 ounces silver per ton; 1 shaft 10 feet deep.

Hidden Treasure. O. Larson, proprietor; office Del Norte; claim 300 by 1,500 feet; discovered 1877; located on Seigle Mountain, 1 mile from Mineral City and 8 miles from Silverton; vertical vein, trending northeast and southwest, width 3 feet, pay vein 18 inches; nature of ore, galena and gray copper; assay value, 82 ounces silver per ton; 1 adit 15 feet, and 1 drift 20 feet in length.

Highland Chief. Owned by W.R. Mitchell, Winfield Scott Stratton, and W.B. Sherman; office Colorado Springs; claim 300 by 1,500 feet; discovered 1876; located on King Solomon Mountain, Animas mining district, 4 miles from Howardsville and 8 miles from Silverton; course of vein, northeast and southeast [sic], width 3 feet, pay vein 8 inches; pitch 15 degrees; nature of ore, galena and gray copper; value, 400 ounces silver per ton; 1 tunnel 60 feet and 1 adit 18 feet in length; yield, $300.

Highland Chief. Gustavus Johnson, proprietor; office Silverton; claim 300 by 1,500 feet; discovered 1875; located in Eureka Gulch, 12 miles from Silverton; course of vein, northeast and southwest, width 7 feet, pay vein 30 inches, pitch 10 degrees; nature of ore, gray copper and galena; value, 200 ounces silver per ton; 1 adit 20 feet, and 1 tunnel 25 feet in length.

Highland Mary. Owned by Edward Innis; discovered July 14, 1874; located in Cunningham Gulch, Animas mining district, 7 miles from Silverton; course of vein, northwest and southeast, width 6 feet, pay vein 30 inches, dip southeast; value of ore, 900 ounces silver per ton; development, 5 levels, aggregating 500 feet, and 1 tunnel 1,300 feet in length. On the surface is a stone engine house, 40 feet square; a large and neatly constructed dwelling house, stables, ore house, post office, etc.

Home Stake Tunnel. Owned by M.J. Condon & Co.; office Eureka; claim 650 by 1,500 feet; discovered 1874; located on Eureka Mountain, near town of Eureka, 8 miles from Silverton; course of tunnel, north and south, length 40 feet.

Humboldt [alternate spelling, Humbolt]. J.W. Strouse & Co., proprietors; office Mineral City; claim 300 by 1500 feet; discovered 1878; located in Poughkeepsie Gulch, 12 miles from Silverton; course of vein, northeast and southwest, width 6 feet, pay vein 2 feet; nature of ore, galena and gray copper; 1 adit 10 feet in length.

Hutchinson. A. Clare & Co., proprietors; office Animas Forks; claim 300 by 1,500 feet; discovered 1874; located on Silver Mountain, Eureka mining district, 3 miles from town of Animas Forks and 11 miles from Silverton; course of vein, northeast and southwest, width 25 feet, pay vein 2 feet; nature of ore, gray copper; assay value, $400 per ton; 1 shaft 10 feet deep, 3 adits 25 feet each.

Hyat Wing. Charles H. and Edward W. McIntire, proprietors; office Ouray; claim 200 by 1,500 feet; discovered 1874; located on Mineral Point Mountain, half mile from Mineral City and 18 miles from Silverton; vertical vein, trending northeast and southwest, width 20 feet, pay vein 30 inches; nature of ore, galena and gray copper; assay value, 100 ounces silver per ton; 2 adits 12 feet each in length.

I.X.L. Charles Thurmond, proprietor; office Animas Forks; claim 300 by 1,500 feet; discovered 1876; located on Houghton Mountain, Eureka mining district, 2 miles from town of Animas Forks and 18 miles from Silverton; course of vein, north and south, width 60 feet, pitch 10 degrees; nature of ore, galena; assay value, 60 to 200 ounces silver per ton; 1 shaft 15 feet deep, 1 adit 15 feet in length.

I.X.L. Tunnel. Owned by the I.X.L Tunnel Co., organized under the laws of the state of Wisconsin, in the year 1877; capital stock, $200,000, in 20,000 shares, $10 each; offices Silverton, Colorado, and Milwaukee, Wisconsin; R.G. Owens, president and treasurer, E. Upson vice president, L.E. Parson, secretary; claim 1,500 by 3,000 feet; discovered 1874; located on Hazelton Mountain, Animas mining district, 3 miles from Silverton; course of tunnel, north and south, length 500 feet.

Idaho. Peter Robertson, proprietor; office Silverton; claim 300 by 3,000 feet; located on Kendall Mountain, Animas mining district, near Silverton; course of vein, southeast and northwest, width 4 feet, pay vein 3 feet; nature of ore, gray copper and galena; value, 84 ounces silver per ton; main shaft 15 feet deep; 2 tunnels, 20 and 140 feet in length, respectively.

Independence. John Lattimore & Co., proprietors; office Lake City; claim 300 by 1,500 feet; discovered 1875; located in Burns Gulch, Eureka mining district, 2 miles from town of Eureka and 10 miles from Silverton; course of vein, northwest and southeast, width 10 feet, pay vein 15 inches; nature of ore, antimonial galena and gray copper; assay value, 130 ounces silver per ton; 1 drift 45 feet in length and 30 feet from surface.

Independent. William Cooper and Co., proprietors; office Lake City; claim 300 by 1,500 feet; discovered 1875; located on Engineer Mountain, 2 miles from Mineral City; course of vein, northeast and southwest, width 18 feet, pay vein 10 inches, pitch 10 degrees; nature of ore, galena and iron pyrites; assay value, 50 ounces silver per ton; 4 adits 10 feet each.

Index. M.W. Dresser & Co., proprietors; office Ouray; claim 300 by 1,500 feet; discovered 1878; located on Silver Mountain, 3 miles from Animas Forks and 11 miles from Silverton; course of vein, northeast and southwest, width 10 feet, pay vein 30 inches, pitch 35 degrees; nature of ore, galena; 1 adit 20 feet in length and 20 feet from surface.

Indian Chief. George Green & Co., proprietors; office Silverton; claim 300 by 1,500 feet; located at the head of Cement Creek, Eureka mining district, 12 miles from Silverton; vertical vein, trending northeast and southwest, width 4 feet, pay vein 12 inches; nature of ore, gray copper and galena; value, 100 to 400 ounces silver per ton; operated by a tunnel 125 feet in length.

Indus. Owned by M. Pray, E.F. Perry, and J.C. Ladd; office Eureka; claim 300 by 1,500 feet; discovered 1877; located on Centennial Mountain, Eureka mining district, 2 miles from town of Eureka and 10 miles from Silverton; vertical vein, trending east and west, width 30 feet; nature of ore, galena and gray copper; assay value, 165 ounces silver per ton; 2 shafts 10 feet each.

Ingleside. C. Paine and I.P. Martell, proprietors; office Animas Forks; claim 300 by 1,500 feet; discovered 1878; located on the Animas River, Eureka mining district, 2 miles from town of Animas Forks and 14 miles from Silverton; course of vein, north and south, width 12 feet, pay vein 20 inches, pitch 15 degrees; nature of ore, galena and gray copper; assay value, 100 ounces silver per ton; 1 adit 12 feet in length.

Ingot. Owned by C. Schofield, of Eureka; claim 300 by 1,500 feet; discovered 1875; located in Niagara Basin, 1 mile from town of Eureka; course of vein, north and south, width 8 feet; nature of ore, galena and sulphurets; assay value, 100 ounces silver per ton; development, 1 shaft 12 feet deep and 2 adits, 10 and 14 feet in length, respectively.

Inter-Ocean Mining Co., of San Juan, Colorado. Incorporated under the laws of the State of Iowa, August 31, 1878; capital stock $1,000,000, in 50,000 shares, $20 each; offices Davenport, Iowa, and Chicago, Illinois; trustees: William Owens, James Reynolds, Robert A. Payne, James M. Pyatt, and John R. Bensley.

Iowa. Owned by C. Gregg, E. Linke, and J. O'Neil; office Silverton; claim 300 by 1,500 feet; discovered 1875; located on Round Mountain, 6 miles from Silverton; course of vein, northwest and southeast, width 40 feet, pay vein 20 inches; nature of ore, free gold and galena; assay value, $250 per ton; 1 shaft 20 feet deep, 1 tunnel 35 feet in length.

Irish Boy. M. Higgins & Co., proprietors; office Silverton; claim 300 by 1,500 feet; discovered 1878; located on Hurricane Peak, Eureka mining district, 12 miles from Silverton; course of vein, northeast and southwest, width 5 feet, pay vein 8 inches; nature of ore, gray copper and galena; 1 adit 10 feet in length.

Iron Crown. George Wilcox & Co., proprietors; office Eureka; claim 300 by 1,400 feet; discovered 1878; located on Silver Mountain, Eureka mining district, 3 miles from town of Eureka and 11 miles from Silverton; vertical vein, trending northwest and southeast, width 24 feet, pay vein 2 feet; nature of ore, galena and gray copper; assay value, 200 ounces silver per ton; 1 adit 13 feet in length.

Isabell. M. Green & Co., proprietors; office Animas Forks; claim 300 by 1,500 feet; discovered 1878; located on Silver Mountain, Eureka mining district, 3 miles from Eureka and 11 miles from Silverton; vertical vein, trending north and south, width 50 feet; nature of ore, gray copper and galena; assay value, 150 ounces silver per ton; 1 shaft 9 feet deep, 1 adit 15 feet in length.

J.L.P. Tunnel. George Green & Co., proprietors; office Silverton; claim 1,500 by 3,000 feet; located on Hazelton Mountain, 2 miles from Silverton; length 400 feet.

Jack-Knife. P. McCay and U.K. White, proprietors; office Del Norte; claim 300 by 1,500 feet; discovered 1876; located on Mineral Point Mountain, 1 mile from Mineral City and 18 miles from Silverton; vertical vein, trending northeast and southwest, width 6 feet; nature of ore, gray copper and sulphurets; assay value, 100 ounces silver per ton; 1 adit, 40 feet in length.

Jarvis. Henry Promise, proprietor; office Animas Forks; claim 300 by 1,500 feet; discovered 1874; located on Tower Mountain, Eureka mining district, 9 miles from Silverton; vertical vein, trending north and south, width 18 feet, pay vein 2 feet; nature of ore, copper pyrites and galena; 1 adit 20 feet in length with a 15-foot face.

Jennie Parker. Owned by Joseph Cuenin, C.J. Rouech, August Tallott, and Henry Guillett; office Del Norte; claim 300 by 1,500 feet, patented; discovered 1875; located on Sultan Mountain, Animas mining district, near limits of Silverton; course of vein, northeast and southwest, width 8 feet, pay vein 3 feet, pitch 10 degrees west; nature of ore, gray copper and galena; average value, 122 ounces silver per ton; main shaft 80 feet deep; 1 tunnel 50 feet in length.

Jessie. R.H. Ferguson & Son, proprietors; office Eureka; claim 300 by 1,500 feet; discovered 1875; located on Crown Mountain, near town of Eureka, 9 miles from Silverton; course of vein, north and south, width 12 feet, pay vein 3 feet, pitch 40 degrees; nature of ore, gray copper and galena; 1 drift 22 feet in length and 20 feet from surface.

John. George W. McIntire, proprietor; office Ouray; claim 300 by 1,500 feet; discovered 1875; located on Middle Creek, half mile from Mineral City and 18 miles from Silverton; course of vein, east and west, width 10 feet, pay vein 4 feet; nature of ore, galena and gray copper; assay value, $100 silver per ton; 3 adits, 10, 25, and 45 feet in length, respectively.

John Franklin. Owned by G.A. McDowell, D.C. Ogsbury, and E.W. Johnstone; office Pueblo; claim 300 by 1,500 feet; discovered 1875; located in Arastra Gulch, Animas mining district, 3 miles from Silverton; course of vein, northeast and southwest, width 3 feet, pay vein 2 feet; nature of ore, free gold; value, $440 per ton; 1 drift 60 feet in length.

Junction City No. 2. W.B. House, proprietor; office Howardsville; claim 300 by 1,500 feet; discovered 1877; located on Tower Mountain, Eureka mining district, 2 miles from town of Eureka and 9 miles from Silverton; course of vein, northeast and southwest, width 12 feet; nature of ore, galena and iron pyrites; 1 tunnel 30 feet in length.

Kansas. G.E. Peters, proprietor; office Animas Forks; claim 300 by 1,500 feet; discovered 1875; located in Picayune Gulch, Eureka mining district, 3 miles from town of Animas Forks, and 11 miles from Silverton; vertical vein, trending northwest and southeast, width 10 feet; nature of ore, brittle silver and gray copper; assay value, 50 to 800 ounces silver per ton; 1 shaft 10 feet deep.

Kansas. Owned by H. Miller and H. Lennon; office Eureka; claim 300 by 1,500 feet; discovered 1877; located in the south fork of Eureka Gulch, 3 miles from town of Eureka and 9 miles from Silverton; vertical vein, trending northwest and southeast, width 6 feet, pay vein, 2 feet; nature of ore, antimony and galena; 1 shaft 14 feet deep.

Keystone. John Egan and William Clary, proprietors; office Silverton; claim 300 by 1,500 feet; discovered 1877; located on Brown Mountain, Eureka mining district, 10 miles from Silverton; vertical vein, trending north and south, width 25 feet; nature of ore, galena; assay value, 60 ounces silver per ton; 1 drift 22 feet in length.

King Solomon. Gillett & Co., proprietors; office Silverton; claim 300 by 1,500 feet; discovered 1876; located on King Solomon Mountain, Animas mining district, 4 miles from Silverton; course of vein northeast and southwest, width 30 feet, pitch 30 degrees east; nature of ore, spar and gray copper; assay value, 70 ounces silver per ton; 1 adit, 25 feet in length and 20 feet from surface.

Lake Park. Colorado Mining and Lake Co., proprietors; office Buffalo, New York; claim 300 by 1,500 feet; discovered 1874; located near Mineral City, 18 miles from Silverton; vertical vein, trending northeast and southwest, width 15 feet, pay vein 8 feet; nature of ore, galena and pay copper; assay value, 70 ounces silver per ton; 1 tunnel 80 feet in length.

Lake View. Henry Jenkins & Co., proprietors; office Mineral City; claim 300 by 1,500 feet; discovered 1877; located on Lake Mountain, Uncompahgre mining district, 2 miles from Mineral City and 20 miles from Silverton; vertical vein, trending northwest and southeast, width 7 feet, pay vein 18 inches; nature of ore, gray copper and galena; assay value, 90 ounces silver per ton; 1 shaft 17 feet deep.

Lake View. Mark Watson, proprietor; office Ouray; claim 300 by 1,500 feet; discovered 1876; located on Mineral Point Mountain, Uncompahgre mining district, 20 miles from Silverton; course of vein, east and west, width 30 feet, pitch 18 degrees; nature of ore, gray copper and galena; assay value, 100 ounces silver per ton; 4 adits 10 feet in length each.

Lamb. P. McEnany and H.F. Ash, proprietors; office Animas Forks; claim 300 by 1,500 feet; discovered 1875; located on Mineral Point Mountain, Uncompahgre mining district, near Mineral City, 18 miles from Silverton; vertical vein, trending northeast and southwest, width 6 feet, pay vein 2 feet; nature of ore, galena and sulphurets; assay value, 300 ounces silver per ton; 1 drift 25 feet in length and 30 feet from surface.

La Plata. R.P. Bukey & Co., proprietors; office Animas Forks; claim 300 by 1,500 feet; discovered 1874; located on Kendall Mountain, Animas mining district, 1 mile from Silverton; vertical vein, trending northeast and southwest, width 4 feet, pay vein 18 inches; nature of ore, galena; average value, 40 ounces silver per ton; 1 shaft 25 feet deep.

Last Chance. Martin C. Clingham and R. Laws; office Silverton; claim 300 by 1,500 feet; discovered 1876; located on Mineral Creek, Animas mining district, 7 miles from Silverton; vertical vein, trending northeast and southwest, width 9 feet, pay vein 30 inches; nature of ore, galena and iron pyrites; assay value, 70 ounces silver per ton; 1 drift 25 feet in length.

Last Chance. C. McCabe & Co., proprietors; office Eureka; claim 300 by 1,500 feet; discovered 1877; located on Cement Creek., Eureka mining district 10 miles from Silverton; course of vein, northwest and southeast, width 24 feet, pitch 40 degrees; nature of ore, galena; assay value, 90 ounces silver per ton; 2 adits, 10 and 15 feet in length, respectively.

Lee. C. Burns and W.L. Guinn, proprietors; office Eureka; claim 50 by 1,500 feet; discovered 1873; located on Missouri Mountain, Eureka mining district, 2 miles from town of Eureka and 10 miles from Silverton; course of vein, northwest and southeast, width 6 feet, pay vein 2 feet, pitch 15 degrees; nature of ore, copper pyrites and gray copper; assay value, 200 ounces silver per ton; 1 drift 25 feet, and 1 adit 20 feet in length.

Legal Tender. George Green & Co., proprietors; office Silverton; claim 300 by 1,500 feet; located on Hazelton Mountain, Animas mining district, 2 miles from Silverton; vertical vein, trending southeast and northwest; value of ore, 40 to 100 ounces silver per ton; 1 shaft 80 feet deep, 1 tunnel 400 feet in length and 250 feet from surface, 1 level 100 feet in length.

Legal Tender. John James and R.C. Muncil, proprietors; office Animas Forks; claim 300 by 1,500 feet; discovered 1876; located on Houghton Mountain, Eureka mining district, 2 miles from town of Animas Forks and 18 miles from Silverton; course of vein, northwest and southeast, width 18 feet; nature of ore, galena and gray copper; assay value, 90 ounces silver per ton; 1 shaft 16 feet deep, 2 adits 10 and 15 feet in length, respectively.

Legal Tender. Thomas Dugan & Co., proprietors; office Silverton; claim 300 by 1,500 feet; discovered 1876; located on Boulder Mountain, Animas mining district, 2½ miles from Silverton; vertical vein, trending north and south, width 4 feet, pay vein 10 inches; nature of ore, galena and gray copper; assay value, 125 ounces silver per ton; 1 adit 15 feet in length with a 12-foot face.

Lelia. Owned by Joseph Ruf and Peter Bicker, of Silverton; claim 300 by 1,500 feet; discovered 1874; located on Jones Mountain, 3 miles from town of Eureka; course of vein, northeast and southwest, width 20 feet, pay vein 6 inches; nature of ore, gray copper and galena; value, 75 ounces silver per ton; development, 1 tunnel 80 feet and 1 level 20 feet in length.

Leviathan. Thomas Johnson proprietor; office Ouray; claim 300 by 1,500 feet; discovered 1876; located in Poughkeepsie Gulch, Uncompahgre mining district, 12 miles from Silverton; vertical vein, trending northwest and southeast, width 20 feet; nature of ore, gray copper and brittle silver; assay value, 250 to 600 ounces silver per ton; 1 tunnel 30 feet in length.

Lily. Con. Burns, proprietor; office Animas Forks; claim 300 to 1,500 feet; discovered 1873; located in Burns Gulch, Eureka mining district, 2 ½ miles from town of Eureka and 10 miles from Silverton; course of vein northwest and southeast, width 4 feet; nature of ore, gray copper and galena; average value, 300 ounces silver per ton; 1 drift 174 feet in length and 180 feet from surface; yield, $700.

Lincoln City. Frank Herrod, proprietor; office Mineral City; claim 300 by 1,500 feet; discovered 1877; located on Abraham Mountain, 14 miles from Silverton; vertical vein, trending northeast and southwest, width 6 feet; nature of ore, copper pyrites and antimony; assay value, 78 ounces silver per ton; 1 shaft 10 feet deep.

Little Bessie. L. Woodbury & Co., proprietors; office Animas Forks; claim 300 and 1,500 feet; discovered 1875; located on Engineer Mountain, 2 miles from Animas Forks and 18 miles from Silverton; course of vein, northwest and southeast, width 8 feet; nature of ore, galena; assay value, 68 ounces silver per ton; 1 adit, 16 feet in length and 30 feet from surface.

Little Maud. Charles H. McIntire, proprietor; office Ouray; claim 300 by 1,500 feet; discovered 1875; located on Seigle Mountain, 1 mile from Mineral City and 18 miles from Silverton; vertical vein, trending northeast and southwest, width 30 feet, pay vein 4 feet; nature of ore, galena; assay value, 80 ounces silver per ton; 1 adit 40 feet in length.

Little Nugget. John H. Pritchard, proprietor; office Del Norte; claim 300 by 1,500 feet; located on the mountain west of the south fork of Mineral Creek, 12 miles from Silverton; vertical vein, tending northeast and southwest, width 4 feet; nature of ore, gray copper and galena; 1 shaft 15 feet deep.

Lizzie Norris. Owned by William Norris; office Eureka; claim 300 by 1,500 feet; discovered 1877; located 2 miles from town of Eureka and 8 miles from Silverton; vertical vein, trending northwest and southeast, width 20 feet; nature of ore, iron pyrites and gray copper; 1 adit 15 feet in length and 10 feet from surface.

London. Barney Mallon & Co., proprietors; office Animas Forks; claim 300 by 1,500 feet; discovered 1877; located on the Animas River, Eureka mining district; 1 mile from town of Animas Forks and 17 mils from Silverton; course of vein, northwest and southeast, width 8 feet, pay vein 2 feet, pitch 10 degrees northwest; nature of ore, galena and gray copper; average value, 100 ounces silver per ton; 1 shaft 30 feet deep; yield $125.

Lone Rock. Johnson Burritt, proprietor; office Silverton; claim 300 by 1,500 feet; discovered 1875; located on Round Mountain, Animas mining district, 4 miles from Silverton; vertical vein, trending northwest and southeast, width 4 feet, pay vein 20 inches; nature of ore, galena and gray copper; value, 100 ounces silver per ton; 1 adit 28 feet and 1 tunnel 40 feet in length.

Louderbach. Owned by J.C. Louderbach of Silverton; claim 300 by 1,500 feet; discovered 1876; located on Mineral Creek, 5 miles from Silverton; course of vein, north and south, width 4 feet, pay vein 20 inches; nature of ore, galena and iron pyrites; value, 72 ounces silver per ton; development, 2 shafts, 10 and 15 feet, respectively.

Lulu. T.S. Barke and H.B. Jenkins, proprietors; office Eureka; claim 300 by 1,500 feet; discovered 1878; located on Missouri Mountain, Eureka mining district, 4 miles from town of Eureka and 12 miles from Silverton;

course of vein, northeast and southwest width 4 feet, pay vein 18 inches; nature of ore, galena and gray copper; assay value, 162 ounces silver per ton; 1 adit 10 feet in length.

M.E. Harrison. J.B. Ross & Co., proprietors; office Eureka; claim 300 by 1,500 feet; discovered 1873; located at the south base of Hurricane Peak, Eureka mining district, 10 miles from Silverton; course of vein, east and west, width 60 feet, pitch 30 degrees; nature of ore, galena and gray copper; average value, 50 to 300 ounces silver per ton; 1 shaft 40 feet deep.

Maggie Bell. Owned by J. Beach, S. Rogers, and W.H. Rogers of Colorado Springs; claim 300 by 1,500 feet; located on Seigle Mountain, Eureka mining district, 14 miles from town of Silverton; course of vein, northeast and southwest, width 12 feet, pitch 45 degrees, pay vein 4 feet; nature of ore, gray copper and galena; average value, 37 ounces silver per ton; 1 drift 100 feet in length.

Maid. R.H. Ferguson & Co., proprietors; office Eureka; claim 300 by 1,500 feet; discovered 1876; located on Tower Mountain, near town of Eureka, 9 miles from Silverton; course of vein, east and west, width 12 feet, pay vein 3 feet, pitch 45 degrees; nature of ore, galena and gray copper; assay value, 140 ounces silver per ton; 1 drift 32 feet in length and 20 feet from surface.

Martha. G. Griffin & Bros., proprietors; office Animas Forks; claim 300 by 1,500 feet; discovered 1876; located on Houghton Mountain, Eureka mining district, 10 miles from Animas Forks and 16 miles from Silverton; vertical vein, running east and west, width 7 feet, pay vein 20 inches; nature of ore, galena and gray copper; assay value, 20 to 200 ounces silver per ton; 2 adits, 12 and 15 feet in length, respectively.

Mary Jane. Owned by T. Byron and J. Tyrrell; office Animas Forks; claim 300 by 1,500 feet; discovered 1877; located on Picayune Mountain, Eureka mining district, 3 miles from Animas Forks and 11 miles from Silverton; course of vein, northeast and southwest, width 20 feet, pay vein 3 feet; nature of ore, gray copper and sulphurets; assay value, 300 to 500 ounces silver per ton; 1 shaft 14 feet deep; 1 adit 20 feet in length and 15 feet from surface.

Mastodon. Owned by the Mastodon Mining Co.; incorporated September 11, 1877; capital stock, $300,000, in 3,000 shares, $100 each; offices Animas Forks, Colorado, and Cleveland, Ohio; trustees: Alfred A.

Hard, J.T. Wamelink, J. Frank Isham, Anthony P. Berghoff, William Selah Chamberlain, Charles A. Brayton, Thomas Maher, Henry W. Heady, and Charles C. Bowen; claim 300 by 1,500 feet; discovered 1874; located on California Mountain, 1 mile from Animas Forks; course of vein, northwest and southeast, width 60 feet, pay vein 8 feet, pitch 10 degrees; nature of ore, gray copper and galena; value, 80 ounces silver per ton; development, 3 shafts, 10, 20, and 35 feet in depth, respectively, and 1 tunnel 160 feet in length.

Mavenau. C. Harding and W. Rummel, proprietors; office Eureka; claim 300 by 1,500 feet; located 1877; located in Minnie Gulch, Eureka mining district, 6 miles from Silverton; vertical vein, trending northeast and southwest, width 14 feet; nature of ore, gray copper and galena; assay value, 180 ounces silver per ton; 1 adit 25 feet in length.

Mayflower. James Thomas & Co., proprietors; office Ouray; claim 300 by 1,500 feet; discovered 1878; located on Engineer Mountain, 2 miles from Mineral City and 20 miles from Silverton; course of vein, northeast and southwest, width 20 feet, pitch 15 degrees; nature of ore, gray copper, iron and copper pyrites; assay value, 90 ounces silver per ton; 1 adit, 18 feet in length.

McClure. A.W. Burrows & Co., proprietors; office Del Norte; claim 300 by 1,500 feet; discovered 1874; located on Mineral Point Mountain, 18 miles from Silverton; vertical vein, trending northeast and southwest, width 75 feet, pay vein 3 feet; nature of ore, galena; assay value, 150 ounces silver per ton; 1 adit 30 feet in length.

McIntire. Colorado Mining and Land Co., proprietors; office Buffalo, New York; claim 300 by 1,500 feet; discovered 1875; located on Houghton Mountain, 1 mile from Mineral City and 17 miles from Silverton; vertical vein, trending northeast and southwest, width 20 feet; nature of ore, gray copper; assay value, $300 silver per ton; 1 shaft 14 feet deep, 1 tunnel 45 feet in length.

McKinnie. J.R. McKinnie & Co., proprietors; office Eureka; claim 300 by 1,500 feet; discovered 1875; located on Niagara Mountain, 2 miles from town of Eureka and 9 miles from Silverton; course of vein, east and west, width 30 feet, pitch 10 degrees; nature of ore, gray copper; assay value, 60 to 600 ounces silver per ton; 1 adit 18 feet in length.

Melrose. William Jackson and N. Brooks, proprietors; office Eureka; claim 300 by 1,500 feet; discovered 1877; located in Ross Basin, Eureka mining district, 10 miles from Silverton; course of vein, north and south, width 40 feet, pitch 18 degrees; nature of ore, galena and iron pyrites, assay value, 20 to 100 ounces silver per ton; 2 shafts, 10 and 12 feet in depth, respectively.

Melville Mining and Reduction Co. Incorporated October 9, 1876; capital stock, $150,000, in 1,500 shares, $100 each; office Silverton; organized by W.G. Melville, G.W. Glick, E. Summerfield, A.P. Clark, Charles C. Duncan and August Poehler. This company owns a half interest in 5 mines, which are located on Sultan Mountain, near limits of town of Silverton, within a radius of 1 mile from their works.

Mineral Point Tunnel Co. Organized under the laws of the State of New York; offices, New York City and Animas Forks, Colorado; Franklin J. Pratt, president, Edward C. Hancock, secretary, E. Steinbach, agent.

Minehaha. G.N. Richardson, proprietor; office Silverton; claim 300 by 1,500 feet; discovered 1876; located on the east fork of Cement Creek, Eureka mining district, 9 miles from Silverton; course of vein, northeast and southwest, width 13 feet, pay vein 2 feet, pitch 10 degrees; nature of ore, galena and gray copper; assay value, 50 to 800 ounces silver per ton; 1 drift 80 feet in length and 45 feet from surface.

Minnehaha. Jack Back & Co. proprietors; office Silverton; claim 300 by 1,500 feet; located 1876; located on Pioneer Mountain, Animas mining district, 7 miles from Silverton; course of vein, northwest and southeast, width 30 inches, pitch 25 degrees; nature of ore, gray copper and galena; assay value, 150 ounces silver per ton; 1 drift 19 feet in length.

Minnie. Owned by E.R. Hopkins, J.J. Isam, and W.B. House; office Howardsville; claim 300 by 1,500 feet; discovered 1876; located on Tower Mountain, Eureka mining district, 2 miles from Silverton; course of vein, northeast and southwest, width 7 feet; nature of ore galena and gray copper; assay value, 500 ounces silver per ton; 1 shaft 12 feet deep, 1 tunnel 60 feet in length and 35 feet from surface.

Minnie. Valentine Goeglin & Bro., proprietors; office Eureka; claim 300 by 1,500 feet; discovered 1877; located on Middle Mountain, Eureka mining district, 2 ½ miles from town of Eureka and 7 miles from Silverton; vertical vein, trending northwest and southeast, width 10 feet, pay vein 2 feet; nature of ore, gray copper; assay value, 75 ounces silver per ton; 1 tunnel 100 feet in length, intersecting vein at a depth of 100 feet.

Mobile. Charles Wickman & Co., proprietors; office Silverton; claim 300 by 1,500 feet; discovered 1875; located in Poughkeepsie Basin, 12 miles from Silverton, 12 miles from Silverton; vertical vein, trending east and west, width 30 feet; nature of ore, galena and sulphurets; assay value, 150 ounces silver per ton; 3 adits, aggregating 50 feet in length.

Modoc. Henry Leroy & Co., proprietors; office Lake City; clam 300 by 1,500 feet; discovered 1876; located on Engineer Mountain, 2 miles from Mineral City and 20 miles from Silverton; vertical vein, trending north and south, width 6 feet; nature of ore, galena and iron pyrites; assay value, 200 ounces silver per ton; 1 shaft 12 feet deep, 1 adit 15 feet in length.

Mogul. D. McGillivray & Co., proprietors; office Eureka; claim 300 by 1,500 feet; discovered 1877; located in Eureka mining district, 1 mile from town of Eureka and 8 miles from Silverton; vertical vein, trending east and west, width 7 feet, pay vein 18 inches; nature of ore, galena and iron pyrites, 1 drift 30 feet in length and 35 feet from surface.

Molas. L.W. Pattison & Co., proprietors; office Silverton; claim 300 by 1,500 feet; discovered 1876; located on Ruby Hill, Animas mining district, 4 miles from Silverton; course of vein, northeast and southwest, width 25 feet, pay vein 14 inches; nature of ore, galena, gray copper, and ruby silver; average value, 200 ounces silver per ton; 2 shafts, 45 and 60 feet in depth, respectively, 1 tunnel 90 feet in length.

Morning Star. Thomas Dunton and C. Price, proprietors; office Eureka; claim 300 by 1,500 feet; discovered 1876; located in Eureka Gulch, Eureka mining district, 1 mile from town of Eureka and 9 miles from Silverton; vertical vein, trending north and south, width 7 feet, pay vein 20 inches; nature of ore, galena and iron pyrites, average value, 20 ounces silver per ton; 1 drift 80 feet in length.

Morning Star. Charles Carlstrom, proprietor; office Silverton; claim 300 by 1,500 feet; discovered 1878; located on Stuart Mountain, Uncompahgre mining district, 14 miles from Silverton; course of vein, northeast and southwest, width 10 feet, pay vein 2 feet, pitch 20 degrees; nature of ore, galena and sulphurets; average value, 224 ounces silver per ton; 1 shaft 11 feet deep.

Mount Vernon. Owned by M. Rich, Frederick Blaisdale, and others; office Lake City; claim 300 by 1,500 feet; discovered 1873; located on the Animas River, near town of Eureka, 9 miles from Silverton; course of vein, north and south, width 5 feet, pay vein 4 to 8 inches, pitch 15 degrees west; nature of ore, gray copper; average value, 170 ounces silver per ton; 2 shafts, 20 and 23 feet in depth, respectively.

Mountain Eagle. Owned by Gustavus Johnson & Co.; office Silverton; claim 300 by 1,500 feet; discovered 1875; located in Ross Basin, 10 miles from Silverton; course of vein, northeast and southwest, width 25 feet, pay vein 2 feet, pitch 10 degrees; nature of ore, galena and gray copper; assay value, 200 ounces silver per ton; 1 shaft 25 feet deep.

Mountain King. Charles Hill, proprietor; office Howardsville; claim 300 by 1,500 feet; discovered 1877; located on Green Mountain, Animas mining district, 9 miles from Silverton; course of vein, east and west, width 125 feet, pay vein 12 feet, pitch 15 degrees; nature of ore, gray copper and galena; assay value, 50 to 1,000 ounces silver per ton; 1 shaft 75 feet deep, 1 drift 75 feet in length; shaft house 20 by 20 feet.

Mountain King. Owned by Charles and Clabe Jones and Henry Warren; office Animas Forks; claim 300 by 1,500 feet; discovered 1878; located on Hurricane Peak, 5 miles from town of Eureka and 12 miles from Silverton; vertical vein, trending northeast and southwest, width 40 feet, pay vein 5 feet; nature of ore, brittle silver and gray copper; assay value, 830 ounces silver per ton; 1 adit 15 feet in length.

Mountain King. C.P. Mellor & Co., proprietors; office Eureka; claim 300 by 1,500 feet; discovered 1877; located in Eureka Gulch, Eureka mining district, 3 miles from town of Eureka and 11 miles from Silverton; course of vein, northwest and southeast, width 35 inches, pay vein 30 inches, pitch 25 degrees; nature of ore, gray copper and galena; assay value, 160 ounces silver per ton; 2 adits 15 feet each.

Mountain Maid. J.M. King, E.S. Finch, and J.C. Haggerty, proprietors; office Ouray; claim 300 by 1,500 feet; discovered 1878; located on Pioneer Mountain, Animas mining district, 8 miles from Silverton; course of vein, east and west, width 18 feet, pay vein 4 feet, pitch 35 degrees; nature of ore, galena and carbonate of lead; average value, 50 ounces silver per ton; 1 adit 10 feet in length.

Mountain Queen. Owned by Frederick B. Beaudry & Co., L. Le Fevre, John H. Maugan, James Raser, Henry Warren, Joseph Eledge and Henry Wolcott; office Lake City; claim 300 by 1,500 feet; discovered 1874; located on California Mountain, at the head waters of the Animas and Cement Rivers, Eureka mining district, 3 miles from Animas Forks and 12 miles from Silverton; vertical vein, trending northeast and southwest, width 20 feet, pay vein 5 feet; nature of ore, gray copper and galena; average value, $75 per ton; 1 shaft 104 feet deep, 2 levels, 50 and 104 feet from surface, respectively, aggregating 150 feet in length; yield, $65,000.

Mountain Pride. C. Plumb and J.S. Roupe, proprietors; office Animas Forks; claim 300 by 1,500 feet; discovered 1875; located on the Animas River, Eureka mining district, 2 ½ miles from town of Animas Forks, and 14 miles from Silverton; course of vein, northeast and southwest, width 45 feet, pitch 20 degrees; nature of ore, galena and iron pyrites; assay value, 30 ounces silver per ton; 1 tunnel 40 feet in length.

Mutual Gold and Silver Mining Association. Incorporated April 18, 1878; capital stock, $50,000, in 5,000 shares, $10 each; offices Silverton, and New York City; trustees: Clint, Lorenzo D., and Almon H. Roudebush, Allen S. Wrightman, and Fred. W. Jones.

My Choice. Owned by Frank Herrod and J.M. Stuart; office Mineral City; claim 370 by 1,500 feet; discovered 1877; located in Poughkeepsie Basin, 12 miles from Silverton; vertical vein, trending northeast and southwest, width 370 feet, pay vein 5 feet; nature of ore, galena and gray copper; assay value, 200 ounces silver per ton; 1 shaft 10 feet deep, 3 adits, each 15 feet from the surface, aggregating 50 feet in length.

N.B. Forrest. G.N. Richardson, proprietor; office Silverton; claim 300 by 1,500 feet; discovered 1878; located on the east fork of Cement Creek, Animas mining district, 9 miles from Silverton; course of vein, northwest and southeast, width 20 feet, pay vein 10 inches, pitch 45 degrees; nature of ore, galena; average value, 50 ounces silver per ton; 1 drift 32 feet in length and 20 feet from surface.

Napoleon Extension. O.O. Larson & Co., proprietors; office Del Norte; claim 300 by 1,500 feet; discovered 1875; located on Houghton Mountain, 1 mile from Mineral City and 18 miles from Silverton; vertical vein, trending northeast and southwest, width 14 feet, pay vein 2 feet;

nature of ore, galena and gray copper; average value, 100 ounces silver per ton; 1 shaft 33 feet deep; shaft house 16 by 18 feet.

Nasby. Owned by Frederick Blaisdale and J.D. Neiswanger; office Eureka; claim 300 by 1,500 feet; discovered 1875; located in town of Eureka, 9 miles from Silverton; vertical vein, trending northeast and southwest, width 3 feet, pay vein 1 foot; nature of ore, iron pyrites and galena; assay value, $75 per ton; 1 drift 15 feet in length and 12 feet from surface.

National. John Hammill & Co., proprietors; office Ouray; claim 300 by 1,500 feet; discovered 1877; located on Wood Mountain, Eureka mining district, 2 miles from town of Animas Forks and 17 miles from Silverton; course of vein, east and west, width 18 feet; nature of ore, galena and iron pyrites; assay value, 60 ounces silver per ton; 1 drift 20 feet in length.

Neptune, Extension of the Mastodon. Owned by M. Rich and Francis A. Cook; office Lake City; claim 300 by 1,500 feet; discovered 1874; located on the left fork of the Animas River, 3 miles from Animas Forks and 15 miles from Silverton; course of vein, northeast and southwest, width 270 feet, pitch 20 degrees south; nature of ore, gray copper, galena, and sylvanite; average value, 60 ounces silver per ton; 3 shafts, aggregating 50 feet in depth.

Nevada. J.M. Donnel & Co., proprietors; office Animas Forks; claim 300 by 1,500 feet; discovered 1874; located on California Mountain, Eureka mining district, 2 miles from Animas Forks and 18 miles from Silverton; course of vein, northeast and southwest, width 5 feet, pay vein 3 feet, pitch 30 degrees; nature of ore, gray copper and brittle silver; assay value, 100 to 10,000 ounces silver per ton; 1 shaft 20 feet deep, 2 adits 12 feet in length each.

New Castle. Henry Carley, proprietor; office Animas Forks; claim 300 by 1,500 feet; discovered 1877; located on Treasure Mountain, Eureka mining district, 1 mile from town of Animas Forks and 17 miles from Silverton; course of vein, northwest and southeast, width 7 feet, pay vein 30 inches; nature of ore, gray copper and galena; average value, 75 ounces silver per ton; 1 adit 25 feet in length with a 15-foot face.

New Era. Owned by George Wilcox, H. Rand, and others; office Eureka; claim 300 by 1,500 feet; discovered 1876; located in Picayune

Gulch, Eureka mining district, 3 miles from town of Eureka and 11 miles from Silverton; course of vein, east and west, width 40 feet, pitch 15 degrees; nature of ore, galena and iron pyrites; assay value, 160 ounces silver per ton; 1 shaft 10 feet deep, 1 adit 15 feet in length.

New York and San Juan Mining Co. Organized under the laws of the State of New York; office Mineral City; A.W. Burrows, agent, Jared F. Harrison, president, Edward C. Hancock, secretary.

Nil Desperandum. Albert Marzetti, proprietor; office Howardsville; claim 300 by 1,500 feet; discovered 1877; located on Tower Mountain, Eureka mining district, 2 miles from town of Eureka and 9 miles from Silverton; course of vein, northwest and southeast, width 11 feet, pay vein, 1 foot; nature of ore, galena and copper pyrites; assay value, 56 ounces silver per ton; 1 adit 21 feet in length.

N.G. J.B. Ross, proprietor; office Eureka; claim 300 by 1,500 feet; discovered 1878; located in Ross Basin, Eureka Mining District, 5 miles from town of Eureka and 10 miles from Silverton; course of vein, northeast and southwest, width 50 feet, pay vein 3 feet; nature of ore, galena and gray copper; 1 drift 50 feet in length and 30 feet from surface.

No Name. R.J. McNutt, proprietor; office Eureka; claim 300 by 1,500 feet; discovered 1873; located in Eureka Gulch, 3 miles from Eureka and 8 miles from Silverton; vertical vein, trending north and south, width 25 feet, pay vein 6 feet; nature of ore, galena; assay value, 40 ounces silver per ton; 1 adit 15 feet in length and 12 feet from surface.

None Such. W. Williams and M.J. Condict, proprietors; office Silverton; claim 300 by 1,500 feet; discovered 1877; located in Ohio Gulch, Animas mining district, 5 miles from Silverton; vertical vein, trending northwest and southeast, width 5 feet, pay vein 3 feet; nature of ore, galena and iron pyrites; assay value, 30 to 200 ounces silver per ton; 1 shaft 10 feet deep, 1 adit 15 feet in length.

North Star. William Williams & Co., proprietors; office Silverton; claim 300 by 1,500 feet; discovered 1876; located on Sultan Mountain, Animas mining district, 1 mile from town of Silverton; vertical vein, trending northwest and southeast, width 6 feet, pay vein 9 inches; nature of ore, galena and gray copper; average value, 200 ounces silver per ton; 1 drift 75 feet in length.

North Star. Owned by Joseph Reef and Crooke Bros., of Silverton; claim 300 by 1,500 feet; discovered 1874; located on King Solomon Mountain, 5 miles from Silverton; vertical vein, trending east and west, width 8 feet, pay vein, 3 feet; value of ore, 200 ounces silver per ton; development, 1 shaft 70 feet and 2 tunnels, 100 and 150 feet in length, respectively; yield, $60,000.

O.K. George A. Castle & Co., proprietors; office Silverton; claim 300 by 1,500 feet; discovered 1876; located on Green Mountain, Animas mining district, 8 miles from Silverton; vertical vein, trending north and south, width 2 feet, pay vein 20 inches; nature of ore, gray copper and copper pyrites; assay value, 300 ounces silver per ton; 1 shaft 10 feet deep, 1 tunnel 45 feet in length, which intersects the vein 60 feet from surface.

O.K. M. Pray & Co., proprietors; office Eureka; claim 300 by 1,500 feet; discovered 1877; located on Treasure Mountain, Eureka mining district, 16 miles from Silverton; vertical vein, trending northeast and southwest, width 10 feet, pay vein 18 inches; nature of ore, galena and gray copper; assay value, 125 ounces silver per ton; 1 adit 18 feet in length.

O.K. Owned by John Meyer, James Dooley, and George Rohwer; office Silverton; claim 300 by 1,500 feet; discovered 1878; located on Poughkeepsie Mountain, 12 miles from Silverton; course of vein, northeast and southwest, width 10 feet, pay vein 10 inches; nature of ore, galena and brittle silver; assay value, 500 ounces silver per ton; 1 drift 20 feet in length and 20 feet from surface.

Oberto. Antony Oberto and John M. Stuart, proprietors; office Silverton; claim 372 by 1,500 feet, patented; discovered 1877; located on Oberto Mountain, Uncompahgre mining district, 12 miles from Silverton; course of vein, northeast and southwest, width 372 feet, pay vein 20 feet; nature of ore, gray copper and galena; average value, 75 ounces silver per ton; 4 drifts aggregating 90 feet in length; yield, $850.

Occidental. T.R. Gammon & Co., proprietors; office Silverton; claim 300 by 1,500 feet; discovered 1877; located on Ruby Hill, Animas mining district, 5 miles from Silverton; course of vein, northwest and southeast, width 4 feet, pay vein 20 inches, pitch 20 degrees; nature of ore, gray copper and galena; assay value, 100 ounces silver per ton; 2 adits, 10 feet each.

Ocean Wave. Andrew Cumtens [Cumtons?], proprietor; office Ouray; claim 300 by 1,500 feet; discovered 1876; located in Grouse Gulch, 3 miles from town of Eureka and 14 miles from Silverton; vertical vein, trending northeast and southwest, width 40 feet; nature of ore, galena; assay value, 30 ounces silver per ton; 1 drift 55 feet in length.

O'Connell. John Egan & Co., proprietors; office Silverton; claim 300 by 1,500 feet; discovered 1876; located on Sultan Mountain, Animas mining district, 3 miles from Silverton; vertical vein, trending east and west, width 4 feet; nature of ore, gray copper and galena; assay value, 200 ounces silver per ton; 1 shaft 20 feet deep.

Onetta. S. Peterson and E. Larson, proprietors; office Eureka; claim 300 by 1,500 feet; discovered 1875; located on Missouri Mountain, Eureka mining district, 2 miles from town of Eureka and 10 miles from Silverton; vertical vein, trending northeast and southwest, width 5 feet, pay vein 18 inches; nature of ore, galena and gray copper; 1 drift 75 feet in length.

Ontario. James T. Rawlings & Co., proprietors; office Denver; claim 300 by 1,500 feet; discovered 1875; located on the Animas River, Eureka mining district, 2 miles from town of Animas Forks and 14 miles from Silverton; course of vein, east and west, width 7 feet, pay vein 18 inches, pitch 20 degrees north; nature of ore, gray copper and galena; 1 drift 30 feet in length and 20 feet from surface.

Othello. Owned by R. Tilton, E.C. Howard, and others; office Silverton; claim 300 by 1,500 feet; discovered 1876; located on Mineral Creek, Animas mining district, 5 miles from Silverton; vertical vein, trending east and west, width 5 feet, pay vein 2 feet; nature of ore, galena and carbonate of lead; assay value, 90 ounces silver per ton; 1 drift 20 feet in length.

Ottawa. P. McEnany and H. T. Ash, proprietors; office Animas Forks; claim 300 by 1,500 feet; discovered 1876; located on Lake Mountain, near Mineral City, 18 miles from Silverton; vertical vein, trending northeast and southwest; width 8 feet, pay vein 30 inches; nature of ore, galena and sulphurets; assay value, 42 to 95 ounces silver per ton; 1 adit 15 feet in length with a 13-foot face.

Ottawa. A.F. Krauss & Co., proprietors; office Ouray; claim 300 by 1,500 feet; discovered 1876; located on Green Mountain, Animas mining district, 8 miles from Silverton; vertical vein, trending northwest and southeast, width 6 feet, pay vein 3 feet; nature of ore, galena and iron pyrites; 1 shaft 22 feet deep.

Pacific. Charles Miller, proprietor; office Eureka; claim 300 by 1,500 feet; discovered 1877; located on Silver Mountain, Eureka mining district, 3 miles from town of Eureka and 9 miles from Silverton; vertical vein, trending north and south, width 10 feet, pay vein 4 feet; nature of ore, galena; assay value, 100 ounces silver per ton; 1shaft 20 feet deep.

Pacific. G. Griffin & Co., proprietors; office Animas Forks; claim 300 by 1,500 feet; discovered 1876; located on California Mountain, Eureka mining district, 3 miles from town of Animas Forks and 17 miles from Silverton; course of vein, northwest and southeast, width 20 feet, pay vein 18 inches, pitch 10 degrees; nature of ore, galena and gray copper; assay value, 100 ounces silver per ton; 2 drifts, 20 feet each.

Pay Master. James Peterson and I. Stoll, proprietors; office Silverton; claim 300 by 800 feet; discovered 1877; located on Kendall Mountain, Animas mining district, 1 mile from Silverton; vertical vein, trending north and south, width 6 feet, pay vein, 18 inches; nature of ore, galena, iron and copper pyrites; assay value, 50 ounces silver per ton; 2 adits, 10 and 15 feet in length, respectively.

Philadelphia. Owned by Niegold Bros., of Niegoldsville; claim 300 by 1,500 feet; discovered 1872; located on Green Mountain, 7 miles from Silverton; development, 300 feet of tunnels; yield, $20,000.

Phoenix. H. Harrington and J. Walsh, proprietors; office Eureka; claim 300 by 1,500 feet; discovered 1877; located in Ross Basin, Eureka mining district, 10 miles from Silverton; vertical vein, trending east and west, width 8 feet, pay vein 28 inches; nature of ore, galena; assay value, 160 ounces silver per ton; 1 shaft 25 feet deep.

Phoenix. Owned by Myer, Dooley & Co.; office Ouray; claim 300 by 1,500 feet; discovered 1878; located on Poughkeepsie Mountain, Uncompahgre mining district, 12 miles from Silverton; course of vein, northwest and southeast, width 6 feet; nature of ore, galena and sulphurets; 1 adit 12 feet in length and 12 feet from surface.

Pioneer Mining and Milling Co. Incorporated October 11, 1875; capital stock, $1,000,000, in 10,000 shares, $100 each; office Silverton;

incorporators: M.D. Harman, M.J. Alkire, W.B. Sherman, William L. Smith, Thomas C. Graden, T.S. White, and W.T. McDonald.

Planet. Owned by Dr. Beiton, of Eureka; claim 300 by 3,000 feet; discovered 1875; located on Tower Mountain, 2 miles from Eureka; vertical vein, trending northwest and southeast, width 24 feet, value of ore, 150 ounces silver per ton; development, 1 drift 60 feet in length and 50 feet from surface.

Pomeroy. Valentine Goeglin and P. Savage, proprietors; office Eureka; claim 300 by 1,500 feet; discovered 1877; located on Middle Mountain, Eureka mining district, 2 ½ miles from town of Eureka and 7 miles from Silverton; vertical vein, trending northwest and southeast, width 4 feet, pay vein 2 feet; nature of ore, iron pyrites and gray copper; 1 adit 15 feet in length.

Pontoon. George N. Propper and A.W. Burrows, proprietors; office Del Norte; claim 300 by 1,500 feet; discovered 1874; located in Burns Gulch, Eureka mining district, 2 miles from town of Eureka and 10 miles from Silverton; vertical vein, trending north and south, width 4 feet, pay vein 20 inches; nature of ore, gray copper and sulphurets; assay value, 50 to 900 ounces silver per ton; 1 drift 40 feet in length and 40 feet from surface.

Port Henry. P. McEnany, proprietor; office Animas Forks; claim 300 by 1,500 feet; discovered 1875; located on Lake Mountain, Uncompahgre mining district, near Mineral City, 18 miles from Silverton; course of vein, northeast and southwest, width 5 feet, pay vein 30 inches, pitch 25 degrees north; nature of ore, brittle silver; assay value, 182 ounces silver per ton; 1 shaft 40 feet deep.

Port Jervis. Henry Jenkins, proprietor; office Silverton; claim 300 by 1,500 feet; located on Sultan Mountain, Animas mining district, 2 miles from Silverton; course of vein, northeast and southwest, width 6 feet, pay vein 20 inches, pitch 15 degrees; nature of ore, gray copper and galena; assay value, 125 ounces silver per ton; 1 adit 30 feet in length.

Poughkeepsie. Owned by R.J. McNutt and G. Howard; office Eureka; claim 300 by 1,500 feet; discovered 1873; located in Poughkeepsie Gulch, Uncompahgre mining district, 8 miles from Ouray; vertical vein, trending east and west, width 40 feet, pay vein 5 feet; nature of ore, gray copper and sulphurets; average value, 150 ounces silver per ton; 2 drifts, 75 and 170 feet from surface, respectively, aggregating 250 in length.

Preston. C. Rugg & Co., proprietors; office Animas Forks; claim 300 by 1,500 feet; discovered 1877; located in Grouse Gulch, Eureka mining district, 2 miles from town of Animas Forks and 12 miles from Silverton; vertical vein, trending northwest and southeast, width 7 feet; pay vein, 3 feet; nature of ore, gray copper and galena; assay value, 100 ounces silver per ton; 1 drift 18 feet in length.

Pride of the West. Owned by Jackson Soward, D.E. Schoelkopf, and others; claim 300 by 1,500 feet; discovered 1871; located on Green Mountain, Animas mining district, 6 miles from Silverton; vertical vein, trending northwest and southeast, width 60 to 80 feet, pay vein 10 to 15 feet; nature of ore, galena and gray copper; average value, 125 ounces silver per ton; development, 250 feet of drifts and tunnels and 1 shaft, 100 feet deep; yield, $30,000.

Prince of Wales. San Juan Bullion Co., proprietors; office Chicago, Illinois; claim 300 by 1,500 feet; discovered 1876; located on Hurricane Peak, Eureka mining district, 3 miles from town of Animas Forks and 12 miles from Silverton; vertical vein, trending north and south, width 5 feet, pay vein 2 feet; nature of ore, gray copper and brittle silver; average value, 168 ounces silver per ton; 1 shaft 35 feet deep, 1 drift 30 feet in length.

Promontory. George N. Propper and A.W. Burrows, proprietors; office Del Norte; claim 300 by 1,500 feet; discovered 1874; located in Burns Gulch, Eureka mining district, 2 miles from town of Eureka, and 10 miles from Silverton; vertical vein, trending northeast and southwest, width 10 feet, pay vein 4 feet; nature of ore, galena; assay value, 100 to 500 ounces silver per ton; 1 adit 30 feet in length, with a 50-foot face.

Prospector. Owned by Joel W. Shackelford and George S. Smith; office; office Denver; claim 200 by 1,500 feet; located on Hazelton Mountain, Animas mining district, 3 miles from Silverton; course of vein, northwest and southeast, width 4 feet, pay vein 4 to 30 inches, pitch 15 degrees south; nature of ore, galena and gray copper; average value, 200 ounces silver per ton; main shaft 120 feet deep, and 2 other shafts, aggregating 150 feet, reached by tunnel at a distance of 600 feet from the mouth an intersected at a depth of 515 feet, aggregate length of levels, 500 feet; 3 shaft houses 15 by 30 feet each; yield, $40,000.

Protection. Charles Carlstrom & Co., proprietors; office Silverton; claim 300 by 1,500 feet; discovered 1877; located on Mineral Point Mountain, 14 miles from Silverton; course of vein, northeast and southwest, width 12 feet, pay vein 10 inches, pitch 30 degrees; nature of ore; gray copper; assay value, 150 ounces silver per ton; 1 adit 15 feet, with a 15-foot face.

Quaker City. John Ryan and Edmund P. True, proprietors; office Silverton; claim 300 by 1,500 feet; discovered 1876; located on Hazelton Mountain, Animas mining district, 5 miles from Silverton; vertical vein, running northwest and southeast, width 4 feet, pay vein 8 inches; nature of ore, galena and gray copper; assay value, 250 ounces silver per ton; 1 drift 22 feet in length.

Queen City. F. Herbst & Co., proprietors; office Ouray; claim 300 by 1,500 feet; discovered 1875; located in Ohio Gulch, Animas mining district, 3 miles from Silverton; course of vein, northeast and southwest, width 5 feet, pay vein 15 inches, pitch 25 degrees; nature of ore, galena; average value, 50 ounces silver per ton; 1 drift 30 feet in length and 20 feet from surface.

Red Cloud. Charles H. McIntire and A.W. Burrows, proprietors; office Del Norte; claim 120 by 1,500 feet; discovered 1874; located on Mineral Point Mountain, Uncompahgre mining district, 18 miles from Silverton; vertical vein, trending northeast and southwest, width 30 feet, pay vein 30 inches; nature of ore, galena and gray copper; assay value, 200 ounces silver per ton; 1 adit 30 feet and 1 drift 50 feet in length.

Red Point. P. McEnany, proprietor; office Animas Forks; claim 300 by 1,500 feet; discovered 1874; located in Placer Gulch; Eureka mining district, 2 miles from town of Animas Forks and 16 miles from Silverton; course of vein, northeast and southwest, width 75 feet; nature of ore, galena; average value, 75 ounces silver per ton; 1 drift 25 feet in length and 20 feet from surface.

Reliable. E.M. King and J.J. Hance, proprietors; office, Silverton; claim 300 by 1,500 feet; discovered 1877; located on Brown Mountain, Eureka mining district, 10 miles from Silverton; course of vein, north and south, width 18 inches; nature of ore, galena and gray copper; 1 drift 20 feet in length and 30 feet from surface.

Republic. S. Jones and W. Hippe, proprietors; office Howardsville; claim 300 by 1,500 feet; discovered 1876; located on King Solomon Mountain, Animas mining district, 6 miles from Silverton; vertical vein, trending northwest and southeast, width 9 feet, pay vein 1 foot; nature of ore, gray copper and galena; assay value, 175 ounces silver per ton; 1 adit 20 feet in length.

Robert Emmet. Owned by James O'Boyle & Bro., of Eureka; claim 300 by 1,000 feet; discovered 1876; located on the Animas River, 4 miles from town of Eureka; course of vein, northwest and southeast, width 8 feet, pay vein 18 inches; nature of ore, gray copper and galena; assay value, 126 ounces silver per ton; development, 1 drift 30 feet in length.

Rock Island. Owned by James J. and Albert Bernard, Daniel J. and James D. McKay, William Forsyth and John M. Ross; office Howardsville; claim 300 by 1,500 feet; discovered 1874; located on Galena Mountain, near Howardsville; course of vein, northeast and southwest, width 4 feet; nature of ore, galena, iron and copper pyrites; average assay value, 75 ounces silver and 3 ounces gold per ton; 1 tunnel 30 feet in length.

Rocky Point. J.H. Crane, W.C. Magee, and R. Magee, proprietors; office Lake City; claim 300 by 1,500 feet; discovered 1875; located on Woods Mountain, Uncompahgre mining district, 3 miles from Mineral City and 19 miles from Silverton; vertical vein, trending north and south, width 8 feet; nature of ore, galena and gray copper; assay value, 70 ounces silver per ton; 1 adit 15 feeet and 1 drift 20 feet in length.

Rollo No. 2. Olaf Sandstone & Co., proprietors; office Silverton; claim 300 by 1,500 feet; located in Poughkeepsie Gulch, Uncompahgre mining district, 12 miles from Silverton; vertical vein, trending northeast and southwest, pay vein 18 inches; nature of ore, gray copper and iron pyrites; assay value, 100 ounces silver per ton; 1 tunnel 25 feet in length and 25 feet from surface.

Roman. H. Harrington & Co., proprietors; office Eureka; claim 300 by 1,500 feet; discovered 1876; located on Silver Mountain, Eureka mining district, 3 miles from Eureka and 11 miles from Silverton; vertical vein, trending east and west, width 5 feet; nature of ore, brittle silver and gray copper; assay value, 800 ounces silver per ton; 1 adit 20 feet in length.

Roman Beauty. R.H. Ferguson & Co., proprietors; office Eureka; claim 300 by 1,500 feet; discovered 1875; located on Crown Mountain, near town of Eureka, 9 mines from Silverton; course of vein, northeast and southwest, width 15 feet, pay vein 5 feet, pitch 40 degrees; nature of ore, gray copper and iron pyrites; assay value, 608 to 3,370 ounces silver per ton; 2 drifts, 28 and 12 feet in length, respectively.

Ross. Owned by James B. and Hugh Ross; office Eureka; claim 300 by 1,500 feet; discovered 1876; located in Ross Basin, Eureka mining district, 5 miles from town of Eureka and 10 miles from Silverton; course of vein, northeast and southwest, width 40 feet, pay vein 30 inches, pitch 10 degrees; nature of ore, copper pyrites and gray copper; assay value, 112 ounces silver per ton; 2 shafts, 10 feet in depth each, 1 drift 50 feet in length and 60 feet from surface.

Royal Gem. Thomas Dugan & Co., proprietors; office Silverton; claim 300 by 1,500 feet; discovered 1877; located on Kendall Mountain, Animas mining district, 1 mile from Silverton; vertical vein, trending east and west, width 6 feet, pay vein 18 inches; nature of ore, galena and iron pyrites; average value, 30 ounces silver per ton; 1 drift 20 feet in length, with a 15-foot face.

Ruby. Owned by A.O. Terry, D.D. Burns, and G.W. Cummings; office St. Louis, Missouri; claim 300 by 1,500 feet; discovered 1875; located on Watson Mountain, near town of Eureka, nine miles from Silverton; course of vein, northeast and southwest, width 7 feet, pay vein 4 feet, pitch 25 degrees; nature of ore, sulphurets, galena and gray copper; average value, 100 ounces silver per ton; 1 shaft 25 deep, 2 tunnels, aggregating 120 feet in length, intersecting vein at a depth of 48 and 60 feet, respectively.

Rudolph. Daniel Wyman, proprietor; office Silverton; claim 300 by 1,500 feet; discovered 1876; located on Mineral Creek, Animas mining district, 4 miles from Silverton; course of vein, north and south, width 4 feet, pay vein 20 inches, pitch 30 degrees; nature of ore, gray copper and iron pyrites; assay value, 80 ounces silver per ton; 1 drift 28 feet in length.

San Francisco. Owned by J. M. Donnel, M.E. Copland, T. Chestnut, and H. Israel; office Animas Forks, claim 300 by 1,500 feet; discovered 1874; located on California Mountain, Eureka mining district, 1 mile from Animas Forks and 18 miles from Silverton; course of vein, northeast and southwest, width 100 feet, pay vein 6 feet, pitch 30 degrees; nature of ore, iron and copper pyrites and galena; average value, 75 ounces silver per ton; 2 adits, 12 and 15 feet in length, respectively.

San Juan Bullion Co. Incorporated under the laws of the State of Wisconsin, September 1878; capital stock, $2,500,000, in 100,000 shares, $25 each; offices Grand Raids, Wisconsin, and Animas Forks, Colorado; trustees: Andrew Forbes, J.P. Kennedy, and others.

San Juan Mining and Development Co. Incorporated December 9, 1875; capital stock, $30,000, in 60 shares, $500 each; offices Chicago, Illinois, and Denver, Colorado; organized by Marcus C. Stearns, Joel W. Shackelford, and John E. Stearns.

Saxon. Owned by James B. Platt & Co.; office Eureka; claim 300 by 1,500 feet; discovered 1877; located in Eureka Gulch, 4 miles from Eureka; vertical vein, trending northeast and southwest, width 15 feet, pay vein 4 feet; average value of ore, 50 ounces silver per ton; development 1 drift 25 feet in length.

Seven-Thirty. Charles Raymond and E. Kelper, proprietors; office Eureka; claim 300 by 1,500 feet; discovered 1876; located on the Animas River, Eureka mining district, 2 miles from Eureka and 10 miles from Silverton; vertical vein, trending east and west, width 7 feet, pay vein 18 inches; nature of ore, galena and iron pyrites; assay value, 80 ounces silver per ton; 1 drift 45 feet in length.

Seventh Son. Charles H. and Edward W. McIntire, proprietors; office Ouray; claim 200 by 1,500 feet; discovered 1874; located on Mineral Point Mountain, half mile from Mineral City and 18 miles from Silverton; vertical vein, trending northeast and southwest, width 15 feet, pay vein 20 inches; nature of ore, galena; assay value, 75 ounces silver per ton; 1 adit 20 feet in length.

Shakespeare. Owned by A.O. Terry, G.W. Cummings, and J.W. Tiernan; office St. Louis, Missouri; claim 300 by 1,500 feet; discovered 1874; located on Watson Mountain, near town of Eureka, 8 miles from Silverton; vertical vein, trending northeast and southwest, width 10 feet, pay vein 3 feet; nature of ore, gray copper and sulphurets; average value, 185 ounces silver per ton; 1 drift 55 feet in length and 60 feet from surface.

Silkstone. George A. Castle, proprietor; office Silverton; claim 300 by 1,500 feet; discovered 1876; located at forks of Cement Creek, Animas mining district, 8 miles from the town of Silverton; vertical vein, trending north and south, width 15 feet, pay vein 30 inches; nature of ore, gray copper and copper pyrites; assay value, 325 to 3,800 ounces silver per ton; 1 shaft 20 feet deep; 1 drift 15 feet in length.

Silver. Charles H. McIntire, proprietor; office Ouray; claim 300 by 1,500 feet; discovered 1874; located on Mineral Point Mountain, 1 mile from Mineral Point Mountain, 1 mile from Mineral City and 18 miles from Silverton; vertical vein, trending northeast and southwest, width 20 feet, pay vein 4 feet; nature of ore, galena; assay value, 80 ounces silver per ton; 2 adits, 12 and 30 feet in length, respectively.

Silver Bell. James Reynolds and Joseph Reynolds, proprietors; office Silverton; claim, 300 by 1,500 feet; discovered in Eureka Gulch, Eureka mining district, 11 miles from Silverton; course of vein, north and south, width 4 feet; nature of ore, gray copper; 1 drift 20 feet in length and 20 feet from surface.

Silver Blossom. J. Larson and Co., proprietors; office Mineral City; claim 300 by 1,500 feet; discovered 1877; located in Poughkeepsie Gulch, Uncompahgre mining district, 15 miles from Silverton; vertical vein, trending northeast and southwest, width 8 feet, pay vein 3 feet; nature of ore, gray copper and galena; assay value, 100 ounces silver per ton; 1 drift 40 feet in length and 50 feet from surface.

Silver Coin. Owned by F. Herrod and G.W. Strouse; office Mineral City; claim 300 by 1,500 feet; discovered 1876; located on Abraham Mountain, Uncompahgre mining district, 12 miles from Silverton; vertical vein, trending northeast and southwest, width 6 feet; nature of ore, copper pyrites and antimony; average value, 46 ounces silver per ton; 2 drifts, 30 and 60 feet from surface, respectively, aggregating 100 feet in length.

Silver Gnome. C.P. Mellor & Co., proprietors; office Eureka; claim 300 by 1,500 feet; discovered 1878; located in Maggie Gulch, Eureka mining district, 2 miles from town of Eureka and 6 miles from Silverton; vertical vein, trending east and west, width 4 ½ feet, pay vein 20 inches; nature of ore, gray copper and galena; assay value, 80 ounces silver per ton; 1 drift 15 feet in length.

Silver Link. M.W. Dresser & Co., proprietors; office Eureka; claim 300 by 1,500 feet; discovered 1878; located on Silver Mountain, 3 miles from Animas Forks and 11 miles from Silverton; course of vein, east and west, width 100 feet, pitch 30 degrees; nature of ore, gray copper and brittle silver; assay value, 40 to 700 ounces silver per ton; 1 adit 20 feet in length and 10 feet from surface.

Silver Plume. John Clark and Charles P. Tallman, proprietors; office Eureka; claim 300 by 1,500 feet; discovered 1875; located on Silver Mountain, Eureka mining district, 1 mile from town of Eureka and 9 miles from Silverton; course of vein, northeast and southwest, width 5 feet, pay vein 7 inches, pitch 15 degrees; nature of ore, gray copper, average value 90 ounces silver per ton; 1 shaft 17 feet deep.

Silver Producing Mining Co. Incorporated October 12, 1877; capital stock, $300,000, in 3,000 shares, $100 each; office Silverton, Edmund Higginbotham, president, Thomas Campau, vice president and treasurer, John G. Heid, secretary; tunnel claim 1,500 by 3,000 feet; discovered 1876; located on Sultan Mountain, Animas mining district, 1 mile from Silverton; course of tunnel, east and west, length 350 feet.

Silver Queen. Robert Hanson, proprietor; office Lake City; claim 300 by 1,500 feet; discovered 1876; located on Treasure Mountain, Eureka mining district, 4 miles from town of Eureka and 12 miles from Silverton; course of vein, northeast and southwest, width 25 feet, pitch 36 degrees; nature of ore, galena and copper pyrites; assay value, 50 ounces silver per ton; 1 adit 16 feet in length and 16 feet from surface.

Silver Wave. H. Goodict & Bro., proprietors; office Ouray; claim 300 by 1,500 feet; discovered 1877; located in Poughkeepsie Gulch, Uncompahgre mining district, 12 miles from Silverton; course of vein, east and west, width 3 feet, pay vein 18 inches, pitch 10 degrees; nature of ore, gray copper; assay value, 30 to 200 ounces silver per ton; 1 adit 30 feet in length.

Silver Wing. Frank Herrod, proprietor; office Mineral City; claim 300 by 1,500 feet; discovered 1876; located on Poughkeepsie Mountain, Uncompahgre mining district, 12 miles from Silverton; course of vein, northeast and southwest, width 8 feet, pay vein 2 feet, pitch 30 degrees northeast; nature of ore, galena; assay value, 123 ounces silver per ton; 1 tunnel, 20 feet in length, intersecting vein at a depth of 20 feet.

Silver Wing Mining and Reduction Co. Incorporated December 9, 1876; capital stock, $1,080,000, in 10,800 shares, $100 each; office Colorado Springs; organized by D. Russ Wood, Joseph S. Reef, Henry McAllister, Jr., William B. Sherman, Matt France, William T. Holt,

William S. Jackson, Nathan S. Culver, and Edward Copley; the property of this company is situated on Jones Mountain, 2 miles from town of Eureka and 10 miles from Silverton; developed by 2 tunnels, 150 and 180 feet in length, respectively.

Sioux City. Owned by Edward M. Brown, B. Wallon, and others; office Animas Forks; claim 100 by 1,500 feet; discovered 1876; located in Burns Gulch, Eureka mining district, 2 miles from town of Eureka and 10 miles from Silverton; course of vein, northwest and southeast, width 6 feet, pay vein, 3 feet, pitch 20 degrees northwest; nature of ore, galena and brittle silver; average value, 200 ounces silver per ton; 2 drifts, 12 and 50 feet in length, respectively; yield, $100.

Sioux City and San Juan Mining Co. Incorporated April 23, 1877; capital stock, $1,000,000, in 10,000 shares, $100 each; offices Del Norte and Animas Forks, Colorado, and Sioux City, Iowa; trustees: C.P. Heath, J.H. Fessenden, A. Bowman, S.W. Raymond, C.A. Orr, J.W. Stewart, J.V. Mellett, John Cleghorn, Sr., W.R. Bowman, A.E. Wilcox, and T.G. Cowgill.

Smuggler. Owned by R.H. Ferguson & Co.; office Eureka; claim 300 by 1,500 feet; discovered 1876; located on Tower Mountain, near town of Eureka, 9 miles from Silverton; course of vein, east and west; width 12 feet, pay vein 5 feet, pitch 45 degrees; nature of ore, gray copper, assay value; 75 ounces silver per ton; 1 adit 15 feet in length and 22 feet from surface.

Snow Drift. T.R. Gammon and S.W. Chinn, proprietors; office Eureka; claim 300 by 1,500 feet; discovered 1877; located in Eureka Gulch, Eureka mining district, 4 miles from town of Eureka and 11 miles from Silverton; vertical vein, trending northwest and southeast, width 8 feet, pay vein 14 inches; nature of ore, galena and gray copper, assay value, 175 ounces silver per ton; 1 adit 20 feet in length.

Snow Flake. Owned by R.V. McKinney & Co., of Eureka; claim 300 by 1,500 feet; discovered 1876; located on Niagara Mountain, 1 mile from town of Eureka; course of vein, north and south, width 5 feet, pay vein 4 feet, assay value of ore, 80 ounces silver pert on; development, 1 adit 18 feet and 1 drift 40 feet in length.

Sonoma. James Raser & Co., proprietors; office Animas Forks; claim 300 by 1,500 feet; discovered 1874; located on Hurricane Peak, 3 miles from Animas Forks and 12 miles from Silverton; course of vein, northeast and southwest, width 2 feet, pay vein 16 inches; nature of ore, galena and gray copper; assay value, $200 per ton; 1 shaft 10 feet deep, 1 adit 15 feet in length.

Sonora. Thomas Johnson and E. Brace, proprietors; office Animas Forks; claim 300 by 1,500 feet; discovered 1876; located on California Mountain, Eureka mining district, 2 miles from town of Animas Forks and 18 miles from Silverton; course of vein, east and west, width 50 feet, pitch 35 degrees; nature of ore, galena and iron pyrites; assay value, 20 to 80 ounces silver per ton; 1 adit 40 feet in length.

Spar. Charles Clase, proprietor; office Eureka; claim 300 by 1,500 feet; discovered 1878; located in Maggie Gulch, 5 miles from town of Eureka and 8 miles from Silverton; course of vein, northwest and southeast, width 5 feet, pitch 30 degrees; 1 adit 10 feet in length and 12 feet from surface.

Spartico. Antonio Oberto and John M. Stuart, proprietors; office Silverton; claim 300 by 1,500 feet; discovered 1878; located on Oberto Mountain, Uncompahgre mining district, 12 miles from Silverton; course of vein, northeast and southwest, width 200 feet, pay vein 2 feet; nature of ore, gray copper and galena; assay value, 75 ounces silver per ton; 2 shafts, 10 and 12 feet, respectively.

Staatsburg Mining Co. Incorporated August 10, 1878; capital stock, $600,000, in 6,000 shares, of $100 each; office Silverton; trustees: J.D. Rollins, J.H. Brink, Charles T. Willcutt, Henry Staats, Alonzo A. Fain, and John M. Stuart.

Star Inn. Joseph Beck & Co. proprietors; office Eureka; claim 300 by 1,500 feet; discovered 1874; located on Brown Mountain, Eureka mining district, 10 miles from Silverton; course of vein, east and west, width 3 feet, pitch 20 degrees; nature of ore, galena; assay value, 100 ounces silver per ton; 1 drift 35 feet in length.

St. Charles. Samuel Ogden, proprietor; office Animas Forks; claim 300 by 1,500 feet; discovered 1876; located on California Mountain, Eureka mining district, 2 miles from town of Animas Forks and 18 mils from Silverton; course of vein, northwest and southeast, width 40 feet, pitch 20 degrees; nature of ore, gray copper and galena; assay value, 60 ounces silver per ton; 1 adit 30 feet in length.

Storm King. Charles Hopkins & Co., proprietors; office Silverton; claim 300 by 1,500; discovered 1875; located on Mineral Creek, Animas mining district, 8 miles from Silverton; vertical vein, trending north and south, width 5 feet, pay vein 30 inches; nature of ore, galena and gray copper; assay value, 150 ounces silver per ton; 1 shaft 11 feet deep, 1 drift 15 feet in length.

Sultan. Leander F. Hollingworth & Co., proprietors; office Silverton; claim 140 by 1,500 feet; discovered 1877; located on Sultan Mountain, Animas mining district, near limits of Silverton; vertical vein, trending northwest and southeast, width 6 feet, pay vein 3 feet; nature of ore, gray copper and galena; assay value, 300 ounces silver per ton; 1 shaft 20 feet.

Sultan Tunnel. John Williams, proprietor; office Silverton; claim 1,500 by 3,000 feet, established 1876; located on Sultan Mountain, Animas mining district, 1 mile form Silverton; length 500 feet.

Summit. Charles S. Roe & Co., proprietor; office Silverton; claim 300 by 1,500 feet; discovered 1876; located on ruby Creek, Animas mining district, 5 miles from Silverton; course of vein, north and south, width 3 feet, pay vein 30 inches, pitch 20 degrees; nature of ore, galena and gray copper; assay value, 250 ounces silver per ton; 1 shaft 15 feet deep.

Summit. Mark Watson and J.O. Feely, proprietors; office Eureka; claim 300 by 1,500 feet; discovered 1877; located in Burns Gulch, Eureka mining district, 3 miles from town of Eureka and 11 miles from Silverton; course of vein, northeast and southwest, width 8 feet, pay vein 10 inches, pitch 30 degrees; nature of ore, gray copper and iron pyrites; assay value, 65 ounces silver per ton; 1 drift 18 feet in length.

Sunbeam. James W. Platt and John Spadling, proprietors; office Eureka; claim 300 by 1,500 feet; discovered 1874; located in Eureka Gulch, Eureka mining district, 9 miles from Silverton; vertical vein, trending north and south, width 10 feet; nature of ore, galena; 1 drift 28 feet in length and 18 feet from surface.

Sunlight. Owned by J. Larson, A. Haddorf, and O. Christenson; office Mineral City; claim 300 by 1,500 feet; discovered 1875; located on Lake Mountain, near Mineral City, 18 miles from Silverton; vertical vein, trending northeast and southwest, width 10 feet, pay vein 4 feet; nature of ore, galena and gray copper; average value, 80 ounces silver per ton; 1 shaft 15 feet deep, 1 drift 40 feet in length.

Sunnyside. Owned by R.J. McNutt and M.M. Engleman; office Eureka; claim 300 by 1,500 feet; discovered 1873; located in Eureka Gulch, 3 miles from town of Eureka and 9 miles from Silverton; course of vein 14 feet, pitch 20 degrees; nature of ore, galena and gray copper; average value, 125 ounces silver per ton; 1 shaft 10 feet deep, 1 drift 30 feet in length and 20 feet from surface; yield, $300.

Sunshine. E.F. Wilkes & Co., proprietors; office Silverton; claim 300 by 950 feet; discovered 1876; located on Pioneer Mountain, Animas mining district, 7 miles from Silverton; course of vein, northwest and southeast, width 15 feet, pitch 30 degrees; nature of ore, galena and carbonate of lead; assay value, 100 ounces silver per ton; 1 tunnel 25 feet in length.

Superior. Benjamin R. Eaton, proprietor; office Silverton; claim 300 by 1,500 feet; discovered 1876; located on Galena Mountain, Animas mining district, 4 miles from Silverton; vertical vein, trending northwest and southeast, width 6 feet, pay vein 20 inches; nature of ore, gray copper and galena; average value, 140 ounces silver per ton; 3 adits aggregating 50 feet in length.

Susquehana. George Ingersoll & Co., proprietors; office Silverton; claim 300 by 1,500 feet; discovered 1873; located on Hazelton Mountain, Animas mining district, 2 miles from Silverton; course of vein, northwest and southeast, width 3 feet, pay vein 4 inches, pitch 45 degrees; nature of ore, gray copper and galena; average value, 375 ounces silver per ton; 1 shaft 160 feet deep, 1 tunnel 90 feet in length, which intersects the vein 160 feet from the surface, 1 level 150 feet in length; yield, $30,000.

Swamp Angel. N. Jones and T.S. Raymond, proprietors; office Eureka; claim 300 by 1,500 feet; discovered 1877; located on Cement Creek, Eureka mining district, 5 miles from Silverton; vertical vein, trending northwest and southeast, width 12 feet, pay vein 2 feet; nature of ore, galena; assay value, 40 ounces silver per ton; 1 drift 15 feet in length.

Tegner. Owned by J.P. Johnson, Carley & Co.; office Eureka; claim 300 by 1,500 feet; discovered 1872; located on Eureka Mountain, near town of Eureka, 8 miles from Silverton; course of vein, east and west, width 8 feet, pay vein 2 feet, pitch 30 degrees; nature of ore, sulphurets and gray copper; average value, $394 per ton; 1 adit 15 feet, 1 tunnel 80 feet, and 1 level 15 feet in length, the later being 80 feet from surface; yield, $50.

Teller. Owned by Alexander N. Knight, Charles Slocum, and Oliver Case; office Silverton; claim 300 by 1,500 feet; discovered 1875; located on Sultan Mountain, Animas mining district, near limits of Silverton; course of vein, northwest and southeast, width 7 feet, pay vein 18 inches, pitch 15 degrees; nature of ore, gray copper and galena; assay value, 200 ounces silver per ton; 1 shaft 26 feet deep, 1 tunnel 50 feet in length.

Teneriffe. S. Richards and F. Kerns, proprietors; office Lake City; claim 300 by 1,500 feet; discovered 1876; located on Seigle Mountain, Uncompahgre mining district, 2 miles from Mineral City, and 18 miles from Silverton; vertical vein, trending north and south, width 15 feet, pay vein 20 inches; nature of ore, galena; average value 40 ounces silver per ton; 1 drift 25 feet in length.

Teneriffe. Johnson Burritt, proprietor; office Silverton; claim 300 by 1,500 feet; discovered 1877; located on Round Mountain, Animas mining district, 4 miles from Silverton; vertical vein, trending northwest and southeast, width 3 feet, pay vein 16 inches; nature of ore, galena and chloride; assay value, 125 ounces silver per ton; 1 adit 18 feet in length.

Terrible. N. Jones & Co., proprietor; office Eureka; claim 300 by 1,500 feet; discovered 1876; located on Brown Mountain, Eureka mining district, 10 miles from Silverton; vertical vein, trending northeast and southwest, width 7 feet, pay vein 30 inches; nature of ore, galena and gray copper; assay value, 100 ounces silver per ton, 1 shaft 24 feet deep.

Three Brothers. Owned by John Stuart, of Silverton; claim 300 by 1,500 feet; discovered 1877; in Poughkeepsie Basin 12 miles from Silverton; course of vein, northwest and southeast, width 60 feet; nature of ore, gray copper; assay value, 190 ounces silver per ton; development, 1 adit 35 feet in length.

Thunderer. C. McCabe & Co., proprietors; office Eureka; claim 300 by 1,500 feet; discovered 1876; located in Eureka Gulch, Eureka mining district, 4 miles from town of Eureka and 11 miles from Silverton; course of vein, east and west, width 5 feet, pitch 10 degrees; nature of ore, gray copper and galena; assay value, 70 ounces silver per ton; 2 shafts, 8 and 10 feet in depth, respectively.

Toledo. R. Tilton & Co., proprietors; office Silverton; claim 300 by 1,500 feet; discovered 1876; located on Pioneer Mountain, Animas mining district, 7 miles from Silverton; course of vein, northwest and southeast, width 8 feet, pay vein 10 inches, pitch 10 degrees; nature of ore, galena and copper pyrites; assay value, 50 ounces silver per ton; 1 shaft 16 feet deep.

Tornado, No. 1. Joseph Bock and Samuel Hunt, proprietors; office Lancaster, Wisconsin; claim 300 by 1,500 feet; discovered 1877; located on South Lookout Mountain, Animas mining district, 8 miles from Silverton; vertical vein, trending east and west, width 12 feet; nature of ore, carbonate of lead; average value, 200 ounces silver per ton; 1 shaft 50 feet deep, 1 tunnel 20 feet, and 1 level 35 feet in length.

Tornado, No. 2. Owned by Alonzo P. Wood and Samuel D. Frazeur; office Silverton; claim 300 by 1,500 feet; located on South Lookout Mountain, Animas mining district, 8 miles from Silverton; course of vein, nearly east and west, width 5 feet, pay vein 30 inches; nature of ore, carbonated of lead; average value 60 ounces silver per ton; 1 shaft 50 feet deep.

Torpedo. Owned by Dr. Small and C. Pratt, of Howardsville; claim 300 by 1,500 feet; discovered 1874; located on Galena Mountain, 1 mile from Howardsville; course of vein, northwest and southeast, pay vein 20 inches; nature of ore, galena and iron pyrites; assay value, 100 ounces silver per ton; development, 4 shafts 10 feet each.

Tower Mountain Mining Co. Incorporated June 12, 1877; capital stock, $400,000, in 8,000 shares, $50 each; offices, Howardsville, Colorado, and Titusville, Pennsylvania; trustees: David Cropley, Samuel G. Stevens, J.D. Angier, F.H. Gibbs, E.O. Emerson, William Barnsdall, Jr., Joel N. Angier, Marcus Bronson, and F. Merrick.

Treasure. W.P. Hartman & Co., proprietors; office Silverton; claim 300 by 1,500 feet; discovered 1877; located on Crown Mountain, Eureka mining district, 2 miles from town of Eureka and 10 miles from Silverton; course of vein, northeast and southwest, width 15 feet, pay vein 2 feet; average value of ore, 35 ounces silver per ton; 1 shaft 15 feet deep.

Treasure Vault. Henry Carley & Co., proprietors; office Animas Forks; claim 300 by 1,500 feet; discovered 1878; located on Hourdin Mountain, Eureka mining district, 2 miles from Animas Forks and 17 miles from Silverton; course of vein, north and south, width 3 feet, pay vein 6 inches; pitch 10 degrees; nature of ore, galena; assay value, 60 ounces silver per ton; 1 shaft 13 feet deep.

Treasury. J. Lattimore & Co., proprietors; office Lake City; claim 300 by 1,500 feet; discovered 1875; located in Poughkeepsie Gulch, Uncompahgre mining district, 12 miles from Silverton; course of vein, northwest and southeast, width 35 feet, pay vein 8 feet; nature of ore, galena and gray copper; assay value, 100 ounces silver per ton; 1 shaft 15 feet deep, 1 adit 30 feet in length.

Tribune. Owned by the Gladstone Smelting Co., of Gladstone; claim 300 by 1,500 feet; discovered 1876; located on Brown Mountain, Eureka mining district, 6 miles from town of Eureka; vertical vein, trending east and west; width 10 feet; nature of ore, gray copper; value, 150 ounces silver per ton; development, 1 tunnel 75 feet in length.

Trinidad. Owned by A.O. Terry, D.M. Edgerton, and F. Johnson; office St. Louis, Missouri; claim 300 by 1,500 feet; discovered 1877; located on Crown Mountain, near town of Eureka, 9 miles from Silverton; course of vein, northeast and southwest, width 4 feet, pay vein 3 feet, pitch 10 degrees; nature of ore, gray copper and galena; average value, 120 ounces silver per ton; 1 shaft 25 feet deep, 1 drift 55 feet in length and 45 feet from surface.

Twilight. Owned by Robert Hanson and H. Melvin; office Lake City; claim 300 by 1,350 feet; discovered 1874; located at the head of Eureka Gulch, 5 miles from town of Eureka and 13 miles from Silverton; course of vein, northeast and southwest, width 12 feet, pitch 20 degrees; nature of ore, copper pyrites and galena; assay value, 60 ounces silver and $28 gold per ton; 1 adit 20 feet in length and 20 feet from surface, 1 tunnel 25 feet in length.

U.P. John Ryan and Peter Hogan, proprietors; office Animas Forks; claim 300 by 1,500 feet; discovered 1877; located on Treasure Mountain, Eureka mining district, 2 miles from Animas Forks and 18 miles from Silverton; course of vein, north and south, width 6 feet, pay vein 30 inches, pitch 30 degrees; nature of ore, galena and gray copper; assay value, 90 ounces silver per ton; 1 shaft 20 feet deep.

U.S. Grant. Owned by William Cooper, Charles H. Cooper, and others; office Lake City; claim 300 by 1,500 feet; discovered 1876; located on Seigle Mountain, 2 miles from Mineral City; course of vein, northeast and southwest, width 8 feet, pitch 40 degrees; nature of ore, galena and gray copper; assay value, 150 ounces silver per ton; 1 shaft 18 feet deep.

Uncompahgre Chief. Edward W. McIntire and Charles H. McIntire, prospectors; office Ouray; claim 300 by 1,500 feet; discovered 1874; located on Lake Mountain, near Mineral City, 18 miles from Silverton; vertical vein, trending northeast and southwest, width 50 feet, pay vein 3 feet; nature of ore, galena; assay value, $74 silver per ton; 1 adit 50 feet in length.

Union. B. Mallon, proprietor; office Animas Forks; claim 300 by 1,500 feet; discovered 1878; located in Burns Gulch, Eureka mining district, 2 miles from town of Eureka and 10 miles from Silverton; course of vein, northeast and southwest, width 5 feet, pay vein 15 inches, pitch 20 degrees; nature of ore, gray copper and iron pyrites; assay value, 100 ounces silver per ton; 1 drift 20 feet in length.

Uniweep. W.R. Long and N.O. Harker, proprietors; office Eureka; claim 300 by 1,500 feet; discovered 1875; located in Eureka Gulch, 4 miles from town of Eureka; vertical vein, trending north and south, width 25 feet; nature of ore, galena, gray copper, iron and copper pyrites; assay value, from 60 to 280 ounces silver per ton; 2 adits, 10 and 15 feet in length, respectively, and 1 tunnel 20 feet in length.

Utah. R.P. Buckeye & Co., proprietors; office Animas Forks; claim 300 by 1,500 feet; discovered 1877; located on California Mountain, 2 miles from Animas Forks and 18 miles from Silverton; course of vein, east and west, width 10 feet, pay vein 14 inches, pitch 10 degrees; nature of ore, galena and gray copper; 1 adit 10 feet in length and 6 feet from surface.

Vera Cruz. Owned by P. Crout and R. Hanson, of Eureka; claim 300 by 1,500 feet; discovered 1875; located on Poughkeepsie Mountain, 13 miles from Silverton; course of vein, east and west, width 8 feet, pay vein 2 feet; nature of ore, brittle silver and copper pyrites; value, $1,600 per ton; development, 1 shaft 30 feet deep.

Victoria. C. Weightman and L.S. Clive, proprietors; office Ouray; claim 300 by 1,500 feet; discovered 1876; located on Mineral Point Mountain, 2 miles from Mineral City and 18 miles from Silverton; vertical vein, trending north and south, width 25 feet; nature of ore, gray copper and galena; assay value, 60 ounces silver per ton; 2 adits, 15 and 25 feet in length, respectively.

W.S. Blakely. W.B. House, proprietor; office Howardsville; claim 300 by 1,500 feet; discovered 1877; located on Tower Mountain, Eureka mining district, 1 mile from town of Eureka and 8 miles from Silverton; course of vein, east and west, width 15 inches; nature of ore, galena; assay value, 140 ounces silver per ton; 1 tunnel 25 feet in length and 14 feet from surface.

Waldron. Charles H. McIntire, proprietor; office Ouray; claim 300 by 1,500 feet; discovered 1874; located on Mineral Point Mountain, near Mineral City, 18 miles from Silverton; vertical vein, trending northeast and southwest, width 40 feet, pay vein 3 feet; nature of ore, galena and gray copper; assay value, 80 ounces silver per ton; 2 drifts, 12 and 20 feet in length, respectively.

War Eagle. Owned by Daniel Wyman & Bro.; office Eureka; claim 300 by 1,500 feet; discovered 1877; located on Brown Mountain, Eureka mining district, 10 miles from Silverton; course of vein, east and west, width 40 inches, pay vein 6 inches, pay vein 6 inches, pitch 15 degrees; nature of ore, galena; assay value, 80 ounces silver per ton; 1 shaft 24 feet deep.

Washington. Thomas Dunton, proprietor; claim 300 by 1,500 feet; discovered 1877; located in Burns Gulch, 2 miles from town of Eureka and 8 miles from Silverton; vertical vein, trending north and south, width 7 feet, pay vein 30 inches; nature of ore, galena and copper pyrites; assay value, 80 ounces silver per ton; 1 adit 30 feet in length and 20 feet from surface.

Waterfall. Owned by J.M. King, E.S. Finch, and others; office Ouray; claim 300 by 1,500 feet, discovered 1878; located on Pioneer Mountain, Animas mining district, 8 miles from Silverton; course of vein, east and west, width 18 feet, pay vein 4 feet, pitch 35 degrees; nature of ore, galena and carbonate of lead; average value, 56 ounces silver per ton; 1 adit 48 feet in length.

Watson. George N. Propper and P.H. Watson, proprietors; office Mineral City; claim 300 by 1,500 feet; discovered 1874; located on Mineral Point Mountain, 1 mile from Mineral City and 18 miles from Silverton; vertical vein, trending northeast and southwest, width 50 feet; nature of ore, galena and gray copper; value, 400 ounces silver per ton; 2 adits, 12 and 20 feet in length, respectively.

Western Enterprise Gold and Silver Mining Co. Incorporated February 1879; capital stock, $3,000,000, in 120,000 shares, $25 each;

offices Animas Forks, Colorado, and Chicago, Illinois; organized by Francis A. Griswold, Edward S. Hunt, Levi N. Woodbury, and others.

Westminster. C. Andrews & Co., proprietors; office Animas Forks; claim 300 by 1,500 feet; discovered 1877; located on California Mountain, Eureka mining district, 2 miles from Animas Forks and 18 miles from Silverton; vertical vein, trending east and west, width 5 feet, pay vein 20 inches; nature of ore, galena and gray copper; assay value, 65 ounces silver per ton; 1 adit, 18 feet in length.

West Virginia. James J. and Albert Bernard, Daniel J. and James D. McKay, William Forsyth, and John M. Ross, proprietors; office Eureka; claim 300 by 1,500 feet; discovered 1874; located on the Animas River, near town of Eureka; course of vein, northeast and southwest, average width 4 feet, pay vein 18 inches; nature of ore, bismuth silver, gray copper, iron and copper pyrites; 1 tunnel 20 feet in length.

West Virginia. Owned by the West Virginia Gold and Silver Mining Co.; organized under the laws of the State of Indiana, in the year 1878; capital stock, $30,000, in 1,000 shares, $30 each; offices Columbus, Ohio, and Eureka, Colorado; J.H. Wasson, president, J.F. Miller, vice president, T.R. Wing, treasurer, C.H. Cole, secretary, E. S. Armstrong, superintendent; claim 300 by 1,500 feet; discovered 1875; located on Jones Mountain, Eureka mining district, near town of Eureka.

White. L. Woodbury & Co., proprietors; office Animas Forks; claim 300 by 1,500 feet; discovered 1875; located on Wood Mountain, Eureka mining district, 2 miles from Animas Forks and 18 miles from Silverton; course of vein, northwest and southeast, width 30 feet, pay vein 7 feet; nature of ore, gray copper and galena; average value, 54 ounces silver per ton; 1 drift 45 feet in length and 20 feet from surface.

White. P. McCay and U.K. White, proprietors; office Del Norte; claim 300 by 1,500 feet; discovered 1877; located on Mineral Point Mountain, 1 mile from Mineral City and 18 miles from Silverton; vertical vein, trending northeast and southwest, width 6 feet; nature of ore, galena, gray copper and tellurium; assay value, 125 ounces silver per ton; 1 adit 18 feet in length.

White Water. E.K. Jessup and N.L. Strong, proprietors; office Mineral City; claim 300 by 1,500 feet; discovered 1875; located on Middle Creek, Uncompahgre mining district, 1 mile from Mineral City and 19 miles from Silverton; course of vein, northeast and southwest, width 15 feet, pay vein 18 inches, pitch 10 degrees; nature of ore, galena; average value, 30 ounces silver per ton; 2 adits, 10 and 25 feet in length, respectively.

Yellow Jacket. Owned by Hann, Cotton, and Grant; office Silverton; claim 50 by 1,500 feet; discovered 1873; located on Hazelton Mountain, Animas mining district, 2½ miles from Silverton; course of vein, northwest and southeast, width 4 feet, pay vein 7 inches; nature of ore, galena and gray copper; average value, 150 ounces silver per ton; 1 shaft 15 feet deep, 1 adit 15 feet in length.

Yellow Jacket. Owned by Frederick Blaisdale and J.W. Neiswanger; office Eureka; claim 300 by 1,500 feet; located 1874; located on Eureka Mountain, 1 mile from town of Eureka and 9 miles from Silverton; course of vein, east and west, pitch 10 degrees; nature of ore, iron pyrites and gray copper; assay value, 45 to 300 ounces silver per ton; 1 shaft 25 feet deep.

San Juan County Ore Mills

Animas Forks Concentration Works. Located in town of Animas Forks, 12 miles from Silverton.

Cameron Smelting Co. Incorporated March 22, 1876; capital stock, $100,000, in 200 shares, $500 each; office located near the middle branch of the lake fork of the Gunnison, near the American Basin, in Adams mining district; incorporators, R.H. Cameron, H.H. Hyatt, Gardner Green, H.P. Carman, and J.M. Fish.

Emma Dean Works. Owned by Winspeare & Co.; located in town of Eureka, 8 miles from Silverton; have a capacity for treating 10 tons of ore daily.

Niegold's Concentration Works. Owned by Niegold Bros.; located on Cement Creek, 6 miles from Silverton; have a capacity for treating 5 tons of ore daily.

Melville Mining and Reduction Co. Incorporated October 9, 1876; capital stock, $150,000, in 15,000 shares, $100 each; organized by William G. Melville, G.W. Glick, E. Summerfield, A.P. Clark, Charles C. Duncan, and August Poehler. The works of this company have a capacity for treating 15 tons of ore in 24 hours; they are run by waterpower and located on Mineral Creek in town of Silverton.

San Juan Reducing Co. Incorporated July 5, 1878; capital stock, $50,000, in 500 shares, $100 each; office at Gladstone, on the forks of Cement Creek; trustees: Lorenzo Dow, W. Broderick Cloeti, T.P. Medley, J.H.E. Waters, and J.H. Hobgon.

Silverton Smelting Works. George Green & Co., proprietors; John L. Pennington, general manager, Thomas E. Bowman, metallurgist. These works have a capacity for treating 20 tons of ore in 24 hours; they are run by waterpower and located on Cement Creek, in town of Silverton.

Abandoned, unidentified mine in Mount Sneffels mining district. The photo was taken in 2001.

This 1903 photo postcard shows miners, prospectors, and travelers in downtown Silverton, Colorado. The man on horseback to the right is laden with mining equipment, a shovel, and a gold pan. He holds his rifle across the front of his saddle, ready for any emergency. A goup of men is standing in front of the J.J. Harris & Co. building. Not surprisingly, there are no women or children in this scene.

Shift leaving Heading, Gunnison Tunnel.

Shift of miners going home after a hard day's work. The supervisor, his watch chain showing on his vest, is standing proudly in front of a car full of workers in the gussison Tunnel. This photo was taken about 1908 or 1909.

438

SUMMIT COUNTY

Thomas B. Corbett reported that Summit County had a population of 3,000 in 1878 and a bullion product that year of $342,000. Mining districts included Bevan, Georgia, Lincoln City, McKay, Peru, Pollock, Silver Lake, Snake River, Ten Mile, and Upper Blue.

Some of the mining camps and towns were deserted when production fell or when the silver crash hit in 1893. Two of these communities, Saints John and Montezuma, are now ghost towns.

Alexander Placer. Owned by Aden Alexander of Denver. This property is located on the Blue River, 6 miles from Breckenridge. It is worked by hydraulic pressure and extends along the river for a distance of 10 miles.

Argenta. James M. Forshey, proprietor; office Leadville; claim 150 by 1,500 feet; located on Fletcher Mountain, Ten Mile mining district, 12 miles from Breckenridge; course of vein, north and south, width 8 feet, pitch 15 degrees west; nature of ore, galena and copper pyrites; value, 2 ounces gold and 50 ounces silver per ton; 2 shafts, 15 and 20 feet in depth, respectively.

Argentine Co. Incorporated October 27, 1875; capital stock, $500,000, in 500 shares, $100 each; offices Montezuma, Colorado, and Chicago, Illinois; incorporators: L. Charles Holland, Jeremiah Laming, Henry M. Payne, James G. Diven, and Abraham V. Hartwell.

Badger. John W. Jacque, proprietor; office Leadville; claim 150 by 1,500 feet; located on Robinson Mountain, Ten Mile mining district, 12 miles from Breckenridge; course of vein, northwest and southeast; nature of ore, galena and carbonate of lead; value, 200 ounces silver per ton; 1 shaft 16 feet deep.

Badger Gold Mining and Fluming Co. Incorporated May 11, 1867; capital stock, $20,000; office Denver; incorporators: M.L. Rood, W.C. Rippey, Robert S. Wilson, and Henry C. Clark.

Ballou Placer. Ballou & Pollock, proprietors; office Breckenridge; claim comprises 800 by 10,000 feet of placer ground, located in French Gulch, McKay mining district, half mile from Breckenridge.

Bartlett Mountain Mining Co., of Colorado. Incorporated 1876; capital stock, $44,000, in 440 shares, $100 each; John Potter, president, Henry McAllister, Jr., secretary and treasurer; claim 2,100 by 10,500 feet; located on Bartlett Mountain, Ten Mile mining district, 12 miles from Breckenridge.

Belle East. Edward C. and Frederick A. Guibor, proprietors; office Empire; claim 150 by 1,500 feet; discovered 1875; located on Glacier Mountain, Snake River mining district, 1 mile from Montezuma and 21 miles from Breckenridge; course of vein, northeast and southwest, width 6 feet, pay vein 16 inches, pitch 20 degrees; nature of ore, galena and gray copper; average value, 100 ounces silver per ton; 3 shafts aggregating 65 feet in depth, 2 drifts aggregating 206 feet in length; yield, $2,000.

Belle West. Edward C. Guibor, proprietor; office Empire; claim 150 by 1,500 feet; discovered 1875; located on Glacier Mountain, Snake River mining district, 1 mile from Montezuma and 21 miles from Breckenridge; course of vein, northeast and southwest, width 6 feet, pay vein 14 inches, pitch 30 degrees; nature of ore, gray copper and galena; assay value, 300 ounces silver per ton; 1 drift 25 feet in length.

Bilk. Owned by James M. Forshey, Frank and Perry Brandon; office Leadville; claim 150 by 1,500 feet; located on Champion Mountain, Ten Mile mining district, 12 miles from Breckenridge; vertical vein, trending east and west, width 3 feet; nature of ore, free gold; value, $140 gold per cord; 1 shaft 16 feet deep.

Black Warrior. E. Lowe & Co., proprietors; office Breckenridge; claim 150 by 1,500 feet; located on Bartlett Mountain, Ten Mile mining district, 12 miles from Breckenridge; vertical vein, trending northeast and southwest, width 12 feet; nature of ore, galena and copper pyrites; value, 90 ounces silver per ton; 1 shaft 25 feet deep.

Blue River Mining Co. Organized under the laws of the State of Missouri, George D. Hall, president, John C. Orrick, secretary, John A. Willoughby, of Breckenridge, Colorado, agent.

Boston and Colorado Gold Placer Mining Co. Organized under the laws of the State of Connecticut; capital stock, $50,000; office Breckenridge, J.A. Willoughby, agent, James B. Potter, president, Levi Nowcomb, secretary.

Boston Mining Co. Incorporated July 27, 1877; capital stock, $500,000, in 50,000 shares, $10 each; offices, Sts. John, Colorado, and Boston, Massachusetts; organized by John R. Brewer, T.H. Perkins, Francis Bartlett, Henry L. Hallett, William L. Candler, and others; John R. Brewer, president, William L. Candler, secretary and treasurer. The property of this company is located on Glacier Mountain and in the town of Saints John, 1 mile from Montezuma and 20 miles from Breckenridge.

C.G. Comstock. William L. and Clarence G. Sampson, proprietors; office Montezuma; claim 150 by 1,500 feet; discovered 1876; located on Glacier Mountain, Snake River mining district, 2 miles from Montezuma and 22 miles from Breckenridge; course of vein, northeast and southwest, width 14 feet, pay vein 8 inches, pitch 35 degrees; nature of ore, gray copper and galena; average value, 220 ounces silver per ton; 3 drifts, 16, 20, and 50 feet in length, respectively; yield, $500.

Carbonate Mining Co. Incorporated November 27, 1878; capital stock, $250,000, in 25,000 shares, $10 each; office Denver and Carbonateville, Colorado; organized by Newton B. Lord, Leonard S. Ballou, John M. Thorn, David A. Gage, Samuel H. Herrick, Alexander M. Lay, and Edward Schenck.

Cashier. D.L. Southworth & Co., proprietors; office Montezuma; claim 150 by 1,500 feet, patented; discovered 1867; located on Teller Mountain, Snake River mining district, 4 miles from Montezuma and 24 miles from Breckenridge; course of vein, northeast and southwest, width 65 feet, pay vein 20 inches, pitch 40 degrees; nature of ore, gray copper; average value, 200 ounces silver per ton; 1 shaft, 100 feet deep; 1 tunnel 450 feet in length, which intersects the vein 200 feet from the surface, 1 level, 40 feet in length; yield, $25,000.

Centennial. Owned by I.W. and F.M. Hibbard; office Alma; claim 150 by 1,500 feet; located on North Star Mountain, Silver Lake mining district, 8 miles from Alma; course of vein, northeast and southwest, width 4 feet, pitch 10 degrees southeast; nature of ore, oxide of iron; average value $50 per ton; developed by 2 tunnels, 50 and 150 feet from surface, respectively, aggregating 250 feet, and 1 drift 40 feet in length; cabin 14 by 16 feet; yield, $20,000.

Centennial. William L. and Clarence G. Sampson, proprietors; office Montezuma; claim 150 by 1,500 feet, patented; discovered 1876; located on Glacier Mountain, Snake River mining district, 2 miles from Montezuma and 22 miles from Breckenridge; course of vein, northeast and southwest, width 40 feet, pay vein 3 feet, pitch 20 degrees; nature of ore, galena and gray copper; average value, 150 ounces silver per ton; main shaft 35 feet deep and 2 other shafts 10 feet each; yield, $500.

Champion. Owned by Edwin S. Platt, John M. Dumont & Co.; office Denver; claim 300 by 1,500 feet, patented; discovered 1873; located on Teller Mountain, Snake River mining district, 5 miles from Montezuma and 25 miles from Breckenridge; course of vein, northeast and southwest, width 20 feet, pitch 45 degrees west; nature of ore, gray copper and galena; developed by 9 shafts, ranging depth from 110 to 300 feet, and 3 levels, 710, 750, and 1,050 feet in length, respectively; on the surface are 3 ore houses, 40 by 70 feet each, 3 blacksmith shops, assay and business offices, saw mill, etc.; yield to date, $27,000.

Champion. Owned by James M. Forshey, Horace and Albert McNeil, office Leadville; claim 150 by 1,500 feet; located on Champion Mountain, Ten Mile mining district, 12 miles from Breckenridge; vertical vein, trending northwest and southeast, width 40 feet; nature of ore, antimonial silver, iron pyrites and sulphurets; assay value, 134 ounces silver per ton; development, 1 shaft 20 feet deep, 1 adit 40 feet, an open cut 15 feet, and a tunnel 10 feet in length.

Chautauqua. Montezuma Silver Mining Co., proprietors; offices Montezuma, Colorado, and New York City; claim 150 by 4,500 feet, patented; located on Glacier Mountain, Snake River mining district, 1 mile from Montezuma, and 21 miles from Breckenridge; course of vein, northeast and southwest, width 6 feet, pay vein 4 feet, pitch 45 degrees; nature of ore, galena and gray copper; assay value, 100 to 800 ounces silver per ton; 1 shaft 140 feet deep, 2 tunnels, 150 and 160 feet in length, respectively.

Chenango Tunnel. Owned by the Silver Gate Mining Co., of Cincinnati, Ohio; claim 1,500 by 3,000 feet; located on Glacier Mountain, Snake River mining district, 1 mile from Montezuma and 20 miles from Breckenridge; length 561 feet.

Clark Placer. C.P. Clark, proprietor; office Lincoln City; claim comprises 2,000 by 10,000 feet of placer ground; located in French Gulch, Bevan mining district, half mile from Breckenridge.

Clark and Cobb Placer. C.P. Clark and J.J. Cobb, proprietors; office Breckenridge; claim comprises 700 by 11,000 feet of ground, located on the Blue River, 1 mile from Breckenridge.

Clinton and Ten Mile Placer. Owned by D.R. Emmett and Edward Lowe. This property is composed of riverbeds, bank and bar claims, and comprises 200 acres of ground. It is located at the mouth of Clinton and Ten Mile Gulches, Ten Mile mining district, 12 miles from Breckenridge and 14 miles from Leadville.

Coaley Extension. Edward C. Guibor & Co., proprietors; office Empire; claim 50 by 350 feet; discovered 1865; located on Glacier Mountain, 1 mile from Montezuma and 21 miles from Breckenridge; vertical vein, trending northeast and southwest, width 4 feet, pay vein 1 foot; nature of ore, galena and gray copper; average value, 250 ounces silver per ton; 1 shaft 65 feet deep; yield, $6,500.

Colorado Springs Prospecting and Mining Co. Incorporated July 24, 1875; capital stock, $25,000; office Colorado Springs; organized by John G. Wilson, Edward Hunt, A.C. Knox, F.P. Lombard, J.M. Stanley, John Z. Potter; R.C. Lyon, president; O.C. Knox, vice president, E.J. Eaton, secretary, L.S. Humiston, superintendent. The property of this company is located on the northeastern slope of Bartlett Mountain, Ten Mile mining district.

Comstock. Owned by the Boston Mining Co.; claim 75 by 1,500 feet, patented; discovered 1875; located on Glacier Mountain, near town of Saints John, 1 mile from Montezuma and 20 miles from Breckenridge; vertical vein, trending northeast and southwest, width 4 feet, pay vein 1 foot; nature of ore, galena and zinc blende; average value, 50 ounces silver per ton; main shaft 250 feet deep, 2 tunnels, 50 and 125 feet from surface, respectively, aggregating 1,300 feet.

Comstock Tunnel. Owned by the Boston Mining Co.; claim 1,500 by 3,000 feet, located on Glacier Mountain, Snake River mining district, near town of Saints John, 1 mile from Montezuma, and 20 miles from Breckenridge. The course is northwest and southeast; it was commenced in 1872, and has cut 9 lodes. It is now 1,500 feet in length and has gained a depth of 850 feet.

Cooper and Yengling Placer. Cooper and Yengling, proprietors; office Breckenridge; claim comprises 40 acres of placer ground, located on the south side of Negro Hill, Upper Blue mining district, half mile from Breckenridge; worked by hydraulic pressure.

Couara Silver Mining Co. Incorporated march 24, 1874; capital stock, $200,000, in 2,000 shares, $100 each; offices Peru and Denver, Colorado; A.B. Robbins, president, H.A. Bagley, secretary, H.J. Bagley, treasurer. The property of this company is located on Ruby and Cooper Mountains, Peru mining district, 5 miles from Montezuma and 22 miles from Breckenridge. It comprises 30 lodes or mineral veins, 150 or 1,500 feet each, a tunnel claim 1,500 by 3,000 feet, a blacksmith and dwelling house.

Croesus Gold and Silver Mining Co. Incorporated September 10, 1877; capital stock, $500,000, in 20,000 shares, $25 each; office Denver; John Dixon, president, Calvin P. Clark, vice president, Calvin P. Clark, vice president, William F. Calloway, secretary; Calvin P. Clark, superintendent. The property of this company is located on Australia Gulch, Bevan mining district, 1 mile from Lincoln City and 3 miles from Breckenridge. It comprises 2 gold lodes, 300 by 1,500 feet each, 8 silver lodes, 300 by 3,000 feet, and a mill site containing 10 acres of ground.

Dexter. F.E. Webster, proprietor; office Montezuma; claim 75 by 1,500 feet; discovered 1869; located on Ruby Mountain, Peru mining district, 4 miles from Montezuma and 22 miles from Breckenridge; course of vein, northeast and southwest, width 2 feet, pay vein 6 inches, pitch 20 degrees; nature of ore, galena; average value, 240 ounces silver per ton; 1 shaft 40 feet deep; yield, $250.

Dora. Owned by Thomas B. Corbett and William F. Kendrick; offices Denver and Alma, Colorado; claim 300 by 1,500 feet; discovered 1878; located on North Star Mountain, Silver Lake mining district, 8 miles from Alma and 14 miles from Fairplay.

Dublin No. 2. Owned by Albert R. Doyle; office Empire; claim 150 by 1,500 feet; discovered 1870; located on Collier Mountain, Snake River mining district, 4 miles from Montezuma; vertical vein, trending north and south, width 7 feet, pay vein 3 feet; nature of ore, iron pyrites and galena;

average value, 60 ounces silver per ton; 1 drift 100 feet in length and 80 feet from surface.

Eclipse. Willard Teller, proprietor; office Denver; claim 50 by 1,600 feet, patented; located on Glacier Mountain, near town of Montezuma; course of vein, northeast and southwest, width 1 foot; nature of ore, galena and gray copper; average value, $200 per ton; main shaft 50 feet and 1 other shaft 40 feet deep, 1 level 60 feet in length; yield, $2,000.

El Paso Mining Co., of Colorado. Incorporated May 24, 1877; capital stock, $10,000, in 500 shares, $20 each; J.H. Kerr, president, W.S. Weed, vice president and treasurer, Charles Ayers, secretary; claim comprises 27,000 linear feet of ground, located in Clinton Gulch, Ten Mile mining district, 12 miles from town of Breckenridge.

Erie, East and West. Montezuma Silver Mining Co., proprietors; office Montezuma; claim 150 by 1,500 feet, patented; located on Glacier Mountain, Snake River mining district, 3 miles from Montezuma and 23 miles from Breckenridge; course of vein, northeast and southwest, width 6 feet, pay vein 1 foot, pitch 20 degrees; nature of ore, galena and gray copper; 1 shaft 18 feet deep, 1 drift 35 feet in length.

Excelsior. James M. Forshey, proprietor; office Leadville; claim 150 by 1,500 feet; located on Fletcher Mountain, Ten Mile mining district, 12 miles from Breckenridge; course of vein, north and south, width 2 feet, pitch 15 degrees west; nature of ore, galena and copper pyrites; value 1 ½ ounces gold and 40 ounces silver per ton; 1 shaft 25 feet deep.

Fletcher, Extension of the Mountain Stake. Owned by James M. Forshey and Albert McNeil; office Leadville; claim 150 by 1,500 feet; located on the west slope of Fletcher Mountain, Ten Mile mining district, 15 miles from Breckenridge; course of vein, 15 degrees west of south, width 4 feet; nature of ore, galena and copper pyrites; average value, $125 per ton; development, 1 surface opening 12 feet in length.

Fletcher Mountain Tunnel. James M. Forshey, proprietor; office Leadville; claim 1,500 by 3,000 feet; located at the base of Fletcher Mountain, Ten Mile mining district, 15 miles from Breckenridge. This tunnel is now 600 feet in length and 1,800 feet from the top of Fletcher Mountain.

Flushing Mining Co.'s Placer. Owned by George M. Clark; office Lincoln City; claim comprises 1,200 square feet of ground, located in French Gulch, Bevan mining district, half mile from Breckenridge.

Fredonia. J.L. Fuller and F. Cromes, proprietors; office Breckenridge; claim 150 by 1,500 feet; discovered 1873; located on Argentine Mountain, Argentine mining district, 5 miles from Breckenridge; course of vein, north and south, width 10 feet, pay vein 3 feet; nature of ore, sulphurets and carbonates of lead; average value, 75 ounces silver per ton; 1 tunnel 60 feet in length, intersecting vein 30 feet from surface.

Fuller and Crome Placer Claim. J.L. Fuller and F. Crome, proprietors; office Breckenridge; claim comprises 27,000 square feet of ground; located on Blue River, Upper Blue mining district, near Breckenridge; it is provided with a flume 400 feet in length and is worked by ground sluicing.

Fuller and Greenleaf Mining and Ditch Co. Incorporated June 16, 1871; capital stock, $30,000, in 60 shares, $500 each; offices Breckenridge Colorado, and Boston and Springfield, Massachusetts; incorporators: T.H. Fuller, A.M. Greenleaf, and J.A. Willoughby.

Fuller Placer Mining Co. Incorporated October 3, 1876; capital stock, $500,000, in 5,000 shares, $100 each; offices Denver, Colorado, and Boston, Massachusetts; organized by Thomas H. Fuller, Mason B. Carpenter, and Gilbert L. Havens. Thomas H. Fuller, president, M.B. Carpenter, secretary and attorney, M.J. Cole, superintendent. The property of this company is located in the valleys of the Swan River and on the tributaries of the Blue River, 6 miles from Breckenridge, the county seat, and 100 miles from Denver. It comprises 3,100 acres of placer ground, over 21 miles of wooden flumes, 25 miles of ditches, 2 miles of sluicing, 9,000 feet of iron piping, 7 little giants, 6 pressure boxes, saw mill, 4 boarding houses, 20 cabins, mining tools, etc. The product of the mines of this company for the year 1876 was $25,000, in 1877 it was $35,000, and in 1878 it rose to $40,000.

G.B.R. Robinson & Co., proprietors; office Leadville; claim 150 by 1,500 feet; located on sheep Mountain, Ten Mile mining district, 12 miles from Breckenridge; course of vein, north and south, width 33 inches, pitch 60 degrees; nature of ore, carbonate of lead; value, 40 ounces silver per ton; 2 shafts 15 feet each.

Gilpin. Owned by James McNassar and John Moon; office Leadville; claim 150 by 1,500 feet; located on the south side of the Continental Divide between Clinton and Mayflower Gulches, 15 miles from Breckenridge; course of vein, 15 degrees east of south, width 4 feet, pitch 15 degrees south; nature of ore, galena and carbonate of lead; average value, $40 per ton; 1 shaft 20 feet deep, surface opening 30 feet in length.

Glacier Mountain Silver Mining Co., Summit County, Colorado. Organized under the laws of the State of New York, July 10, 1877; offices Montezuma, Colorado, and New York City; capital stock, $5,000,000, in 50,000 shares, $100 each; trustees: James B. Craig, John O. Bradford, Edward T. Bradford, Hugh J. Begley, George K. Clark, William H. Rand, and Robert L. Martin.

Gold Run Gold Washing Co. Incorporated April 16, 1878; capital stock, $500,000, in 50,000 shares, $10 each; office Denver; trustees: Lelon Peabody, Henry D. Steele, Oliver A. Whittemore, D.G. Blass, and N.W. Riker.

Gold Run Gravel Washing Co. Incorporated February 11, 1878; capital stock, $500,000, in 50,000 shares, $10 each; office Denver; trustees: Benjamin G. Blass, Nathan W. Riker, and Edward E. Hartwell.

Golden Bill. John W. Jacque, proprietor; office Leadville; claim 150 by 1,500 feet; located on Fletcher Mountain, Ten Mile mining district, 12 miles from Breckenridge; vertical vein, trending northeast and southwest, width 8 feet, pay vein 4 feet; nature of ore, free gold and iron pyrites; value, $100 gold per ton; 2 shafts, 10 and 15 feet in depth, respectively.

Golden Cord. Frank Probasco, proprietor; office Leadville; claim 150 by 1,500 feet; located on Fletcher Mountain, Ten Mile mining district, 12 miles from Breckenridge; course of vein, north and south, width 5 feet, pitch 15 degrees west; nature of ore, galena and copper pyrites; value $100 silver per ton; 2 adits, 15 and 20 feet in length, respectively.

Golden Eagle. John W. Jacque & Co., proprietors; office Leadville; claim 150 by 1,500 feet; located at head of Mayflower Gulch, Ten Mile mining district, 12 miles from Breckenridge; vertical vein, trending northeast and southwest, width 4 feet; nature of ore, free gold and iron pyrites; value, $100 gold per ton; 1 shaft 15 feet deep.

Golden Gate. John Fitzgerald, proprietor; office Leadville; claim 150 by 1,500 feet; located on Fletcher Mountain, Ten Mile mining district, 12 miles from Breckenridge; course of vein, north and south, width 4 feet, pitch 15 degrees west; nature of ore, galena and copper pyrites; value $100 silver per ton; 1 adit, 20 feet in length.

Great Republic. Owned by F.E. Webster, J.A. Conwell, and J.J. McKinney; office Montezuma; discovered 1877; located on Bear Mountain, Snake River mining district, 3 miles from Montezuma and 23 miles from Breckenridge; course of vein, north and south, width 3 feet, pay vein 7 inches, pitch 45 degrees; nature of ore, galena and copper pyrites; value, 92 ounces silver per ton; 2 tunnels, 70 and 115 feet in length, respectively, 1 level 65 feet in length; shaft house 25 by 25 feet.

Gray's Peak Mining and Milling Co. Incorporated November 6, 1878; capital stock, $1,000,000, in 10,000 shares, $100 each; office Montezuma; organized by Harvey Quicksell, Marcus A. Root, T. Norval Jeffries, and E.J. Mathews.

Griffin Gold and Silver Mining Co. Incorporated November 10, 1869; capital stock, $1,000,000, in 10,000 shares, $10 each; office Shawhunvill; trustees: Adam L. and John Shock, and Ziba Surles.

Herman. Edward C. Guibor and Herman Silver, proprietors; office Empire; claim 150 by 1,500 feet; discovered 1875; located on Glacier Mountain, Snake River mining district, 1 mile from Montezuma and 21 miles from Breckenridge; course of vein, northeast and southwest, width 5 feet, pay vein 18 inches, pitch 30 degrees; nature of ore, gray copper; average value, 160 ounces silver per ton; 1 shaft 46 feet deep, 1 level 87 feet in length; shaft house 18 by 42 feet; yield, $2,000.

Hoosier. John W. Jacque, proprietor; office Leadville; claim 150 by 1,500 feet; located on Robinson Mountain, Ten Mile mining district, 12 miles from Breckenridge; course of vein, northwest and southeast, width 4 feet, pitch 45 degrees southwest; nature of ore, galena and carbonate of lead; value, 200 ounces silver per ton; 1 shaft 15 feet deep.

Hoosier Gulch Placer. Owned by Henry M. Bostwick and William Bemrose; claim comprises 240 acres of placer ground; located in Hoosier Gulch, 10 miles from Breckenridge. It is worked by hydraulic pressure, and has a flume 1,000 feet in length; yield to date, $25,000.

Humbug Gulch Placer. Henry Schuster and Henry Stahl, proprietors; office Lincoln City; claim comprises 700 by 800 feet of ground; located near Georgia Gulch, Georgia mining district, 6 miles from Breckenridge.

Ingleside. Colorado Springs Prospecting and Mining Co., proprietors; office Colorado Springs; claim 150 by 1,800 feet; discovered 1875; located on Bartlett Mountain, Ten Mile mining district, 12 miles from Breckenridge; vertical vein, trending north and south, width 6 feet, pay vein 4 feet; nature of ore, galena and copper pyrites; value, 105 ounces silver per ton; 1 drift 120 feet in length.

Ingleside Mining Property. Owned by the Colorado Springs Prospecting and Mining Co.; office Colorado Springs; claim comprises 11 lodes; 150 by 1,500 feet each; located on the northeastern slope of Bartlett Mountain, Ten Mile mining district, 15 miles from Breckenridge.

Invincible Mining Co.'s Placer. Joseph P. Stell, T.P. Goodman, and William S. Cooper; office Breckenridge; claim comprises 30,600 square feet of placer ground, relocated 1878; located in French Gulch, Blue River mining district, 1 mile from Breckenridge; it is worked by hydraulic pressure, has 500 inches of water, and is furnished with a flume 500 feet in length.

Justice. John W. Jacque, proprietor; office Leadville; claim 150 by 1,500 feet; located on Robinson Mountain, Ten Mile mining district, 12 miles from Breckenridge, course of vein, northwest and southeast, width 4 feet, pitch 45 degrees southeast; nature of ore, galena and carbonate of lead; value, 200 ounces silver per ton; 1 shaft 11 feet deep.

La Plata. O.C. Knox, proprietor; office Colorado Springs; claim 150 by 1,500 feet; located on Bartlett Mountain, Ten Mile mining district, 12 miles from Breckenridge; vertical vein, trending north and south, width 4 feet, pay vein 3 feet; nature of ore, galena and copper pyrites; 1 adit 15 feet in length.

Legal Tender. Colorado Springs Prospecting and Mining Co., proprietors; office, Colorado Springs; claim 150 by 1,500 feet; located on Fletcher Mountain, Ten Mile mining district, 12 miles from Breckenridge; course of vein, north and south, width 5 feet, pay vein 3 feet, pitch 10 degrees west; nature of ore, galena; 1 shaft 15 feet deep, 1 tunnel 35 feet in length.

Little Annie. Griffith G. Jones, proprietor; office Montezuma; claim 150 by 1,500 feet; located on Glacier Mountain, near Montezuma, 20 miles from Breckenridge; course of vein, east and west, width 2 feet, pitch 20 degrees; nature of ore, galena; average value, 50 ounces silver per ton; 1 tunnel 40 feet in length, intersecting vein 30 feet from surface, 1 level 208 feet in length.

Little Frank. John W. Jacque, proprietor; office Leadville; claim 150 by 1,500 feet; located on Robinson Mountain, Ten Mile mining district, 12 miles from Breckenridge; course of vein, northwest and southeast, width 4 feet, pitch 45 degrees southeast; nature of ore, galena and carbonate of lead; value, 200 ounces silver per ton; 1 shaft 15 feet deep.

Little Giant. Robinson & Co., proprietors; office Leadville; claim 150 by 1,500 feet; located on Robinson Mountain, Ten Mile mining district, 12 miles from Breckenridge; course of vein, north and south, width 15 feet, pay vein 20 inches, pitch 45 degrees; nature of ore, carbonate of lead, value, 110 ounces silver per ton; 2 shafts 15 feet each.

Louise. Joseph B. Steel & Co., proprietors; office Breckenridge; claim 150 by 1,500 feet; discovered 1878; located on Gold Run, Gold Run mining district, 2 miles from Breckenridge; vertical vein, trending northeast and southwest, width 5 feet, pay vein 8 inches; nature of ore, galena; assay value, 30 ounces silver per ton; 1 shaft 15 feet deep.

Lucky. Owned by E. Lowe & Co.; office Breckenridge; claim 150 by 1,500 feet; located on Bartlett Mountain, Ten Mile mining district, 16 miles from Breckenridge; vertical vein, trending northeast and southwest, width 3 feet, pay vein 8 inches; nature of ore, galena; average value, $50 per ton; 1 adit 20 feet in length, with a 15-foot face.

Malabar. Discovered 1866; located on Ruby Mountain, Peru mining district, 25 miles from Breckenridge; course of vein, northeast and southwest, width 10 feet, pay vein 6 feet, pitch 20 degrees; nature of ore, galena; average value, 60 ounces silver per ton; aggregate length of drifts, 100 feet.

Manitou. Colorado Springs Prospecting and Mining Co., proprietors; office Colorado Springs; claim 150 by 1,500 feet; located on Bartlett Mountain, Ten Mile mining district, 12 miles from Breckenridge; vertical vein, trending north and south; nature of ore, galena and carbonate of lead; 3 adits 20 feet each.

Marxon. William L. and Clarence G. Sampson, proprietors; office Montezuma; claim 150 by 1,500 feet; discovered 1876; located on Glacier Mountain, Snake River mining district, 2 miles from Montezuma and 22 miles from Breckenridge; course of vein, northeast and southwest, width 40 feet, pay vein 3 feet, pitch 20 degrees; nature of ore, gray copper and galena; assay value, 60 to 800 ounces silver per ton; 2 shafts, 10 and 25 feet in depth, respectively.

May Bell. Owned by Frank and Perry Brandon; office Leadville; claim 150 by 1,500 feet; located on the west slope of Fletcher Mountain, Ten Mile mining district, 14 miles from Leadville and 16 miles from Breckenridge; course of vein, northwest and southeast, width 4 feet, pitch 10 degrees west; nature of ore, galena and iron pyrites; average value, 40 ounces silver per ton; 1 shaft 15 feet deep, surface opening 15 feet in length.

McNulty Gulch Placer. Owned by James McNassar, and Frank and Perry Brandon; located on Ten Mile creek, Ten Mile mining district, 15 miles from Breckenridge, comprises 60 acres of placer ground; worked by hydraulic pressure.

Metal Milling and Mining Co. Incorporated June 3, 1878; capital stock, $100,000, in 10,000 shares, $10 each; offices Montezuma, Colorado, and New York City; trustees: John A. Leslie, Edgar Harriott, Robert L. Martin, Ebenezer Rich, and N.H. Davis.

Minnie. Samuel F. Cary, proprietor; office Breckenridge; claim 150 by 1,500 feet; discovered 1877; located on Mineral Hill, Lincoln City mining district, 3 miles from Breckenridge; vertical vein, trending northeast and southwest, width 5 feet, pay vein 8 inches; nature of ore, galena; assay value, 90 ounces silver per ton; 2 shafts, 20 and 40 feet in depth, respectively.

Modoc. Bressler & Co., proprietors; office Montezuma; claim 150 by 1,500 feet; discovered 1869; located on Collier Mountain, 2 miles from Montezuma and 22 miles from Breckenridge; course of vein, northeast and southwest, width 1 foot; nature of ore, galena and bismuth silver; average value, 500 ounces silver per ton; 1 shaft 60 feet deep, 1 tunnel 100 feet in length; shaft house 14 by 16 feet.

Monte Rosa Mining Co. Incorporated January 27, 1876; capital stock, $10,000, in 1,000 shares, $10 each; office Colorado Springs;

incorporators, W.H. Slack, J.H. Mathers, R.C. Bristol, H.E. Freeman, George L. King, J.D. Crane, J.W. Leighton, and D.L. Welch.

Montezuma Mining and Smelting Co. Incorporated April 10, 1875; capital stock, $50,000, in 500 shares, $100 each; office Montezuma; incorporators: George L. and J.D. Scott and William Andrews.

Montezuma Silver Mining Co., Colorado. Organized under the laws of the State of New York, November 9, 1877; capital stock, $5,000,000, in 50,000 shares, $100 each; offices Montezuma, Colorado, and New York City; James B. Craig, president, J.O. Bradford, treasurer, Edwin T. Bradford, secretary; Hugh J. Begly, J.H. McChesney, Edwin S. Fowler and Robert L. Martin, trustees; Ebenezer Rich, agent.

Mount Ruby Tunnel. Owned by the Couara Mining Co.; office Denver; claim 1,500 by 3,000 feet; located on Ruby Mountain, Peru mining district, 5 miles from Montezuma and 23 miles from Breckenridge. This tunnel will intersect 20 lodes; it was commenced in 1876 and is now 250 feet in length.

Mount Ruby Tunnel Co. Incorporated May 5, 1877; capital stock, $1,000,000, in 100,000 shares, $10 each; office Denver; trustees: A. B. Robbins, Avery Gallup, Samuel Watson, Henry A. Bagley, B.R. Ragland and Herman J. Bagley. The property of this company is located on Mount Ruby, Peru mining district, 5 miles from Montezuma and 22 miles from Breckenridge. The tunnel has cut 2 lodes, is now 250 feet in length, and has for its object the intersection of all lodes running at right angles.

Mountain Pride. F.E. Webster, proprietor; office Montezuma; claim 75 by 1,500 feet; discovered 1869; located on Ruby Mountain, Peru mining district, 4 miles from Montezuma and 22 miles from Breckenridge; course of vein, north and south, width 4 feet, pay vein 8 inches; nature of ore, galena; average value, 60 ounces silver per ton; 1 shaft, 15 feet deep.

Mountain Stake. James M. Forshey, proprietor; office Leadville; claim 150 by 1,500 feet; located on Fletcher Mountain, Ten Mile mining district, 12 miles from Breckenridge; course of vein, north and south, width 4 feet, pitch 75 degrees west; nature of ore, carbonate of lead, value, 2 ounces gold and 100 ounces silver per ton; 1 shaft 25 feet deep, 1 adit 15 feet in length.

Napoleon. St. Lawrence Silver Mining Co., proprietors; office Canton, New York; claim 50 by 1,500 feet; discovered 1867; located on Glacier Mountain, near Montezuma, 20 miles from Breckenridge; vertical vein, trending northeast and southwest, width 5 feet, pay vein 1 foot; nature of ore, galena and zinc blende; average value, 150 ounces silver per ton; 1 drift 350 feet in length and 150 feet from surface; yield, $1,500.

National Treasury. Discovered 1865; located in Snake River Gulch, Peru mining district, 25 miles from Breckenridge; course of vein, northeast and southwest, width 5 feet, pay vein 6 inches, pitch 20 degrees; nature of ore, galena; average value, 175 ounces silver per ton; 1 shaft 80 feet deep.

No. 1. Owned by the Couara Silver Mining Co; offices Peru and Denver, Colorado; claim 150 by 1,500 feet; discovered 1875; located on the southwestern slope of Ruby Mountain, Peru mining district, 7 miles from Montezuma and 23 miles from Breckenridge; course of vein, northeast and southwest, width 5 feet, pay vein 3 inches, pitch 45 degrees northwest; nature of ore, galena and sulphurets; average value 480 ounces silver per ton; 1 surface opening 20 feet in length.

No. 2. Owned by the Couara Silver Mining Co., claim 150 by 1,500 feet; discovered 1875; located on Ruby Mountain, Peru mining district, 7 miles from Montezuma and 23 miles from Breckenridge; course of vein, northeast and southwest, width 12 feet, pay vein 6 inches, pitch 35 degrees northwest; nature of ore, galena and zinc blende; assay value, 155 ounces silver per ton; 1 drift 12 feet in length, reached by the Mount Ruby Tunnel, at a distance of 170 feet from the mouth, and intersected at a depth of 90 feet.

No. 4. Silver Gate Mining Co., proprietors; office Cincinnati, Ohio; claim 150 by 1,500 feet; located on Glacier Mountain, 1 mile from Montezuma, and 21 miles from Breckenridge; vertical vein, trending north and south, width 7 feet, pay vein 3 feet; nature of ore, zinc blende and galena; assay value, 86 ounces silver per ton; intersected by the Chenango Tunnel, 200 feet from the surface, 1 level 60 feet in length.

No. 5. Owned by the Boston Mining Co.; claim 150 by 1,500 feet, patented; discovered 1875; located on Glacier Mountain, near town of Saints John, 1 mile from Montezuma and 20 miles from Breckenridge; vertical vein, trending northeast and southwest, width 4 feet, pay vein 1 foot; nature of ore, galena and zinc blende; intersected by the Comstock Tunnel, at a depth of 650 feet, drifts aggregating 2,100 feet in length.

No. 6. Silver Gate Mining Co., proprietors; office Cincinnati, Ohio; claim 150 by 1,500 feet; located on Glacier Mountain, 1 mile from Montezuma and 21 miles from Breckenridge; vertical vein, trending north and south, width 2 feet, pay vein 7 inches; nature of ore, galena; average value, 42 ounces silver per ton; intersected by Chenango Tunnel 225 feet from surface.

No. 7. Owned by the Boston Mining Co.; offices Saints John, Colorado, and Boston, Massachusetts; claim 150 by 1,500 feet, patented; discovered 1875; located on Glacier Mountain, 1 mile from Montezuma and 20 miles from Breckenridge; course of vein, northeast and southwest, width 5 feet, pay vein 2 feet, pitch 40 degrees; nature of ore, galena and zinc blende; intersected by the Comstock Tunnel at a depth of 700 feet, aggregate length of drifts 1,400 feet.

Norse. Owned by the Couara Silver Mining Co.; offices Peru and Denver, Colorado; claim 150 by 1,500 feet; discovered 1875; located on the northeast slope of Mount Ruby, Peru mining district, 7 miles from Montezuma and 23 miles from Breckenridge; course of vein, northwest and southeast, width 3 feet, pay vein 6 inches, pitch 10 degrees east; nature of ore, sulphurets and ruby silver; assay value, 20 to 26,000 ounces silver per ton; 1 shaft 40 feet deep, 1 adit 90 feet in length.

North Star. D.W. Wiley & Co., proprietors; office Montezuma; claim 150 by 1,500 feet; discovered 1864; located on Collier Mountain, 1 mile from Montezuma and 21 miles from Breckenridge; course of vein, northeast and southwest, width 6 feet, pay vein 10 inches, pitch 20 degrees; nature of ore, galena and gray copper; average value, 300 ounces silver per ton; main shaft 30 feet deep and 2 other shafts aggregating 30 feet, 3 levels 20, 25, and 60 feet in length, respectively; yield, $500.

North Star. John Fitzgerald & Co., proprietors; office Leadville; claim 150 by 1,500 feet; located on Fletcher Mountain, Ten Mile mining district, 12 miles from Breckenridge; course of vein, north and south, width 2 feet, pitch 15 degrees west; nature of ore, galena and copper pyrites; value, $100 silver per ton; 1 adit 25 feet in length with a 20-foot face.

Occidental. Owned by the Colorado Springs Prospecting and Mining Co.; office Colorado Springs; claim 150 by 1,500 feet; located on Bartlett Mountain, Ten Mile mining district, 16 miles from Breckenridge; course of vein, southwest and northeast, width 6 feet, pay vein 3 feet, pitch 10 degrees east; nature of ore, galena and gray copper; 1 shaft 10 feet deep; worked through a tunnel from an adjoining claim.

Olsen Placer. Henry Shuster and M.J. Cole, proprietors; office Lincoln City; claim comprises 8,000 square feet of placer ground; located on South Swan River, Georgia mining district, 6 miles from Breckenridge.

Ondawa Mining Co. Incorporated July 26, 1876; capital stock, $50,000, in 1,000 shares, $50 each; office Denver; incorporators: Henry K. Steele, Harvey P. Platt, Edwin T. Platt, Charles B. Phillips, and John Fitch.

Orion. Owned by the Colorado Springs Prospecting and Mining Co.; claim 150 by 1,500 feet; located on Bartlett Mountain, Ten Mile mining district, 16 miles from Breckenridge; course of vein, north and south, width 3 feet; nature of ore, galena and copper pyrites; 1 tunnel, 40 feet in length.

Oro-Fino. Owned by J.H. Follett & Co., office Breckenridge; claim 150 by 1,500 feet; located on Surles Point near the mouth of Surles Gulch, 14 miles from Breckenridge; course of vein, north and south, width 12 feet, pitch 25 degrees northwest; nature of ore, honeycomb quartz and iron pyrites; average value, $30 per ton; 1 tunnel 40 feet in length.

P.G. F.E. Webster, proprietor; office Montezuma; claim 150 by 1,500 feet; discovered 1875; located on Collier Mountain, 3 miles from Montezuma and 23 miles from Breckenridge; vertical vein, trending northeast and southwest, width 4 feet, pay vein 18 inches; nature of ore, galena and ruby silver; average value, 150 ounces silver per ton; 1 drift 40 feet in length.

Pacific. Owned by J.H. Follett & Co.; office Breckenridge; claim 150 by 1,500 feet; located on Surles Point, near the mouth of Surles Gulch, Ten Mile mining district, 14 miles from Breckenridge; course of vein, north and south, width 12 feet, pitch 75 degrees; nature of ore, honeycomb quartz and iron pyrites; average value $30 per ton.

Pacific. William Bemrose, proprietor; office Alma; claim 150 by 1,500 feet; discovered 1876; located on North Star Mountain, Pollock mining district, 10 miles from Breckenridge; course of vein, northeast and southwest, width 6 feet, pitch 15 degrees; nature of ore, oxide of iron; average value, $20 gold per ton; 1 drift 40 feet in length and 25 feet from surface; yield, $500.

Pilgrim's Folly. Grigg & Co., proprietors; office Montezuma; claim 150 by 1,500 feet; located on Tariff Mountain, Peru mining district, 5 miles from Montezuma and 23 miles from Breckenridge; vertical vein, trending east and west, width of pay vein, 4 inches; nature of ore, galena and ruby silver; 1 shaft 25 feet deep.

Potosi. Edward C. Guibor & Co., proprietors; office Empire; claim 50 by 800 feet, patented; discovered 1865; located on Glacier Mountain, 2 miles from Montezuma and 22 miles from Breckenridge; course of vein, northeast and southwest, width 12 feet, pitch 45 degrees; nature of ore, sulphurets and ruby silver; average value, 230 ounces silver per ton; 2 shafts, 14 feet in depth each.

Pride of the West. James M. Forshey, proprietor; office Leadville; claim 150 by 1,500 feet; located at head of Clinton Gulch, Ten Mile mining district, 12 miles from Breckenridge; course of vein, north and south; width 4 feet, pitch 75 degrees south; nature of ore, carbonate of lead; 2 shafts, 18 and 30 feet in depth, respectively.

Radical and Radical Jr. Montezuma Silver Mining Co., proprietors; office Montezuma; claim 150 by 3,000 feet; located on Teller Mountain, 2 miles from Montezuma and 22 miles from Breckenridge; course of vein, northeast and southwest, width 8 feet, pay vein 3 feet; nature of ore, galena and gray copper; assay value, 1,320 ounces silver per ton; 2 shafts, 80 and 110 feet deep, respectively, 1 tunnel 55 feet in length, aggregate length of levels 185 feet.

Ready Pay. D.L. Southworth & Co., proprietors; office Montezuma; claim 150 by 1,500 feet, patented; discovered 1867; located on Teller Mountain, 4 miles from Montezuma and 24 miles from Breckenridge; course of vein, northeast and southwest, width 65 feet, pay vein 2 feet, pitch 40 degrees; nature of ore, gray copper and galena; average value, 200 ounces silver per ton; 1 tunnel 25 feet, and 1 level 15 feet in length; shaft house 12 by 14 feet.

Register. Willard Teller, proprietor; office Denver; claim 50 by 1,500 feet, patented; located on Glacier Mountain, near town of Montezuma, average width of vein 3 feet, pay vein 1 foot; nature of ore, galena; average value, $80 per ton; main shaft 140 feet deep.

Resumption. Owned by S.C. Emery & Co.; office Leadville; claim 300 by 1,500 feet; discovered 1878; located in Mayflower Gulch, Ten Mile mining district, 12 miles from Breckenridge; course of vein, north and south, width 15 feet, pay vein 3 feet, pitch 20 degrees east; nature of ore, galena and sulphurets; average value, 160 ounces silver per ton; 1 drift 60 feet in length and 30 feet from surface.

Rising Sun. Owned by Recen, Allison, and others; claim 50 by 1,500 feet; discovered 1878; located on Surles Point, Ten Mile mining district, 12 miles from Breckenridge; course of vein, north and south, pitch 70 degrees; nature of ore, carbonate of lead; assay value, 28 ounces silver per ton; 1 drift 60 feet in length.

Rustler. William L. and Clarence G. Sampson, proprietors; office Montezuma; claim 150 by 1,500 feet; discovered 1875; located on Glacier Mountain, Snake River mining district, 2 miles from Montezuma and 22 miles from Breckenridge; course of vein, northeast and southwest, width 12 feet, pay vein 5 inches, pitch 25 degrees; nature of ore, gray copper and brittle silver; assay value, 200 to 12,000 ounces silver per ton; 1 drift 60 feet in length and 40 feet from surface.

Ryan Gulch Placer. Roby & Silverthorn, proprietors; office Breckenridge; claim comprises 800 by 2,800 feet of placer ground located on the Blue River, 13 miles from Breckenridge.

Sampson. William L. and Clarence G. Sampson, proprietors; office Montezuma; claim 150 by 1,500 feet; discovered 1877; located on Glacier Mountain, 2 miles from Montezuma and 22 miles from Breckenridge; course of vein, northeast and southwest, width 8 feet, pay vein 8 inches, pitch 30 degrees; nature of ore, gray copper and galena; average value, 100 ounces silver per ton; 1 drift 120 feet in length and 60 feet from surface; yield, $400.

Seventy-Six. Owned by Frank and Perry Brandon; office Leadville; claim 150 by 1,500 feet; located at the head of Clinton Gulch, on the northwest slope of Bartlett Mountain, 15 miles from Breckenridge; vertical vein, trending northeast and southwest, width 4 feet; nature of ore, iron pyrites and brittle silver; value, $300 per ton; 1 drift and 1 surface opening, 10 and 15 feet in length, respectively.

Seventy-Eight. Robinson & Co., proprietors; office Leadville; claim 150 by 1,500 feet; located at base of Robinson Mountain, Ten Mile mining district, 12 miles from Breckenridge; course of vein, north and south, width

4 feet, pitch 45 degrees; nature of ore, carbonate of lead; value, $200 silver per ton; 1 shaft 15 feet deep.

Silver Blossom. Frank Probasco and Calvin Blossom, proprietors; office Topeka, Kansas; claim 150 by 1,500 feet; located at head of Mayflower Gulch, Ten Mile mining district, 15 miles from Breckenridge; course of vein, north and south, width 4 feet, pitch 15 degrees east; nature of ore, galena and carbonate of lead; value, $100 silver per ton; 1 adit 13 feet in length.

Silver Lake. William Bemrose, proprietor; office Alma; claim 150 by 1,500 feet; discovered 1878; located on North Star Mountain, 10 miles from Breckenridge; course of vein, northeast and southwest, width 4 feet, pay vein 2 feet, pitch 15 degrees; nature of ore, oxide of iron; assay value, 1 to 7 ounces gold and 80 ounces silver per ton; 1 shaft 15 feet deep.

Silver Queen. John W. Jacque, proprietor; office Leadville; claim 150 by 1,500 feet; located at base of Robinson Mountain, Ten Mile mining district, 12 miles from Breckenridge; course of vein, northwest and southeast, width 4 feet, pitch 45 degrees; nature of ore, galena and carbonate of lead; value, 200 ounces silver per ton; 1 shaft 15 feet deep, 1 adit 25 feet in length.

Silver Spike. Colorado Springs Prospecting and Mining Co., proprietors; office Colorado Springs; claim 150 by 1,500 feet; discovered 1875; located on Bartlett Mountain, Ten Mile mining district, 12 miles from Breckenridge; course of vein, northwest and southeast, width 6 feet, pay vein 1 foot, pitch 10 degrees; nature of ore, galena and copper pyrites; value, 50 ounces silver per ton; 1 drift 20 feet in length.

Silver Wing. St. Lawrence Silver Mining Co., proprietors; office Canton, New York; claim 50 by 3,000 feet, patented; discovered 1867; located on Glacier Mountain, Snake River mining district, near Montezuma, 20 miles from Breckenridge; course of vein, northeast and southwest, width 5 feet; pay vein 2 feet, pitch 15 degrees; nature of ore, zinc, galena, and ruby silver; average value, 150 ounces silver per ton; 1 shaft 50 feet, 1 tunnel 100 feet, and 1 drift 400 feet in length; yield, $10,000.

Sisler Placer. John Sisler, proprietor; office Breckenridge; claim comprises 1,600 by 2,400 feet of placer ground; located in French Gulch, Upper Blue mining district, 1 mile from Breckenridge.

Speckled Quail. Joseph P. Steel, proprietor; office Breckenridge; claim 150 by 1,500 feet; discovered 1878; located in Lincoln City mining district, 5 miles from Breckenridge; vertical vein, trending northeast and southwest, width 6 feet, pay vein 14 inches; nature of ore, galena; 1 shaft 30 feet deep.

Springfield Mining, Ditching and Fluming Co. Incorporated June 23, 1871; capital stock, $20,000, in 40 shares, $500 each; offices Breckenridge, Colorado, and Springfield, Massachusetts; incorporators: John A. Willoughby, A.M. Greenleaf, and T.H. Fuller.

St. Cloud. D.W. Willey and T.E. Clery, proprietors; office Montezuma; claim 150 by 1,500 feet; discovered 1872; located on Glacier Mountain, 2 miles from Montezuma and 22 miles from Breckenridge; course of vein, east and west, width 10 feet, pay vein 2 feet, pitch 15 degrees; nature of ore, gray copper and galena; assay value, 500 ounces silver per ton; 2 shafts, 10 and 25 feet in depth, respectively; shaft house 12 by 14 feet.

St. Louis. John W. Jacque, proprietor; office Leadville; claim 150 by 1,500 feet; located on Robinson Mountain, Ten Mile mining district, 12 miles from Breckenridge; course of vein, northwest and southeast, width 4 feet, pitch 45 degrees southeast; nature of ore, galena and carbonate of lead; value 200 ounces silver per ton; 1 shaft 15 feet deep.

Star of the West. Owned by Frank Bunholzer, F. Aicher, and J. Harvat; office Georgetown; claim 150 by 1,500 feet; discovered 1873; located on Taylor Mountain, Snake River mining district, 23 miles from Georgetown; vertical vein, running north and south, width 15 feet, pay vein 5 feet; nature of ore, sulphurets; average value, 15 ounces gold and from 200 to 600 ounces silver per ton; development, 1 adit 125 feet in length.

Stillson Patch Placer. John Sisler, proprietor; office Breckenridge; claim comprises 800 by 4,400 feet of placer ground; located in French Gulch, Upper Blue mining district, 1 mile from Breckenridge.

Sukey. William W. Webster, proprietor; office Montezuma; claim 50 by 3,000 feet; discovered 1864; located on Glacier Mountain, Snake River mining district, near Montezuma, 20 miles from Breckenridge; course of vein, northeast and southwest, width 10 feet, pay vein 3 feet, pitch 10 degrees; nature of ore, galena and zinc; average value, 175 ounces silver per ton; 1 drift 110 feet in length and 50 feet from surface; yield, $5,000.

Swan River Placer Mining Co. Incorporated May 1, 1878; capital stock, $100,000, in 1,000 shares, $100 each; office Central City; incorporators, Otto Eckhardt, Frederick Wendt, Israel B. Ward, George Ebert, and Frederick Nieninger.

Teller. Montezuma Silver Mining Co., proprietors; office Montezuma; claim 150 by 1,600 feet, patented; discovered 1876; located on Glacier Mountain, 1 mile from Montezuma and 21 miles from Breckenridge; course of vein, northeast and southwest, width 40 feet, pay vein 3 feet, pitch 20 degrees; nature of ore, galena and gray copper; assay value, 60 to 800 ounces silver per ton; 1 shaft 30 feet deep.

Thirty Per Cent. Colorado Springs Prospecting and Mining Co., proprietors; office Colorado Springs; claim 128 by 1,500 feet; discovered 1875; located on Bartlett Mountain, Ten Mile mining district, 12 miles from Breckenridge; course of vein, northwest and southeast, width 4 feet, pitch 10 degrees east; nature of ore, galena and gray copper; 1 drift 36 feet in length.

Tiger. Owned by D.W. Willey, T.E. Clery, and others; office Montezuma; claim 150 by 1,500 feet; discovered 1864; located on Glacier Mountain, 1 mile from Montezuma and 21 miles from Breckenridge; vertical vein, trending northeast and southwest, width 5 feet, pay vein 8 inches; nature of ore, galena, ruby and brittle silver; average value, 500 ounces silver per ton; 6 shafts aggregating 90 feet in depth; shaft house 12 by 14 feet.

Tiger Extension. D.W. Willey & Co., proprietors; office Montezuma; claim 150 by 1,500 feet, patented; discovered 1864; located on Glacier Mountain, 1 mile from Montezuma and 21 miles from Breckenridge; vertical vein, trending northeast and southwest, width 10 feet, pay vein 6 feet; nature of ore, galena and gray copper; average value, 150 ounces silver per ton; 1 shaft 20 feet deep, 1 drift 52 feet in length and 40 feet from surface; yield, $1,000.

Troy Bar Diggings. Owned by B.F. and J.H. Follett; office Breckenridge; claim comprises 260 acres of placer ground worked by hydraulic pressure and located on Ten Mile Creek, Ten Mile mining district, 13 miles from Breckenridge.

Tunnel Lode No. 1. Owned by the Couara Silver Mining Co.; offices Peru and Denver, Colorado; claim 150 by 1,500 feet; discovered 1875;

located on Ruby Mountain, Peru mining district, 7 miles from Montezuma, and 23 miles from Breckenridge; course of vein, northeast and southwest, width 4 feet, pay vein 6 inches, pitch 30 degrees northwest; nature of ore, galena; average value, 25 ounces silver per ton and 20 per cent lead; 1 drift 20 feet in length; reached by the Mount Ruby Tunnel at a distance of 41 feet from the mouth and intersected at a depth of 25 feet.

Vandemore. Owned by Middleton W. and W.K. Smith and Jasper Ward; office Alma; claim 150 by 1,500 feet; located on North Star Mountain, Silver lake mining district, 8 miles from Alma; course of vein, northeast and southwest, width 6 feet, pitch 10 degrees southeast; nature of ore, oxide of iron; development, 1 shaft 25 feet deep.

Vanderbilt. William King and J.D. Rankin, proprietors; office Breckenridge; claim 150 by 1,500 feet, patented; discovered 1870; located in Pollock mining district, 9 miles from Breckenridge; course of vein, northeast and southwest, width 9 feet; nature of ore, oxide of iron; average value, $36 per ton; 1 drift 20 feet in length and 100 feet from surface.

Whitesides. G.G. Jones and Co., proprietors; office Montezuma; claim 150 by 1,500 feet; discovered 1874; located on Glacier Mountain, 2 miles from Montezuma and 22 miles from Breckenridge; course of vein, northeast and southwest, width 4 feet, pay vein 15 inches, pitch 40 degrees; nature of ore, galena; average value, 50 ounces silver per ton; 1 shaft 25 feet deep, 1 tunnel 25 feet in length.

Wild Wagoner. Lucius Bradshaw and Perry Brandon, proprietors; office Leadville; located 1874; located on Fletcher Mountain, Ten Mile mining district, 12 miles from Breckenridge; vertical vein, trending northeast and southwest, width 3 feet, pay vein 30 inches; nature of ore, galena and oxide of iron; value, 1½ ounces gold and 15 ounces silver per ton; 1 shaft 20 feet deep, 1 adit 30 feet in length.

Wisconsin. Owned by Griffith G. Jones, E. Jones, and J. Thomas; office Montezuma; claim 150 by 1,500 feet; discovered 1874; located on Glacier Mountain, 2 miles from Montezuma and 22 miles from Breckenridge; course of vein, northeast and southwest, width 20 inches, pitch 25 degrees; nature of ore, galena and gray copper; average value, 90 ounces silver per ton; 1 shaft 25 feet deep.

Wool Sack. Joseph P. Steele, proprietor; office Breckenridge; claim 150 by 1,500 feet; discovered 1878; located in Lincoln City mining district, 5 miles from Breckenridge; vertical vein, trending northeast and southwest, width 3 feet, pay vein 5 inches; nature of ore, galena; 2 shafts 30 feet each.

Summit County Ore Mills

Boston Mining Co.'s Mill. Located in town of Saints John, 1 mile from Montezuma and 20 miles from Breckenridge; has a capacity for treating 25 tons of ore in 24 hours, and contains an engine, Blake and Dodge crushers, Cornish rollers, 10 stamps, jigs, tables, furnaces, etc.

Sisapo Mill. John H. Yonley, of Montezuma, proprietor; located in Montezuma; has a capacity for treating 20 tons of ore in 24 hours.

Sukey Mill. William W. Webster, of Montezuma, trustee; located near Montezuma; has a capacity for treating 3 tons of ore daily.

WELD COUNTY

The 1879 directory listed the following coalmines and corporations in Weld County.

Boulder Valley Coal Co. Capital stock, $3,000,000; John D. Perry, president, John W. Hannah, superintendent; offices Erie and Denver, Colorado. This company owns 5,120 acres of land, located in town of Erie, on Boulder Valley Railroad, 24 miles from Denver. The mine is developed by 2 inclines, 200 feet each, and 8 entries aggregating 3,600 feet.

Co-operative Mining Co. Incorporated April 7, 1875; capital stock, $5,000, divided into 18 shares; office Erie, Colorado; organized by John Mackin, Joseph L. Griffith, John McDowell, Hiram Marfell, Joseph George, William Loughlin, Robert Loughlin, James McDowell, John R. Thomas, and others.

Greeley Coal Mining Co. Incorporated May 28, 1870; capital stock, $10,000, in 1,000 shares, $10 each; office Greeley; organized by Russell Fisk, William B. Plato, William Inderlied, Thomas Rush, George Pyburn, David Boyd, Samuel F. French, Joseph S. Williams, and Edward T. Nicholls.

Mitchell Coal Mine. Joseph Mitchell, proprietor; property comprises 80 acres of coal land, located 1 mile south of Erie, on Boulder Valley Railroad, 24 miles from Denver; developed by a main shaft 200 feet, an air shaft of the same depth, and 4 entries aggregating 900 feet in length; total yield, 35,000 tons.

St. Vrain Mining Co. Organized under the laws of the State of NewYork, March 22, 1877; capital stock, $20,000, in 2,000 shares, $100 each; offices Denver, Colorado, and 33 Wall Street, New York; trustees: R.S. Grant, George N. Wheeler, and John Child; I.H. Grant agent.

INDEX

ABBEY, Charles 35
ABBOT, C.S. 79
ABBOTT, E.C. 295 298 Jacob J. Jr. 242 251 James W. 242 251 William H. 261
ABEEL, Jacob H. 202
ABERTO [OBERTO?], Antonio 376
ABODIE, E.R. 94
ADAMS, -- 156 364 Frank 239 Henry 239 251 J. H. 84 R.F. 367 368 W.A. 241 243 248 250
ADUDDEL, Dr. Robert G. 165 188 194
AGGERS, George L. 25 65
AICHER, F. 459
AIRY, Joseph 65
AKERS, I.N. 233
ALBRIGHT, Ferdinand C. 19 41 62
ALBRO, O.M. 205 217
ALDER, J.W. 176
ALDERMAN, Lamartine 57
ALDERSON, John 18 42
ALEXANDER, Aden 44 263 266 270 271 281 289 354 439
ALKIRE, M.J. 301 307 418
ALLEN, Daniel B. 351 Harry 351 S. 391 Samuel P. 78 130
ALLISON, -- 457 F.H. 84
ALMER, Harry 95
ALSEBROOK, -- 229
ALVATER, Henry 214
ALVORD, C.C. 61 334
AMES, Frances E. 51 George C. 63 Oliver 67 R.A. 138
AMM, Philip T. 359
ANDERSON, Charles W. 216 Henry 100 J.C. 316 T.J. 233 W.H. 79 William 452
ANDRE [ANDREWS?], Francis L. 79 130 139
ANDREWS, C. 296 435 Francis 91

J.B. 312
ANGIER, H.G. 345 J.D. 245 431 Joel N. 245 431
ANGLE, C.W. 288 H.G. 5 354
ANKENY, -- 23 71
ANKNEY, Dr. 372
ANTHONY, T.D. 281
ANTISELL, Thomas 79
ARCHER, James 274
ARMOR, William 94
ARMSTRONG, -- 23 71 E.S. 390 435
ARNETT, Anthony 8 11 13 16 21 22 30 33 34 41 Mary G. 64
ARNOLD, Mrs. A. 53 George 102 J.L. 305 313
ASH, H.F. 403 H.T. 416
ASHLEY, Eli M. 80 128 211
ATKINS, Horace H. 146 175 192 210 215
ATKINSON, H.M. 198
AUGUST, A.J. 138 Anthony J. 114
AUSTIN, James A. 10 M.A. 160 O.O. 7 S.S. 313
AUTRIE, George 302
AVERY, Alfred D. 157
AYER, Charles 232 268
AYERS, Charles 445 Cyrus B. 13 40
BABBITT, John F. 79
BACK, Jack 409 John 259
BACON, Corbit 216
BAEDER, Samuel 19 31 38 63 66 70
BAER, Thomas 107
BAGLEY, H.A. 444 H.J. 444 Henry A. 80 452 Herman J. 452
BAILEY, D.S. 144 John W. 160 Melvin 43
BAKER, A.H. 167 A.P. 204 Charles 155 Daniel 153 J. 168 231 Jasper 153 L.C. 364 Thomas A. 15

INDEX

Thomas H. 7 363 William 87 101 103 118 123 142
BALDWIN, Charles P. 80 86 95 102 112 113 145 William H. 80 102 113
BALES, Charles 377
BALL, David J. 81 130
BALLARD, -- 192 A.J. 159 O. Jr. 92
BALLOU, -- 440 Franklin 149 150 Leonard S. 441
BAMBERRY, Charles 188
BANGS, -- 159
BARBER, -- 74
BARCKHAMER, A.M. 14
BARDINE, Joseph 341 361
BARKE, T.S. 406
BARKER, C.S. 294 James W. 111 William J. 168 179 182 210
BARLOW, G.S. 157
BARNARD, A.W. 124 William H. 263 290
BARNES, David 194 William H. 256
BARNSDALL, William Jr. 431
BARNUM, T.J. 245
BARRETT, G.H. 84 George H. 84 J.J. 239 240 244 J.N. 242 J.S. 109
BARROWS, Henry A. 52 J.P. 316 317 322
BARTLETT, E.N. 268 289 Francis 441 George W. 338 J.C. 310 L.H. 51
BARRON, P.H. 356
BASEY, Robert 232
BASH, A.W. 10
BASS, -- 344 Elisha 337 341
BASSICK, Edmund C. 156
BATCHELDER, George H. 22 27 51 67
BATES, -- 168 Joseph E. 2 Julius E. 397

BAUM, J. 238
BAXTER, -- 168 Enos K. 37 61 82 147 180 Richard 108
BAYLE, Hugh E. 167
BEACH, E.C. 213 Edwin N. 57 J. 407
BEAN, A.J. 12 43
BEARCE, J.F. 87
BEASLEY, Andrew J. 71
BEATTIE, Harry 318 J.A. 325
BEATY, H.C. 248
BEAUCHAMP, Thomas 342 Thomas B. 340 351
BEAUDRY, Frederick B. 412
BEAUPRE, Pierre 103
BEBEE, M.F. 144 219 Mark F. 181 221
BECK, Isaac T. 160 Joseph 331 427 William 364 371 William E. 60
BECKER, Henry 188 Theodore H. 176 203
BEDAL, Isaac 63 J. 5
BEDEN, Cecil 91
BEEBE, Joseph A. 78
BEEKER, Peter 364
BEERS, W. 334 William M. 244
BEGLEY, Hugh J. 447 452
BEGSBEE, Joseph 67
BEIDLE, Adam 310
BEITON, Dr. 418
BELCHER, A.J. 289
BELDEN, -- 167 169 170 183 187 193 206 214 David D. 169 170 206 242 251
BELL, J.T. 371 Robert W. 308 W. 392
BELLAMY, Charles T. 111
BEMROSE, William 337 448 455 458
BENBOW, T.A. 339
BENDER, -- 139
BENEDICT, James F. 109

INDEX

BENNE, A.S. 96
BENNET, R. 333
BENNETT, A.S. 102 Erastus S. 2
   Lucy J. 2 Stephen D.N. 231
BENSLEY, John R. 400
BENSON, James 272
BENTON, J. 333
BERDELL, -- 267 Charles P. 47
BERGEN, Benjamin F. 34
BERGER, Bros. 59 George B. 275
   William B. 63 74
BERGHOFF, Anthony P. 408
BERGMAN, John 287
BERNARD, Albert 239 332 377
   387 391 421435 Ambrose W. 129
   Benjamin 273 279 283 James J.
   239 332 377 387 391 421 435
   William R. 245 251 252
BERNSTEIN, Levy 107
BERRY, George 119 123 238
BERRYMAN, John R. 222
BEST, John D. 340 341 356
BETTMAN, B. 279 B. 291
BETTS, James A. 344
BICKER, Peter 405
BIEDELL, Mark 370 376
BIEGER, H. 353
BIGGER, Andrew M. 261
BIGGS, John A. 91
BIGLOW, Charles E. 146 Joshua R. 146
BILL, Charles K. 344 352
BILLINGS, John 300 311
BILLS, Albert 22 25 28 71
BILYIEN, C.J. 371
BINCKLEY, George M. 366
BINFORD, Edward J. 33 53 73
BIRCH, William W. 39
BISHOP, S.B. 85 Samuel P. 99
BIXBY, Amos 22 Anne 48
BLACKMAN, August 117
BLAIR, Thomas 372 Woodbury 263

290
BLAISDALE, Frederick 370 411
   413 436
BLAKE, -- 220 Frank O. 20 32 39
   H. 308 Henry 51 Henry T. 160
   John C. 38 51 Orris 20 28 32 37
   39 51 William 40
BLAKESLEE, H.B. 16
BLASS, Benjamin G. 447 D.G. 447
BLATCHFORD, S.M. 91
BLEVIN, Giles 39
BLODGETT, A.D. 84
BLOSS, Benjamin G. 63
BLOSSOM, Calvin 458 Charles F. 155
BLOWER, Richard 33
BLUNT, George P. 238
BLYTHE, H.F. 324
BOARD, -- 256
BOCK, Joseph 236 431
BODEN, Austin 38 William 38
BOGGS, George 240
BOH, Joseph 106
BOHM, A.V. 275
BOHRAM, J. 183
BOIES, J.K. 159
BOLEY, A.S. 232 D.C. 232
BOLTHOFF, Henry 138 196 203 204 217
BOND, Hiram G. 2 217 N.J. 353
   Richard 214
BONTECON [or, BONTICON],
   Elijah W. 340 359
BOONE, W.C. 131
BOOT, J.W. 93
BORDEN, -- 266 267 270 278 282 289
BORDERS, John S. 340 342 344 351 354
BOREHAM, J.S. 88 Jeremiah 168
BORGESON, John 40
BORMAN, A.H. 84

467

INDEX

BOSTWICK, Henry M. 448 Joseph H. 194
BOTHSELL, -- 160
BOTTOMS [BUTTON?], Granville M. 47
BOULWARE, M. 388
BOUN, J.H. 96
BOUTON, J.C. 242
BOWDEN, Edward 82 144 Edwin 117 127
BOWEN, A. 426 Charles C. 408 Sayles J. 79 Thomas M. 238 365 366 367
BOWKER, Dr. Seth B. 61
BOWMAN, Thomas E. 437 W.R. 426
BOYCE, George 153 157 162
BOYD, Alexander 43 Col. 160 David 463 James H. 12 72
BOYER, Henry 139
BRACE, E. 427
BRADBURY, C.C. 46 51
BRADEN, William 94
BRADFORD, Edward T. 447 452 J.O. 452 John M. 266 John O. 447
BRADLEY, E.C. 321 James B. 168 Lemuel M. 168 S.H. 169
BRADSHAW, Lucius F. 265 277 287 289 461 Royal C. 378
BRADY, Thomas 23
BRAGG, W.C. 167
BRAINARD, Wesley 14
BRANCH, Harry A. 159
BRANDON, Frank 440 451 457 Perry 440 451 457 461
BRANDT, Ferdinand Henry 364
BRANTLEY, George A. 308
BRAUN, Theodore F. 158
BRAYTON, Charles A. 408
BREASLEY, Elias 272
BRECKINRIDGE, Thomas A. 393 Thomas E. 331

BREECE, Estate 354 Sullivan D. 282
BREESE, Edward 273
BRENNAN, James 348
BRESSLER, -- 451
BRETT, Matthew D. 57
BREWER, John R. 67 441 Perry 59
BREWSTER, W.J. 372
BRIEGLIEB, C. 272 Conrad 339 Max 272
BRIGGS, Bros. 191 227 Charles H. 172 212 George W. 172 J.L. 232
BRIGHT, Richard J. 144 147
BRINK, J.H. 330 397 427
BRISTOL, R.C. 452 Richard C. 381
BRITTENSTEIN, Henry 273 278 284 288
BRITTON, J. 295 297
BROAD, W.E. 259
BROADBENT, Samuel 188
BROCK, C. 294 T.V. 389
BROCKETT, J.W. 242
BROLASKIE, E. 303
BRONSON, H. Marcus 39 100 132 Marcus 431
BROOKS, James G. 354 N. 409
BROOKES, M.V. 379 Max 286
BROOME, James E. 344
BROWN, Albert 30 Arthur A. 91 B. Peyton 111 D. 393 D.F. 26 28 Edward M. 426 F.H. 49 G. 301 George W. 24 Henry C. 24 91 J. Warren 90 John W. 91 P. 318 R.T. 326 Robert 53 Squire L. 381 W.N. 231
BROWNELL, J.L. 113
BROWNLOW, John T. 337 345 351 356
BRUMFIELD, A.W. 250
BRUNER, Ely 18 W. 302
BRUNK, -- 342 343 George W. 337 341 345 346 347 348 350 351 352

INDEX

BRUNNEL, Frank A. 168
BUCHANAN, James 85
BUCHARDEE, Frederick C. 67
BUCKEYE, R.P. 433
BUCKINGHAM, Charles B. 20 34 39 41 66
BUDD, B.S. 65
BUELL, Bela S. 172 173 198 200 221 226 J.S. 382
BUKEY, R.P. 387 404
BULLIS, A.D. 140
BUNHOLZER, Frank 138 459 Frederick 92
BURCH, Thomas R. 327 W.W. 55
BURCHARD, A.T. 121
BURCHINELL, W.K. 339 346
BURDELL, -- 18
BURDETT, Walter W. 110
BURDSAL, Charles W. 114 120 138
BURKE, William 17 176
BURLEIGH, Charles H. 86
BURNETT, C.C. 99
BURNHAM, C. 345
BURNS, C. 404 Con. 405 D.D. 422 Henry 285
BURR, W.A. 134
BURRELL, Charles F. 137 158
BURRIS, C.W. 377
BURRITT, Johnson 406 430
BURROWS, A.W. 373 375 378 385 386 395 408 414 418 419 420 John R. 261 363 364
BURTON, Arthur 364
BUSH, J.J. 266 William H. 90 166 177 200 217 218
BUSHNELL, A.R. 236
BUSKIRK, M.J. 326
BUSS, J. 279 291
BUTTON, Granville M. 8 54 H.O. 81 95 J.S. 95 Schuyler 81 95

BUTZEL, John L. 42 65 L.J. 25
BUZBY, Albert G. 188 220
BYERS, William N. 2
BYFIELD, Charles 272
BYRON, T. 392 394 407
CADY, -- 343
CAIN[CAINE?], John 120
CAINE, John 84
CAKE, H.L. 124
CALDER, I.S. 377
CALDWELL, W. 329 330
CALELY, W. 320
CALKINS, Leonard G. 117
CALLOWAY, William F. 444
CAMERON, Dr. E.F. 342 344 347 H.C. 42 James H. 216 R.H. 436
CAMPAU, Thomas 387 425
CAMPBELL, A. 260 Albert 237 241 243 248 250 E.F. 346 George 249 H. 249 Henry 268 John 64 Robert A. 194
CANDLER, W.L. 121 William I. 441
CANFIELD, Samuel W. 91 100
CANINS, Frank 160
CANNON, George L. 77 82 86 108 112 127 146
CARLETON, Thomas 319
CARLEY, -- 429 Henry 380 387 413 431 R.J. 384
CARLISLE, James 361
CARLSTROM, Charles 410 420
CARMAN, H.P.
CARNAHAN, J.M. 62 S. 267
CARPENTER, A.C. 86 A.S. 132 Mason B. 234 247 285 446 William O. 101 148
CARR, E.T. 83 130 144 218 James A. 58 Lyman A. 255
CARRERA, J.C. 288
CARRUTH, R. 96
CARSTENS, Alexander 188

INDEX

CART, Timothy 331
CARTER, Harry 56 64 T.J. 256
CARVELL, Jacob 35
CARY, Samuel F. 451
CASE, F.M. 281 J.D. 316 Joseph 206 Oliver 430
CASEY, James H. 238
CASS, Jas. B. 348
CASSEL, William H. 65
CASSIDAY, Alexander M. 11 158 D.R. 11 158
CASTLE, George A. 385 415 423
CATREN, B.C. 147
CATTELL, Alexander G. 44
CAVENDER, Charles 278
CAZIN, Francis M.F. 2 Fred 267
CESSNA, J.O. 371
CHACKFIELD, -- 200 John 169
CHADBURNE, G.W. 280
CHAFFEE, --- 281 Hon. Jerome B. 12 72 73 174 177 208
CHAMBERLAIN, William Selah 408
CHAMBERLAND, Onesine 103
CHAMBERS, -- 246 George W. 62 Joseph 234 246 Wesley 321
CHANDLER, W. 297
CHAPIN, H.C. 268 Howard C. 120
CHAPMAN, John J. 264 276 291 Wilmot H. 255
CHAREST, Emile 307
CHARIST, -- 335
CHARLES, John Q. 256 Levin C. [Charles Levin?] 2
CHATILLON, Henry 168 193 195
CHEATLEY, John F. 182 William H. 94 212 222
CHEESMAN, Walter S. 354 358
CHENEY, O.F. 267 268
CHERNEY, Jacob 32 39
CHESTNUT, T. 422
CHILCOTT, George M. 361

CHILD, John 463
CHILDS, George P. 242
CHINN, R.W. 89 102 112 113 126 137 S.W. 426
CHISOLM, A.R. 43 Alexander R. 352
CHITTENDEN, L.R. 30
CHMAL, Henry 28 43
CHRISTENSON, O. 428
CHRISTIAN, Charles J. 22 40 54 64 William 272 357 William A. 21 22 39 40 45 54 60 64
CHURCH, Bros. 82
CLARE, A. 314 398
CLARK, -- 315 316 A.B. 177 A.P. 409 436 B. Franklin 100 C.P. 443 Calvin P. 444 Charles M. 446 Edward M. 272 Frederick A. 343 348 George R. 447 George T. 286 H. 302 Henry B. 364 Henry C. 439 J.T. 58 James 172 James F. 21 John 378 425 John G. 236 L. 319 O.E. 105 Philip 307 R.N. 153 157 162 163 William 110 119 123 285 William A. 306 William M. 98 111 124
CLARY, William 403
CLASE, Charles 375 427
CLAY, L.B. 233 259 297 303 313 319 327 Samuel Jr. 32 49
CLAYPOOL, Benjamin F. 144 147 Edward F. 144 147
CLAYTON, William P. 5
CLEGE, Thomas 371
CLEGHORN, John Sr. 426
CLEMMER, J.H. 29
CLERY, T.E. 459 460 Thomas 200
CLEVELAND, Sylvester 261
CLIFTON, -- 178
CLINGHAM, M.C. 333 Martin C. 404
CLINN, William 90

INDEX

CLINTON, Edward F. 192
CLIVE, L.S. 433
CLOETI, W. Broderick 437
CLOUGH, John A. 275 William 261
COALEY, -- 178 189
COBB, E. Winslow 42 191 J.J. 443
COBLE, W.W. 267 268
COBURN, John 95
COCHRANE, Barney 193
COCKEY, J.C. 176
CODDINGTON, Matthew O. 112
COE, R.H. 95
COFFEE, Henry N. 33
COFFMAN, Louis 38
COHEN, -- 273
COLE, C.H. 390 435 J. 334 Judson E. 272 M.J. 446 455 Seth B. 91 100
COLEMAN, -- 192 Thomas 10 11 13
COLLIER, Joseph 37 John 59 88 William 63
COLLINS, W. 141 142
COLLOM, Charles 90 142 148 John 90 142
COLLYER, Samuel E. 327
COLT, E. Boudinot 345
COLVIN, Albert 300
COMBS, -- 155
CONDICT, M.J. 414
CONDON, M.J. 374 398
CONGDEN, George E. 144
CONGER, Samuel P. 10 61
CONKLIN, Charles H. 59
CONNELY, Robert 57 94 326
CONNER, William W. 261
CONNERS, John A. 256
CONVERSE, J.W. 1 224
CONWELL, J.A. 448
COOGAN, C. 315 329 333
COOK, A.P. 249 Francis A. 413

George D. 66 Henry D. 263 Low P. 63 Lyman A. 222 W.P. 352
COOKE, G.W. 271 George 379 Henry D. 290
COOMBES, F. 332
COON, John D. 272
COOPER, -- 444 Albert 265 266 285 343 Albert A. 341 356 C.N. 381 Charles H. 432 Henry Y. 234 Isaac 34 J. 331 James T. 167 Kemp G. 29 93 136 M.D. 318 326 William 399 432 William S. 449
COPE, E.B. 312
COPLAND, M.E. 422
COPLEN, J. 300 328 J.D. 324 L. 300
COPLEY, Edward 426
CORBETT, M.S. 295 316 320 336 Thomas B. 444
COREY, W.S. 328 330
CORK, G.F. 308
CORNING, George C. 17 34
CORNWELL, Lafayette 163
CORSER, John 56
COTTON, -- 436
COWENHOVEN, H.P. 210
COWGILL, T.G. 426
COWLES, Alfred 177 G.W. 91 106 109 126 H.C. 107 141 H. Clinton 139
CRAGIN, Aaron H. 351
CRAIG, J.B. 300 James B. 447 452
CRANE, J.D. 452 J.H. 421 J.J. 333 M. 310
CRARY, Harry C. 25, 62
CRAWFORD, C.C. 331 D.C. 256 David C. 89 112 113 137 John S. 16 R.J. 79 R.R. 232
CREE, Edward 248 369
CRIM, E.L. 309
CRISSY, J.M. 376
CRITTENDEN, Thomas T. 79

INDEX

CROMES, F. 446
CROOK, A.B. 339 346 Albert B. 354
CROOKE, -- 244 250 Bros. 415 John J. 237 247 252 364 Lewis 252 364 367 368 Robert 252
CROOKS, William L. 28
CROPLEY, David 431
CROPSEY, Andrew H. 85
CROSBY, F.W. 117
CROSS, C. 107 D. 107 M.J. 378 Robert S. 378
CROUT, Louis 253 P. 388 393 433
CROW, Henry 2 90 130 339 343 353 354 M.D. 297
CROZER, Peter W. 255
CRUMMEY, George 237 239 George W. Sr. 238
CRYDER, A.C. 243 246 J.B. 243
CUENIN, Joseph 325 401
CULVER, Allen M. 242 252 Dr. N.S. 51 66 Nathan S. 426
CUMMINGS, G.W. 422 423 George M. 84
CUMTONS [or, CUMTENS], Andrew 312 416
CUNNINGHAM, R. 113
CURRAN, John 320 M. 241 Mark 320
CURRY, John R. 329 376
CURTICE, William J. 339
CURTIN, A.G. 44 Constance 44 John J. 44
CURTIS, Frank 242 246 249 Henry H. 43 Rodney 133 146
CUSHMAN, Estate 84 144 Rodney 160 Samuel 174 William H. 98 117
CUSICK, P. 312 317 386
CUSTER, John M. 385
CUSTIS, Thomas J. 185 W. Tracy 185

CUTHBERT, Mayland 163
CUTSHAW, Leonard 16 64
CYPHER, Jacob F. 144
DAILEY, James 117
DAMARIN, Louis C. 122 W.N. 354
DANA, J.F. 283 J.W. 283
DANFORD, A. 238 Thomas 18 54 63
DANFORTH, A. 300
DANIELS, Bros. 155 Jacob J. 159
DARBY, Thomas L. 269
DARWIN -- 275
DAVEY, John R. 72
DAVIDSON, C.C. 93 156 Calvin C. 80 David 70 James R. 2 Samuel 352 Thomas 183 William 49 54 71 William A. 15 50 63 74
DAVIS, Abner Y. 248 318 Charles 38 Charles S. 18 Frank M. 350 George W. 272 J.L. 268 J.N. 55 N.H. 451 Peter 14 34 42 Professor 3 Samuel S. 75 Sidney 204 Thomas 261
DAVY, George 58
DAWLEY, Isaac N. 2
DAY, F.C. 364 George W. 58 231 P.O. 282 S.H. 319 393
DE KAY, Drake 91
DE LANO, John 113
DE WITT, David M. 83 92 122
DEAVER, -- 216
DECATUR, Stephen 124
DECKER, James H. 53
DEFEBAUGH, Charles 348
DEITZ, Henry 10 11 13 20 28 39 45
DEPP, George 254
DERVY, C.W. 277 L.M. 277 S.M. 277
DEVERE, James F. 89 112 113 137
DEVOTIE, B. 107 125
DEXTER, -- 343 James V. 337 343 345 351

INDEX

DIAMOND, Daniel 95
DICK, Theodore 243
DICKERSON, William N. 87
DICKEY, S.J. 316
DIEHL, Harry 315 318 335
DILLINGHAM, -- 163
DILLON, Henry C. 256 James S. 350 Richard 264 271 William 182
DIMICK, E.H. 10
DINGLE, Simon 117
DISBROW, Park 93 109
DIVEN, James G. 439
DIX, George W. 97 112 115 129 John D. 97 112 115 129
DIXON, John 444
DIXWELL, John J. 67
DODDS, John F. 233 236 239 241 247
DODGE, Le Grand 364 367 368
DOE, William H. 87 William H. Jr. 186 William H. Sr. 199 224
DOLPHIN, M. 304
DONALD, William 61
DONALDSON, James 277 286
DONNEL, J.M. 384 385 413 422
DONNELL, H.C. 131
DONNELLY, Charles 282
DONOVAN, Cornelius 368 Thomas 366
DOOLEY, -- 417 James 391 415
DOREMUS, R. Ogden 65
DORL, Frederick C. 13
DORY, John A. 112
DOUGAN, Dr. D.H. 339 340 341 355
DOUGHERTY, E.R. 327
DOW, Lorenzo 437
DOWNEND, Frank L. 89 121 124 141 200
DOWNER, N.J. 318 319 326
DOYLE, Albert R. 444 W.A. 107 140 W.H. 118

DOZIER, Joseph 268
DRAKE, Alonzo 223 Eugene 222 223 Lester 222 223
DRESSER, Joseph C. 75 M.W. 302 308 400 424
DRIGGS, Edmund 190
DRUMMOND, Daniel 4 9 41 96 118
DUBOIS, -- 119 269 J.J.B. 267 270 Justin E. 84 117
DUDLEY, Edward 44 Judson H. 286 348 350 351 Thomas H. 44
DUGAL, Louis 281 354
DUGAN, Thomas 405 422
DUGGAN, George 16
DUHEM, Constant 2
DULANEY, Clara 110 David 110
DUMONT, John M. 103 120 135 442
DUN, R.G. 50
DUNBAR, A.D. 70
DUNCAN, -- 156 C.G. 1 72 225 Charles C. 409 436
DUNHAM, W.B. 373
DUNN, John C. 394 L.C. 374
DUNTON, H. 327 Thomas 390 410 434
DUPUIS, John 103
DURFEE, D. 312 D.D. 247
DUSTON, Samuel L. 31
DWYER, Henry 18
DYER, Judge Elias 269 270 Rev. John Lewis 270
EADOR, Jacob 57
EARLE, George 24
EASLEY, John C. 96
EATON, Benjamin H. 21 39 45 60 Benjamin R. 371 429 E.J. 443 G.N. 56
EBERHARDT, Frederick 68
EBERT, George 460
ECKHARDT, Otto 460

INDEX

ECKLIN, James 238
EDDY, C.C. 6 Clem C. 6 F. 6 John 323
EDGERTON, D.M. 245 432
EDONS, Stephen A. 159
EDWARDS, -- 154 John J. 153 160 John W. 81 99 133 R.J. 157 161 Robert 91 100
EDMUNDSON, W. 194
EGAN, John 403 416
EGBERT, Benjamin F. 122
EGGERT, Frederick 30
ELBERT, F.J. 320 Samuel H. 2
ELDER, Clarence P. 42 211
ELDRIDGE, E.S. 321
ELEDGE, Joseph 412
ELKINS, Richard S. 78
ELLERY, -- 223
ELLET, John A. 7 17 23
ELLINGHAM, John J. 33 73
ELLIOT, B.R. 99 John 377 W.H. 7
ELLIS, Charles W. 58 Peter 129 Thomas W. 129
ELLISON, -- 54 Jacob 68
ELSNER, John 355
ELWOOD, Alonzo 251
ELY, Samuel P. 63
EMERINS, John 2 Bertha 2
EMERSON, Charles H. 196 E.O. 431 John L. 196
EMERY, Andrew M. 182 S.C. 456
EMIGH, W. 156
EMMETT, D.R. 443
EMRICK, A.J. 56
ENGLEMAN, M.M. 394 429
EPLEY, John J. 368
ESCHLER, C. 54
ESTABROOK, George W. 201 213
EVANS, C.B. 238 David 16 173 John 358 John G. 61
EVARTS, Charles O. 236 J.D. 236
EVENS, S.S. 330

EVEREST, H.P. 45
EVERETT, F.E. 256
EWING, Mary R. 241 Simon 203 206 W.N. 236 252 William N. 233
EYSTER, C.S. 337 350 352
EZEKIEL, David I. 355
FAGALY, F. 261
FAGAN, James C. 183 195 206 222
FAIN, A.A. 330 Alonzo A. 427
FAIRBURN, George 45
FAIRCHILD, D.L. 251
FANNER, George 34
FARMER, James M. 37
FARRAR, B.F. 65 Fiske 234
FARRELL, Thomas 139
FARWELL, C.B. 52 John V. 148 278
FAY, Joseph B. 378 Joseph R. 378 Lyman 341 349 355
FEE, Eva 31
FEELY, J.O. 428
FENNELLY, William 184
FENNEY, Thomas 147
FERGUSON, G. 373 391 R.H. 402 407 421 426
FERNEON, Aaron 51
FERNSWORTH, William 272 339
FESSENDEN, J.H. 426
FIDLER, John 341
FIELD, Mrs. Francis 170 180 207 209 226
FILLIUS, Jacob 90
FINCH, E.S. 382 411 434
FINERTY, Peter 264 268 269 271 282 285
FINK, W.S. 321 322 323 332
FINLEY, Henry 240
FISCHER, C.F.A. 2 Herman 25
FISH, Charles R. 26 84 90 94 102 124 128 135 J.M. 436 John A. 91 96

474

INDEX

FISHER, G.W. 280 George R. 278 William 216 William H. 53
FISK, -- A. 186 A.C. 355 Archie C. 366 394 C. 327 Russell 463
FITCH, Arthur J. 85 Charles R. 236
FITZGERALD, John 448 454 John E. 275
FLAGG, Wilbur W. 196
FLANAGAN, James F. 337
FLEMING, Alexander 331
FLETCHER, -- 282 W.M. 114 William A. 269 270 286
FLEURY, Charles 342
FLOWERS, W.T. 259
FLUKE, O.P. 88
FLYKIGER, John 54
FLYNN, Daniel 64 Michael 199 Patrick 199
FODDE, F. 295 302 322
FOGARTY, M. 124
FOLLETT, B..F. 460 J.H. 455 460
FOLSOM, Becker 326
FONDA, A.P. 288 W.B. 309 325 327 William B. 238 306
FOOT, John 345
FORBES, Albert R. 26 Andrew 376 423
FORCE, Oscar B. 50 Samuel 35
FORD, L. 26 345 348 351 353
FORMAN, H.W. 227
FORNEY, J.W. 124
FORREST, W.T. 16 234
FORRESTER, Nicholas C. 186
FORSHEY, James M. 439 440 442 445 452 456
FORSYTH, Elbridge 32 40 William 239 332 377 387 391 421 435
FOSS, F.H. 281
FOSTER, D. Ernest 123 George J. 37 59 Gorham 48 J.P. Geraud 216 219 Robert 153 154 159 162 163
FOUNTAIN, Ashburton 205 Charles 205
FOWLER, Edward P. 352 Edwin S. 452 William B. 53
FRANCE, Matt. 203 425
FRANCISCO, J.M. 253
FRANKLIN, -- 243 F. 390
FRANKS, J.C. 187
FRASER, James 166
FRAZER, -- 343 James 179 William 296 305 315
FRAZEUR, Samuel D. 431
FREEMAN, H.E. 452 J.M. 109 Jacob 271 S. 240 Thomas J. 354
FRENCH, John H. 261 Rodney 216 Samuel 463 Stephen B. 175
FRICK, Conrad 246 F. 384
FRIEND, William M. 235 248 314 324
FRISBEE, B.H. 37
FRITS, Ezra D. 178 179
FROHM, Nils 88 134
FRYE, Charles B. 311 J. 335 396
FULLAN, H. 50
FULLER, Hiram 34 44 J.L. 446 T.H. 446 459 Thomas H. 446
FULLERTON, W. 226 William 50 192 227
FUNK, M. 323 M.N. 312 334
FURNALD, -- 223 Alonzo 169
GAFF, James W. Jr. 163
GAGE, David A. 441
GAILLARDON, Jacques 255
GAINES, C.C. 304
GAISER, Louis 106
GALBRAITH, -- 388 Charles 397
GALLUP, Avery 452 Francis 85 Thomas P. 65
GALT, Ernest 311
GAMMON, T.R. 415 426
GARBUTT, Frank C. 7 15 234 239 242 247 251 335 336
GARDNER, F. 306

INDEX

GARRET, Samuel 14 46 William 46
GARRETT, N. 372
GAY, Edward B. 63
GAYNOR, E.J. 259
GEIST, A.W. 361
GENTH, Professor F.A.3
GEORGE, Joseph 463
GEORKE, J. 233 389
GERMAIN, William 181
GERMANSEN, C. 315
GERRY, M.B. 242
GHOST, A.M. 355 Helen M. 361
GIBBONS, C.J. 388
GIBBS, F.H. 431
GIBSON, James 288 John 364
GILBERT, James I. 84 92 95 103 104 108 117 Jarvis 17 John W. 113 Joseph I. 85 103
GILBERTSON, Iner 263
GILCHRIST, -- 298 William J. 77
GILDERSLEEVE, J.W. 278
GILL, A.W. 277 347 Andrew 242 Andrew W. 15 47 286 350 Ely 156
GILLASPIE, George W. 10 H.B. 26 John N. 10 30 John W. 10
GILLESPIE, James G. 267 277 286
GILLETT, -- 403 Charles F. 366 James 372
GILLETTE, J. 379
GILMAN, John A. 28
GILMORE, A. 8 Alexander 57 James 57 Mrs. Phoebe 138
GIRARDIN, William 373 374
GIVEN, H.H. 342
GLADSTONE, -- 432
GLAIZE, -- 139
GLASGOW, John 260
GLAVE, Paul C. 25
GLENN, John 90
GLICK, G.W. 409 436

GODDARD, C. 5
GODIN, J. 293
GOEGLIN, Valentine 409 418
GOERKE, -- 160 Paul 158 161
GOETZEL, Mathias 91 96 106 107
GOLDMAN, Elias 169 182
GOODICT, H. 425
GOODIN, John 371
GOODMAN, James B. 101 148 T.P. 449
GOODWIN, Dr. Harrison 39 John N. 237
GOOGEN, C. 321
GOOKIN, Samuel H. 185
GORHAM, Arthur 89
GORIS, R.J. 274
GORSLINE, William 124
GORTMANS, -- 2
GOSS, Carver J. 99 107 133 146
GOULD, James B. 62
GOVE, Aaron, 80 81 G.D. 307
GRACE, W.N. 247
GRADEN, Thomas C. 418
GRAHAM, Benjamin 348 Thomas J. 8 12 35 36 46 60
GRANT, -- 436 I.H. 463 James 290 James B. 290 Orville L. 54 R. 376 R.S. 463
GRAVELLE, Alex. L. 48
GRAVES, A.M. 37
GRAY, -- 222 E.C. 221 Eli 108 George H. 167 Horace A. 358
GREDIG, Jacob 243
GREEN, David 81 G.B. 5 Gardner 436 George 331 372 374 378 395 400 401 404 437 Leland W. 42 M. 372 377 383 401 Stephen 57
GREENLEAF, A.M. 345 446 459 E.B. 371
GREENOUGH, Edward P. 320
GREENWOOD, Fred 320
GREGORY, -- 245 256

INDEX

GREGG, C. 400 George W. 50
GRIFFIN, G. 392 407 417 H.M. 93
   111 115 116 134 136 T.J. 36
GRIFFITH, D.T. 278 David T. 120
   G.W.E. 93 111 116 134 136 J.D.
   278 Joseph L. 463
GRIGG, -- 456 George W. 62
GRISWOLD, Francis A. 435
GROENMAYER, H. 207
GROSS, Carver J. 85 146 I. 390
GROUPE, E. N. 301
GROUT, Joseph M. 55 William T. 55
GROVE, K.P. 142
GUIBOR, Edward C. 115 120 440
   443 448 456 Frederick A. 440
GUILLETT, Henry 401
GUINN, M. 370 W.L. 392 404
GUISE, James H. 22
GUMP, E.J. 288
GUNNEL, A.T. 245 251 252
GURLEY, Charles D. 231 232
GWINN, John 289
HAAK, L.T. 5
HAARDT, William 8 9
HADDORF, A. 428
HADDOX, B.B. 307 Benjamin B. 332 334
HADLEY, William L. 119
HAFER, Joseph B. 173 188
HAFFORD, George H. 157 161 Joseph 167
HAGGERTY, J.C. 411
HAHN, Frederick L. 320
HAIGHT, E. Graham 355 R.S. 185
HAINES, J.L. 295 310
HALE, B.E. 65 Horace M. 165 J.A. 223
HALL, -- 342 343 Assyria 337 338
   341 345 346 347 348 350 351 357
   358 Charles L. 357 G.W. 91 135
   149 George D. 440 Horace F. 28

69 J.S. 397 John R. 129 Moses
   338 355 Robert D. 257 Wright W. 28 69
HALLETT, Henry L. 441 Moses 53
HALLOCK, Nelson 265 269 278
   285 352 354
HALM, M. 208
HAMBEL, John R. 111 128 138
HAMER, Joseph 235 246
HAMILL, Charles W. 10 11 John
   413 Hon. William A. 91 143 150
HAMILTON, Oliver P. 13 T.S. 32
HAMLEN, Robert D. 382
HAMMILL, John 380
HAMMOND, W.A. 47
HANCE, J.J. 384 420
HANCOCK, Edward C. 409 414
   William 90 148 255
HANN, -- 436
HANNAH, James W. 208 223 229
   463 William 204
HANRAHAN, James 261 356
HANSBROUGH, P.M. 110
HANSON, R. 388 433 Robert 425 432
HARBOTTLE, -- 243 244 245 250
HARBOUR, J.V. 11
HARD, Alfred A. 407 408 Bros. 240
HARDENBROOK, William A. 35 53 65
HARDENSTEIN, -- 264
HARDIN, Claude P. 326
HARDING, C. 236 408 Theodore M. 154 163
HARKER, N.O. 433 O.H. 37 59 S.B. 20
HARKINS, Hugh 5 Thomas F. 5
HARLEY, Henry 261
HARMAN, M.D. 418
HARPER, John 63
HARRINGTON, H. 417 421 Henry

INDEX

C. 128
HARRIOTT, Edgar 451
HARRIS, -- 344 Daniel 220 Dennis B. 326
HARRISON, E. 100 117 132 Edwin 264 287 290 Jared F. 414
HART, A.B. 397 C. 388 W. 102 W.R. 327
HARTMAN, Casper R. 256 W.P. 391 431
HARTON, C. 231 232
HARTSELL, W.W. 341
HARTWELL, Abraham V. 439 Edward E. 85 354 447
HARTZEL, H. 112
HARVAT, J. 459
HASSELBACHER, John 8 9
HATCH, Franz 206 Israel B. 70
HATHAWAY, Charles G. 346 352
HAUSER, J.H. 221
HAUSLE, John 256
HAVENS, Gilbert L. 446
HAWKE, Jacob 21 39 45 60
HAWKES, Daniel 109 F.T. 54 J.A. 92
HAWKINS, E.M. 33 William A. 351
HAWLEY, Henry J. 193 210 219 Sidney B. 193
HAYDEN, F. 277 Lewis 289 Nathaniel 57 S.G. 357
HAYES, William A. 67
HAYMAN, Benjamin 39
HAYNES, John 280 Michael 280
HAYNOR, Charles 20
HAZARD, Horatio E. 78 88 174 190 199 224 Morris 78 88 174 190 199 224
HEADY, Henry W. 408
HEATH, C.P. 426
HEATON, George W. 217
HECKENDORF, August 246
HEDGES, Rufus W. 2
HEID, John G. 425
HEISER, Mat. 38
HELMER, Henry 143 William 143
HELMICK, William 110 111
HENDERSHOT, Thomas F. 36 54
HENDERSON, -- 66 Edward W. 29 174 193 215 216 219 222 224 George L. 269 George S. 287 H. 121
HENDRICK, D.D. 3 31
HENDRICKS, M. 302
HENDRIE, Charles F. 138 203 204 217
HENRY, E. 319 I.N. 143 John C. 51
HENSE, Estate 194 Heirs of John 217 Joseph 245
HENSHALL, James 129
HEPBURN, J.C. 242 Mary 179
HEPNER, George 20 Jacob 20
HERBST, -- 318 F. 294 302 306 378 420
HEREFORD, A.P. 278
HERR, A.J. 159 Hiero B. 159 Theodore W. 159
HERRICK, Samuel 441
HERROD, F. 424 Frank 384 405 412 425
HERZINGER, J.L. 20
HESS, E.L. 300 W.C. 331
HEWITT, Henry C. 272
HIBBARD, F.M. 441 Horace W. 94 I.W. 441 Russell A. 366
HIBSCHLE, Herman 264
HICOCK, W.C. 81 105 111 137 142
HIGBEE, Frederick ("Fred") L. 21 56
HIGGINBOTHAM, Edmund 387 425
HIGGINS, L.L. 337 M. 401 Richard E. 247
HIGHT, G.W. 304 318 323

INDEX

HILL, A. 316 Britton A. 56 Charles 303 411 David 75 79 George 207 Hiram 197 J.M. 239 240 244 James L. 263 290 364 Jerry N. 61 Julius D. 2 Nathaniel P. 1 172 184 224 225 269 353 Sylvester 31 33 64
HINMAN, Porter T. 9 W.S. 11
HIPPE, W. 421
HIRSCH, Emanuel 121 124
HOAGLAND, George W. 85
HOBGON, J.H. 437
HOBSON, William 32
HOCKADAY, Charles N. 32
HOCKER, R.W. 49
HODGES, E.W. 383 Henry W. 350
HOES, E.V.B. 239 240 244
HOFFARD, -- 154
HOFFER, J. George 191
HOGAN, M. 392 394 Peter 432
HOIL, Benjamin 342 346 347
HOLBROOK, J.J. 234 247 James J. 233
HOLCOMBE, John M. 242 251
HOLLAND, Charles 340 359 Dwight G. 340 359 L. Charles 439 Park 340 359
HOLLIDAY, C.K. 233
HOLLINGWORTH, Leander F. 428
HOLLISTER, Marshall 101
HOLMAN, J.W. 167 Joseph W. 175 176 182
HOLMES, E.C. 261 George 276 291 John 5 100
HOLT, -- 240 G. 304 M.J. 9 33 Matthew 4 William T. 425
HOME, Howard 325
HOOD, F. 125 Frank 107
HOOPER, George L. 344
HOPKINS, Charles 301 428 E.R. 409
HORN, A. 88

HORNER, G.C. 313 J.E. 321 J.W. 120
HORRIGAN, Edward 101 William 101
HOSANG, Martin 68
HOSKINS, L.E. 196
HOUGH, John S. 240 241 243 245 248
HOUGHTON, E.W. 7 Jacob 351 Peter 237
HOUSE, W.B. 402 409 433
HOUSEN, Herman 138
HOUX, -- 284
HOWARD, E.C. 416 Edward P. 381 F. 110 Flodvardo 111 G. 418 Gains C. 381 George 394
HOWCUTT, E. 370
HOWE, William 101
HOWELL, Cornelius C. 29 30 William 25
HOYLE, Stephen Z. 342
HOYT, -- 159 161 Arthur W. 88 George A. 170 180 Henry S. Jr. 364
HUBBARD, E.L. 42 Emerson 231
HUBBELL, Jay A. 263 290
HUBERTY, Henry 33
HUDDLESTON, S.F. 30
HUDSON, Joseph 263 273 274 280
HUFF, A. 259 G.O. 104 T.J. 141 142 W.F. 104
HUG, M.M. 16
HUGHES, -- 282 Bela M. 120 143 George 319 J.H. 269 270 286 J.W. 244 250 Owen 89 106 190 Richard Owen 191 T.J. 348 358 T.W.B. 354
HUMASTON, Charles 260 394 L.S. 443
HUMMEL, C. 259 Jacob 19 31 63
HUMPHREYS, Horace 216 255
HUNDLEY, Elisha E. 340 356

479

INDEX

HUNT, -- 74 A. Cameron 2 Edward 443 Edward S. 435 Harry 324 Samuel 431
HUPPER, E.A. 19 45 59
HURD, A.H. 251 Charles R. 157 Nathan S. 138 140 William S. 157
HURLBUT, W.H. 288
HUTCHINSON, D.J. 154 163 George 212 George E. 166
HYATT, H.E. 205 217 436
HYNDMAN, John 223 M.B. 201
IFINGER, John 350
ILIFF, William 58
INGERSOLL, George 429 W. 312
INGOLS, Augustus B. 43
INNIS, Edward 398
IRWIN, Dick 153 Richard 155
ISAM, J.J. 409
ISHAM, J. Frank 408
ISRAEL, H. 422
JACKSON, -- 347 J.G. 261 U. 260 W.U. 383 William 375 409 William S. 426
JACOBS, George W. 195 227
JACOBSON, Eugene P. 2
JACQUE, John W. 439 447 448 449 450 458 459
JAMES, A.F. 114 John 404 John W. 6 William H. 356
JAMESON, E.R. 324 O.W. 304 320 333
JEBB, J.G. 359
JEFFRIES, Howard B. 324 Lorenzo A. 324 S.G. 372 T. Norval 448
JENKINS, -- 219 H.B. 392 406 Henry 403 418
JENNISS, Edward 261 Walter B. 16
JERVEY, Theodore D. 352
JESSUP, E.K. 435
JOHNS, Frank W. 122 Thomas C. 122

JOHNSON, -- 370 382 A.C. 190 210 213 Albert 111 116 122 B.F. 111 C.J. 88 Charles 309 330 Charles W. 231 Clase [Chas.?] J. 134 E.F. 272 Egbert 111 F. 432 Gustavus 395 397 398 411 H.A. 178 226 H.B. 104 H.D. 238 H.L 91 Howard N. 320 J.B. 16 105 J.H. 289 J.P. 384 429 John 194 L.A. 172 M.W. 317 Mary 207 Peter C. 207 217 R.A. 178 Theron W. 16 Thomas 405 427 W.H. 104 William U. 353
JOHNSTONE, E.W. 402
JONES, -- 263 264 276 Aaron M. 194 Amanzo L. 257 Charles 260 381 411 Clabe 260 381 411 E. 461 Edward 89 106 190 191 Evan 257 Fred. W. 412 G.G. 461 Griffith G. 450 461 Harry C. 241 J.H. 28 162 191 261 J. Harvey 43 120 J.T. 321 James H. 2 6 James K. 17 19 49 Joab 165 John P. 263 290 N. 429 430 S. 421 William 89 106 110 190 191
JORDEN, C.C. 304
JORDIE, John 68
JOSEPHS, Charles 318 325
JOSLIN, Anthony 84 J.B. 38 J. Jay 39 43 83 212
JOYCE, Stephen 278
JUDD, Norman B. 117
JUNKINS, -- 159
KAFKA, Louis 239
KALBAUGH, Perry 86 92 Zadock 87 103 130 145
KALEY, Frank 354
KANE, Stephen K. 79
KATZENMAYER, E.J. 41
KEARNEY, H.S. 101 110 125 148
KEARNS, R. 311
KEELER, Joseph 241

INDEX

KEHLER, Frederick 17
KEITH, C.B. 117 Edson 177
KELLEY, C. 96
KELLOGG, Aaron W. 40 C.C. 278 George A. 244 246 250
KELLY, -- 183 H.N. 157 James 189 John 269
KELPER, E. 423
KELSO, Fletcher 143
KELTY, -- 165 181 215
KENDALL, J.W. 259
KENDRICK, William F. 344 345 444
KENNEDY, J.P. 423 Silas S. 21 39 45 60 109 Thomas 199 William R. 238
KENNEY, Robert D. 59
KEPHART, E.B. 324
KERCHER, C.F. 88
KERNS, F. 430
KERR, J.H. 268 445
KERWAN, Robert 184
KESLER, John 54
KETTLE, George E. 124
KEYES, D. 388 Estate of E.W. 274
KIEFER, John 25 350
KILBOURN, -- 219 A. Asahel 219 198 F. 198 Frank 219
KILBY, L.C. 320
KILHAM, Leonard C. 297 299 300 309 311 322 323 326
KIMBALL, Charles H. 233 249 F.E. 272 339 G. 324
KIMBER, -- 166 222 A.J. 30 Job V. 94 181 191 192 226 227 228
KIMBERLY, Peter 373
KING, David L. 78 E.M. 384 420 Francis G. 297 299 300 309 311 322 323 326 George L. 452 J.H. 297 299 300 309 311 322 323 326 J.M. 382 411 434 John H. 91 John W. 24 O.H. 131 R.C. 309 Theodore 128 139 345 William 461
KINKEAD, James 79 93 127 130
KIRK, Charles 304 323
KIRKER, Robert A. 337 338 339 345 347 349 353 356 357 358
KIRKLAND, J.H. 121
KIRTLEY, Jeremiah 117
KISSLER, Marion 43
KLINE, George J. 188 213 H.M. 94 J.F. 188 213
KLINGENSMITH, A.A. 325
KNIGHT, -- 389 397 Abel 201 Albert M. 381 Alexander N. 430 Estate of Henry 356 W.L. 155
KNOX, A.C. 443 George W. 288 O.C. 339 443 449
KOCH, Barney 194 Joseph 107
KORNER, Gustave 25
KOUNTZE, George 54 63 74
KRAFT, William 289
KRAMER, Joseph 188
KRAUSS, A.F. 416 O.F. 386
KREUSS, A.F. 293
KROHN, Louis 279 291
KRUG, Gotfried 188
KRUSE, Frederick 169 Gustave 188 Henry 214 Henry J. 214 John 188
KUHNSTER, Frederick 206
KUSLICK, Robert 240
KUSZ, Charles L. Jr. 266 267 268
LA GRAVE, F.A. 264 Francis H. 70
LA PAUGH, Harry A. 187
LACEY, Robert S. 110 111
LACOME, Joseph E. 306 331 334
LADD, J.C. 400
LAFFER, Henry 21
LAFFOON, J.B. 307 L.B. 332
LAFRENZ, John H. 171 214
LAHAYE, George M. 16 29
LAKE, Henry W. 170 178 180 192 207 209 R.N. 394

INDEX

LAMBERT, William E. 277
LAMING, Jeremiah 340 358 439
LANDIS, -- 253
LANDON, John 42 R.B. 381
LANE, H.E. 90
LANGE, George 102
LANGFORD, Augustine G. 75 Nathaniel P. 75
LANGHOFF, John 271
LANGHORNE, James C. 2 269
LARSEN, Nils 85 Thomas H. 261
LARSON, E. 416 J. 424 428 O. 388 391 397 O.O. 412 S. 391
LASHUER, Stephen 64
LATHROP, M.A. 10 16 56 Samuel P. 144
LATSHAW, William H. 93
LATSON, M.W. 85 Morton W. 144 147
LATTIMORE, J. 432 John 377 399
LAW, John 277
LAWRENCE, Charles E. 159
LAWS, R. 404
LAWSON, Alexander 85 125 132 142 James 80 93 L.M. 91
LAWTON, Andrew L. 232
LAY, Alexander 441
LE CLAIR, H. 325 Henry 370
LE FEVRE, L. 412 Owen E. 43
LEACH, Hiram S. 2 Samuel 353 354
LEARNARD, Heber A. 24 27 67 Horace F. 24 27 67
LEARNED, Samuel 148 William H. 222
LEAS, George W. 345
LEDUC, Napoleon 103
LEE, George S. 242 252 Green 223 N.J. 317 W.E. 275
LEEDON, S.W. 37
LEHMAN, Charles W. 246 H.C. 159

LEIGHTON, Frederick 144 221 J.W. 452
LEIMER, Charles F. 191 Rudolph A. 261
LEITER, Levi Z. 265 269 273 275 278 285 287
LELAND, Charles M. 207
LENNON, H. 402
LEONARD, J.E. 238 William M. 261
LEPORIEU, Frederick 107
LEPOSIN, Frederick 96 106
LEROY, H. 376 Henry 410
LESLIE, John A. 451
LESSIG, William H. 21 39 45 60
LESTER, C.H. 267 James H. 247 305 307
LETT, H.C. 198
LEWIS, -- 229 L.W. 357
LIGHTFOOT, J. 277
LILLEY, John G. 2 W.H. 296 326
LIMBERG, Charles 272
LINCOLN, Henry E. 220 S.W. 144
LINDERMAN, Rufus 57
LINDGREN, John R. 337 341
LINDSLEY, Taylor 202 221 William 221
LINKE, E. 400
LINSCOTT, Samuel 100
LIONS, Henry 43
LISSBURGER, Marks 110
LISTANCE, George K. 65
LIVINGSTON, Johnston 364
LOCK, Bradford H. 226
LOCKE, Isaac H. 67
LOCKWOOD, J.S. 277 Radcliffe B. 65
LOESCHER, Emil 267
LOESCHIGK, William O. 252
LOGAN, John A. 135
LOKER, James R. 280
LOMBARD, F.P. 443

INDEX

LONG, J. 392 J.F. 271 275 276 J.W 271 275 276 R.F. 336 Robert T. 326 W.R. 433
LOOMIS, M. 125 William H. 343
LOPER, Isaac N. 48
LORAH, Samuel I. 216
LORD, John M. 57 94 326 Newton B. 441
LOTHROP, Wilbur C. 211
LOTT, J.D. 251
LOUDERBACH, J.C. 406
LOUGHEED, Rev. Samuel 253
LOUGHLIN, Robert 463 William 463
LOVE, -- 318 N.H. 294 Peter 288
LOVELAND, William A.H. 256
LOWE, -- 244 E. 440 450 Edward 443
LOWERRE, J.W. 16
LOYE, Peter 282
LUCAS, Joseph 247
LUCE, Frederick R. 6 47 John A. 326 Joseph 255
LUESLEY, R.C. 382 383
LUNSFORD, Iswold 21
LUNT, S.P. 348 358
LUTTRELL, James 357
LYLES, D.C. 278
LYNCH, Edward L. 96 106 James M. 285
LYNN, William P. 117 120 131
LYON, Henry E. 211 R.C. 268 443
LYONS, William A. 157
MAAS, John B. 9 23 34
MABEE, George W. 185 194 228
MACK, Henry 33 Jacob 169 182 201
MACKEY, Andrew J. 42 58 65 Richard 173 192 197 226 227
MACKIN, John 463
MAGEE, R. 421 W.C. 421
MAGRUDER, John R. 263 290

MAHAN, D. 6 N.V. 16
MAHANY, Jeremiah 122 M.A. 344 M.H. 341
MAHER, John 270 281 Patrick 199 Thomas 408
MAIR, Charles A. 177
MAJORS, R. 88
MALLON, B. 433 Barney 406
MALLORY, D.D. 297 299 300 309 311 322 323 326
MALTBY, George E. 320 336
MANGOLD, Otto 356
MANN, William 41 William E. 24 William J. 24 50
MANNION, William 349
MANSUR, Alva 189
MARBURY, W.H. 54
MARFELL, Hiram 463
MARKEL, J.H. 231
MARKERT, Samuel 45
MARKS, W.N. 235
MARLOW, F. 385
MARR, William 37
MARRH, C.J. 397
MARSH, George E. 77
MARSHALL, Charles H. 91 115 146 F.J. 120 123 135 Frank J. 85 J.G. 347 James M. 372 379 William S. 320
MARTELL, I.P. 400
MARTIN, -- 253 F.L. 51 253 G.H. 302 George 94 182 212 222 James 13 74 John 378 R.H. 182 R.L. 93 Robert 94 Robert L. 451 452 William 45 222
MARTINE, Charles A. 117 202
MARZETTI, Albert 381 414
MASON, -- 253
MATER, -- 267 Charles 264
MATHER, J.C. 361
MATHERS, C.W. 209 J.H. 339 452
MATHEWS, E.J. 448 James F. 149

INDEX

Oscar L. 368
MATTICE, Benjamin 158
MAUDE, John B. 164 280
MAUGAN, John H. 412
MAUGHAM, J.H. 238 John H. 236
MAXWELL, -- 154 James P. 10 56
 John 14 John M. 23
MAY, C. 293 John 57 William H.
 163
MAYER, G.F. 56 William 289
McALLISTER, Henry Jr. 425 440
 Thomas 203 206
McANDREWS, -- 356
McARTHUR, Duncan 134
McBARNES, S. 58
McBREEN, William 200
McBRIDE, H.D. 247 Robert A. 28
 W.P. 247
McCABE, C. 404 430
McCADON, J. 19
McCALL, Alexander 58
McCANN, T.J. 149
McCARTHY, C.C. 356
McCARTY, E. 293 317 370 395
 Leander 43
McCAY, P. 401 435
McCHESNEY, J.H. 452
McCLELLAN, Catharine 268
McCLELLAND, -- 252 Erskine 87
 130 William F. 350
McCLOSKEY, J.D. 289
McCLURE, A.H. 55 Charles Y. 17
 Joseph E. 278 William 43
McCOMB, -- 154
McCOMBE, John 283
McCONNAUGHEY, William 37
McCONNELL, D. 383
McCOOK, Edward M. 124 125
 Major General 20
McCORMICK, David 31 J.M. 295
 M. 310 315 332 335 M.C. 295
McCRELLIS, P.M. 114

McCUNNIFF, Thomas 143
McDERMITH, W.J. 267
McDONALD, Alexander 96 146
 J.F. 238 J.H. 253 James 361 John
 282 288 John A. 364 Miles 349
 Patrick 96 W.T. 418
McDOWELL, G.A. 402 James 463
 John 463
McELHENNEY, A.M. 156
McELROY, John 295
McENANY, P. 392 403 416 418
 420
McFALL, John B. 14 John W. 46 47
McFARLAND, James 368 Samuel
 368
McFARLANE, William 227
McFERRAN, J.W. 234 James H.B.
 353 359
McGILLIVRAY, D. 393 410
McGINNIS, E. 305
McGONIGAL, Daniel 196
McGOONEY, A.A. 289
McGRAIN, James 79
McGRAW, Hugh D. 35 M. 371
McGRUDER, Alfred 199
McGUIGAN, M. 304
McGUIRE, P. 248
McINTIRE, Charles H. 369 378 380
 385 386 396 399 406 420 423
 424 433 434 Charles W. 374 Mrs.
 Clara 381 Edward W. 369 399
 423 433 Ellen M. 387 389 George
 381 George W. 402 W. 332 391
McINTOSH, James 112 John 79
 Joseph P. 28
McKAY, Daniel J. 198 239 332 377
 387 391 421 435 James C. 332
 James D. 239 377 387 391 421
 435 John B. 198 William 21
McKEE, Charles 83 130 191 218
 Robert 66
McKEEN, William R. 144 147

INDEX

McKENNEY, L.C. 245
McKENZIE, D. 293 294 Neil D. 53
McKINLEY, Charles 13 M.J. 102
McKINNEY, Andrew 256 J.J. 448 R.V. 426
McKINNIE, J.R. 376 390 408
McKINZIE, B.F. 369 J. 397
McKNIGHT, Sumner T. 212 William 13
McLAUGHLIN, Felix 344
McLEOD, Alexander 198
McMILLAN, E.C. 89 121 141 Samuel 348
McMORRIS, Thomas A. 255
McMURDY, Estate 95 96 99 115
McNASSER, James 194 198 204 344 352 447 451
McNEIL, Albert 442 445 Horace 42
McNULTY, Patrick 84 116 120
McNUTT, R.J. 414 418 429
McPHERSON, Burrell 48 Daniel A. 42 Joseph 11
McWILLIAMS, John 261 Samuel 320
MEARS, Otto 234 244
MECHLING, John 354
MEDARY, Charles 91
MEDLEY, A. 117 T.P. 437
MEEK, J. 307 John 301
MELLETT, J.V. 426
MELLOR, C.P. 411 424 John 226 Samuel 226
MELVILLE, W.G. 321 409 William G. 436
MELVIN, -- 160 H. 432 J.A. 162 James A. 155
MENDENHALL, William 92 144
MERALLS, William A. 35 52
MERCER, R.M. 308 334 William 316
MERRIAM, George 274

MERRICK, F. 431
MERRILL, J.Warren 1 224
MESLER, -- 235 236 237 248 251 A.A. 249 Alexander 234 O.A. 241 249
MESSINGER, Frank H. 170 171 180 188 195 206 207 220 224
METCALF, Charles A. 111
MEUGG, Joseph 171
MEYER, -- 270 August R. 263 266 268 279 280 283 285 291 Henry G. 279 291 John 391 415
MEYERS, Andrew 135 David 105 145 George 105 145
MEYRING, Henry 43 65
MICHAEL, Francis 267
MICHAELS, J.P. 247
MIDDLETON, Reuben S. 90
MILES, -- 149 John 238
MILLER, Charles 417 Christopher 65 Christopher C. 206 Frank 125 G. 385 H. 402 J.F. 390 435 J.M. 47 James J. 54 John 36 54 Paul P. 46 Thomas 254 William G. 167
MILLS, -- 243 244 245 J. 108 127 250 J.W. 303 Joseph K. 124 Milton 144 S.B. 5 Stiles E. 222 Sydenham 271 277 345 350 William H. 158
MILLSBAUGH, Mrs. William 57
MILNER, G.C. 235
MINDENHALL, F.M. 307
MINER, John 54 Joseph S. 2
MINFORD, Thomas I. 178
MISH, G. 315
MISHLER, Samuel 16 21 34 53 181
MITCHELL, -- 267 David H. 135 Frank G. 277 G. 328 John 9 56 John P. 34 Joseph 463 W. 305 328 334 W.R. 375 398 Walter R. 277
MIZE, Edward 78

INDEX

MOFFAT, David H. Jr. 15 358 R.W. 297 314
MOLANDER, Jonas O. 352
MONELL, Henry 4 42 Ira F. 4
MONK, George W. 161
MONTIETH, W.R. 261
MONTREUIL, Alexander G. 18
MOODY, Moses S. 212
MOON, John 447
MOORE, George 21 Henry 67 John F. 253 John S. 26 R.C. 114 Samuel 59 W.H. 85 140 William 94 100 106 110 William H. 136 146
MOREHOUSE, Philip E. 26 92 146 321
MOREY, F. 329 395 Henry 235
MORGAN, -- 276 Charles N. 63 Samuel 263 281 289 W. 308 William 397
MORLEY, Henry 241
MORRELL, D.J. 124
MORRILL, Charles J. 100 Wiley E. 100
MORRIS, -- 149 Austin W. 85 Benjamin 97 Benjamin F. 112 115 129 C. 319 Charles 299 302 Charles H. 26 97 112 115 129 Frank E. 177 J.G. 369 John E. 102 Nelson 14 Robert 102
MORRISON, -- 335 H.S. 99 H.W. 307 John 17 27 70 M.J. 117 R. 223 Robert M.D. 208 223 Robert S. 100
MORSE, Harley B. 94 184 195 196 199 202 203 N.J. 118
MORSEMAN, Ruben 123
MORTIMORE, Dr. D. 51 58
MORTON, -- 253 E. 313 394 George W. 253
MOSLEY, -- 192
MOSS, Drury R. 168 N.J. 107 140

MOULTON, H.S. 231 232 William W. 78
MOYLE, -- 379 380
MOYNAHAN, James 349 352 357
MULHOLLAND, J.M. 305 313 William 372
MULLEN, -- 298 343 Charles M. 358 J.K. 240 Thomas 215
MULLET, Alfred B. 79
MULLIN, Loudon 25
MULOCK, Ira 277 349 Joshua 349
MUNCIL, R.C. 404
MUNFORD, James E. 50 M. 50
MUNN, Bros. 307 David A. 306 309 332 334 John 306 334 John C. 325 Joseph A. 125 148
MURDOCK, Hugh 357
MURPHY, D.M. 256 J.T. 79 John D. 374
MURRAY, C.H. 236 James 269
MUSGROVE, H. 240 William E. 351
MYER, -- 417
MYERS, Charles H. 113 John C. 157 Joseph H. 350
NALLE, B.F. 54
NAPHEYS, Benjamin F. 98 143 149
NAPIER, Barnett T. 34
NASH, David B. 84
NATHAN, Adolph 236 Joseph 236
NAYLOR, Charles T. 59 Edward Y. 143 Frank M. 143
NEEDHAM, John 144 221
NEGUS, Timothy G. 100 202
NEIDIG, A.H. 89
NEIKIRK, -- 17 32 Henry 43
NEILL, J.A. 261
NEISWANGER, J.D. 413 J.W. 436 John 370
NELSON, August 134
NESCHKE, William A. 96

INDEX

NETTER, Gabriel 42
NEUMEYER, -- 293 A.W. 294 297
NEWCOMB, George 4 64 Henry E. 100
NEWELL, Samuel V. 194
NEWMAN, John 139 150
NICHOLS, Edward H. 91 Edward T. 463 Foster 194 215 H.C. 156 John 256 Sylvester 178 184 194 W.H.J. 16 64 William H. 261
NICHOLSON, E. 268 William 190
NIEGOLD, Bros. 417 436
NIENINGER, Frederick 460
NILES, A. 289
NILSON, August 88 Sophie 110
NOBLIT, Dell Jr. 44
NORRIS, A.E. 380 William 380 406
NORTH, Charles 37
NORTON, James B. 181
NORWOOD, Talmadge 93
NOTT, S.W. 144
NOWCOMB, Levi 441
NOXON, Abram M. 112
NOYES, George E. 129
NUTT, Emory C. 38 55 71
NUTTING, -- 245 M.P. 303 311
NYCE, George 275
O'BOYLE, James 308 421
O'BRIEN, John 382
O'CONNELL, D. 389
O'CONNOR, J.B. 139
O'NEIL, Dennis 248 Edward 30 J. 400
OBERTO, Antony [or, Antonio] 415 427
OFFENBACHER, W.A. 155
OGDEN, Samuel 427
OGLE, -- 382 C. 319
OGLESBY, John P. 327
OGSBURY, D.C. 402
OHIO, Edward 326

OHLWILER, Jacob 321
OLD, Robert O. 117 Thomas C. 97
OLDS, A.M. 261
OLIVER, Thomas 120
OLLIVER, James R. 357
OLMSTED, L.F. 150
OLNEY, Henry C. 233
OLSON, Andrew G. 26 36
ORAHOOD, Harper M. 218 221
OREAR, Henry 278
ORR, C.A. 426 J. 329 395
ORRICK, John C. 440
OSBORN, George W. 394 John M. 102 148
OWENS, A. 320 Charles J. 320 R.G. 399 William 400
OYLER, Thomas J. 30 144 221
PADDOCK, August 6 47
PAGE, W. 328
PAINE, C. 400
PALMER, A.L. 140 A.S. 85 John 200 William J. 358
PAMPERIN, H. 235
PARKER, Charles M. 261 George W. 57 Horatio G. 88 James D. 163 Orville 382
PARRISH, Thomas C. 155 163
PARROTT, R.D. 320 336 W. 294
PARSON, L.E. 399
PARSONS, Albro L. 58 Charles H. 79 Curtis R. 58
PARTRIDGE, -- 364 J.S. 364
PASCO, Joseph 42
PATRICK, James M. 291 W.F. 291
PATT, J.H. 364
PATTEN, George A. 144 219 221
PATTERSON, -- 253 Charles B. 128 E.H.N. 124 J.W. 311 John W. 393 O.P. 18
PATTISON, L.W. 410 W.L. 231
PAUL, Estate of J. Marshall 283 Henry 340 356 Henry N. 283 J.

INDEX

Marshall 282 283 John Rodman 283
PAYNE, Henry M. 439 Robert A. 400
PEABODY, Jacob L. 350 Lelon 353 447
PEARCE, Joseph 265 274 R. 224 Richard 1 Thomas 117 William 265
PEASE, Charles F. 21 56 George E. 281 354 M. Granville 50 62 Stephen 281 Stephen H. 354
PECK, A. 307 329 Arthur 289 339 C. 307 Fred C. 240 James 81 97 100 106 123 128 139 William 234
PEIRCE, J.W. 386
PELL, James 25 32 William 25 32 William G. 3 6 31
PELTON, Benjamin H. 249
PENDLETON, Ira W. 10 Joy H. 78
PENISTON, Rienzi E. 238
PENN, Elijah G. 51 S.M. 51
PENNINGTON, John L. 437
PERKINS, T.H. 441 William F. 80
PERLEY, J.A. 183
PERRIN, Edward S. 211
PERRY, E.F. 400 John d. 463
PETER, Thomas J. 233
PETERS, Daniel 90 142 148 255 G.E. 402 George 294
PETERSON, -- 370 382 August 88 134 Charles 176 J. 379 J.C. 370 James 417 James E. 232 N. 305 N.C. 15 N.D. 21 P. 332 335 Pear J. 363 364 365 367 S. 395 416
PEVEY, C.K. 218
PEYTON, C.L. 248 Clark 241
PFLUM, Fidel 188
PHELPS, M. 295 297
PHILIP, George G. 6
PHILLIPS, Charles B. 455 Edward

A. 17 27 Ivers 77 86 James 238 Richard 130 William 240
PICKENS, A.C. 267
PIERCE, E.W. 28 57 71 Edward W. 394 John 56 64 L.H. 341 343 356 W.W. 13
PIERSON, E.R. 26 J. Fred 215 Oliver H. 344 352
PINE, B.F. 42
PIPPIN, D. 213
PITKIN, Hon. F.W. 238
PLANT, James 344
PLATO, William B. 463
PLATT, -- 18 Edwin S. 103 442 Edwin T. 455 Harvey P. 455 James B. 423 James W. 316 372 380 428
PLUMB, C. 412
PLUMMER, Daniel 350
POCOCK, Frederick 18 42
POEHLER, August 409 437
POGUE, Americus Loudy 339 340 341 355 Christopher Columbus 339 Loudy J. 339
POHLE, Julius G. 78 83 89 96 99 104 105 107 115 116 118 120 122 123 126 130 133 135 136 137 145 149 150
POLLARD, C.E. 145
POLLOCK, -- 440 Irving J. 139
POMEROY, J.V. 7 36 74 James 14
POOR, W.H. 139 141
POPE, B. 385 F.A. 120
PORTER, G.R. 318 N.T. 90
POSEY, -- 244 364 A.H. 277
POTTER, James B. 441 James W. 245 251 252 John 440 John Z. 443 Thomas H. 179 184 195 211 219 227 228
POWELL, Robert 157 160 161 W.P. 29
POWERS, -- 154 E.H. 259 Horace

INDEX

H. 368 Joseph 123
PRATT, C. 431 Daniel H. 256
  Franklin J. 409 J. 108
PRAY, M. 388 400 415
PRESCOTT, -- 154 A.K. 308 392
PRESSLER, Jacob 188 196 Philip
  188
PRICE, C. 410 J.L. 397 Mark B.
  233
PRIDDY, C.N. 232
PRINCE, H.J. DeBruyn 29
PRINDLE, James R. Polk 122
PRITCHARD, A.L. 207 John H.
  406
PROBASCO, Frank 447 458
PROMISE, Henry 386 387 401
PROPPER, George N. 369 373 375
  385 394 395 418 419 434
PROTHEROE, John 220
PUGH, Samuel A. 110 111 Ulysses
  16 34 61
PUGHE, John 35
PURCELL, George 374 386
PURCELLS, G. 312
PUSKS, T. 346 348 351 353
PUTNAM, J.B. 266
PYATT, James M. 400
PYBURN, George 463
QUICKSELL, Harvey 448
QUINN, D.P. 309 L. 324 384
RAGLAN, B.R. 452
RAINEY, F.M. 350
RALSTON, J.H. 57 James H. 26 70
  71 288
RAMPAGE, William 217
RAMSDELL, Thomas 335 336
RAND, H. 413 William H. 447
RANDOLPH, George E. 165 194
  202 209 220 228 Peyton 54
RANKIN, J.C. 312 J.D. 461
RANSOM, S.M. 299 322
RAPP, A. 247

RASER, James 412 426
RASIN, Mifflin 218
RAWLING, James F. 377
RAWLINGS, James T. 396 416
RAY, -- 275 E. 372 379 James 79
  John 261 305 Robert 286
RAYMOND, Alfred 260 Charles
  381 423 S.W. 426 T.S. 429
REYNOLDS, F.A. 158 John O. 228
RAYNOR, W.C. 221 William 317
READ, Thomas B. 63
RECEN, -- 457
REDMAN, James 342 William 342
REED, Clinton 136 220 Crawford
  350 James H. 179 James K. 332
  John 350
REEF, Joseph 415 Joseph S. 364
  425
REEMER, George 56
REES, James H. 356
REESE, D. 372 Dempsey 307 James
  H. 341
REGAN, S. 385
REID, James K. 303 John S. 7 26 5
  L. Wilbur 54 Leonard B. 368 W.
  326
REILEY, William 259
REILLY, Thomas 28
REILY, Richard C. 93
REINHART, Charles S. 150
REMINE, L.M. 261 W.W. 261
RENWICK, Thomas 102
RESOR, J. 309 J.C. 320 J.D. 393
  J.S. 305 T. 294 325
REYNOLDS, James 400 424 Joseph
  84 92 95 108 117 424 William T.
  86 90 97
RHEINHOLDT, Rudolph 279 291
RHODES, Abram S. 247
RICE, E.C. 89 121 J.E. 57 P.S. 266
RICH, Ebenezer 451 452 M. 245
  318 370 373 411 413

489

INDEX

RICHARDS, Gomer 272 J. 396 Joseph 287 S. 430 William 239
RICHARDSON, A.W. 309 G.N. 409 412 John 9
RICHMAN, Thomas I. 177 218 228
RICKARD, Richard H. 217
RIED, James H. 189 190 200
RIETHMANN, John J. 55
RIFE, Edward H. 232
RIGGS, F.F. 365
RIGGINS, Benjamin 273 Benjamin L. 276 291 J.R. 264
RIKER, N.W. 447 Nathan W. 63 447
RILEY, Edward 147 J.E. 234 246 William 113
RINEHART, A.E. 16
RINKER, J. 298 J.W. 303
RIPPEY, W.C. 439
RISCHE, August 158
RISLEY, John E. 91
RITTER, Jacob 96
ROBBINS, -- 18 31 379 380 A.B. 444 452 Aaron B. 80 98
ROBERTS, G.O. 315 329 333 J.N. 117 John G. 26 Mrs. Martha 321 322 323 332 William C. 327
ROBERTSON, M.N. 120 Peter 237 399
ROBESON, Solomon 135
ROBINS, Prof. C.E. 364 Charles E. 363 365 367
ROBINSON, -- 159 446 450 157 Charles E. 96 104 109 134 147 149 Daniel A. 17 James H. 55 James W. 55 67 John B. 376 M.N. 120 S.C. 268
ROBISON, J.F. 313 321 371
ROBY, -- 457
ROCKEFELLOW, -- 388
ROCKWELL, Lewis C. 98 201
ROE, C.S. 384 Charles S. 428

ROGERS, Andrews N. 166 170 171 173 180 197 199 205 212 224 226 228 J. 137 J.H. 81 105 James 111 142 Platt 17 S. 407 Theodore F. 320 W.H. 316 328 407
ROHM, R.L. 143
ROHWER, George 391 415
ROLFE, J.N. 307
ROLLINS, David J. 14 31 53 Ellen 14 J.D. 427 John Q.A. 190 210 228
RONK, R. 334
ROOD, M.L. 439
ROOS, J. 387
ROOT, G.D. 342 H. Walling 80 81 Leonard S. 65 M.H. 197 223 Marcus A. 448
ROPER, John L. 320 336 Joseph Jr. 27
ROSE, Charles [Corydon?] 249 W. Willet 150 William W. Jr. 150
ROSENCRANS, Isaac H. 16
ROSS, C.W. 390 Edward G. 79 Hugh 422 J.B. 380 407 414 James B. 422 John M. 198 239 332 377 387 391 421 435 Morris M. 85 Norman M. 85 110 W.W. 379
ROTCH, Benjamin S. 185 William J. 185
ROUDEBUSH, Almon H. 412 Clint 412 Lorenzo D. 412
ROUECH, C.J. 325 401
ROUILLIARD, Jean B 16 64
ROUNDS, J.R. 312
ROUPE, J.S. 412
ROURK, J.J. 261
ROUSSEAU, David 78
ROUTT, John L. 143 281
ROUX, Victor 28 43 52 69
ROWAN, James J. 246 247
ROWE, T.R. 271
ROWLAND, Capt. Ebenezer 3 6 9

INDEX

23 29 31 34 49 58 68 69
ROWLEY, Henry A.
ROWORTH, William M. 175 193 197 212 213
ROY, Antoine 103
RUDDLE, James 79
RUF, Joseph 405
RUGG, C. 386 419
RUGGLES, Henry 245
RULE, William M. 191
RUMMEL, George H. 357 W. 236 408
RUOSSOU [ROUSSOU?], Otto 366
RUSH, M. 301 Thomas 463
RUSSELL, -- 282 B.O. 175 J. 300 324 328 William 269 270 286 William B. 211
RUST, George W. 74
RUSTIN, L.J. 296
RUTAR, Gustave 27
RUTTER, James G. 22
RUVERA [or, RUVERO], Angelo 48 66
RYAN, John 420 432
SAFFORD, Jacob 233 Sylvestus A. 354
SAGENDORF, George H. 139 141
ST. CLAIR, Joel T. 41
SALE, H.H. 79 William H. 79
SALISBURY, Edward L. 202 219 J.H. 42 William H. 12
SALOMON, A.Z. 109
SALTIEL, Emanuel H. 355 361
SALTONSTALL, Francis G. 177
SAMPSON, Clarence G. 441 442 451 457 William L. 441 442 451 457
SANBORN, Corydon W. 12 17
SANDERS, -- 267 J.F. 286
SANDERSON, H. 245 J.L. 245
SANDSTONE, Olaf 374 421
SANFORD, B.N. 120 E.M. 247

SAPPINGTON, Thomas 260
SARGEANT, -- 155 M.L. 233
SAUNDERS, J.F. 51 Theodore O. 7 22
SAVAGE, P. 418
SAYER, Daniel 144 Windham S. 352
SAYLER, D.H. 42 56
SAYRE, Hal 64 125 192 215
SCHAEFER, Charles 243 249 William 160
SCHAFFTER, P.P. 135 Peter 135 Theobald 135
SCHALL, John H. 260
SCHELLENGER, John L. 20 60
SCHENCK, Edward 441
SCHEYER, -- 2 163
SCHIEDLER, Frank M. 239
SCHIEFENDECKER, C.C. 106
SCHIFFER, Alfred 241
SCHIRMER, Professor 3
SCHMIDT, Oscar E. 252
SCHMIT, F. 161
SCHNEIDER, Deitmar 188
SCHNELL, Ferdinand August 267
SCHOELKOPF, D.E. 419
SCHOFIELD, C. 400
SCHREINER, John 236
SCHULTZ, William 303
SCHUSTER, Henry 449
SCHWARZ, T.E. 127
SCITZ, J. 324
SCOBEY, Jason E. 175 183
SCOTT, -- 382 Bros. 19 S. 231 232 George L. 452 J.D. 452 R.E. 296 298 Robert N. 286
SCUDDER, John 94 192 217 Miss Mary T. 192
SEAL, A.D. 321
SEARS, -- 315 C.W. 272 Charles B. 232 J.H. 307 L. 306 310 330 Nathan 223 T.D. 169

INDEX

SEARY, Thomas 178 187
SEELEY, John E. 277
SEIFRIED, Henry 100
SEIWELL, Ed. A. 144 221
SELKIRK, Estate 39
SELLARS, W. 14
SELLERS, William 31 46 53
SELLIER, M. 255
SENEY, George R. 272
SHACKELFORD, Joel W. 285 419 423
SHAFFENBURG, Marc A. 12
SHALLCROSS, Mrs. 344
SHANKS, Charles A. 288 289 H.F. 306 330 John S. 288 289 William F. 57 94 326
SHANLEY, John Sr. 268 T. 241
SHAW, -- 263 264 E.J. 237 Edwin J. 238 Mrs. J.B. 62 J.H. 240 John H. 238 Joseph 19 61 Judson B. 67 70 T. 301 376 William S. 220
SHEAHAN, D. 347
SHEDD, William G. 14 46 51
SHEHAN, William 101
SHELDON, W.C. 207
SHELHAMER, -- 347
SHEPERD, S.A. 162
SHEPPARD, R.C. 364
SHERIDAN, Phil 53
SHERMAN, Elijah C. 186 W.B. 375 398 418 William 238 William B. 425
SHERRY, John 175
SHERWOOD, Clifford E. 57
SHIELDS, -- 343 J. 81 105 John 111 137 142
SHILLITO, John 81
SHIPMAN, C.N. 78
SHIRES, Thomas 34 44 50
SHIVEL, S.L. 65
SHIVELY, H.M. 231 Peter 114
SHOCK, Adam L. 448 John 448

SHOUP, Dr. 23
SHOWMAKER, Benjamin F. 340
SHUMATE, James 288
SHUSTER, Henry 455
SHUTTS, B.B. 327
SIDELL, George 345
SIEMS, F. 237
SIGAFUS, William 351 354
SILVER, Herman 448
SILVERTHORN, -- 457
SIMMONS, John H. 249 T.F. 81
SIMPSON, John 56 Robert 72 William V. 150
SINGER, Lawrence 42 68
SISLER, John 458 459
SIVYER, George J. 60
SKELLEY, Lawrence 272 Thomas 200 William 220
SKELTON, Thomas 13
SKINNER, A.P.W. 16 R 391
SLACK, W.H. 452
SLADE, Charles P. 57
SLATER, M.H. 286
SLEIGHT, W.W. 63
SLOCUM, -- 389 397 Charles 381 430
SLOSSON, William B. 133 146
SMALL, Dr. 431
SMEDLEY, William 394
SMILEY, George W. 74
SMITH, Angus 371 Azor A. 176 B.F. 27 59 B.J. 233 238 Benjamin F. 19 23 52 C.S. 351 354 Charles W. 67 David N. 98 114 131 137 145 E.F. 356 E.M. 319 Eben 12 73 Ed. P. 157 Ellis 368 Frank A. 157 G.H. 302 308 George S. 419 H.F. 246 352 Henry 173 199 Homer H. 35 71 J. 172 J. Alden 3 5 22 33 60 89 J.J. 140 J.S. 93 158 James D. 22 40 54 64 John H. 347 John W. 35 46 162 343 353

INDEX

354 Jonathan S. 63 74 L.C. 364
L.K. 61 Lewis C. 363 364 M.W.
351 354 Middleton W. 461 R.P.
364 S. 161 331 S.I. 268 352 356
W.K. 351 354 461 W.S. 125
William 282 William H. 237
William L. 418 William S. 107
SMOCK, Robert B. 169 206
SNIDER, George E. 160 Jacob 145
SNODGRASS, J.F. 132 John B. 139
SNOOK, E.F. 242
SNOW, William C. 129
SNOWDEN, B. 235 J. 247
SNYDER, -- 293 L.C. 169 175 185
220 Nicholas S. 326
SOULE, Albert G. 17
SOUTHGATE, J.N. 383
SOUTHWORTH, D.L. 441 456
SOWARD, Jackson 419
SPADLING, John 370 428
SPALDING, C.J. 66 H.M. 56
SPARLING, James 244
SPEER, Robert H. 361
SPENCER, B.D. 24 155 163 George
45 George W. 310 J.W. 376
SPILKER, L.H. 245 251 252
SPLANE, John 263 264
SQUIRES, -- 17 32 Frederick A. 11
12 17 27 51
STAATS, H. 330 Henry 427
STACEY, John A. 80 102 113 L.
333
STAFFORD, B.S. 108
STAHL, Henry 449
STALLCUP, John C. 278
STANDLEY, Joseph 174 177
STANLEY, J.M. 443 Luke 261
STANTON, John Jr. 170 171 180
Thomas 331
STAPLES, E. 99
STARR, Thomas 278 280 287 291
STATTLER, J.B. 89 121

STEAD, Charles M. 351
STEARNS, John E. 423 John R. 57
Marcus 423
STEBBINS, G.I. 53 W.R. 53
STEEL, Joseph P. 459
STEELE, Henry D. 447 Henry K.
455 Joseph B. 450 Joseph P 461
R.W. 139
STEINBACH, E. 409
STELL, Joseph P. 449
STEPHENS, Amos P. 150 Andrew
140 W.V. 127 William H. 269
STEPPLER, Joseph 4 7 44 67 70
Joseph Jr. 44 67
STETSON, D.S. 44
STEVENS, -- 140 216 278 E.W.
216 George E. 288 James 7 22 48
Samuel G. 431 W.H. 266 343
William H. 265 273 275 282 285
287 340 353 356 357
STEWARD, James O. 122 V.E. 120
STEWART, -- 160 C.A. 50 L.T. 16
J.W. 426 James Oscar 114 138
STIMPSON, John H. 261
STINE, Simon J. 124
STINGLEY, Baxter 272 H.W. 324
STINSON, Charles N. 30 T.R. 30
W.S. 325
STODDARD, Orange B. 368
STOIBER, Felix 120
STOLL, I. 417
STONE, Daniel 233 George W. 13
STOREY, W. 305 307
STOTTS, Israel 109
STOVER, Berty 231 232 Daniel C.
61 231 232 E.S. 233 James H.
231 232
STOWELL, Caleb S. 56
STRAIGHT, W.W. 394
STRATTON, Robert M. 212
Winfield Scott 375 398
STREETER, Eli S. 143

INDEX

STRELONG, August 160
STRIKER, John 177 Thomas 177
STRINGHAM, Charles A. 366
STROCK, C.C. 71 Cyrus 9 David 71
STRONG, C. 120 Cyrus 55 N.L. 435 Romeo D. 112
STROUSE, G.W. 424 J.W. 398 O.W. 313
STROUT, Frederick W. 7 22 35 W.H. 320 336
STRUBY, R. 397
STUART, C.A. 41 J.M. 330 412 John 24 144 379 430 John M. 377 384 415 427
STURNS, Frederick E. 16
SUERRERI, P. 288
SULLIVAN, D. 171 198 205 S. 173 184 212 220 227 229 Dennis 194
SUMMERFIELD, E. 409 436
SUMMERS, H. 56
SUMNER, Edward C. 231
SURLES, Ziba 448
SUTFON, Daniel 11
SWANSEY, N.M. 245
SWANSON, J. 369
SWARENGER, Z. 157
SWEEDER, Henry 169
SWEENEY, John 270 281
SWEM, James M. 275
SWETT, Samuel 112
SYMES, George G. 143
TABOR, -- 266 267 270 278 281 282 289 H.A.W. [Horace] 265 269 270
TALBOTT, John H. 269
TALCOTT, Mancel 14
TALLMAN, Charles P. 425 John 142 Nelson 142 R.W. 118
TALLOTT, August 401
TALMADGE, Van Nest 35
TANKERSLEY, Charles W. 366

TANNER, George 14 42
TAPPAN, John G. 189
TARVIN, E.M. 61
TASCHER, Jacob 166 228
TAWDEY, Edgar 64
TAYLOR, -- 200 Alexander 184 199 218 219 Charles H. 24 Charles M. 353 Cyrus 26 24 57 Ferdinand D. 160 Frank 90 246 Frank M. 100 143 148 202 H. Augustus 90 148 Henry 22 25 28 71 J.W. 246 M.S. 240 N.C. 319 R. 241 Theodore 246 W.W. 141 William 167 173 William E. 260 261
TEAL, George 94 128 139
TELLER, Henry M. 173 179 180 184 189 191 201 203 209 210 211 215 221 Willard 169 173 203 212 213 214 445 456 William M. 104 136 165
TEMPLETON, Andrew 327
TENNAL, -- 167 169 170 183 187 193 206 214
TERRELL, Col. 241
TERRILL, -- A. 301 314 A.T. 373
TERRY, A.O. 385 422 423 432 Allen J. 18 28 Noah 11 18 28
TEVIS, Joseph C. 304 310 L.W. 297 324
THACKERAY, John W. 101 Joseph A. 179
THALLER, John 171
THATCH, Thomas H. 79 93 127 130
THATCHER, J.A. 353 Joseph A. 178 Stephen D. 288
THAYER, E. 293 321
THEODORE, -- 316
THIES, Michael 188
THOMAS, -- 235 George 23 58 314 Henry 301 J. 461 James 408

494

INDEX

James M. 257 John 321 John R. 463 William H. 89
THOMASON, James A. 48
THOMASSON, Tower 160
THOMPKINS, B. 54
THOMPSON, Alba R. 242 251 E.P. 38 66 George W. 374 H.C. 38 66 Henry 100 146 Henry C. 39 41 55 James 75 James H. 214 John 88 John P. 17 Peter 205 Richard S. 340 359 S.W. 26 Samuel T. 248 William 26 101
THORN, John M. 441
THORNTON, Alexander 153 157 162 J. 324 James 392
THORPE, William 79 91 100
THURBER, Orrin 193
THURMOND, -- 154 Charles 298 399
THYNG, -- 161
TIERNAN, J.W. 385 423
TIERNEY, John 17 176
TILTON, R. 416 430 W.F. 272
TINKER, G.A. 397 Wesley 397
TITUS, -- 252
TOBIN, Samuel 88
TOLL, Charles H. 366
TOLLIVER, James 301
TOMAY, John 82
TOMLINSON, -- 165 181 215 Joseph B. 228
TOMPKINS, Charles C. 54 George V. 252
TOOKER, -- 119 269
TOPPING, Clark S. 350
TOSSER, Jacob 188
TOUSER, Frank 68
TOWER, H.F. 259
TOWN, R.B. 303
TOWNSEND, Albert 100 139 George 135
TRAEGER, Charles M. 268 Charles W. 268
TRAILOR, W.W. 142
TRAVERSE, S. 383
TRELLINGER, A. 390 Daniel 270 282 288
TREVAN, -- 357
TREVELLIAN, James 142
TRIMBLE, George W. 278
TRIPP, T.M. 261
TRUE, Edmund P. 420
TRUITT, F.M. 244 322 334 S.P. 244 322 334
TUBB, John 105 119 132
TUCKER, -- 284 Alfred 42 Wenzel 38
TUERKE, Henry W. 271 Herman 271
TUFFS, G.E. 224
TUGO, Charles T. 177
TUMBLESON, Silas T. 9 55
TURCK, John 101 133 148
TURNBULL, -- 143
TURNER, Professor Harry 65 John 200 W. 299 322 William H. 242 252
TUTTLE, H.H. 66 Nye 245
TYLER, E.W. 204 Sidney W. 167 196
TYRE, Simon 104
TYRRELL, J. 407
UNDERLIED, William 463
UPDEGRAF, -- 166
UPSON, E. 399
UTTER, Charles H. 120
VALLE, Nerii 264
VAN BUREN, G.M. 106
VAN DEMOER, John J. 2 352
VAN DEREN, Archibald J. 34
VAN DEVENTER, A.J. 166
VAN DIEST, P.H. 40 49 59 60 164
VAN DOREN, J. 293
VAN FLEET, Charles G. 3 4 6 63

495

INDEX

74
VAN GIESON, W.H. 363
VAN KEUREN, Cornelius 35 37 63
VAN NEST, John 166 173 199
VAN PELT, E. 308 G. 319
VAN WAGENEN, Theodore F. 277
VAN WENDT, Alex 7
VAN WETERING, H. J. 15 21 40 49 59 60
VANCE, David 371 J.W. 324
VANDERBILT, Henry S. 351 W.W. 100
VANDERVORT, Dr. 104 142
VEALE, Richard 46
VER PLANK, Phillip W. 47
VERDEW, John 67
VINCENT, Levi 261
VIRGINIA, Saline 103
VIVIAN, G. 127 S.H. 127
VOLKMAN, H.E. 304 320 333
VON SCHULZ, A. 1 224
VON WENDT, Alexander 27
VOORHIES, John H.P. 377 386
VOSE, Rufus C. 233 249
WADE, S. 328 Samuel 234
WAIN, William 221
WAKEFIELD, Gilbert 104 W.G. 96
WALCOTT, Henry R. 1
WALDO, W. 384 385
WALDRON, David 40 Jeremiah 16
WALKER, Edward S. 51 Harlan P. 27 36 73 W.A. 259 296 William S. 2 43
WALLACE, Edward F. 27 67 J.W. 384 Thomas 302 W.H. 261
WALLON, B. 426
WALSH, J. 417
WALTERS, L.B. 267
WALTON, F.B. 302 George 67
WAMELINK, J.T. 408
WARD, -- 293 Alfred 350 C.A. 299 322 Israel B. 460 J.B. 294 297

Jasper 461
WARDWELL, Samuel 177
WARE, Richard B. 350
WARING, J.D. 328
WARNER, Ed. B. 172 212 J.A. 243
WARREN, Henry 411 412
WARWICK, T. 132 William 144
WASSON, J.H. 390 435
WATERMAN, Herbert 186
WATERS, J.H.E. 437
WATKINS, Horace 111 John Q. 164
WATSON, J. 384 385 James E. 321 James F. 12 Mark 403 428 P.H. 434 Samuel 98 369 305 307 335 452
WEARE, John W. 19 Nathan 32 49 73
WEAVER, E.S. 117 Elijah S. 202
WEED, W.S. 445
WEBB, -- 156 H.N. 155 157 Henry S. 153 J.L. 155 157 John W. 276
WEBBER, H. 242 John H. 355
WEBSTER, F.E. 444 448 452 455 W.N. 95 96 118 William W. 459 462
WEESE, John 34
WEIGHTMAN, C. 433
WEINRICH, William H. 368
WEIR, W.H. 342
WEISE, A.V. 290
WEIST, Frank J. 14
WELCH, C.C. 256 Charles C. 63 75 89 112 113 137 D.L. 452 Frank L. 75 Luther J. 75
WELLBORN, James F. 144
WELLER, Benjamin S. 79
WELLEY, Thomas T. 357
WELLS, B.F. 136 B.T. 170 185 207 209 228 Thomas S. 270 285 289
WEMPLE, Jay C. 90
WENDELL, S. 235

INDEX

WENDT, Frederick 460
WERLEY, P.J. 6
WERTZBERGER, Herman 245
WESSELL, C.H. 100
WEST, A.R. 6 Henry T. 75 James 317 William 255
WESTBROOK, W.E. 63
WESTERN, J.H. 92
WESTLEY, Joseph 381
WESTMAN, George 79 124 342
WESTON, Elijah 100 John H. 146 W. 298 320
WETHERELL, -- 267
WETMORE, T.C. 288
WHEELER, George N. 463 J.H. 275 J.O. 213 Theodore W. 225 229 William R. 268
WHILSTON, R.B. 347
WHIPPLE, -- 66 185 Lewis 234
WHITCOMB, Truman 15 V 91
WHITE, C.H. 359 Elmer 67 G.G. 117 J.A. 312 374 386 Nathan 80 265 Robert J.S. 24 S.H. 116 129 T.S. 418 U.K. 401 435
WHITEHEAD, Augustus H. 139
WHITLEY, R.H. 51
WHITNEY, F.H. 329 Norman 32
WHITTEMORE, Oliver A. 246 447
WHITTIER, D. 34
WICKMAN, Charles 369 410
WILCOX, A.E. 426 A.M. 100 George 401 413
WILDE, Anthony 68
WILDER, A. 42 Edward P. 242 251 Eugene 22 48 George 285
WILEY, D.W. 454
WILKES, E.F. 429
WILKINS, Frederick 52
WILLARD [WILLIARD?], Ann 134 Mrs. Louisa A. 201 Z.H. 134
WILLCOTT, Charles T. 427
WILLEY, Alfred 348 D.W. 459 460

Thomas 337 348 350 352
WILLIAMS, -- 357 A.J. 162 191 Alpheus S. 79 Annie 191 Benjamin M. 75 C.M. 136 Charles M. 93 Edward W. 89 106 190 191 Edwin E. 261 Elizabeth 191 Henry 1 225 339 340 341 359 Henry B. 228 John 4 9 11 64 239 428 Joseph S. 463 M. 304 333 O.M. 190 W. 237 396 414 W.B. 20 W.C. 135 Dr. W.H. 286 W.S. 286 Williams 414
WILLIAMSON, Frederick R. 90 148 George R. 29 J. 370 383 396
WILLIARD [WILLARD?], Z.A. 74
WILLIS, B. 85
WILLOUGHBY, J.A. 441 446 John A. 440 459
WILSON, Robert S. 439
WIMBER, M.L. 322
WING, Samuel 300
WINSPEARE, -- 436
WILSON, Almerson 27 David G. 176 186 Daniel H. 354 Elihu C. 348 George 240 J.C. 279 289 J.L. 112 J.M. 16 James A. 97 James L. 53 John G. 443 Joseph C. 324 M. 16 T.B. 54 W.L. 338
WINCHESTER, -- 364 Eliza S. 364 Frank W. 364 Locke W. 90 Lucius A. 364
WING, H.O. 310 T.R. 390 435
WINNE, Peter 13 57 71
WINTERMUTE, James 251
WISE, William O. 75
WISWAL, Charles E. 47
WITMORE, Hill 251
WITTER, Estate 281 354
WOEBER, Adam 256
WOLCOTT, Henry 412 Henry R. Horace [H.R.] 16 52 224 343 345 351 353

INDEX

WOLD, Andrew 132
WOLFE, J.S. 359
WOOD, A.B. 282 343 Albinus B. 266 Alonzo P. Wood 431 B. Frank 47 Benjamin F. 48 D. Russ 425 George 315 373 Henry 70 Henry E. 33 James D. 174 Jared 58 John H. 47 John R. 48 R.H. 126 Samuel 19 Thomas L. 47 48 William L. 232
WOODBURY, L. 389 405 435 Levi N. 435 R.W. 259
WOODRUFF, William L. 352
WOODS, John L. 245 R. 131 William 61
WOODSIDE, A.J. 158
WOODWARD, Benjamin F. 2 George Jr. 266 P. Henry 335 336 Richard W. 335 336 Sarah C. 335 336
WORDEN, L.J. 321
WORLEY, Nathan Jr. 89
WORTHING, W. 14 34 42
WRIGHT, Dr. A.E. 265 280 282 Harvey 242 James C. 112 O. 304 318 R.H. 357
WRIGHTMAN, Allen S. 412
WRIGLEY, W.W. 87 103 124 145
WULSTEN, Carl 154 160 162
WYKE, -- 54
WYMAN, Daniel 422 434
YANKEE, W.H. 266
YENGLING, -- 444
YONLEY, John H. 462
YOUNG, Frank C. 194 228 John H. 363 William 371 William B. 255
ZANTZ, George 243 248
ZIMMERMAN, Uriah 247

www.ingramcontent.com/pod-product-compliance
Lightning Source LLC
Chambersburg PA
CBHW060909300426
44112CB00011B/1402